GROWTH HORMONE AND RELATED PEPTIDES

GROWTH HORMONE AND RELATED PEPTIDES

Proceedings of the IIIrd International Symposium,
Milan, September 17-20, 1975

Editors:

A. PECILE and E.E. MÜLLER

Institute of Pharmacology
University of Milan

1976

EXCERPTA MEDICA, AMSTERDAM - OXFORD
AMERICAN ELSEVIER PUBLISHING CO., INC., NEW YORK

© EXCERPTA MEDICA 1976

All rights reserved. No part of this publication may be reproduced, stored in a retrieval system or transmitted, in any form or by any means, electronic, mechanical, photocopying, recording or otherwise, without permission in writing from the publisher.

INTERNATIONAL CONGRESS SERIES NO. 381

ISBN EXCERPTA MEDICA 90 219 0312 1
ISBN AMERICAN ELSEVIER 0 444 15214 8

Library of Congress Cataloging in Publication Data

International Symposium on Growth Hormone and Related
   Peptides, 3d, Milan, 1975.
   Growth hormone and related peptides.

   (International congress series; no. 381)
   Includes index.
   1. Somatotropin--Congresses. 2. Peptides--Congresses.
I. Pecile, Antonio. II. Müller, E.E.
III. Title. IV. Series.
QP572.S6157   1976   612.6   76-3752
ISBN 0-444-15214-8

Publisher:
Excerpta Medica
305 Keizersgracht
Amsterdam
P.O. Box 1126

Sole Distributors for the USA and Canada:
American Elsevier Publishing Company, Inc.
52 Vanderbilt Avenue
New York, N.Y. 10017

Typeset by Fototekst bv, Naarden
Printed in The Netherlands by Verenigde Grafische Industrie bv, Rijswijk.

INTERNATIONAL SYMPOSIUM ON GROWTH HORMONE AND RELATED PEPTIDES

PRESIDENT
C.H. Li, *U.S.A.*

LOCAL CHAIRMEN
E. Trabucchi, *Italy*
C. Ricci, *Italy*

ORGANIZING COMMITTEE
A. Pecile, *Italy*
E.E. Müller, *Italy*
P. Neri, *Italy*

PROGRAMME COMMITTEE

R.M. Blizzard, *U.S.A.*
A.T. Cowie, *U.K.*
W.H. Daughaday, *U.S.A.*
F.C. Greenwood, *U.S.A.*
J.B. Josimovich, *U.S.A.*
C.H. Li, *U.S.A.*

R. Luft, *Sweden*
E.E. Müller, *Italy*
P. Neri, *Italy*
A. Pecile, *Italy*
A. Prader, *Switzerland*
A.E. Wilhelmi, *U.S.A.*

SECRETARY
Maria Luisa Pecile, *Milan, Italy*

PARTICIPANTS OF THE SYMPOSIUM

FOREWORD

Many important advances on somatotropin research have been made since the Second International Symposium was held in 1971. Most noteworthy is the demonstration that biologically active fragments can be prepared by limited digestion of human somatotropin (HGH) with human plasmin. Perhaps next in importance is the understanding of the various types of heterogeneity found in nearly all somatotropin preparations. In addition, the complete amino acid sequences of the ovine and bovine somatotropin molecules have been proposed. It is of interest to note that 169 amino acid residues out of 191 residues occupy either identical or homologous positions between the human and bovine/ovine hormones, and yet animal hormones are not active in man.

The isolation of human prolactin in highly purified form has finally been achieved and its properties are distinctly different from human somatotropin. A hypothalamic peptide with 14 amino acids possessing an inhibiting effect on somatotropin release from the pituitary has been identified and synthesized. This tetradecapeptide, called somatostatin, is also found to inhibit the secretion of insulin and glucagon. The somatotropin-dependent serum sulfation factors, which are known as somatomedins, have been highly purified and partially characterized.

Finally, the extra-somatotropic effects of HGH have been observed in diverse cases of clinical interest. For instance, HGH exhibits beneficial effects on patients with bleeding ulcers. Together with calcitonin, HGH is effective in treatment of idiopathic osteoporosis. The human hormone is also capable of reducing the serum cholesterol level in hypercholesterolemic subjects.

This volume is a collection of the complete texts of invited papers presented at the Third International Symposium on Growth Hormone in September of 1975. These papers include some of the subjects mentioned above, as well as up-to-date developments on somatotropin, prolactin, choriomammotropins, somatostatin and somatomedins. I am sure that Professors Trabucchi and Ricci join me in thanking the members of the Programme and Organizing Committees both for their advice and for undertaking the task of organizing the Symposium. We also thank Dr. Maria Luisa Pecile for her untiring skill in making the Symposium a success.

Choh Hao Li

San Francisco, California
February, 1976

ACKNOWLEDGEMENT

The President of the Symposium and the Editors gratefully acknowledge the valuable assistance extended by Professor F.J. Ebling and Dr. I. Henderson of Sheffield in the linguistic revision of the papers written by authors whose mother tongue is not English.

CONTENTS

I. GROWTH HORMONES
Chemical studies. Modification during the secretory process. Nature in plasma. Synthesis in cell-free systems.

Studies on the chemistry of bovine and rat growth hormones   Michael Wallis and Richard V. Davies: 1
Studies on plasmin-modified human growth hormone and its fragments   Choh Hao Li and Thomas A. Bewley: 14
The nature of fragments of human growth hormone produced by plasmin digestion   Jack L. Kostyo, John B. Mills, Charles R. Reagan, Daniel Rudman and Alfred E. Wilhelmi: . 33
In vivo and in vitro actions of synthetic part sequences of human pituitary growth hormone   J. Bornstein: 41
Synthesis of fragments with aminoacid sequences of growth hormones   F. Chillemi, A. Aiello, A. Pecile and V.R. Olgiati: 50
Human growth hormone: a family of proteins   U.J. Lewis, R.N.P. Singh, S.M. Peterson and W.P. VanderLaan: 64
Studies on the nature of plasma growth hormone   S. Ellis, R.E. Grindeland, T.J. Reilly and S.H. Yang: 75
Studies of growth hormone synthesis in cultured rat pituitary cells and in cell-free systems   F. Carter Bancroft, Phyllis M. Sussman and Robert J. Tushinski: 84

II. GROWTH HORMONES AND GROWTH FACTORS:
Biological activities

Cellular mechanisms of the acute stimulatory effect of growth hormone   K. Ahrén, K. Albertsson-Wikland, O. Isaksson and J.L. Kostyo: 94
Growth hormone action on thymus and lymphoid cells   G.P. Talwar, S.N.S. Hanjan, Z. Kidwai, P.D. Gupta, N.N. Mehrotra, R. Saxena and Q. Bhattarai: 104
Growth hormone and insulin secretion   Leslie L. Bennett and Donald L. Curry: 116
Radioreceptor assay of plasma NSILA-s in man: basal and stimulated levels in normal and pathologic states   Klara Megyesi, C. Ronald Kahn, Jesse Roth, Phillip Gorden and David M. Neville Jr: 127
Mitogenic factors from the brain and the pituitary: physiological significance   D. Gospodarowicz, J.S. Moran and H. Bialecki: 141

III. SOMATOMEDIN:
Chemistry. Regulation of generation. Biological effects.

Somatomedins A and B. Isolation, chemistry and in vivo effects   L. Fryklund, A. Skottner, H. Sievertsson and K. Hall: 156
Regulation of somatomedin generation   William H. Daughaday, Lawrence S. Phillips and Adrian C. Herington: 169

Contents

Studies on the regulation of somatomedins A and B Kerstin Hall, Kazue Takano, Gösta Enberg and Linda Fryklund: 178
Specificity, topography, and ontogeny of the somatomedin C receptor in mammalian tissues A. Joseph D'Ercole, Louis E. Underwood, Judson J. Van Wyk, Charles J. Decedue and Doretha B. Foushee: 190
Evidence for a role of adenosine 3':5'-monophosphate in growth hormone-dependent serum sulfation factor (somatomedin) action on cartilage Harold E. Lebovitz, Marc K. Drezner and Francis A. Neelon: 202

IV. SOMATOSTATIN AND THE ENDOCRINE PANCREAS

Somatostatin, a hormone of the α_1-cells of the pancreatic islets and its influence on the secretion of insulin and glucagon S. Efendić, T. Hökfelt and R. Luft: 216

V. GROWTH HORMONES. CLINICAL INVESTIGATIONS:
Control of growth hormone secretion

Serotoninergic control of human growth hormone secretion: the actions of L-dopa and 2-bromo-α-ergocryptine George A. Smythe, Paul J. Compton and Leslie Lazarus: ... 222
Neuroendocrine control of growth hormone secretion: experimental and clinical studies A. Liuzzi, A.E. Panerai, P.G. Chiodini, C. Secchi, D. Cocchi, L. Botalla, F. Silvestrini and E.E. Müller: 236
Growth hormone release in Huntington's disease Stephen Podolsky and Norman A. Leopold: 252

VI. GROWTH HORMONES. CLINICAL INVESTIGATIONS:
Selected topics

Human growth hormone changes with age T.L. Bazzarre, A.J. Johanson, C.A. Huseman, M.M. Varma and R.M. Blizzard: 261
Plasma somatomedin. Clinical observations J.L. Van den Brande: 271
Interrelations of the effects of growth hormone and testosterone in hypopituitarism M. Zachmann, A. Aynsley-Green and A. Prader: 286
Intermittent versus continuous HGH treatment of hypopituitary dwarfism Zvi Laron and Athalia Pertzelan: 297
Metabolic and therapeutic studies on 246 patients with acromegaly treated with heavy particles John H. Lawrence: 312

VII. HUMAN CHORIONIC SOMATOMAMMOTROPIN:
Isolation and characterization. Biological effects. Clinical investigations

Isolation and characterization of bovine and ovine placental lactogen Robert E. Fellows, Franklyn F. Bolander, Thomas W. Hurley and Stuart Handwerger: 315
Chemical structure and biologic and immunologic activity of 'big' human placental lactogen Arthur B. Schneider, Kazimierz Kowalski and Louis M. Sherwood: 327
Biological action of human chorionic somatomammotropin during pregnancy — its lipolytic action and fetal growth S. Tojo, M. Mochizuki, H. Morikawa and Y. Ohga: 334
Effects of human chorionic somatomammotropin on the male reproductive apparatus of rodents and on placental steroids during human pregnancy P. Neri, C. Arezzini, C. Fruschelli, E.E. Müller, P. Fioretti and A.R. Genazzani: 345
Correlation between plasma levels of human chorionic somatomammotrophin (HCS), sexual steroids and their precursors in normal and pathological pregnancies G. Magrini, J.P. Felber, F. Méan and A. Curchod: 369

Contents

Clinical application of human chorionic somatomammotrophin A.T. Letchworth: 380

VIII. PROLACTIN:
Selected topics

Polypeptide hormone production by cells cultured on artificial capillaries Richard A. Knazek and Jay S. Skyler: 386
Prolactin and fertility control in women C. Robyn, M. Vekemans, P. Delvoye, V. Joostens-Defleur, A. Caufriez and M. L'Hermite: 396
The relationship between relaxin and prolactin immunoactivities in various reproductive states: radioimmunoassay using porcine relaxin Wayne A. Chamley, Roger D. Hooley and Gillian D. Bryant: 407
The relationship between relaxin and prolactin immunoactivities in various reproductive states: physical-chemical and immuno-biological studies Simon C.M. Kwok, John P. McMurtry and Gillian D. Bryant: 414
Mammotrophic hormones in ruminants Isabel A. Forsyth and I.C. Hart: 422
Prolactin binding to plasma membranes, and its effect on monovalent cation transport in mammary alveolar tissue — a possible mechanism of action Ian R. Falconer: 433
Inhibition of prolactin secretion by dopamine and Piribedil (ET-495) Robert M. MacLeod, Hiroko Kimura and Ivan Login: 443

Author Index 455
Subject Index 457

I. Growth hormones

STUDIES ON THE CHEMISTRY OF BOVINE AND RAT GROWTH HORMONES

MICHAEL WALLIS and RICHARD V. DAVIES*

Biochemistry Laboratory, School of Biological Sciences, University of Sussex, Falmer, Brighton, Sussex, United Kingdom

Pituitary growth hormone (GH) is a protein of moderate molecular weight (about 20,000 in most species) containing a single polypeptide chain. The chains tend to aggregate, and this has led to considerable difficulties in interpreting the structure of the hormone. There is considerable species specificity among the growth hormones, which can be detected by chemical, immunological and biological properties.

This paper will be confined mainly to a review of our studies on the growth hormones from 2 species; the ox and the rat, but an attempt will be made to integrate our results with those of others working on these hormones and on growth hormones and related proteins from other species.

BOVINE GROWTH HORMONE

Bovine (ox) growth hormone (BGH) tends to aggregate, and in neutral solution and at moderate concentrations dimers tend to predominate. This fact (which led to molecular weight estimates of about 50,000), and the observation that most preparations of BGH contain 2 N-terminal residues (phenylalanine and alanine) led to some confusion in the early work on the structure of BGH, and a model was proposed in which 2 (dissimilar) polypeptide chains were joined covalently. Subsequent studies on the molecular weight of the hormone, using dissociating conditions or gel filtration, showed that the monomer molecular weight is only 20-26,000 (Andrews and Folley, 1963; Dellacha et al., 1966; Ellis et al., 1966; Wallis and Dixon, 1966). Studies on the amino acid sequence of BGH, particularly at the amino and carboxyl ends (Santomé et al., 1966; Wallis, 1966, 1969) confirmed the idea that the hormone contains single polypeptide chains of molecular weight about 20,000.

Amino acid sequence studies

The early work on the chemistry of BGH suggested that studies on the amino acid sequence could be fruitfully carried out. Several groups (Santomé et al., 1971, 1973; Fellows et al., 1972; Fellows, 1973; Wallis, 1973a; Gráf and Li, 1974a) have now

---

* Present address: Biochemistry Department, A.R.C. Institute of Animal Physiology, Babraham, Cambridge, U.K.

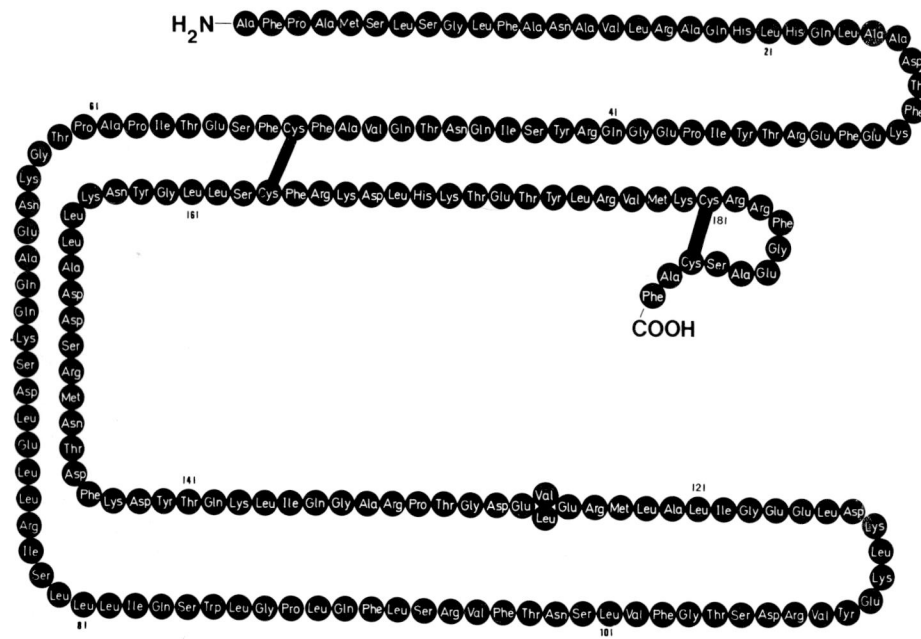

Fig. 1. The amino acid sequence of bovine growth hormone. The sequence shown is from Wallis (1973a). Fellows (1973) reported an identical structure (though a few amide residues were not assigned). Santomé et al. (1973) have a substantially different sequence between residues 84 and 94, and different amide assignments at residues 69 and 168. The sequence shown here is confirmed by Gráf and Li (1974a). Heterogeneity at residues 1 and 127 is discussed in the text.

carried out such studies. Our sequence for the hormone (Wallis, 1973a) is presented in Figure 1. The results are in complete agreement with those of Fellows (1973) and have been confirmed in part by Gráf and Li (1974a). There is some disagreement between our results and those of Santomé et al. (1973) — the details are indicated in the legend of Figure 1. The hormone contains 191 residues, 2 disulphide bridges, and a single tryptophan.

N-terminal heterogeneity

Preparations of BGH show heterogeneity at various points, and the sequence work which has now been completed allows this to be assessed.

Heterogeneity at the N-terminus of BGH was recognized by Li and Ash (1953) who reported the presence of 2 N-terminal groups, phenylalanine and alanine. In some preparations of BGH the N-terminal heterogeneity is more extensive (Ellis, 1961; Fellows et al., 1972) probably as a result of enzymic degradation of the hormone during isolation. The nature of this N-terminal heterogeneity was elucidated by investigating the N-terminal sequence (Wallis, 1969). It became clear that there were 2 N-terminal sequences which differed only by the presence or absence of an extra N-terminal alanine. Subsequent studies on the sequence of the remainder of the BGH molecule indicate that the alanyl and phenylalanyl chains do not differ elsewhere in the sequence.

Ellis et al. (1972) showed that the alanyl and phenylalanyl chains of BGH can be separated by isoelectric focussing. The significant difference in isoelectric points which this revealed suggested that there must be differences between the 2 chains other than the presence or absence of an alanyl residue. However, it seemed possible that the difference in pKa between the α-amino group of an N-terminal alanyl residue and that of an N-terminal phenylalanyl residue might be sufficient to explain the difference in isoelectric points (Wallis, 1971). Evidence in support of this idea was provided by experiments showing that alanyl and phenylalanyl peptides derived from the N-terminus of BGH could be separated by high voltage electrophoresis, owing to substantial differences in the pKa's of their α-amino groups (Wallis, 1973b). Lorenson and Ellis (1975) have recently shown that the different chains of BGH can be separated by polyacrylamide gel electrophoresis, and that the electrophoretic behaviour could, again, be explained in terms of differences between the pKa's of the α-amino groups. It thus seems probable that the separations of the alanyl and phenylalanyl chains of BGH that have been achieved are due solely to differences at the N-termini of the chains. It is interesting that Ferguson and Wallace (1961) reported that BGH could be separated electrophoretically into 2 components which may well have been the alanyl and phenylalanyl chains, though this was not demonstrated at that time.

The *origin* of the N-terminal heterogeneity remains in some doubt. Genetic causes have been largely ruled out. The possibility that the difference results from allelic polymorphism was tested by preparing BGH from a substantial number of individual pituitaries (Peña et al., 1969; Wallis, unpublished data); every gland contained both the alanyl and phenylalanyl chains, ruling out allelic polymorphism. That the 2 chains arise as a result of gene duplication also seems very unlikely, as indicated by the existence and behaviour of an allelic variation at residue 127 (see below).

The alternative explanation for the alanyl and phenylalanyl chains is that they arise as a result of enzymic degradation. This could occur during extraction and isolation of the hormone, but such an explanation is made less likely by the fact that the 2 chains are always obtained in approximately equal amounts (at least, in those extraction conditions where more extensive degradation is avoided — see below). An alternative explanation is that the N-terminal heterogeneity results from the processing of a growth hormone precursor. Thus a precursor with the sequence:

$$\cdots X - Y \underset{b}{\overset{a}{\updownarrow}} Ala \underset{\uparrow}{-} Phe - Pro - Ala \cdots$$

might be converted to BGH by cleavage at either point a, or point b (the specificity of the converting enzyme enabling either cleavage to occur). Once cleavage at a has occurred, however, the enzyme (an endopeptidase) would be unable to remove the N-terminal alanine by a second cleavage at b. Bancroft and Sussman (1975) have recently demonstrated the occurrence of a precursor of rat growth hormone, and this encourages us to favour this last explanation for the origin of the 2 BGH chains. It is noteworthy that most mammalian growth hormones (except sheep and ox) do not show a corresponding N-terminal heterogeneity, and possess only phenylalanine as N-terminal residue. It may be that in these cases the sequence of the GH precursor is such as to restrict the processing to cleavage at position b. Clearly, sequence studies on the GH precursor should be able to resolve some of these problems.

In some preparations of BGH there is considerably more N-terminal heterogeneity than has been discussed here. It seems probable that this is a result of enzymic degradation occurring during extraction. Such degradation is most pronounced when extraction is performed at neutral or mildly acidic pHs (Ellis, 1961). When the initial extraction is performed at alkaline pH it is minimized, and such an alkaline extraction has been used for preparation of most of the BGH used in our work (Wallis and Dixon, 1966).

Allelic polymorphism

At residue 127 in the BGH molecule an ambiguity was noticed during the sequence work (Fellows and Rogol, 1969; Santomé et al., 1971, 1973; Wallis, 1964, 1973a). Both valine and leucine are found at this position, in a ratio of about 1:2. Seavey et al. (1971) prepared growth hormone from individual beef pituitaries, and showed that, in an individual gland, residue 127 could be all leucine, all valine, or both. They thus demonstrated that variation at this position is due to allelic polymorphism in the cows from which the hormone is derived. The fact that individuals that contain only valine or only leucine can be found provides good evidence against the idea that there has been a duplication of the GH gene in cows (see previous section).

Electrophoretic heterogeneity

Nearly all preparations of BGH show some heterogeneity on gel electrophoresis (in addition to that due to N-terminal heterogeneity). Much work has gone into elucidating the basis for this, and it seems likely that it derives from 2 sources — loss of amide groups and proteolytic cleavage of the chain. Such variations are difficult to analyse in detail, and it remains in doubt whether they are of physiological significance or whether they result entirely from degradation during extraction.

Modification of methionine residues in bovine growth hormone

Progress with studies on the primary structure of growth hormones from various species has facilitated studies on modification of the hormone. Interesting results have been obtained from enzymic digestion of GH (e.g. Sonenberg et al., 1972; Gráf and Li, 1974b; Reagen et al., 1975). Modification of specific residues by chemical means has also been used to produce GH derivatives with altered properties. We have studied the modification of methionine residues in BGH (Wallis, 1972).

BGH contains 4 methionine residues (Fig. 1). These can be modified by alkylation with iodoacetic acid at low pH (pH 3.5). At this pH other residues do not react; methionine is converted to carboxymethyl methionine:

$$\begin{array}{c} CH_3 \\ | \\ +S-CH_2COO^- \\ | \\ (CH_2)_2 \\ | \\ ---NH-CH-CO--- \end{array}$$

The modified methionine possesses no net charge but, as can be seen, exists as a zwitterion so that considerable alteration of the properties of the 'new' residue must be expected.

BGH was completely carboxymethylated, using a 50-fold molar excess of iodoacetic acid, in the presence or absence of 8 M urea. The product was completely inactive when tested for growth-promoting activity in pituitary dwarf mice (Table 1).

Reaction of BGH with iodoacetic acid was also used to investigate the relative reac-

TABLE 1

Growth-promoting activity of carboxymethyl-bovine growth hormone in dwarf mice

Sample	Number of animals	Weight gain (g)	(± S.E.M.)
Assay I			
1. saline controls	5	+ 0.41	(± 0.36)
2. 2 µg BGH/day	6	+ 1.51*	(± 0.31)
3. 10 µg BGH/day	5	+ 2.90**	(± 0.40)
4. 10 µg urea treated BGH/day	4	+ 2.62**	(± 0.20)
5. 20 µg carboxymethyl-BGH/day	6	+ 0.14	(± 0.25)
Assay II			
6. saline controls	6	+ 1.01	(± 0.17)
7. 2 µg BGH/day	6	+ 1.85**	(± 0.15)
8. 5 µg BGH/day	6	+ 2.73***	(± 0.28)
9. 40 µg carboxymethyl-BGH/day	6	+ 1.14	(± 0.21)
10. 5 µg BGH + 40 µg carboxymethyl-BGH/day	6	+ 3.40***	(± 0.36)

The weight gain shown was the mean increase in weight for each group of mice after 21 (Assay I) or 22 (Assay II) daily injections of the test solution indicated. The volume injected was 0.1 ml/day in every case, and all the samples were dissolved in 0.9% NaCl and adjusted to pH 7-9. For the difference between samples and the appropriate controls: *p < 0.05, **p < 0.01, ***p < 0.001. The differences between 1 and 5, 3 and 4, 6 and 9, and 8 and 10 are not significant. The assay method used is described in Wallis and Dew (1973). The data shown are from Wallis (1972).

TABLE 2

Reactivity of the methionine residues of bovine growth hormone

Residue (see Fig. 1)	Specific activity of carboxymethyl-Met residue*	
	No denaturant	8 M urea
Met[5]	2,910	3,370
Met[124]	1,030	1,400
Met[149]	5,920	5,250
Met[179]	4,550	5,090

* Specific activities are expressed in dpm/nmole. BGH was reacted with ^{14}C-labelled iodoacetic acid in the presence or absence of 8 M urea. Data from Wallis (1972).

tivities of the methionine residues in BGH. When a small quantity (2 mole/mole of BGH) of labelled [^{14}C]iodoacetate was reacted with BGH, the methionine residues competed for the alkylating reagent. Subsequent complete carboxymethylation with excess unlabelled iodoacetate, followed by identification of those residues which had been labelled and the amount of radioactivity associated with them, enabled the relative reactivities of the 4 methionine residues to be determined (Table 2). It is clear that the methionine at residue 124 is much less reactive than the other 3 methionines.

Glaser and Li (1974) have shown that 3 out of 4 methionine residues in BGH can be oxidized with hydrogen peroxide at pH 8.5 without any significant loss of biological activity. In this case it was Met[179] that was unmodified. The difference between these results and our own presumably reflects the different conditions and reagents used. It could be a result, as Glaser and Li (1974) pointed out, of differences of conformation in the BGH molecule in acidic and basic conditions.

RAT GROWTH HORMONE

Extensive comparative studies have been described for growth hormones from many vertebrate groups, but detailed chemical information is available only for those from a few mammals — man and several ungulates (ox, sheep, pig, horse). With a few exceptions, a similar situation exists for most anterior pituitary hormones — the range of species for which detailed chemical information is available is restricted to man and a few domestic animals. These are the species, of course, for which substantial amounts of material can be prepared, but they are not necessarily those of sole interest to the endocrinologist or biochemist. There is an urgent need for chemical information from a wider range of species (especially in view of the specific variation seen in the GHs), which in turn implies a need for more sensitive chemical methods, so that chemical information, particularly amino acid sequences, can be obtained with small quantities of material.

We have attempted to develop such methods, and have done so in connection with studies on rat growth hormone (RGH). This is of particular interest because the rat is the most widely used test animal for studies on the synthesis, secretion, metabolism, and mode of action of growth hormone. (Moderate quantities of RGH have been generously made available to us by the NIAMDD-NIH rat pituitary hormone programme.)

Two approaches have been used in studying the amino acid sequence of rat growth hormone:

1. Peptides have been prepared from peptide maps (after detection with very weak ninhydrin) and then used for sequence studies using the sensitive dansyl method for N-terminal determinations, in combination with the Edman degradation. This type of approach has been used previously by ourselves (Wallis, 1973a) and others (e.g. Seavey et al., 1973) and allows a considerable amount of structural information to be obtained from 0.1-0.2 μmole of material.

2. Radioactive [^{14}C] RGH has been prepared by incubating pituitaries in vitro with ^{14}C amino acids in a suitable buffered salts solution (Davies and Wallis, 1971). The labelled hormone was purified by polyacrylamide gel electrophoresis and/or gel filtration (Fig. 2) and subsequently used for structural studies. Amino acids, peptides and their derivatives were detected by virtue of their radioactivity rather than by conventional (chemical) means, which leads to a considerable increase in sentivity, and means that detection of peptides (by autoradiography, for example) is not accompanied by loss of material due to chemical degradation.

This second approach has several potential advantages. In particular, it should eventually be applicable to structural determinations on species for which only a few pituitaries are available (provided these can be obtained fresh). It also has the advantage that, once the hormone has been purified from other (radioactive) pituitary proteins, contamination with (non-radioactive) amino acids, peptides and proteins is not a problem — indeed addition of unlabelled material as a carrier is generally to be recommended. The use of radioactive proteins for sequencing has some disadvantages, however, the chief of which is the problem of getting sufficient label into the protein of interest. Growth hormone is a particularly favourable case, since a high proportion

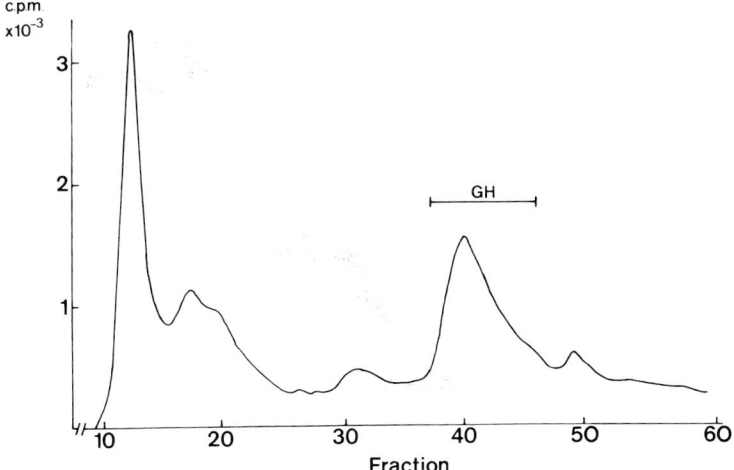

Fig. 2. Purification of labelled rat growth hormone. Rat anterior pituitaries were incubated for 20 hours in the presence of [^{14}C] amino acids. They were then extracted and the extract was applied to a preparative polyacrylamide gel electrophoresis column. 1-ml fractions were collected, and 5 μl of each was taken for scintillation counting. The peak labelled GH was identified as RGH by analytical polyacrylamide gel electrophoresis and immunoprecipitation.

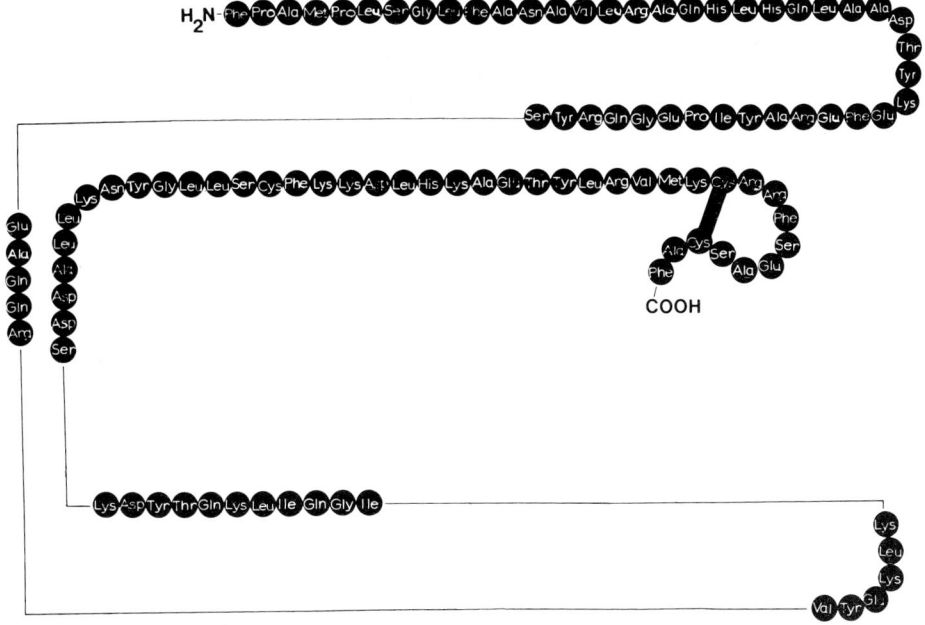

Fig. 3. Tentative, partial amino acid sequence of rat growth hormone. The sequences determined have been positioned by comparison with the sequence of bovine growth hormones. The solid line indicates regions of the sequence which have not yet been determined. See 'Note added in proof'.

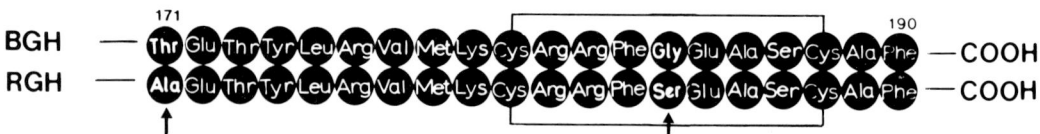

Fig. 4. Comparison of the C-terminal sequences of rat and bovine growth hormones. See 'Note added in proof'.

of the label incorporated into pituitary proteins goes into this hormone (30-40%; Betteridge and Wallis, 1973). Associated with this problem is the fact that the specific activities of the amino acids in the labelled hormone are very variable, owing to the varied size of amino acid pools, and the varied metabolic possibilities for amino acids within the pituitary. Non-essential amino acids (especially glycine and alanine) are incorporated with much lower specific activities than are essential amino acids. It may be that material synthesized in vitro by a cell-free system will prove more satisfactory as a starting material for structural studies (such a possibility is being investigated); nevertheless, a considerable amount of structural information has been obtained using labelled RGH prepared by incubating intact pituitary glands in vitro, and the method has considerable promise as a general way of investigating GH sequences.

The amino acid sequences which have been determined for rat growth hormone are shown in Figure 3. Homology with the bovine hormone has been used in aligning some of the tryptic peptides. The sequences which have been determined currently account for about 55% of the total sequence, and although the information obtained so far is only tentative and incomplete it is already sufficient to enable useful comparisons to be made with GHs from other species. Comparison of the C-terminal sequences of rat and bovine GHs is made in Figure 4.

MOLECULAR EVOLUTION OF GROWTH HORMONES AND RELATED HORMONES

The amino acid sequences that have been determined so far enable some preliminary conclusions to be drawn about the evolution of the mammalian GHs (Wallis and Davies, 1974; Wallis, 1975). Comparison of the amino acid sequences of GHs from various mammals is shown in Figure 5. From such a comparison several conclusions can be drawn.

The sequences of ovine (sheep) and bovine GHs are very similar: they differ at only 3 positions, residue 99 (Asp in ovine GH; Asn in bovine GH), residue 127 (the allelic polymorphism seen in BGH does not occur in the sheep, where only Leu is found in this position) and residue 130 (Gly in BGH, Val in ovine GH). Whether the difference at residue 99 is a real one is doubtful, since one would expect it to be reflected in different electrophoretic mobilities for bovine and ovine GHs and these are not seen.

The sequence of human growth hormone is very substantially different from that of

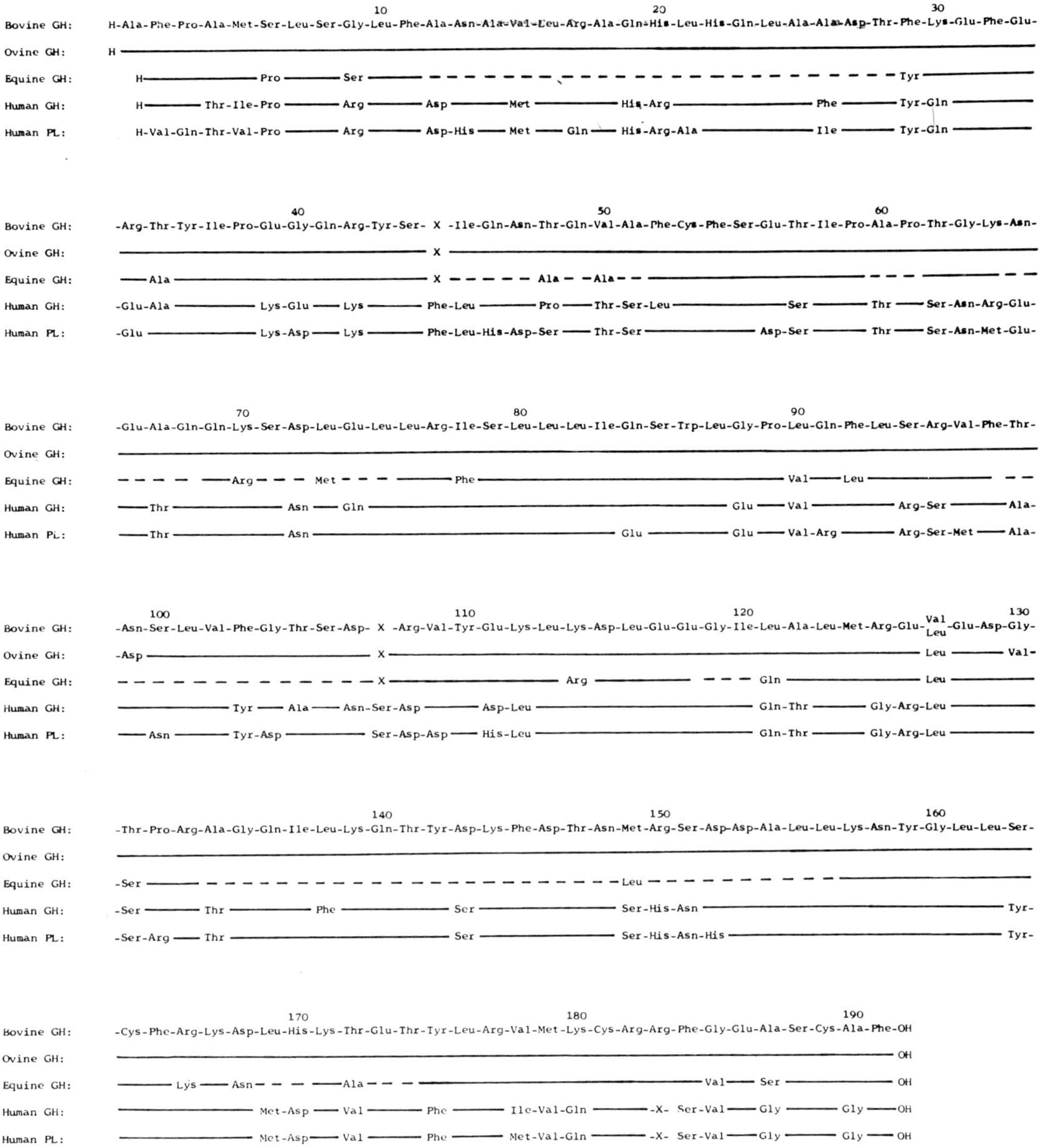

	10	20	30
Bovine GH:	H-Ala-Phe-Pro-Ala-Met-Ser-Leu-Ser-Gly-Leu-Phe-Ala-Asn-Ala-Val-Leu-Arg-Ala-Gln-His-Leu-His-Gln-Leu-Ala-Ala-Asp-Thr-Phe-Lys-Glu-Phe-Glu-
Ovine GH:	H——
Equine GH:	H————————— Pro ——————— Ser —— Tyr ————————
Human GH:	H——————— Thr-Ile-Pro ——————— Arg ————— Asp ———— Met ———————— His-Arg ———————————— Phe ———————— Tyr-Gln
Human PL:	H-Val-Gln-Thr-Val-Pro ——————— Arg ——————— Asp-His ———— Met ——— Gln ——— His-Arg-Ala ——————— Ile —————— Tyr-Gln

	40	50	60
Bovine GH:	-Arg-Thr-Tyr-Ile-Pro-Glu-Gly-Gln-Arg-Tyr-Ser- X -Ile-Gln-Asn-Thr-Gln-Val-Ala-Phe-Cys-Phe-Ser-Glu-Thr-Ile-Pro-Ala-Pro-Thr-Gly-Lys-Asn-
Ovine GH:	———————————————————————————— X ———
Equine GH:	———— Ala ————————————————— X — — — — ———— Ala — —Ala — — ————————————— — — — — — —
Human GH:	-Glu-Ala ——————————— Lys-Glu —————— Lys ————— Phe-Leu ———————— Pro ——— Thr-Ser-Leu ————————— Ser ————— Thr ——— Ser-Asn-Arg-Glu-
Human PL:	-Glu ——————————— Lys-Asp ————— Lys ————— Phe-Leu-His-Asp-Ser ——————— Thr-Ser ———————————— Asp-Ser ————— Thr ——— Ser-Asn-Met-Glu-

	70	80	90
Bovine GH:	-Glu-Ala-Gln-Gln-Lys-Ser-Asp-Leu-Glu-Leu-Leu-Arg-Ile-Ser-Leu-Leu-Leu-Ile-Gln-Ser-Trp-Leu-Gly-Pro-Leu-Gln-Phe-Leu-Ser-Arg-Val-Phe-Thr-
Ovine GH:	——
Equine GH:	— — — — ——— Arg — — — Met — — — ——————— Phe ————————————————————— Val ——— Leu ——————— — —
Human GH:	———— Thr ——————————— Asn ——— Gln ———————————————————————— Glu ——— Val ——————— Arg-Ser ————————— Ala-
Human PL:	———— Thr ——————————— Asn ———————————————————————————————— Glu ——— Glu ——— Val-Arg ————— Arg-Ser-Met ——————— Ala-

	100	110	120	130
Bovine GH:	-Asn-Ser-Leu-Val-Phe-Gly-Thr-Ser-Asp- X -Arg-Val-Tyr-Glu-Lys-Leu-Lys-Asp-Leu-Glu-Glu-Gly-Ile-Leu-Ala-Leu-Met-Arg-Glu-Val-Glu-Asp-Gly- Leu
Ovine GH:	-Asp———————————————————— X —————————————————————————————————————— Leu ———————— Val-
Equine GH:	— — — — — — — — — — — — — X ——————————————— Arg ————————————— — — — Gln —————————— Leu ———————
Human GH:	—————— Tyr ——— Ala ——— Asn-Ser-Asp ————— Asp-Leu ————————————————————————— Gln-Thr ——————————— Gly-Arg-Leu ————
Human PL:	————— Asn ———————— Tyr-Asp ——————— Ser-Asp-Asp ——— His-Leu ————————————————— Gln-Thr ——————————— Gly-Arg-Leu ————

	140	150	160
Bovine GH:	-Thr-Pro-Arg-Ala-Gly-Gln-Ile-Leu-Lys-Gln-Thr-Tyr-Asp-Lys-Phe-Asp-Thr-Asn-Met-Arg-Ser-Asp-Asp-Ala-Leu-Leu-Lys-Asn-Tyr-Gly-Leu-Leu-Ser-
Ovine GH:	——
Equine GH:	-Ser ————— — — — — — — — — — — — ————————————————————————— Leu ———————————————— — — — — — —
Human GH:	-Ser ————— Thr —————— Phe ——————————— Ser ——————————— Ser-His-Asn ——————————————————— Tyr-
Human PL:	-Ser-Arg ——— Thr ———————————————————————— Ser ————————— Ser-His-Asn-His ——————————————————— Tyr-

	170	180	190
Bovine GH:	-Cys-Phe-Arg-Lys-Asp-Leu-His-Lys-Thr-Glu-Thr-Tyr-Leu-Arg-Val-Met-Lys-Cys-Arg-Arg-Phe-Gly-Glu-Ala-Ser-Cys-Ala-Phe-OH
Ovine GH:	—— OH
Equine GH:	—————— Lys ——— Asn — — — ——— Ala — — — ————————————————— Val ——— Ser ———————— OH
Human GH:	————————————————— Met-Asp ——— Val ——— Phe ———————— Ile-Val-Gln ——— -X- Ser-Val ——— Gly ————————— Gly ——— OH
Human PL:	———————————————— Met-Asp ——— Val ——— Phe ————— Met-Val-Gln ——— -X- Ser-Val ——— Gly ————————— Gly ——— OH

Fig. 5. A comparison of the primary structures of growth hormones from several species and human placental lactogen. The sequence of bovine growth hormone is shown in full. For the other sequences, a solid line indicates identity with bovine growth hormone; differences are written out in full. -X- indicates a 'gap'.

The sequence for the bovine hormone is from Wallis (1973a) (see Fig. 1). The sequence of ovine GH is from Li et al. (1972); it is largely confirmed by Davies (1974), Fernandez et al. (1972) and Bellair (1972). The sequence of equine (horse) GH is from Zakin et al. (1973); dashes indicate incompletely sequenced regions. In some places, where amide assignments have not been completed, the assignment giving maximum homology with bovine GH has been assumed. The sequence of human GH is from Niall (1972) and Bewley et al. (1972). These authors differ about 1 or 2 amide assignments, and in these cases the sequence of Bewley et al. (1972) has been used. The sequence of human placental lactogen (human PL) is from Bewley et al. (1972), Niall (1972) and Sherwood et al. (1971). Again, in the case of a few disagreements about amide assignments, the sequence of Bewley et al. (1972) has been used.

Disulphide bridges join Cys^{53} to Cys^{164} and Cys^{181} to Cys^{189}. (From Wallis, 1975, by courtesy of the Editors of *Biological Reviews*.)

BGH (or any of the other mammalian GH sequences) — it differs at about 35% of all residues. The sequences of pig and horse GHs are similar to BGH — incomplete sequence determinations make it difficult to give a precise estimate of the extent of the differences here, but they are probably about 10% of all residues.

It seemed clear from such comparisons that the rate of evolution of the mammalian GHs had been rather variable, and it was of particular interest therefore to include the information now available for RGH into the comparison. It is clear from Figures 3 and 4 that RGH is rather similar to BGH, and probably differs at only about 11% of all residues. It is well established that the rodent, artiodactyl and primate lines separated at about the same time in the evolution of the mammals. The fact that human and bovine (or rat) GHs differ by about 35% of all residues while rat and bovine GHs differ by only 11% confirms that the rates of evolution in the growth hormone family have been very variable and demonstrates that the evolution of human GH (and presumably that of other primates) has been much more rapid than that of the other mammalian GHs that have been studied. This evidence from amino acid sequences agrees rather well with other evidence which suggests that the biological and immunological properties of human growth hormone differ substantially from those of other mammalian GHs.

Another protein which must be considered when the evolution of GHs is being discussed is human placental lactogen. This has a sequence very similar to that of human GH (differing at about 15% of all residues). It is much more like human GH than any of the other GHs for which sequences are available, and this suggests rather strongly that its appearance (presumably as the result of a gene duplication and subsequent divergence) must have occurred after the divergence of the main groups of placental mammals, during the evolution of the primate line. In this case, of course, the precise relationship between the human placental lactogen and the placental lactogens which have been reported from other (non-primate) mammalian species is thrown in doubt — it may be that placental lactogens in the various mammalian groups had more than one origin (Wallis, 1971; Bewley et al., 1972). The resolution of this problem will have to await further studies on the characterization and structure of non-primate placental lactogens.

The growth hormones also show homology with prolactin (e.g. Bewley et al., 1972; Wallis, 1971, 1974, 1975). The homology is distant, but distinct (about 25% of all residues are identical) and indicates that growth hormone and prolactin were originally derived from a common ancestor by gene duplication and then divergence. GH and prolactin are known to be distinct hormones in most fish groups, however, so the point at which the gene duplication giving rise to the hormone family occurred must have been very far back in evolutionary time.

The various structural relationships between the proteins of the growth hormone-prolactin family can be usefully summarized by means of an evolutionary tree (Fig. 6), constructed on the lines discussed by Dayhoff (1972). This tree is slightly different from that presented previously (Wallis, 1975); the divergence of bovine, rat and human GHs is assumed to have occurred at about the same time, a feature justified by the fossil record, but not by the sequence data alone.

Within the sequence of any member of the growth hormone-prolactin family, evidence for internal duplications can be discerned (Niall et al., 1971; Fellows et al., 1972). Proposals have also been made for homologies between members of this family and other proteins, but these are inevitably more speculative.

The comparative sequence studies discussed here have been based on sequences of a limited range of mammalian hormones. There is clearly a need for more sequence information, particularly from lower vertebrates. Studies on the characterization of

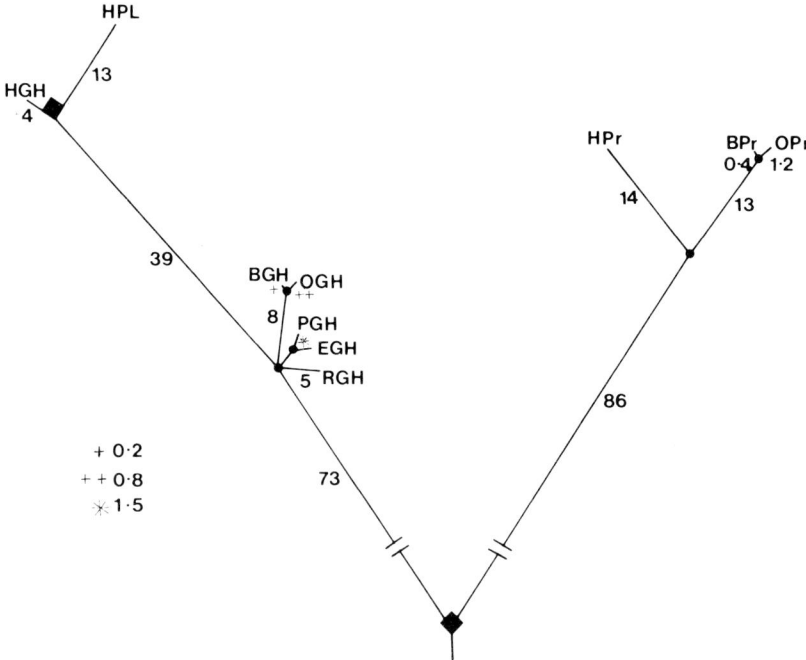

Fig. 6. 'Evolutionary tree' for the growth hormone-prolactin family. A diamond indicates gene duplication; a circle indicates species divergence. The figures on the branches ('branch lengths') indicate the distance, in accepted point mutations per 100 residues (Dayhoff, 1972) between the 2 nearest nodes. They are only approximate, because of incompleteness of some sequences, and disagreements about others. The tree was constructed from the corresponding amino acid sequences, using a matrix method. Dayhoff (1972) has also prepared a tree showing evolutionary relationships within this family.

The tree shown here differs from an earlier one (Wallis, 1975) in that the main groups of placental mammals are here shown as originating at the same point. This is based on evidence from the fossil record, and not amino acid sequence data; it emphasizes the point made in the text that the rate of evolution of growth hormone has varied from one group to another. The unexpectedly close resemblance of pig GH to horse GH is based on the partial sequences available for GHs from these species (Mills and Wilhelmi, 1972; Zakin et al., 1973); it may be reduced when the full sequences become available.

Abbreviations: BGH, bovine (ox) growth hormone; OGH, ovine (sheep) growth hormone; PGH, porcine (pig) growth hormone; EGH, equine (horse) growth hormone; RGH, rat growth hormone; HGH, human growth hormone; HPL, human placental lactogen; HPr, human prolactin; BPr, bovine prolactin; OPr, ovine prolactin.

reptilian, bird, amphibian and elasmobranch GHs have been reported (e.g. Lewis et al., 1972; Hayashida et al., 1973; Farmer et al., 1974) and we can clearly hope for extensive sequence data from these species to augment that which is available so far from mammals.

The comparative studies of growth hormone sequences have thrown much light on structural relationships within this protein family. They have been less successful however in illuminating structure-function relationships. Thus, we still have no clear idea why human placental lactogen, which is much more similar to growth hormones than it is to prolactins, has lactogenic activity but very low growth-promoting activity. Similarly, no structural basis is apparent for the possession of lactogenic activity by human GH, but not by the non-primate GHs (in most assay systems) — they are equally similar to prolactin. Clearly, the resolution of such problems lies at the level of 3-dimensional structure, and may have to await the results of X-ray crystallographic analysis.

Note added in proof

The sequence at the C-terminus of rat GH, shown in Figures 3 and 4, should read:

--- Phe — Ala — Glu — Ser — Ser — Cys — Ala — Phe — COOH

ACKNOWLEDGEMENTS

We thank the Medical and Agricultural Research Councils and the Royal Society for research grants and the NIAMDD Pituitary Hormone Distribution Program for gifts of various pituitary hormones. We are grateful to Mrs. Jenny Dew, Miss Margaret Elms, Mr. David Watson and Mr. Peter Dew for skilful technical assistance.

REFERENCES

Andrews, P. and Folley, S.J. (1963): *Biochem. J., 87,* 3P.
Bancroft, F.C. and Sussman, P.M. (1975): In: *Abstracts, The Endocrine Society, 57th Annual Meeting,* p. 111.
Bellair, J.T. (1972): *Biochem. biophys. Res. Commun., 46,* 1128.
Betteridge, A. and Wallis, M. (1973): *Biochem. J., 134,* 1103.
Bewley, T.A., Dixon, J.S. and Li, C.H. (1972): *Int. J. Peptide and Protein Res., 4,* 281.
Davies, R.V. (1974): *Studies on Sheep and Rat Growth Hormone.* Thesis, University of Sussex.
Davies, R.V. and Wallis, M. (1971): *Biochem. J., 125,* 65P.
Dayhoff, M.O. (1972): *Atlas of Protein Sequence and Structure, 1972, Vol. 5.* National Biomedical Research Foundation, Washington, D.C.
Dellacha, J.M., Enero, M.A. and Faiferman, I. (1966): *Experientia (Basel), 22,* 16.
Ellis, G.J., Marler, E., Chen, H.C. and Wilhelmi, A.E. (1966): *Fed. Proc., 25,* 348.
Ellis, S. (1961): *Endocrinology, 69,* 554.
Ellis, S., Lorenson, M., Grindeland, R.E. and Callahan, P.X. (1972): In: *Growth and Growth Hormone,* p. 55. Editors: A. Pecile and E.E. Müller. ICS 244, Excerpta Medica, Amsterdam.
Farmer, S.W., Papkoff, H. and Hayashida, T. (1974): *Endocrinology, 95,* 1560.
Fellows, R.E. (1973): *Recent Progr. Hormone Res., 29,* 404.
Fellows, R.E. and Rogol, A.D. (1969): *J. biol. Chem., 244,* 1567.
Fellows, R.E., Rogol, A.D. and Mudge, A. (1972): In: *Growth and Growth Hormone,* p. 42. Editors: A. Pecile and E.E. Müller. ICS 244, Excerpta Medica, Amsterdam.
Ferguson, K.A. and Wallace, A.L.C. (1961): *Nature (Lond.), 190,* 629.
Fernandez, H.N., Peña, C., Poskus, E., Biscoglio, M.J., Paladini, A.C., Dellacha, J.M. and Santomé, J.A. (1972): *FEBS Letters, 25,* 265.
Glaser, C.B. and Li, C.H. (1974): *Biochemistry, 13,* 1044.
Gráf, L. and Li, C.H. (1974a): *Biochem. biophys. Res. Commun., 56,* 168.
Gráf, L. and Li, C.H. (1974b): *Biochemistry, 13,* 5408.
Hayashida, T., Licht, P. and Nicoll, C.S. (1973): *Science, 182,* 169.
Lewis, U.J., Singh, R.N.P., Seavey, B.K., Lasker, R. and Pickford, G.E. (1972): *Fishery Bull., 70,* 933.
Li, C.H. and Ash, L. (1953): *J. biol. Chem., 203,* 419.

Li, C.H., Dixon, J.S., Gordon, D. and Knorr, J. (1972): *Int. J. Peptide and Protein Res., 4,* 151.
Lorenson, M.G. and Ellis, S. (1975): *Endocrinology, 96,* 833.
Mills, J.B. and Wilhelmi, A.E. (1972): In: *Growth and Growth Hormone,* p. 38. Editors: A. Pecile and E.E. Müller. ICS 244, Excerpta Medica, Amsterdam.
Niall, H.D. (1972): In: *Prolactin and Carcinogenesis,* p. 13. Editors: A.R. Boyns and K. Griffiths. Alpha Omega Alpha Publishing, Cardiff.
Niall, H.D., Hogan, M.L., Sauer, R., Rosenblum, I.Y. and Greenwood, F.C. (1971): *Proc. nat. Acad. Sci. (Wash.), 68,* 866.
Peña, C., Paladini, A.C., Dellacha, J.M. and Santomé, J.A. (1969): *Biochim. biophys. Acta (Amst.), 194,* 320.
Reagan, C.R., Mills, J.B., Kostyo, J.L. and Wilhelmi, A.E. (1975): *Proc. nat. Acad. Sci. (Wash.), 72,* 1684.
Santomé, J.A., Dellacha, J.M., Paladini, A.C., Peña, C., Biscoglio, M.J., Daurat, S.T., Poskus, E. and Wolfenstein, C.E.M. (1973): *Europ. J. Biochem., 37,* 164.
Santomé, J.A., Dellacha, J.M., Paladini, A.C., Wolfenstein, C.E.M., Peña, C., Poskus, E., Daurat, S.T., Biscoglio, M.J., De Sesé, Z.M.M. and De Sangüesa, A.V.F. (1971): *FEBS Letters, 16,* 198.
Santomé, J.A., Wolfenstein, C.E.M., Biscoglio, M.J. and Paladini, A.C. (1966): *Arch. Biochem., 116,* 19.
Seavey, B.K., Singh, R.N.P., Lewis, U.J. and Geschwind, I.I. (1971): *Biochem. biophys. Res. Commun., 43,* 189.
Seavey, B.K., Singh, R.N.P., Lindsay, T.T. and Lewis, U.J. (1973): *Gen. comp. Endocr., 21,* 358.
Sherwood, L.M., Handwerger, S., McLaurin, W.D. and Lanner, M. (1971): *Nature New Biol., 233,* 59.
Sonenberg, M., Yamasaki, N., Kikutani, M., Swislocki, N.I., Levine, L. and New, M. (1972): In: *Growth and Growth Hormone,* p. 75. Editors: A. Pecile and E.E. Müller. ICS 244, Excerpta Medica, Amsterdam.
Wallis, M. (1964): *Studies on Ox Growth Hormone.* Thesis, University of Cambridge.
Wallis, M. (1966): *Biochim. biophys. Acta (Amst.), 115,* 423.
Wallis, M. (1969): *FEBS Letters, 3,* 118.
Wallis, M. (1971): *Biochem. J., 125,* 54P.
Wallis, M. (1972): *FEBS Letters, 21,* 118.
Wallis, M. (1973a): *FEBS Letters, 35,* 11.
Wallis, M. (1973b): *Biochim. biophys. Acta (Amst.), 310,* 388.
Wallis, M. (1974): *FEBS Letters, 44,* 205.
Wallis, M. (1975): *Biol. Rev., 50,* 35.
Wallis, M. and Davies, R.V. (1974): *Biochem. Soc. Trans., 2,* 911.
Wallis, M. and Dew, J.A. (1973): *J. Endocr., 56,* 235.
Wallis, M. and Dixon, H.B.F. (1966): *Biochem. J., 100,* 593.
Zakin, M.M., Poskus, E., Dellacha, J.M., Paladini, A.C. and Santomé, J.A. (1973): *FEBS Letters, 34,* 353.

STUDIES ON PLASMIN-MODIFIED HUMAN GROWTH HORMONE AND ITS FRAGMENTS*

CHOH HAO LI AND THOMAS A. BEWLEY

Hormone Research Laboratory, University of California, San Francisco, Calif., U.S.A.

The use of plasmin for the partial digestion of somatotropins was first described by Ellis et al. (1968). Chrambach and Yadley (1970) reported an 8-fold enhancement of tibial-line bioactivity of human somatotropin (HGH) produced by plasmin at the 1970 Endocrine Society Annual Meeting. Later, Yadley and Chrambach (1973) described in detail that digestion of HGH with plasmin leads to a progressive transformation of the hormone to various components of higher electrophoretic mobility and increased prolactin activity. Reagan et al. (1973) and Mills et al. (1973) reported that hydrolysis of HGH with plasmin did not impair metabolic effects in the rat or in man. A detailed chemical and physical characterization of the cleaved products of HGH by plasmin digestion was not given in any of the studies.

Li and Gráf (1974) reported the isolation and characterization of 2 biologically active fragments from plasmin digests of HGH. It was shown that these 2 fragments [Cys(Cam)^{53}-HGH-(1-134) and Cys(Cam)^{165,182,189}-HGH-(141-191)] were derived from cleavage of the Arg-Thr (positions 134-135) and the Lys-Gln (positions 140-141) bonds of the hormone molecule (see Fig. 1). The isolation and biological characterization of Cys(Cam)^{53}-HGH-(1-134) have also been described by Reagan et al. (1975). Both peptide fragments possess immunoreactivity in complement fixation and radioimmunoassay experiments (Clarke et al., 1974). In this paper, we report our recent studies on plasmin-modified HGH and its derivatives and fragments.

PREPARATION AND CHARACTERIZATION OF CYS(CAM)^{53}-HGH-(15-125)

A hendekakaihekaton peptide fragment has been prepared by cyanogen bromide cleavage of Cys(Cam)^{53}-HGH-(1-134). Cyanogen bromide cleavage was carried out by

---

Abbreviations used are: HGH, human pituitary growth hormone; PL-HGH, HGH modified by limited digestion with human plasmin; Cys(SH)^{182,189}-PL-HGH, the product formed by reduction of one disulfide bond in PL-HGH; Cys(Cam)^{182,189}-PL-HGH, the product formed by alkylation of Cys(SH)^{182,189}-PL-HGH with α-iodoacetamide; Cys(SH)^{53,165,182,189}-PL-HGH, the product formed by reduction of both disulfide bonds in PL-HGH; Cys(Cam)^{53,165,182,189}-PL-HGH, the product formed by alkylation of Cys(SH)^{53,165,182,189}-PL-HGH with α-iodoacetamide; Cys(Cam)^{53}-HGH-(1-134), the 134 residue N-terminal fragment prepared from Cys(Cam)^{53,165,182,189}-PL-HGH; Cys(Cam)^{165,182,189}-HGH-(141-191), the 51 residue C-terminal fragment prepared from Cys(Cam)^{53,165,182,189}-PL-HGH; REOX-PL-HGH, reduced and auto-oxidized PL-HGH; DTT, dithiothreitol; CD, circular dichroism; ODC, ornithine decarboxylase.

* This work was supported in part by the American Cancer Society and the Allen-Geffen Fund.

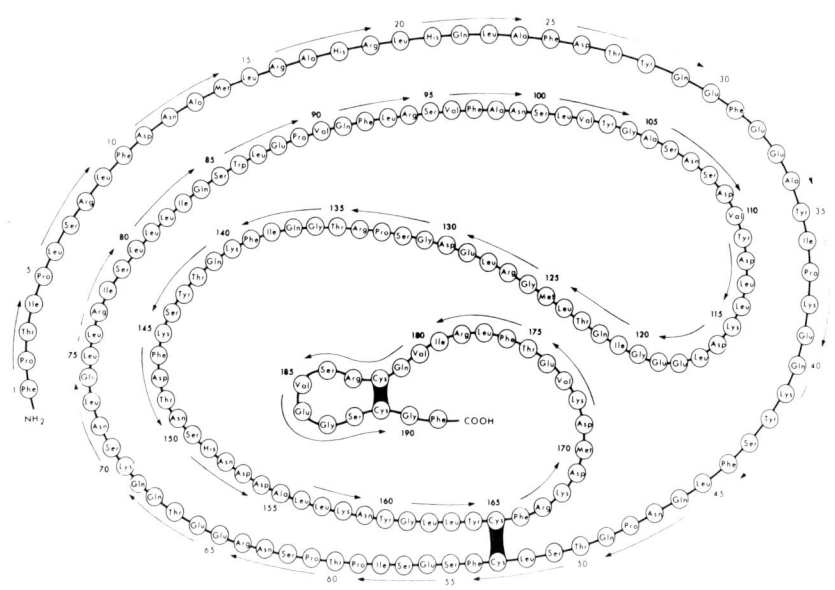

Fig. 1. Amino acid sequence of the HGH molecule.

the procedure of Gross and Witkop (1962) as described previously (Li et al., 1970). The CNBr fragments (50 mg) were fractionated by gel filtration on Sephadex G-75 in 20% acetic acid, and the effluents were examined by an automatic fluorescence monitor (Böhlen et al., 1975) as well as by absorbance at 280 nm. The contents in the 2 main peaks were lyophilized and yielded 35 and 10 mg, respectively. Amino acid (Moore, 1972) and end group (Gray, 1967) analyses revealed that the first peak contained the expected product, Cys(Cam)[53]-HGH-(15-125), whereas the second peak consisted of 2 peptide fragments with compositions corresponding to HGH-(1-14) and HGH-(126-134). The amino acid compositions of the isolated Cys(Cam)[53]-HGH-(15-125) fragment after acid hydrolysis and total enzymatic digestion are in agreement with theoretical values. The NH_2-terminal sequence of the fragment was found to be Leu-Arg-Ala-His-Arg-Leu-, while treatment with carboxypeptidase gave Hsl, Leu, and Thr.

The chromatographic and sedimentation behavior of the fragment in comparison with HGH, plasmin-treated HGH, and Cys(Cam)[53]-HGH-(1-134) on Sephadex G-100 in 0.01 M NH_4HCO_3 indicated that it is an aggregated form with a Ve/Vo value of 1.25 (Table 1). The fragment migrates as a single band with the same mobility as Cys(Cam)[53]-HGH(1-134) in disc electrophoresis experiments (Fig. 2). In radioimmunoassay, this fragment and Cys(Cam)[53]-HGH-(1-134) both gave inhibition curves with guinea pig antiserum to HGH which were parallel to those of HGH as shown in Figure 3. However, the fragment was much less effective than Cys(Cam)[53]-HGH-(1-134), requiring much higher antigen concentrations to give the same degree of inhibition. Rabbit antiserum to Cys(Cam)[53]-HGH-(1-134) at a dilution of 1:130 reacted with 50 ng of the antigen to fix about 91% complement, whereas the fragment's reaction was slightly lower (Fig. 4).

The CD spectra (Fig. 5) of the 2 peptides, indicated that Cys(Cam)[53]-HGH-(1-134) possesses 40% α-helix, whereas the smaller fragment appears to contain virtually no

TABLE 1

Relative elution volume and sedimentation coefficients of HGH and its fragments

Preparation	Ve/Vo[a]	$S_{20,w}$[b]
HGH	1.89	2.50
Plasmin-treated HGH	1.98	2.41
Cys(Cam)[53]-HGH-(1-134)	1.67	3.81
Cys(Cam)[53]-HGH-(15-125)	1.25	4.55

[a] Sephadex G-100 column (2.7 × 78 cm); 0.01 M, NH_4HCO_3, pH 8.2.
[b] 0.1 M Tris buffer, pH 8.20; protein conc. 1 mg/ml.

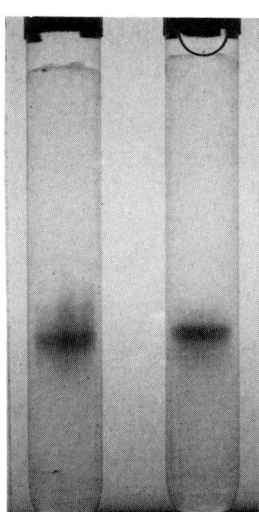

Fig. 2. Electrophoresis of Cys(Cam)[53]-HGH-(15-125) (left) and Cys(Cam)[53]-HGH-(1-134) (right) on polyacrylamide gel (7%) at pH 4.5 (50-μg samples).

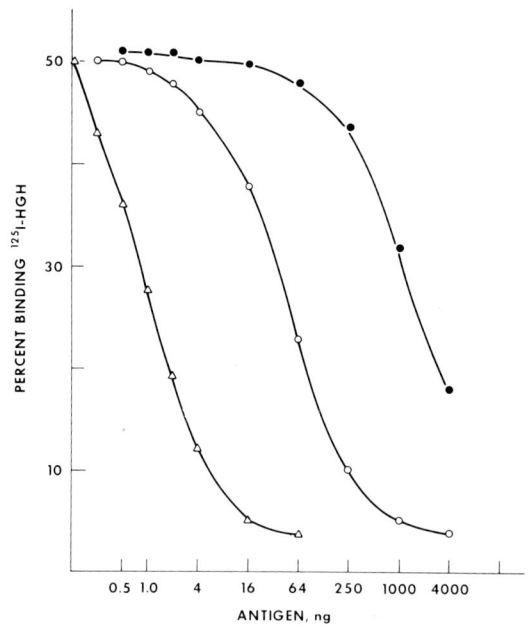

Fig. 3. Competition of HGH (△—△), Cys(Cam)[53]-HGH-(1-134) (o—o), and Cys(Cam)[53]-HGH-(15-125) (●—●) in the HGH radioimmunoassay systems. Final dilution of guinea pig antiserum was 1/160,000.

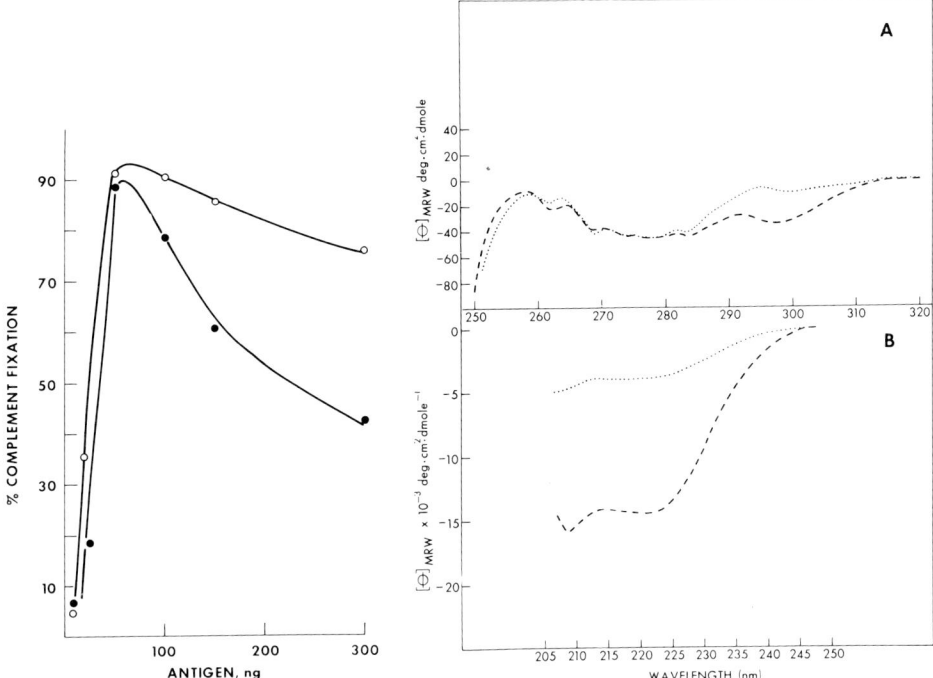

Fig. 4. Microcomplement fixation curves obtained with Cys(Cam)[53]-HGH-(1-134) (o—o) or Cys(Cam[53]-HGH-(15-125) (●—●) and rabbit antiserum to Cys-(Cam)[53]-HGH-(1-134).

Fig. 5. CD spectra in 0.1 M Tris buffer of pH 8.2. A. Side-chain spectra. B. Amide bond spectra: Cys(Cam)[53]-HGH-(1-134) (---) and Cys(Cam)[53]-HGH-(15-125) (....).

α-helix. In the region of side-chain absorption, the fragment is almost devoid of indole dichroism.

Table 2 summarizes the results of the biological activity in the pigeon crop sac and rat tibia tests. It may be noted that the prolactin activity of the fragment is about 50% that of Cys(Cam)[53]-HGH-(1-134). In comparison with HGH, the fragment retains about 3% of the prolactin potency. In the rat tibia assay, the fragment gave a non-parallel dose-response relationship when compared with the data obtained with either HGH or Cys(Cam)[53]-HGH-(1-134). Hence, although it is not possible to estimate its potency, it is nevertheless evident that the fragment has measurable growth-promoting activity.

Bovine somatotropin is known to stimulate ornithine decarboxylase activity in the liver (Jänne and Raina, 1969; Russell et al., 1970). Recent data of Rao et al. (1975) show that HGH, PL-HGH, and Cys(Cam)[53]-HGH-(1-134) stimulate hepatic ornithine decarboxylase activity in 21-day-old male rats (see Table 3). HGH produced a significant stimulation at a dose of 0.14 nmole and the response was proportional to the dose of the hormone. Plasmin-treated HGH appears to be more potent than HGH. The NH$_2$-terminal fragment was found to possess 10% of the potency of the native hormone. The COOH-terminal fragment was inactive at the highest dose tested (120 nmole). Table 4 presents the ornithine decarboxylase stimulating activity of

TABLE 2
Biological activity of Cys(Cam)^{53}-HGH-(15-125)

Preparation	Tibia test			Crop sac assay[d]	
	Total dose (μg)	Response	p-Value[a]	Total dose (μg)	Response
Saline	0	166 ± 2.1		0	10.3 ± 0.4
HGH	60	267 ± 5.2		3	24.6 ± 1.0
				6	32.6 ± 0.1
Cys (Cam)^{53}-HGH-(1-134)	300	234 ± 4.9	0.005-0.001[b]	60	26.6 ± 0.8[e]
				120	35.4 ± 1.6
Cys (Cam)^{53}-HGH-(15-125)	300	219 ± 3.9	< 0.001[b]	60	19.3 ± 0.7[f]
			0.005-0.001[c]	120	27.3 ± 1.6

[a] Mean ± S.E. in microns; 10 rats in each group.
[b] In comparison with the response of saline controls.
[c] In comparison with the response of Cys(Cam)^{53}-HGH-(1-134).
[d] Dry mucosal weight in mg; mean ± S.E.; 3 birds in each group.
[e] Relative potency to HGH, 6.1%, with 95% confidence limit of 5.1-7.6 and λ = 0.061.
[f] Relative potency to HGH, 3.2%, with 95% confidence limit of 2.4-3.9 and λ = 0.063.

TABLE 3
Stimulation of hepatic ornithine decarboxylase activity by HGH and its derivatives

Preparation	Dose (nmole)	Response[a]
Saline	0.0	81.2 ± 8.4 (27)
HGH	0.14	100.8 ± 14.8 (9)
	0.42	161.2 ± 26.7 (19)
	1.26	274.5 ± 11.0 (10)
PL-HGH[b]	0.14	137.0 ± 11.0 (5)
	0.42	274.0 ± 44.0 (5)
Cys (Cam)^{53}-HGH-(1-134)[c]	3.10	109.7 ± 10.2 (10)
	9.30	233.5 ± 45.2 (10)
Cys (Cam)$^{165, 182, 189}$-HGH-(141-191)	40.0	97.7 ± 24.6 (5)
	120.0	82.0 ± 32.0 (5)

[a] Values (pmole $^{14}CO_2$/mg protein/hr) are the mean ± S.E. The number of animals is given in parentheses. For experimental conditions, see footnote 'a' of Table 4.
[b] Relative potency to HGH, 130% with confidence limit of 34-1300 and λ = 0.45.
[c] Relative potency to HGH, 10% with confidence limit of 5-41 and λ = 0.33.

TABLE 4

Hepatic ornithine decarboxylase stimulating activity[a] of HGH and Cys(Cam)[53]-HGH-(15-125)

Preparation	Dose (nmole)	Response[b]
Saline	0	189 ± 26 (8)
HGH	0.42	361 ± 31 (6)
	1.26	617 ± 38 (6)
Cys(Cam)[53]-HGH-(15-125)[c]	7.75	443 ± 59 (6)
	23.27	598 ± 27 (6)

[a] Male rats, 21 days of age, of the Sprague-Dawley strain were injected intraperitoneally with 0.2 ml of the hormone preparation or saline. Four hours later, the animals were sacrificed, and the livers removed and homogenized at 4° in 0.25 mM Tris buffer of pH 7.5 containing 0.1 mM disodium EDTA and 5 mM dithiothreitol. The homogenate was centrifuged at 20,000 × g for 60 minutes at 4° and an aliquot of the supernatant was assayed for ornithine decarboxylase activity.
[b] pmoles $^{14}CO_2$/mg protein/hr; mean ± S.E. (number of animals).
[c] Relative potency to HGH, 6.9% with a confidence limit of 4.1-11.9 and λ = 0.25.

Cys(Cam)[53]-HGH-(15-125) in the rat liver. It was found that the 111 residue fragment possesses 7% potency of the natural hormone.

It is evident from the CD spectra (Fig. 5) that the fragment does not possess distinct, 3-dimensional structures. Apparently the removal of the NH$_2$-terminal 14 amino acids and the COOH-terminal 9 amino acids from Cys(Cam)[53]-HGH-(1-134) causes an extensive loss of α-helical content and a marked change in the rigid tertiary structure. However, the absence of the secondary and tertiary structures in the 111 residue fragment does not abolish all the biological activities. The ODC stimulating activity of the fragment appears to be slightly lower than the 134 residue fragment (Rao et al., 1975). It is entirely possible that the fragment has 3-dimensional structures distinct from HGH which are not detectable in the CD spectra.

REGENERATION OF BIOLOGICAL ACTIVITY BY MIXING THE TWO PLASMIN FRAGMENTS OF PL-HGH

A mixture was prepared by adding 1.34 mg (0.087 μmole) of Cys(Cam)[53]-HGH-(1-134) to 0.54 mg (0.089 μmole) of Cys(Cam)[165,182,189]-HGH-(141-191) in 1.0 ml of pH 8.4 Tris buffer (0.1 M, 2% butanol). The turbid solution was kept at room temperature (23°C) for 5 hours and then transferred to the refrigerator (4°C). After 240 hours, the solution appeared to have become clear and was submitted for bioassays.

It may be seen in Table 5 that the solution containing the 2 fragments had much higher growth-promoting activity in comparison with the potency of the 134 residue fragment in the tibia test. At the dose tested, the potency of the 51 residue fragment was almost nil. The regeneration of lactogenic activity was also evident in the pigeon crop sac assay as shown in Table 6. It has not been determined whether the regeneration of biological activities is due to the repair of the original conformational state of PL-HGH, or due to synergistic action of the fragments when mixed together under these experimental conditions.

TABLE 5

Growth-promoting activity of $[Cys(Cam)^{53}$-HGH-$(1$-$134)]$ plus $[Cys(Cam)^{165, 182, 189}$-$HGH$-$(141$-$191)]$ in the tibia test

Preparation	Total dose (nmole)	Response[a]
PL-HGH	0.93	241.0 ± 2.4
	2.79	267.2 ± 7.7
Cys (Cam)^{53}-HGH-(1-134)[b, c] +	2.27	252.5 ± 10.2
Cys (Cam)$^{165, 182, 189}$-HGH-(141-191)	6.81	281.2 ± 6.4
Cys (Cam)^{53}-HGH-(1-134)[d]	3.26	207.2 ± 1.5
	9.78	235.5 ± 1.5
Cys (Cam)$^{165, 182, 189}$-HGH-(141-191)	9.78	186.7 ± 4.4
Saline	0	176.5 ± 4.5

[a] Tibia width in micra; Mean ± S.E.; 4 animals in each group.
[b] Relative potency to PL-HGH, 60% with a confidence limit of 40-153 and $\lambda = 0.24$.
[c] Relative potency to Cys(Cam)^{53}-HGH-(1-134), 828% with a confidence limit of 441-3077 and $\lambda = 0.17$.
[d] Relative potency to PL-HGH, 8% with a confidence limit of 4-12 and $\lambda = 0.12$.

TABLE 6

Prolactin activity of $[Cys(Cam)^{53}$-HGH-$(1$-$134)]$ plus $[Cys(Cam)^{165, 182, 189}$-$HGH$-$(141$-$191)]$ in the pigeon crop sac assay

Preparation	Total dose (nmole)	Response[a]
PL-HGH	0.093	24.6 ± 0.8
	0.279	28.8 ± 1.7
Cys (Cam)^{53}-HGH-(1-134)[b, c] +	0.48	16.8 ± 0.7
Cys (Cam)$^{165, 182, 189}$-HGH-(141-191)	1.44	20.7 ± 1.4
Cys (Cam)^{53}-HGH-(1-134)[d]	0.65	9.6 ± 1.8
	1.95	12.4 ± 2.7
Cys (Cam)$^{165, 182, 189}$-HGH-(141-191)	20.33	13.4 ± 1.2
Saline	0	10.5 ± 0.2

[a] Dry mucosal weight in mg; Mean ± S.E.; 4 birds in each group.
[b] Relative potency to PL-HGH, 23% with a confidence limit of 0.3-62 and $\lambda = 0.27$.
[c] Relative potency to Cys (Cam)^{53}-HGH-(1-134), approximately 947% with $\lambda = 0.46$.
[d] Relative potency to PL-HGH, approximately 1% with $\lambda = 0.46$.

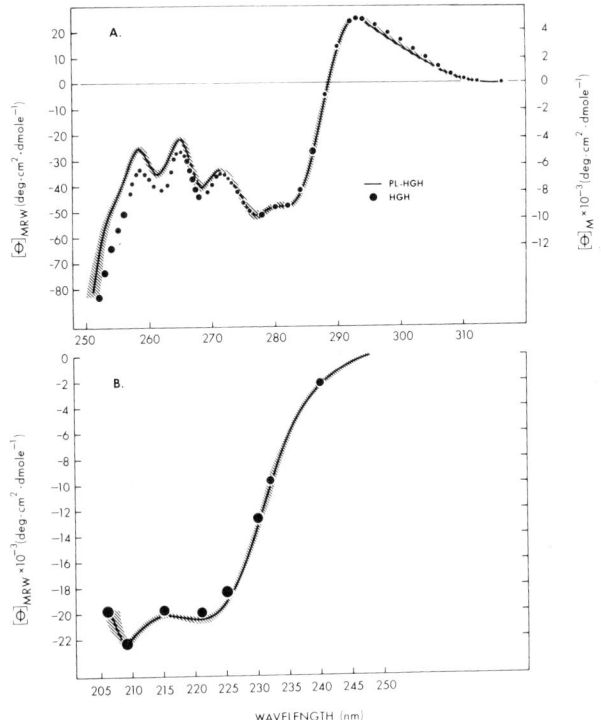

Fig. 6. Circular dichroism spectra in the region of side-chain absorption (A) and the region dominated by amide bond absorption (B) of PL-HGH (———) and HGH (....) in 0.1 M Tris-Cl buffer, pH 8.2. In (A) the spectrum of PL-HGH is the mean of 10 separate spectra using 3 preparations of the protein. The shaded area represents the S.E.M. for these spectra as a function of wavelength. Similarly, the diameter of the circles represents the S.E.M. for 7 spectra of HGH. In (B), 3 spectra were used for PH-HGH and 4 for HGH.

CIRCULAR DICHROISM STUDIES OF PL-HGH, DERIVATIVES AND FRAGMENTS

Figure 6 shows the CD spectrum of PL-HGH compared with the spectrum of native HGH. It can be seen that except for very small differences above 294 nm and below 266 nm, the 2 spectra are essentially identical. The positive asymmetric band above 289 nm has been previously assigned to the tryptophan residue at position 86 in HGH (Aloj and Edelhoch, 1972; Bewley et al., 1972; Bewley and Li, 1972). The negative bands at 282 and 277 nm have been assigned predominantly to tyrosine residues, with the 2 negative bands at 268-269 and 261-262 nm being assigned to phenylalanine (Aloj and Edelhoch, 1972; Bewley and Li, 1972). The same assignments may be applied to the bands in PL-HGH. From Figure 6B, it can be seen that in the region dominated by amide bond absorption, PL-HGH and HGH again display identical CD spectra within the limits of experimental error. The α-helix content of PL-HGH is therefore estimated to be 55 ± 5% as previously reported for native HGH (Bewley et al., 1969). The entire CD spectra of both these proteins are unchanged on standing at room temperature for periods up to 100 hours. The CD spectra of PL-HGH presented here clearly

TABLE 7

Exclusion chromatographic and optical properties of HGH, PL-HGH, and reduced-carbamidomethylated derivatives and fragments of PL-HGH[a]

Preparation	Ve/Vo	K_{av}[k]	s (Å)[k]	Fluorescence emiss. max. (nm)	$E_{1\,cm,\,277\,nm}^{0.1\%}$
HGH	2.12	0.537	23.8	336[b]	0.931[c]
PL-HGH	2.13	0.541	23.6	336-337[e]	0.931[d]
Cys (Cam)[182, 189]-PL-HGH	2.13	0.541	23.6	–	0.931[d]
Cys (Cam)[53, 165, 182, 189]-PL-HGH	2.12	0.537	23.8	–	0.931[d]
Cys (Cam)[53]-HGH-(1-134)[h]	1.75	0.360	32.3	343-344[e]	0.799[g]
Cys (Cam)[53]-HGH-(1-134)[j]	1.75	0.359	32.4	334-335[e]	0.799[g]
Cys (Cam)[165, 182, 189]-HGH-(141-191)[h]	3.10	–	–	305-306[f]	0.657[g]
Cys (Cam)[165, 182, 189]-HGH-(141-191)[j]	3.08	–	–	307-308[f]	0.657[g]

[a] Measured in 0.1 M Tris-Cl buffer, pH 8.2.
[b] Data from Bewley et al. (1972), excitation at 285 nm.
[c] Data from Bewley et al. (1969)
[d] Assumed to be the same as HGH.
[e] Excitation at 294 nm.
[f] Excitation at 280 nm.
[g] Calculated from amino acid composition.
[h] Fragment prepared in this study.
[j] Fragment prepared by Li and Gráf (1974).
[k] Calculated as described by Laurent and Killander (1964).

demonstrate that the loss of the hexapeptide comprising residues 135-140 does not produce any significant alterations in either the secondary or tertiary structure of the molecule. This fact is also evidenced by the equivalent exclusion chromatographic behavior and fluorescence emission maxima of these 2 proteins (Table 7).

Cys(Cam)[53]-HGH-(1-134) and Cys(Cam)[165,182,189]-HGH-(141-191) were prepared by a modified procedure which will be described in the next section. As shown in Figure 7, CD spectra of these two fragments are different from that of earlier samples of Li and Gráf (1974). The new preparation of Cys(Cam)[53]-HGH-(1-134) exhibits a very intense negative CD band at 298 nm as well as an increased negative dichroism in the far UV region. The 2 preparations of Cys(Cam)[165,182,189]-HGH-(141-191) appear to be about the same with regard to all properties measured except for the appearance of β-structure in the sample of Li and Gráf (1974).

REDUCTION AND ALKYLATION OF PL-HGH WITH DTT

2.5-mole excess DTT

The side-chain CD spectrum of PL-HGH after 2×10^3 seconds in the presence of a 2.5-mole excess of DTT is shown in Figure 8 along with the spectrum of the starting material for comparison. The positive indole band is unchanged but there is a signif-

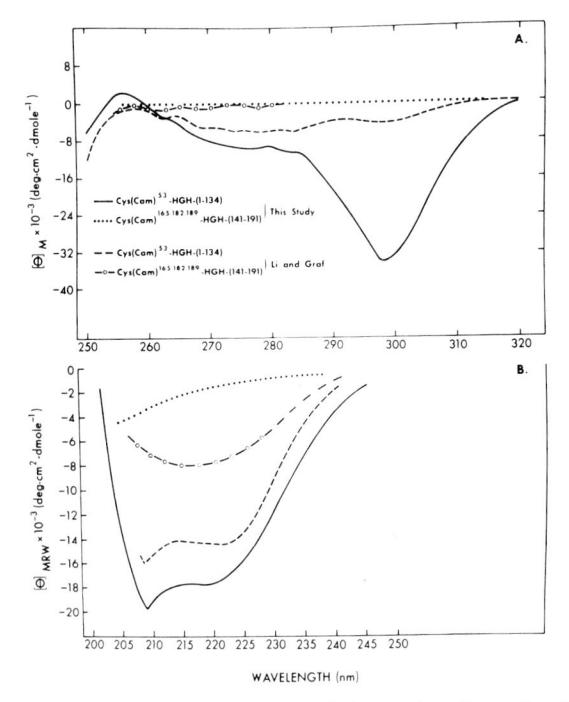

Fig. 7. Circular dichroism spectra in the region of side-chain absorption (A) and the region dominated by amide bond absorption (B) of Cys(Cam)^{53}-HGH-(1-134) as prepared in this study (———) and as prepared by Li and Gráf (1974) (----), and Cys(Cam)^{165,182,189}-HGH-(141-191) as prepared in this study (....) and as prepared by Li and Gráf (- o -). All spectra were taken in the Tris-Cl buffer.

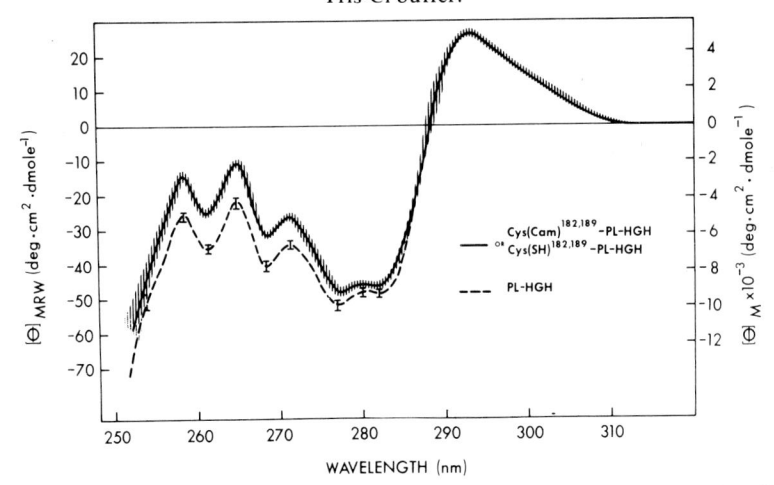

Fig. 8. Circular dichroism spectra in the region of side-chain absorption of Cys(SH)^{182,189}-PL-HGH (———) and PL-HGH (----). Spectra of the reduced form were taken after 2×10^3 seconds in the presence of a 2.5-mole excess of DTT; the shaded area represents the S.E.M. for 4 preparations. The S.E.M. for the PL-HGH spectrum is the same as in Figure 6, but shown here as vertical bars. All spectra were taken in the Tris-Cl buffer. The spectrum of Cys(Cam)^{182,189}-PL-HGH is identical with that of Cys(SH)^{182,189}-PL-HGH.

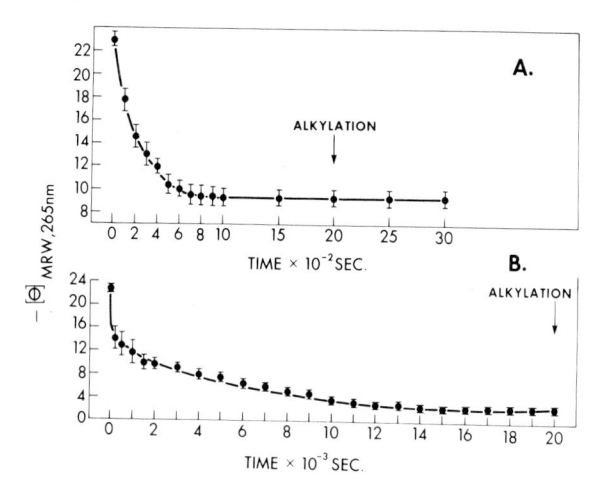

Fig. 9. The rate of change in the ellipticity at 265 nm for PL-HGH treated with a 2.5-mole excess of DTT (A) and a 25-mole excess (B) of DTT. In both curves, the filled circles represent the mean values for 4 experiments; S.E.M. values are shown as vertical bars. Reactions were carried out at 27°C in the Tris-Cl buffer.

icant loss of negative dichroism in the reduced protein. This loss increases from 286 to about 260 nm. Below 260 nm the 2 spectra appear to converge again with no significant differences appearing between them at wavelengths below 240 nm. Due to rapidly increasing sample absorptivity, producing an unfavorable signal-to-noise ratio, the data from 250 to 245 nm are of limited accuracy. Figure 9A shows the rate of loss of negative dichroism at 265 nm. The reaction is more than 50% complete after the first 5 minutes with an apparently stable state being reached in about 15 minutes. This state remains unchanged for the next hour. The spectrum of this product, shown in Figure 8, was taken between 2 and 3 × 10^3 seconds after adding the reducing agent. During this period, there was no time dependence at any wavelength between 320 and 200 nm. Alkylation of the reduced product was performed after the final CD scan had been completed. A monomeric form of the alkylated protein was obtained from Sephadex G-100 in about 95% yield (see Fig. 10D). Amino acid analysis of the product indicated that only one of the 2 -S-S- bonds had been modified. The chromatographic behavior of the purified monomer was unchanged on standing at room temperature for periods up to 150 hours (Fig. 10D). Within experimental error, the entire CD spectrum of the Cys(Cam)[182,189]-PL-HGH was indistinguishable from that of the Cys-(SH)[182,189]-PL-HGH shown in Figure 8. This includes both the near and far-UV regions (see also Table 8). Moreover, within experimental error, the far-UV CD of both these derivatives are indistinguishable from either PL-HGH or native HGH (Table 8). Both the near and far-UV CD spectra of Cys(Cam)[182,189]-PL-HGH were found to be stable at room temperature for periods up to 150 hours.

25-mole excess DTT

Figure 11 shows the side-chain CD spectrum of PL-HGH after 20 × 10^3 seconds in the presence of a 25-mole excess of DTT. A small but significant loss of positive dichroism is seen above 289 nm, with an even greater loss of negative dichroism below 285 nm

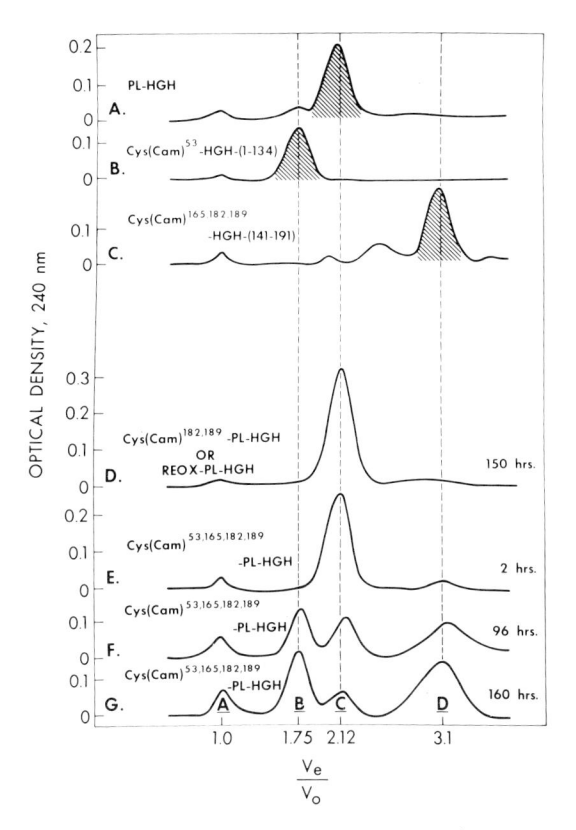

Fig. 10. Exclusion chromatography of PL-HGH, derivatives and fragments on Sephadex G-100 (column 1.5 × 59 cm) with ascending elution at a flow rate of 4.88 ml/hr. The times indicated at the right of patterns D, E, F, and G are the intervals during which the samples were incubated at room temperature before chromatography. The samples producing patterns A, B and C were all prepared as described by Li and Gráf (1974).

than found in the case of Cys(SH)$^{182-189}$-PL-HGH. Again the spectrum of the starting material and this reduced form converge below 260 nm. The rate of loss of ellipticity at 265 nm is shown in Figure 9B. The reaction is about 50% complete in the first 2 minutes. At this point a slow but progressive change in ellipticity continued to occur for the next 12-15 × 10^3 seconds, with a stable state being achieved by 20 × 10^3 seconds. The total change in ellipticity at 265 nm is almost twice that seen for Cys-(SH)^{182,189}-PL-HGH. Despite this greater difference in the side-chain CD, the far-UV CD of Cys(SH)^{53,165,182,189}-PL-HGH is unchanged from that of the starting material (Table 8). Although the CD changes brought about by reduction of the protein are indeed small, it is clear that they are significant and can be measured with accuracy.

Alkylation of the reduced protein was performed at 20 × 10^3 seconds and the alkylated protein immediately submitted to exclusion chromatography on Sephadex G-100. A monomeric product was obtained in approximately 90% yield (Fig. 10E). Amino acid analysis indicated that both disulfide bonds had been quantitatively modified. Although the far-UV CD spectrum of this tetracarbamidomethylated product was not markedly changed from that of the PL-HGH (or Cys(SH)^{53,165,182,189}-PL-

TABLE 8

Mean residue weight ellipticities and estimated α-helix contents of PL-HGH derivatives and fragments[a]

Preparation	220 nm[b]	209 nm[b]	α-helix content[c]
Cys (SH)[182, 189]-PL-HGH	−21,390	−23,390	55
Cys (Cam)[182, 189]-PL-HGH	−21,300	−23,400	55
Cys (SH)[53, 165, 182, 189]-PL-HGH	−20,370	−21,350	55
Cys (Cam)[53, 165, 182, 189]-PL-HGH[d]	−21,500	−23,000	55
REOX-PL-HGH	−20,200	−22,650	55
Cys (Cam)[53]-HGH-(1-134)[e]	−17,900	−19,700	50
Cys (Cam)[53]-HGH-(1-134)[f]	−14,500	−16,000	40
Cys (Cam)[165, 182, 189]-HGH-(141-191)[e]	−1,600	−3,500	0[g]
Cys (Cam)[165, 182, 189]-HGH-(141-191)[f]	−7,500	−7,000	0[h]

[a] Measured in 0.1 M Tris-Cl buffer, pH 8.2.
[b] In deg · cm^2 · dmole^{-1}
[c] Estimated to the nearest 5% as described by Bewley et al. (1969).
[d] Measured on a freshly thawed, frozen sample.
[e] Fragment prepared in this study.
[f] Fragment prepared as described by Li and Gráf (1974).
[g] This fragment appears to be a random coil with a negative maximum at 195-197 nm.
[h] This fragment appears to contain ≈ 40% β-structure with a negative maximum at 215-217 nm.

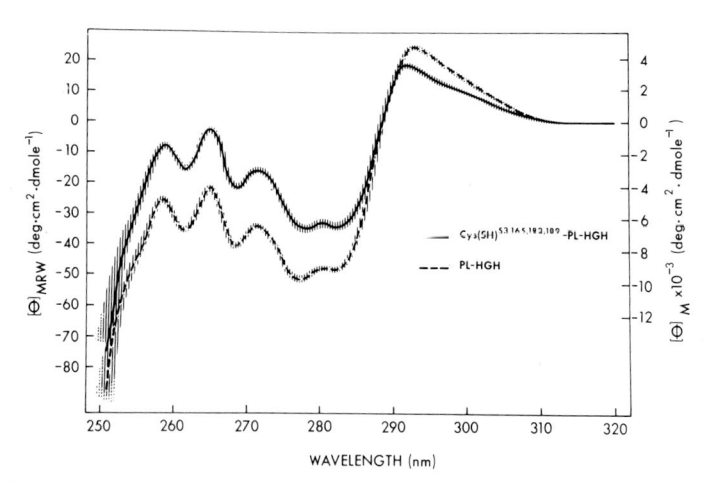

Fig. 11. Circular dichroism spectra in the region of side-chain absorption of Cys(SH)[53,165,182,189]-PL-HGH (—) and PL-HGH (----). The spectrum of the reduced form is the mean of 8 experiments begun after 20×10^3 seconds in the presence of a 25-mole excess of DTT in the Tris-Cl buffer. The shaded area represents the S.E.M. for these experiments. The spectrum of PL-HGH is the same as in Figure 6A.

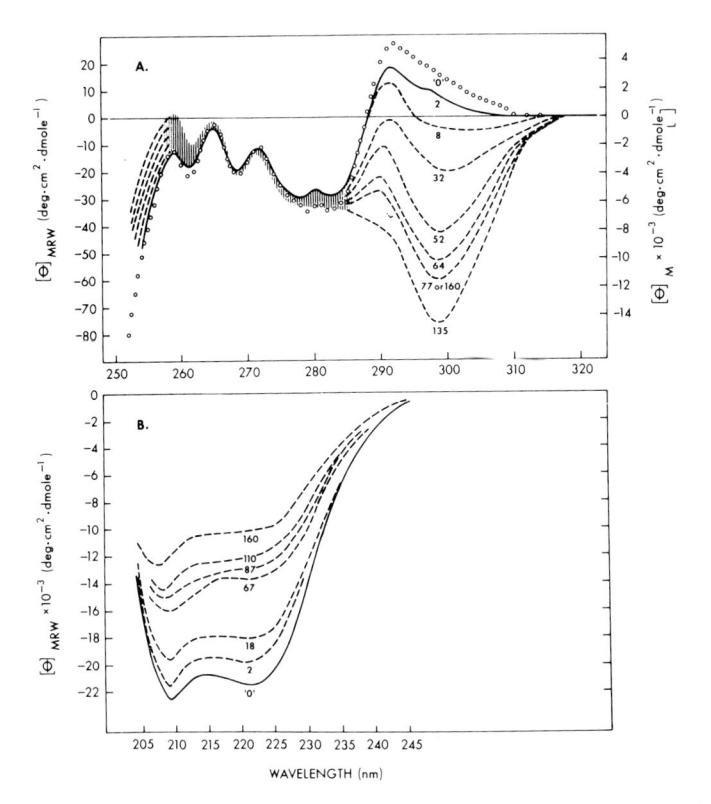

Fig. 12. The effect of incubation in the Tris-Cl buffer, at room temperature, on the circular dichroism spectra of Cys(Cam)[53,165,182,189]-PL-HGH in the region of side-chain absorption (A) and the region dominated by amide bond absorption (B). The time of incubation is indicated just below each spectrum. The '0' time sample is the spectrum of a freshly thawed sample of this derivative, frozen for a period of 3 months. In (A), the spectrum taken at 160 hours was virtually identical to the spectrum taken at 77 hours.

HGH), the side-chain CD showed a somewhat increased loss in positive dichroism above 289 nm, relative to that seen in the 4-SH form. The amount of the additional change appeared to depend on how much time had elapsed between the alkylation and the time the CD was taken. It was found that if the G-100 step is eliminated, the CD spectrum of the Cys(Cam)[53,165,182,189]-PL-HGH, taken approximately 2 hours after alkylation, was only very slightly different from that of the 4-SH form. However, in the early stages of the study, the product was chromatographed and the monomeric Cys(Cam)[53,165,182,189]-PL-HGH immediately frozen in several aliquots at 1.0-1.5 mg/ml for future use. When thawed up to 3 months later, the side-chain CD of the 'frozen' 4-Cam derivative surprisingly exhibited an indole dichroism identical in all respects to either PL-HGH or native HGH. The remainder of the spectrum was unchanged from that of the freshly prepared but unchromatographed samples. The CD spectra of these thawed samples were quite stable at room temperature for periods up to 24 hours. After this time they too began to lose their positive dichroism above 289 nm.

Figure 12A shows the effect of standing at room temperature on the side-chain CD of Cys(Cam)[53,165,182,189]-PL-HGH. The positive indole dichroism is slowly lost, being replaced by a stronger negative band centered at 298 nm, which may also be assigned to the indole chromophore. The negative dichroism of the 298 nm band continued to increase for approximately 100-150 hours. Beyond this point, the 298 nm band begins to decrease in intensity so that the spectrum taken at 160 hours is not significantly different from that taken at 77 hours. Between 280 and 255 nm, very little change occurs at any time. However, as shown in Figure 12B, the far-UV spectra exhibit a continuous, marked decrease in negative ellipticity throughout the 160-hour period. We have not extended our studies beyond 160 hours as yet. Chromatography of these samples during the same period (Fig. 10E-10G) dramatically demonstrates that the Cys(Cam)[53,165,182,189]-PL-HGH is slowly dissociating into its 2 components, Cys-(Cam)[53]-HGH-(1-134) and Cys(Cam)[165,182,189]-HGH-(141-191). Amino acid analyses of the 'B' and 'D' fractions of the dissociated form are consistent with the identification of these 2 fragments.

The fact that PL-HGH dissociates only after the second disulfide bond has been modified strongly suggests that this must be the bond linking Cys-53 to Cys-165 (the only covalent link holding PL-HGH together). Thus, the bond modified in Cys-(Cam)[182,189]-PL-HGH must be the one forming the smaller, COOH-terminal loop between Cys-182 and Cys-189. This is consistent with earlier reports of the selective reduction of the COOH-terminal disulfide bonds in both HGH (Gráf et al., 1971) and BGH (Gráf et al., 1975).

Bioassay and radioimmunoassay of PL-HGH and derivatives

The growth-promoting potency of PL-HGH, Cys(Cam)[53,165,182,189]-PL-HGH and REOX-PL-HGH* are compared with native HGH in Table 9. It is evident that all 4 proteins are essentially equipotent in this assay. It should be noted that the Cys(Cam)[53,165,182,189]-PL-HGH sample assayed was an aliquot of the 'frozen' material, thawed just prior to assay. Table 9 also demonstrates that PL-HGH and derivatives are at least as potent as HGH when measured in the pigeon crop sac assay. The possibility that the REOX-PL-HGH is slightly more potent than native HGH in both assay systems is currently under further investigation. The potency of PL-HGH, Cys(Cam)[182,189]-PL-HGH, Cys(Cam)[53,165,182,189]-PL-HGH, and REOX-PL-HGH in a radioimmunoassay against anti-HGH is also shown in Table 9. PL-HGH, Cys(Cam)[182,189]-PL-HGH, and REOX-PL-HGH appear to react equally with this antiserum, being only slightly less potent than native HGH. However, a freshly thawed sample of the frozen Cys(Cam)[53,165,182,189]-PL-HGH exhibits a definitely reduced potency.

THE DISULFIDE BOND CD IN PL-HGH

Since in the 2 reduced-carbamidomethylated forms of PL-HGH there is no evidence for any pronounced changes in the CD bands assigned to the aromatic and amide bond chromophoric groups and chemically, the only difference between PL-HGH and these 2 modified proteins is the absence of 1 and 2 moles of disulfide bonds, respectively, the loss in negative dichroism between 285 and 240 nm has been assigned to the

---

* Exclusion chromatography of 2 preparations of REOX-PL-HGH prepared by allowing the 4-SH form to auto-oxidize for 48 hours in the presence of the 25-mole excess of DTT gave a 95% yield of monomeric material (Ve/Vo = 2.0-2.1). Both the near and far-UV CD spectra of the reoxidized protein were identical with PL-HGH.

TABLE 9

Biological activity of PL-HGH and derivatives as measured in the rat tibia assay

Preparation	Total dose (μg)	Tibia width (μm)[a]
HGH	20	245 ± 8 (5)
	60	275 ± 10 (5)
PL-HGH	20	239 ± 3 (5)
	60	273 ± 9 (5)
Cys(Cam)[53, 165, 182, 189]-PL-HGH[b]	20	248 ± 9 (5)
	60	278 ± 12 (5)
REOX-PL-HGH	20	243 ± 7 (4)
	60	327 ± 13 (4)

Biological activity of PL-HGH and derivatives as measured in the pigeon crop sac assay

Preparation	Total dose (μg)	Dry mucosal weight[a]
HGH	2	22.9 ± 3.5 (4)
	6	26.2 ± 4.3 (4)
PL-HGH	2	24.6 ± 0.8 (4)
	6	28.9 ± 1.8 (4)
Cys(Cam)[53, 165, 182, 189]-PL-HGH[b]	2	20.8 ± 3.9 (4)
	6	27.1 ± 3.7 (4)
REOX-PL-HGH	2	28.4 ± 2.2 (4)
	6	34.9 ± 3.5 (4)

Potency of PL-HGH and derivatives as measured in a radioimmunoassay against purified anti-HGH[c]

Preparation	Potency (relative to HGH)	95% Confidence limits
HGH	100	—
PL-HGH	95	83-109
Cys(Cam)[182, 189]-PL-HGH	91	79-108
Cys(Cam)[53, 165, 182, 189]-PL-HGH[b]	62	54-71
REOX-PL-HGH	92	82-103

[a] Mean response ± S.E.M. The number of test animals is shown in parentheses.
[b] Measured on a freshly thawed, frozen sample.
[c] Rabbit antiserum R51-UCS as described by Aubert et al. (1974).

optical activity of the disulfide bonds linking Cys-182 to Cys-189 and Cys-53 to Cys-165. We also believe the reaction kinetics obtained by following the change in ellipticity at 265 nm (Fig. 9) to be a direct measurement of the rate of reduction of these bonds. The circular dichroism contributed by the COOH-terminal disulfide is obtained by subtracting the CD spectrum of Cys(Cam)[182,189]-PL-HGH from that of PL-HGH. Similarly, the difference between the spectrum of Cys(Cam)[182,189]-PL-HGH and 'frozen' Cys(Cam)[53,165,182,189]-PL-HGH represents the optical activity of the disulfide

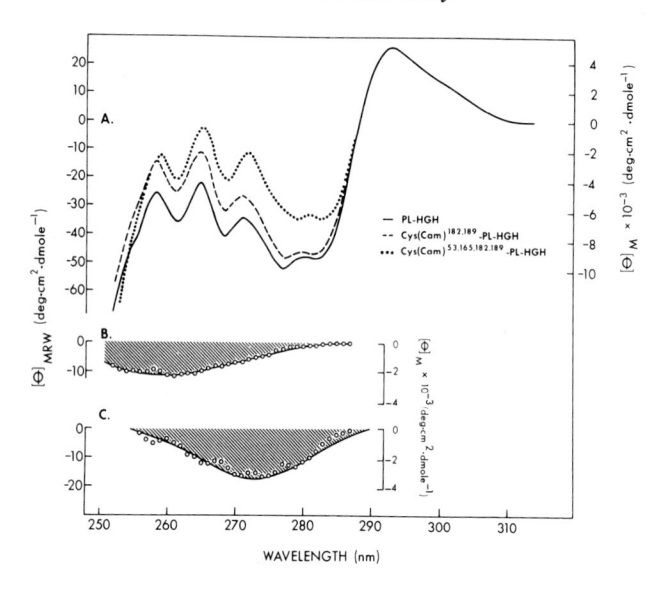

Fig. 13. A: Circular dichroism spectra in the region of side-chain absorption of PL-HGH (———), Cys(Cam)[182,189]-PL-HGH (---), and Cys(Cam)[53,165,182,189]-PL-HGH (....). B: The CD band generated by subtracting the spectrum of Cys(Cam)[182,189]-PL-HGH from that of PL-HGH. C: The CD band generated by subtracting the spectrum of Cys(Cam)[53,165,182,189]-PL-HGH from that of Cys(Cam)[182,189]-PL-HGH. The spectra in A are taken from Figure 8 and the '0' time spectrum from Figure 12.

bond linking Cys-53 and Cys-165. In this case we have used the spectrum of the 'frozen' 4-Cam derivative because it is the only spectrum on this derivative in which the other chromophoric groups appear to be contributing the same dichroism as in the starting material. These 2 CD bands are shown in Figure 13. The CD band contributed by the small COOH-terminal disulfide is very similar in shape, intensity and spectral position to the band assigned to the disulfide bond CD in L-cystine and other small molecular weight disulfides (Beychok, 1966; Coleman and Blout, 1968; Yamashiro et al., 1975). In contrast, the CD band contributed by the other disulfide is redshifted, suggesting significant differences in the local environments and dihedral angles of the 2 disulfide bonds (Coleman and Blout, 1968; Linderberg and Michl, 1970; Yamashiro et al., 1975).

SUMMARY

The action of human plasmin on HGH has briefly been reviewed. Reduction and carbamidomethylation of the plasmin-modified hormone gives rise to 2 biologically active fragments: Cys(Cam)[53]-HGH-(1-134) and Cys(Cam)[165,182,189]-HGH-(141-191). Among these 2 fragments, the NH_2-terminal possesses higher biological activity. It is now possible to show that a mixture of these 2 fragments regenerates the growth-promoting activity from an initial value of 10%, largely due to the NH_2-terminal fragment, to 60% in the tibia test.

A 111 residue fragment has been prepared by cyanogen bromide cleavage of Cys-(Cam)[53]-HGH-(1-134). The fragment, corresponding to amino acid residues 15-125 in the HGH molecule, possesses hepatic ornithine decarboxylase stimulating, lactogenic, and somatotropic activity. It has immunoreactivity in microcomplement fixation and radioimmunoassay experiments. The circular dichroism data indicate that the 111 residue peptide fragment is largely devoid of secondary and tertiary structure.

It has also been shown by circular dichroism studies that PL-HGH is nearly equivalent to native HGH with regard to its conformation. Moreover, the reduction and carbamidomethylation of the COOH-terminal disulfide bond causes no pronounced changes in this structure. From the data it is apparent that this bond may be modified with considerable selectivity. The reduction of the second disulfide bond causes a series of slow, temperature dependent conformational changes which ultimately result in the dissociation of the molecule into 2 fragments. These changes may be prevented if the Cys(Cam)[53,165,182,189]-PL-HGH derivative is frozen soon after preparation. For approximately 24 hours after thawing these 'frozen' preparations exhibit no significant conformational changes from native HGH. We feel that the retention of *full* biological activity in these derivatives is intimately associated with the retention of the conformation of native HGH although large fragments may still be capable of eliciting partial activities in some assay systems.

ACKNOWLEDGMENTS

We thank Dr. T. Hayashida and Dr. M.L. Aubert for radioimmunoassay data, and Dr. A.J. Rao for ODC data. We also thank D. Gordon, J. Knorr, J.D. Nelson, P. Geffen and K. Hoey for technical assistance.

REFERENCES

Aloj, S. and Edelhoch, H. (1972): *J. biol. Chem., 247,* 1146.
Aubert, M.L., Bewley, T.A., Grumbach, M.M. and Kaplan, S.L. (1974): In: *Advances in Human Growth Hormone Research,* pp. 434-462. Editor: S. Raiti. DHEW Publ. No. (NIH) 74-612.
Bewley, T.A., Brovetto-Cruz, J. and Li, C.H. (1969): *Biochemistry, 8,* 4701.
Bewley, T.A., Kawauchi, H. and Li, C.H. (1972): *Biochemistry, 11,* 4179.
Bewley, T.A. and Li, C.H. (1972): *Biochemistry, 11,* 885.
Beychok, S. (1966): *Science, 154,* 1288.
Bölen, P., Stein, S., Stone, J. and Udenfriend, S. (1975): *Analyt. Biochem.,* in press.
Chrambach, A. and Yadley, R.A. (1970): In: *Abstracts, 52nd Endocrine Society Meeting,* Abstract No. 151.
Clarke, W.C., Hayashida, T. and Li, C.H. (1974): *Arch. Biochem., 164,* 571.
Coleman, D.L. and Blout, E.R. (1968): *J. Amer. chem. Soc., 90,* 2405.
Ellis, S., Nuenke, J.M. and Grindeland, R.E. (1968): *Endocrinology, 83,* 1029.
Gráf, L., Barát, E., Borvendég, J., Patthy, A. and Cseh, G. (1971): In: *Abstracts, II International Congress on Growth Hormone and Related Peptidés, Milan 1971,* Abstract No. 40. ICS 236. Excerpta Medica, Amsterdam.
Gráf, L., Li, C.H. and Bewley, T.A. (1975): *Int. J. Peptide and Protein Res.,* in press.
Gray, W.R. (1967): *Meth. Enzymology, 11,* 469.
Gross, E. and Witkop, B. (1962): *J. biol. Chem., 237,* 1856.
Jänne, J. and Raina, A. (1969): *Biochim. biophys. Acta (Amst.), 174,* 769.
Laurent, T.C. and Killander, J. (1964): *J. Chromatography, 14,* 317.
Li, C.H., Dixon, J.S., Lo, T.B., Schmidt, K.D. and Pankov, Y.A. (1970): *Arch. Biochem., 141,* 705.
Li, C.H. and Gráf, L. (1974): *Proc. nat. Acad. Sci. (Wash.), 71,* 1197.
Linderberg, J. and Michl, J. (1970): *J. Amer. chem. Soc., 92,* 2619.

Mills, J.B., Reagan, C.R., Rudman, D., Kostyo, J.L., Zachariah, P. and Wilhelmi, A.E. (1973): *J. clin. Invest., 52,* 2941.
Moore, S. (1972): In: *Chemistry and Biology of Peptides,* pp. 629-653. Editor: J. Meienhofer. Ann Arbor Science Publishers.
Rao, A.J., Ramachandran, J. and Li, C.H. (1975): *Life Sci.,* in press.
Reagan, C.R., Kostyo, J.L., Mills, J.B. and Wilhelmi, A.E. (1973): *Fed. Proc., 32,* Abstract No. 294.
Reagan, C.R., Mills, J.B., Kostyo, J.L. and Wilhelmi, A.E. (1975): *Endocrinology, 96,* 625.
Russell, D.H., Snyder, S.H. and Medina, V.J. (1970): *Endocrinology, 86,* 1414.
Yadley, R.A. and Chrambach, A. (1973): *Endocrinology, 93,* 858.
Yamashiro, D., Rigbi, M., Bewley, T.A. and Li, C.H. (1975): *Int. J. Peptide and Protein Res., 7,* 389.

THE NATURE OF FRAGMENTS OF HUMAN GROWTH HORMONE PRODUCED BY PLASMIN DIGESTION*

JACK L. KOSTYO, JOHN B. MILLS, CHARLES R. REAGAN, DANIEL RUDMAN and ALFRED E. WILHELMI

Departments of Physiology, Biochemistry and Medicine, Emory University, Atlanta, Ga., U.S.A.

The digestion of human growth hormone (HGH) with the enzyme plasmin removes a hexapeptide, residues 135-140, from the large disulfide loop of the hormone molecule. The cleaved hormone, which consists of amino acid residues 1-134 attached to residues 141-191 by the disulfide bond between residues 53 and 165, retains the biological activities of the intact hormone in a variety of animal test systems and in man (Mills et al., 1973; Li and Gráf, 1974; Reagan et al., 1975a). This large fragment has been split into 2 smaller peptides by reduction of the disulfide bridges and S-carbamidomethylation of the resultant half-cystine residues (Reagan et al., 1973; Li and Gráf, 1974). The N-terminal peptide consisting of residues 1-134 is quite active in vitro in stimulating protein synthesis in the isolated rat diaphragm and glucose oxidation in isolated rat adipose tissue (Reagan et al., 1975a). If injected intravenously, peptide 1-134 also stimulates thymidine incorporation into costal cartilage of hypophysectomized rats. However, it has only weak activity in the weight gain and tibia assays in hypophysectomized rats (Li and Gráf, 1974; Reagan et al., 1975a), tests which involve the subcutaneous or intraperitoneal administration of the hormone. From these observations it would appear that the preparations of peptide 1-134 made to date are quite labile in vivo. The C-terminal peptide consisting of amino acid residues 141-191 is essentially inactive.

Since the biological activities of the HGH molecule are retained by peptide 1-134, it was of considerable interest to find alternative and more efficient methods to prepare substantial amounts of this peptide for further fragmentation studies. It was also hoped that a product could be prepared that would not have the high in vivo lability of the earlier preparations. The method chosen involved the digestion of reduced and S-carbamidomethylated HGH (RCAM-HGH) with plasmin. The assumption was made that plasmin would have the same initial mode of attack on RCAM-HGH as it has on native HGH, i.e. that it would cleave 2 peptide bonds resulting in the removal of hexapeptide 135-140. In that event, peptide 1-134 would be generated.

For this work, RCAM-HGH was prepared by reduction of highly purified HGH with dithiothreitol at pH 8.6 and alkylation of the half-cystine residues with iodoacetamide (Reagan et al., 1975b). The extent of reduction and alkylation of the product was determined by amino acid analysis and, in every case, found to be complete. Various batches of RCAM-HGH prepared in this manner were tested for growth-promoting

---

* This research was supported by grants from the National Institutes of Health (HD 04485, HD 01231 and RR 00039) and The Kroc Foundation.

activity in hypophysectomized rats using the 9-day weight gain test (Parlow et al., 1965). The potencies of these materials relative to the International Growth Hormone Standard, Bovine, ranged from 0.6-2.7 IU/mg, with most preparations having a relative potency above 1 IU/mg (Reagan et al., 1975b). Since the highly purified native HGH from which the RCAM-HGH preparations were derived has a relative potency approximating 2 IU/mg in the weight gain test, reduction and S-carbamidomethylation, in some instances, may cause a moderate reduction in potency. RCAM-HGH was also quite active in stimulating thymidine incorporation into costal cartilage of hypophysectomized rats. Also, when added in vitro, RCAM-HGH stimulated glucose oxidation by isolated adipose tissue of hypophysectomized rats, the effects obtained being similar to those produced with comparable amounts of native HGH. The reduced and alkylated hormone was also tested for activity in 3 growth hormone-deficient boys aged 12, 13 and 15, using a metabolic balance procedure described previously (Mills et al., 1973). When given 0.17 IU of RCAM-HGH/kg body weight$^{3/4}$ per day for 7 days, each subject showed positive balances (relative to the 7-day control period) of nitrogen, phosphorus, potassium, sodium and chloride and a gain in body weight. The effects obtained, particularly on the retention of nitrogen, were comparable to those produced by an equivalent daily dose of native HGH. Thus, these results demonstrate that RCAM-HGH is highly active in both animals and humans and complement the earlier finding of Dixon and Li (1966) indicating that RCAM-HGH is active in the tibia test, and the work of Ceraci et al. (1972) and Connors et al. (1973) showing that RCAM-HGH has metabolic activity in human subjects.

Digestion of RCAM-HGH with human plasmin was carried out overnight at 37°C (Reagan et al., 1975b). Disc gel electrophoresis of the digests revealed that approximately 95% of the starting material was converted to 2 products that are more acidic than the starting material (Fig. 1). Of these 2 products, the major electrophoretic component has been typically the less acidic of the 2.

When tested for growth-promoting activity in hypophysectomized rats, 6 different digests of RCAM-HGH were found to have relative potencies ranging from 0.9-2.1 IU/mg, with 5 of the 6 having potencies of 1.2 IU/mg or better. Further, these digests were very active when injected intraperitoneally in stimulating thymidine incorporation into the costal cartilage of hypophysectomized rats. Effects were produced that were comparable to those obtained with equivalent doses of native HGH. The digests also stimulated glucose oxidation by isolated rat adipose tissue when added in vitro to concentrations ranging from 0.1-0.5 μg/ml of incubation medium. The effects obtained, however, appeared to be somewhat smaller than those produced with comparable concentrations of native HGH. Whether these digests do indeed have reduced insulin-like activity remains to be firmly established. In any event, the high activity of these digests in all of the above assay systems cannot be attributed to the small amount of undigested RCAM-HGH remaining in the preparations. One plasmin digest of RCAM-HGH was also tested for metabolic activity in the 3 human subjects referred to earlier. Each boy received a daily dose of 0.17 IU of digest/kg body weight $^{3/4}$ per day for 7 days. In each case, treatment with the digest produced positive balances of nitrogen, phosphorus, potassium, sodium and chloride and a gain in body weight. The responses obtained were generally of the magnitude produced with an equivalent dose of native HGH.

From the above results it is clear that plasmin digests of RCAM-HGH retain a high degree of activity in animals and in growth hormone-deficient human subjects. Of particular interest is the fact that these digests have high activity in in vivo assays involving administration of test materials by the subcutaneous, intraperitoneal or intramuscular routes. It will be recalled that peptide 1-134 produced by plasmin digestion

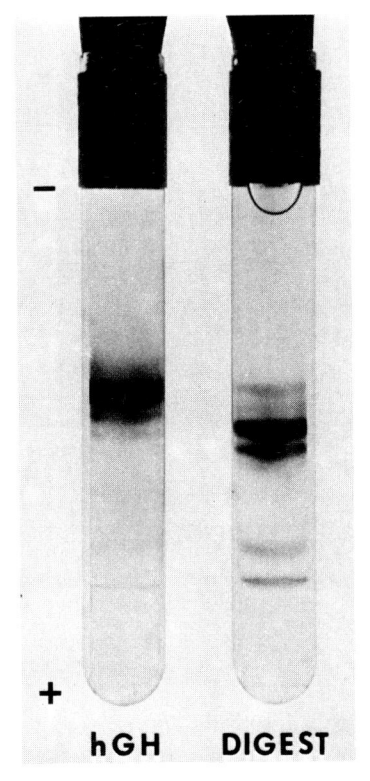

Fig. 1. Disc gel electrophoresis of precursor HGH and a plasmin digest of RCAM-HGH (100-μg samples). RCAM-HGH gives an electrophoretic pattern identical to native HGH.

of native HGH followed by reduction and alkylation of the cleavage product had little activity in vivo unless administered intravenously. Clearly, the plasmin digests of RCAM-HGH do not share this apparent in vivo lability. This raised the possibility that either the active component in the digests was not peptide 1-134 or that the fewer chemical manipulations involved in producing these digests resulted in a more stable version of peptide 1-134.

As a first step toward the chemical characterization of the biologically active component(s) in the digests of RCAM-HGH, efforts were made to separate major peptide components from minor cleavage products. Lyophilized digests were dissolved in 1 M propionic acid and subjected to gel filtration on Sephadex G-50 fine in 0.01 N HCl. Five absorbance peaks (R1-R5) were typically obtained (Reagan et al., 1975b). Peak R1 contained aggregated material and undigested precursor. Peak R2 consisted of the 2 major peptide components present in the whole digest (as judged by gel electrophoresis of R2). Peaks R3-R5 contained mixtures of small peptides that have not been characterized with the exception of peptide 135-140 which was present in peak R5. Since this hexapeptide was found in the digests, it would appear that the action of plasmin on RCAM-HGH is similar to its action on native HGH in that residues 135-140 are cleaved from the molecule. The digestion would appear to be more exten-

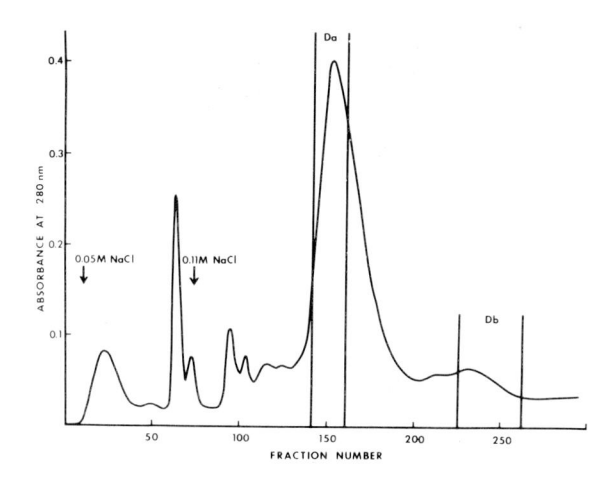

Fig. 2. Elution pattern of a plasmin digest of RCAM-HGH (ca. 200 mg) applied to a 2.7 × 21.5 cm column of DEAE-32 cellulose (Whatman) equilibrated with 0.05 M NaCl in 0.01 M Tris, pH 8.0. Chromatography was carried out at 4°C. Five-ml fractions were collected. Buffer was changed to 0.11 M NaCl, 0.01 M Tris, pH 8.0 at the point indicated by the arrow. The pairs of vertical lines bracket the fractions that were pooled to obtain preparations Da and Db.

sive, however, since a fragment of the size of peptide 141-191 was not recovered in the digests.

Several preparations of peak R2 were tested for growth-promoting activity in hypophysectomized rats and found to have relative potencies (1.1-1.9 IU/mg) similar to those of the digests from which they were derived. Peak R2 also had a high degree of activity in stimulating thymidine incorporation into rat costal cartilage and glucose oxidation by rat adipose tissue in vitro (Reagan et al., 1975*b*). Hence it would appear that the biological properties of the plasmin digests of RCAM-HGH are due primarily to the 2 large peptide components that they contain.

An attempt was made to separate these 2 major components by chromatography of the whole plasmin digest on DEAE-cellulose. Various amounts of whole digest ranging from 60-200 mg were adsorbed to a column of DEAE-cellulose that had been equilibrated with 0.05 M NaCl in 0.01 M Tris, pH 8.0. Elution was then carried out with 0.11 M NaCl in 0.01 M Tris, pH 8.0. A typical elution pattern is shown in Figure 2. By pooling selected fractions from the major absorbance peak (at 280 nm), it was possible to obtain the individual major peptide components in relatively pure form. This can be seen in Figure 3 which shows the disc gel electrophoretic patterns of such pools from a typical chromatographic run. Pool Da contains the major, less acidic component of the digest, and pool Db contains the minor, more acidic component. Table 1 gives the yields of Da and Db obtained in 4 such chromatographic separations. The yield of Da averaged about 15% of the starting material and that of Db about 7%. Perhaps this difference in yield, like the gel electrophoretic pattern, reflects the relative abundance of these 2 components in the whole digest. In any event, it would be noted that these yields are conservative, since an effort was made to select only a limited number of fractions for pools Da and Db to achieve the highest purity possible.

Only pool Da has been subjected to extensive biological characterization. Two Da preparations have been assayed in the weight gain test in hypophysectomized rats and

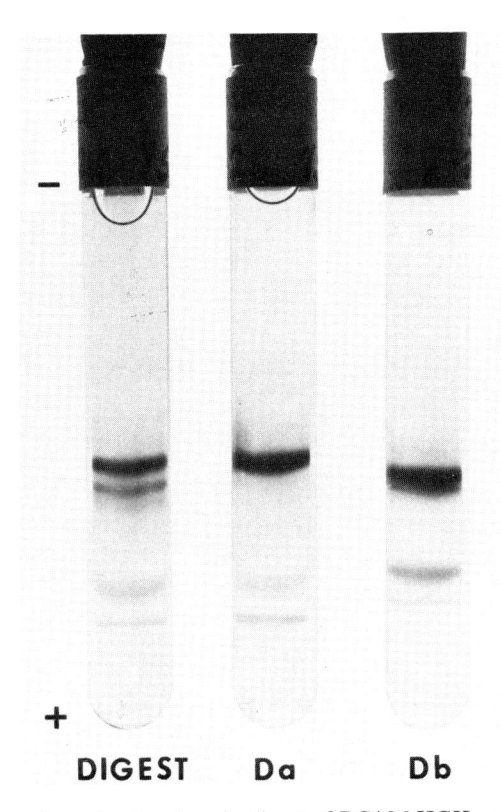

Fig. 3. Disc gel electrophoresis of a plasmin digest of RCAM-HGH and pools Da and Db from DEAE-cellulose chromatography of the whole digest (100-μg samples).

found to possess growth-promoting activity comparable to that of highly purified native HGH [M-33-89 = 2.2(1.6-2.9) IU/mg and M-12-2 = 3.7(2.2-7.2) IU/mg]. The data in Table 2 demonstrate that Da also retains the ability to stimulate thymidine incorporation into the costal cartilage of hypophysectomized rats when administered intraperitoneally. Da also had a high degree of in vitro activity on glucose oxidation by isolated rat adipose tissue as the results in Table 3 indicate. However, like the whole plasmin digest from which Da was derived, its activity in this assay on a weight basis appears to be only 20-40% that of native HGH. Only one preparation of pool Db (M-33-13) was assayed for growth-promoting activity, and a high degree of activity was found [1.6(1.1-2.2) IU/mg]. It is conceivable that Db is a chemically modified (deamidated?) version of Da.

Some preliminary information has been obtained regarding the chemical identity of Da. Table 4 shows the average amino acid composition of the 4 preparations of Da listed in Table 1. The low content of S-carboxymethyl cysteine indicates that the carboxy terminal region of HGH, which is relatively rich in cysteine (residues 165, 182 and 189), is not present in Da. However, the composition analysis does compare favorably with that of the amino terminal portion of the molecule. This is evident from the comparison of the composition of Da with the known compositions of HGH peptides 1-134 and 9-134 shown in Table 4. Amino terminal analysis of Da with dinitrofluoro-

TABLE 1

Yields of pools Da and Db obtained by DEAE-cellulose chromatography of plasmin-digested RCAM-HGH

Amount of digest used (mg)	Da preparation	Amount of Da recovered (mg)	Db preparation	Amount of Db recovered (mg)
71.2	M-184-6	12.7	M-184-8	7.4
71.2	M-12-2	12.5	M-12-4	6.1
68.3	M-22-2	8.1	M-22-6	4.7
213.2	M-33-89	35.6	M-33-13	12.7

TABLE 2

Effects of intraperitoneal injection of fraction Da on ^{3}H-thymidine incorporation into rat costal cartilage in vitro*

Preparation	Dose (μg/day)	^{3}H-thymidine incorporation (dpm/mg protein) Control	^{3}H-thymidine incorporation (dpm/mg protein) Treated
HGH	2.5	882 ± 68 (16)	1211 ± 114 (8)
	5	882 ± 68 (16)	2078 ± 469 (8)
Da	2.5	882 ± 68 (16)	1847 ± 377 (15)
	5	882 ± 68 (16)	2628 ± 563 (15)

* Hypophysectomized rats were injected i.p. with saline or hormone 48 and 24 hours prior to sacrifice. Cartilage was incubated for 3 hours with ^{3}H-thymidine. (For details of method see Mills et al., 1973.)

TABLE 3

In vitro effects of Da on glucose oxidation by rat adipose tissue*

Preparation	Concentration (μg/ml)	No. of rats	$^{14}CO_2$ production (dpm/mg/hr) Control	Hormone	M.D. ± S.E.
HGH	0.1	8	35.7	91.1	55.5 ± 4.4
	0.5	8	35.7	124.1	88.5 ± 10.4
Da	0.1	8	35.7	56.4	20.8 ± 3.8
	0.25	8	35.7	94.9	59.3 ± 7.9
	0.5	8	35.7	104.3	68.7 ± 3.8

* Epididymal adipose tissue of hypophysectomized rats was incubated for 1 hour with ^{14}C-glucose and test materials. (For details of method see Mills et al., 1973.)

TABLE 4
Amino acid composition of Da*

Amino acid	Found (moles %)	Theoretical (1-134) (moles %)	Theoretical (9-134) (moles %)
Lys	4.8	3.0	3.2
His	1.9	1.5	1.6
Arg	5.5	6.0	5.6
Trp	0.5	0.8	0.8
Asp	10.8	9.0	9.5
Thr	4.4	4.5	4.0
Ser	7.8	10.5	10.3
Glu	15.9	16.4	17.5
Pro	4.3	6.0	4.8
Gly	4.3	3.0	3.2
Ala	4.8	4.5	4.8
Val	3.7	3.0	3.2
Met	1.5	1.5	1.6
Ile	3.5	4.5	4.0
Leu	14.2	15.7	15.9
Tyr	5.0	3.7	4.0
Phe	6.8	6.0	5.6
CM Cys	0.9	0.8	0.8

* Average of analyses of the 4 Da preparation listed in Table 1.

benzene showed phenylalanine (32% of DNP-amino acids detected, uncorrected), leucine (45%) and glutamic acid (19%). Reaction of Da with cyanogen bromide to cleave the polypeptide chain at methionine residues resulted in the release of peptide 126-134. This was not unexpected, since, as noted earlier, hexapeptide 135-140 was found in the digests of RCAM-HGH, indicating that plasmin had cleaved the reduced hormone between residues 134 and 135. Another peptide was also produced by the cyanogen bromide cleavage. It has not been characterized fully, but its amino acid composition indicates that it does not contain any S-carboxymethyl cysteine, suggesting that it may be an N-terminal peptide derived from the cleavage of Da at the methionine residue at position 14. Further, analysis of Da by SDS gel electrophoresis indicated the presence of 2 components with approximate molecular weights of 16,250 and 13,050. All of the above data, taken together, suggest the possibility that Da is a mixture of peptides 1-134 (15,570 daltons) and 9-134 (14,659 daltons). The presence of amino terminal glutamic acid in the preparation is not in accord with this hypothesis. However, the finding of N-terminal glutamic acid may indicate that Da is contaminated with an adherent peptide possibly derived from the 141-191 sequence of the molecule (residue 141 is glutamine, and plasmin cleaves RCAM-HGH between residues 140 and 141).

In summary, the results of this study indicate that it is possible by plasmin cleavage of RCAM-HGH to produce highly active peptide fragments that retain many of the biological properties of native HGH and which are not labile in the whole animal. If the tentative identification of the structure of Da is correct, this work would indicate that the full growth-promoting property of HGH resides in the N-terminal 134 amino acid residues of the hormone molecule.

REFERENCES

Cerasi, E., Li, C.H. and Luft, R. (1972): *J. clin. Endocr., 34,* 644.
Connors, M.H., Kaplan, S.L., Li, C.H. and Grumbach, M.M. (1973): *J. clin. Endocr., 37,* 499.
Dixon, J.S. and Li, C.H. (1966): *Science, 154,* 785.
Li, C.H. and Gráf, L. (1974): *Proc. nat. Acad. Sci. (Wash.), 71,* 1197.
Mills, J.B., Reagan, C.R., Rudman, D., Kostyo, J.L., Zachariah, P. and Wilhelmi, A.E. (1973): *J. clin. Invest., 52,* 2941.
Parlow, A.F., Wilhelmi, A.E. and Reichert, L.E. (1965): *Endocrinology, 77,* 1126.
Reagan, C.R., Kostyo, J.L., Mills, J.B. and Wilhelmi, A.E. (1973): *Fed. Proc., 32,* 265 (abstract).
Reagan, C.R., Mills, J.B., Kostyo, J.L. and Wilhelmi, A.E. (1975a): *Endocrinology, 96,* 625.
Reagan, C.R., Mills, J.B., Kostyo, J.L. and Wilhelmi, A.E. (1975b): *Proc. nat. Acad. Sci. (Wash.), 72,* 1684.

IN VIVO AND IN VITRO ACTIONS OF SYNTHETIC PART SEQUENCES OF HUMAN PITUITARY GROWTH HORMONE*

J. BORNSTEIN

Department of Biochemistry, Monash University, Clayton, Victoria, Australia

Previous communications reviewed by Bornstein et al. (1973) have reported in vitro and in vivo actions of 2 polypeptide fractions obtained from growth hormone hydrolysates, and have shown that these actions could be duplicated by the synthetic sequence H-Phe-Pro-Thr-Ile-Pro-Leu-Ser-Arg-Leu-Phe-Asp-Asn-Ala-Met-Leu-OH representing amino acids 1-15 of the sequence of human growth hormone (HGH 1-15) (Niall, 1971) in the case of insulin-like actions; and by the sequence H-Arg-Lys-Asp-Met-Asp-Lys-Val-Glu-Thr-Phe-Leu-Arg-Ile-Val-Gln-Cys-Arg-Ser-Val-Glu-Gly-Ser-Cys-Gly-Phe-OH representing amino acids 167-191 (HGH 167-191) of human growth hormone which reproduced the anti-insulin actions of the natural peptides. Later work (Ng et al., 1974) showed that in the case of the N-terminal sequence the synthetic peptides HGH 1-20, HGH 1-15, HGH 1-13, HGH 3-13 and HGH 6-13 were active both in vivo and in vitro whereas HGH 1-10 and HGH 6-11 were inert. The actions of HGH 1-15 and HGH 167-191 are summarised in Table 1.

This paper describes investigations of the actions of the polypeptides HGH 6-13 (i.e., H-Leu-Ser-Arg-Leu-Phe-Asp-Asn-Ala-OH) and HGH 176-180 (i.e., H-Phe-Leu-Arg-Ile-Val-OH) on the glucose tolerance of mice both immediately following intravenous injection and following intraperitoneal injection for 4 days in potentiating insulin action on the rat diaphragm and on release of insulin by isolated islets taken 24 hours following the injection of the peptides. Further, the effect of the peptide HGH 6-13 has been investigated on the binding of insulin to specific receptors on lipocytes and hepatocytes.

The effect of the peptides HGH 176-191, HGH 177-191 and HGH 180-191 has been investigated on the level of glucose in 16-hour fasted mice following intravenous injection.

MATERIALS

Polypeptide HGH 6-13 was a gift of Institut Choay, Paris, France. All other polypeptides were synthesized in the laboratory by Dr. C.A. Browne and Mrs. Claire Pullin by the solid phase technique (Stewart and Young, 1969). Only limited experiments could be performed with the HGH 176-191-type peptides due to difficulties in synthesis arising from poor coupling of Gly (190) to Cys (189).

---

* The results reported in this paper have been obtained by a number of staff and students, and are acknowledged in the text.

TABLE 1

	HGH 167-191	HGH 1-15
In vivo	Not tested	Potentiates action of insulin. Long-term effect on glucose disposal
In vitro (tissues)	Inhibits glucose uptake, fatty acid synthesis, protein synthesis. Accelerates lipolysis[2]	Accelerates glucose uptake[1], fatty acid synthesis[1], protein synthesis[1]. Inhibits lipolysis
In vitro (enzymes)	Inhibits glyceraldehyde 3-phosphate dehydrogenase (GAPD), pyruvic dehydrogenase[3], glycogen synthase[4], acetyl CoA carboxylase[5]	Reverses HGH 167-191 inhibition of GAPD

1. Action occurs only in the presence of insulin.
2. No change in cyclic AMP levels (Fajgman and Gould, unpublished data).
3. This finding is an inhibition of activation (Aylward et al., 1974) rather than a direct inhibition of the enzyme.
4. Data of Aylward et al. (1975) – inhibition of activation.
5. Data of M. Dobos – inhibition of activation.
4 and 5 obtained with natural peptide fraction.

All animals used were of standard laboratory stock and were fasted for 16 hours prior to use, in vivo experiments being done under pentobarbital anaesthesia.

Lipocytes were prepared according to the technique of Rodbell (1964); hepatocytes according to the technique of Howard et al. (1973); and pancreatic islets according to the technique of Ashcroft et al. (1972). Insulin binding was done according to the technique of Kono and Barham (1971). Radioimmunoassays of insulin were done by Dr. Frances Stephenson according to the technique of Hunter and Greenwood (1964).

All peptides were monitored for activity in the GAPD assay as described by Bornstein et al. (1968). Dosage of peptides was: HGH 6-13, 3 µg/g body weight; HGH 176-180, 6 µg/g body weight; HGH 176-191, 1 ng/g body weight; HGH 177-191, 5 ng/g body weight; and HGH 180-191, 10 ng/g body weight.

EXPERIMENTAL

All peptides were initially tested as a screening procedure in the GAPD assay. The results (Table 2) show that the peptide HGH 176-180, a part of the structure of the inhibitory peptide, reverses or prevents inhibition, pointing to competition between similar structures for a single binding site (Bornstein, 1972). Accordingly, it appeared that reversal of GAPD inhibition by other pituitary peptides was not a likely mechanism of action, and the peptides were investigated separately.

Investigation of the actions of HGH 6-13 and HGH 176-180

The effect of these peptides was initially investigated by measuring the fall in blood glucose between the peak value and values up to 90 minutes. Anaesthetised mice were given a single intravenous injection of peptide (3 µg/g body weight) which was

TABLE 2
Effect of synthetic peptides on the activity of glyceraldehyde 3-phosphate dehydrogenase

HGH 176-191	Inhibits
HGH 177-191	Inhibits
HGH 180-191	Not active
HGH 6-13	Reverses inhibition
HGH 176-180	Reverses inhibition

TABLE 3
Fall in blood glucose between 15 and 90 minutes in mg/100 ml (mean ± S.D. 4 animals in each group)

Time (min)	Controls	HGH 6-13	HGH 176-180
15-30	15 ± 11.75	34 ± 5.6	17 ± 16.15
-45	45.5 ± 8.6	68 ± 6.4	43.25 ± 27.6
-60	71.25 ± 9.6	97 ± 7.8	65 ± 31.9
-90	102.75 ± 11.0	122.75 ± 9.1	103 ± 27.5

No significant difference for controls and HGH 176-180

For controls and HGH 6-13 at:

	n	Δ	T	P
15-30	6	19	2.92	< 0.05
-45	6	29	5.42	< 0.01
-60	6	25.75	4.16	< 0.01
-90	6	20	2.80	< 0.05

TABLE 4
Fall in blood glucose between 15 and 90 minutes in mg/100 ml (mean ± S.D.)

Time (min)	Controls (4)	HGH 6-13 (4)	HGH 176-180 (4)
15-30	8.75 ± 12.4	28.10 ± 3.5	25.50 ± 13.7
-45	33.25 ± 12.5	52.25 ± 10.7	58.75 ± 17.8
-60	42.25 ± 18.0	76.00 ± 10.9	89.50 ± 22.6
-90	88.00 ± 17.8	129.00 ± 13.7	126.00 ± 22.7

For controls and HGH 176-180 at:

	n	Δ	T	P
15-30	6	16.75	1.81	> 0.1
-45	6	25.50	2.34	< 0.1
-60	6	47.25	3.27	< 0.02
-90	6	38.00	2.63	< 0.05

For controls and HGH 6-13 at:

	n	Δ	T	P
15-30	6	19.35	3.00	< 0.05
-45	6	19.00	2.31	> 0.1
-60	6	33.75	3.21	< 0.02
-90	6	41.00	3.65	< 0.02

TABLE 5

Glucose uptake by rat diaphragm (mg/g wet weight/90 min ± S.D.)

Number of experiments	Control	HGH 6-13	Number of experiments	Control	HGH 176-180
8	5.2 ± 0.74	6.8 ± 0.81	7	4.8 ± 0.45	4.7 ± 0.64

For control and HGH 6-13, n = 14 t = 4.12 P = 0.001
For control and HGH 176-180 n.s.d.
Peptide concentration 0.25 µg/ml. Insulin concentration 0.1 mµ/ml.

Data by J. Bornstein, Miss D. Pakenham and Miss B. Farrell.

immediately followed by a glucose load of 250 µg/g. Blood was taken for glucose estimations at zero time, 15, 30, 45 and 90 minutes after the injection of glucose.

The results of the experimental series are given in Table 3. Thus it is seen that polypeptide 6-13 has an immediate action in potentiating the uptake of glucose from the blood, whereas polypeptide 176-180 is inactive in this system.

However, when the polypeptides were administered for either 3 or 4 days and the test was carried out on the next day, the result was as observed in Table 4.

It is clearly seen that, following chronic administration, both peptides accelerate the removal of a glucose load, thus indicating that the immediate and long-term actions could be mediated by different mechanisms. This viewpoint was given further weight when the 2 peptides were tested for their ability to potentiate glucose uptake by the isolated rat diaphragm in the presence of insulin, which had been shown to be an obligatory requirement for this effect by Ng et al. (1974).

Table 5 shows that, whereas during a 90-minute incubation HGH 6-13 significantly potentiated glucose uptake by the diaphragm, HGH 176-180 was inactive in this regard. As these data tended to confirm that reversal of GAPD inhibition was not a physiologic phenomenon, HGH 186-190 having no immediate activity, the effect of the rapidly acting peptide HGH 6-13 was investigated on specific insulin binding by adipocytes and hepatocytes.

Experiments with adipocytes showed no effect of HGH 6-13 at concentrations of 5, 50 and 500 ng/ml on the binding of insulin to its receptors either as a direct effect or following preincubation with the peptide, although previous studies (Bornstein et al., 1975a) had shown that although HGH 1-15 had no effect on insulin binding by collagenase treated lipocytes, this peptide accelerated the recovery of binding sites following trypsin treatment of the cells according to Kono and Barham (1971), and that this acceleration was abolished by actinomycin D and cycloheximide. However, experiments using hepatocytes showed that although the peptide had a marginal immediate action on insulin binding, preincubation of HGH 6-13 with the cells greatly retarded the loss of insulin binding sites during the preincubation (1 hour at 22°) (Table 6).

If the mechanism of the action of these peptides (i.e., HGH 1-15 and active analogues) is to maintain insulin binding at optimum levels, then the very puzzling long-term action must be due to other mechanisms.

In order to examine this possibility, the effect of the injection of HGH 6-13 and HGH 176-180 on the response of the islets to glucose stimulation was examined 24 hours after a single injection of peptide and 24 hours after chronic treatment by Dr. L.C.C. Weerasinghe and Mr. C. Moodie. The results (Table 7) showed that although

TABLE 6
Insulin binding (counts/m/mg dry cells)

Experiment	Control	HGH 6-13	% change
1	8541	9262	+ 8%
2	11790	14132	+18%
Following 1 hour preincubation:			
1	3551	4890	+38%
2	4227	7027	+66%

Concentration HGH 6-13, 1 µg/ml. Experiments with HGH 176-180 are at present in progress.
Data by Dr. L.C.C. Weerasinghe and Mr. J. Landsberger.

TABLE 7
Insulin release by isolated islets after injection of HGH 6-13 and HGH 176-180

µU insulin/islet/hr ± S.D. (n = 5)

	0 mM	Glucose level 10 mM	20 mM
24 hours after single injection of peptide:			
Control	4.8 ± 2.4	73.9 ± 22.6	—
HGH 6-13	2.8 ± 1.9	128.6 ± 22.6	—
HGH 176-180	4.5 ± 2.8	93.0 ± 21.0	—
For control and 6-13 (10 mM)	n = 8	T = 3.61	$P < 0.01$
For control and 176-180 (10 mM)	n = 8	T = 1.2	$P > 0.2$
After 4 days pretreatment with peptide:			
Control	2.8 ± 1.3	45.8 ± 11.6	170.8 ± 21.9
HGH 6-13	11.8 ± 6.3	147.2 ± 35.6	142.8 ± 32.7
HGH 176-180	4.2 ± 1.3	74.0 ± 13.6	147.2 ± 35.6
For control and 6-13 (10 mM)	n = 8	T = 4.37 $P < 0.01$	At 20 mM, n.s.d.
For control and 176-180 (10 mM)	n = 8	T = 3.53 $P < 0.01$	

both HGH 1-20 and HGH 1-15 had been shown by Mr. A. Burton to have no immediate effect on insulin release at a concentration of 100 ng/ml, following treatment by the peptide islets were more responsive to glucose at the concentration 10 mM, but there was no difference when they were maximally stimulated by 20 mM glucose. It is clearly seen from this table that a single injection of HGH 6-13 is sufficient to sensitise β cells to glucose and that although chronic injection of HGH 176-180 will sensitise the islets, the effect is not as large.

Investigations in regard to peptides HGH 176-191, 177-191 and 180-191

Early investigations (Bornstein et al., 1975b) showed that peptide HGH 176-191 and the similarly active fractions from blood and growth hormones produced an early

TABLE 8*

Change in blood glucose levels (experimental minus controls) following i.v. injection of HGH 176-191, HGH 177-191 and HGH 180-191

Time after injection (min)	Change in blood glucose in mg/100 ml ± S.D. (n = 6)			
	(1) 176-191	(2) 176-191 + AGS**	(3) 177-191	(4) 180-191
5	+ 2.0 ± 1.5	− 1.0 ± 1.0	+ 7.8 ± 3.16	− 4.8 ± 4.17
10	+ 7.8 ± 1.1	+ 2.1 ± 2.0	+ 7.9 ± 3.11	− 6.7 ± 7.18
15	+ 7.6 ± 1.3	+ 3.0 ± 1.9	+12.4 ± 2.16	− 3.8 ± 4.42
20	+13.7 ± 3.3	+ 3.0 ± 1.2	+17.5 ± 2.65	− 7.0 ± 4.72
30	+11.0 ± 2.6	+ 4.0 ± 4.1	+22.5 ± 2.89	− 3.3 ± 3.40
45	+18.0 ± 3.9	+ 8.1 ± 4.2	+27.0 ± 5.83	− 6.2 ± 4.20
60	+21.0 ± 6.1	+10.0 ± 6.1	+29.0 ±10.36	− 3.7 ± 4.10

* Due to shortage of peptides the data in this table is preliminary and requires further investigation both in respect to the effects of antiglucagon sera and radioimmunoassay of glucagon in the portal blood.
** 0.1 ml antiglucagon serum 16 hours prior to injection of peptide.
Peptide administered: (1) HGH 176-191, 1 ng/g body weight; (2) HGH 176-191, 1 ng/g body weight; (3) HGH 177-191, 5 ng/g body weight; (4) HGH 180-191, 10 ng/g body weight.
(1) All differences significantly different from control ($P < 0.05$) from 10-60 minutes.
(1) and (2) Differences from 10-45 minutes significant ($P < 0.05$).
(2) No significant difference from control until 60 minutes.
(3) All differences from control significant ($P < 0.05$).
(4) No significant difference from control.
Data by Misses D. Pakenham and B. Farrell.

inhibition of glucose uptake under the conditions of the glucose tolerance test and that the peptide was capable of raising the blood sugar by glycogenolysis. Table 8 shows the effect of the above peptides on the blood glucose of mice following intravenous injection in nanogram dosages.

Both HGH 176-191 and HGH 177-191 produced an immediate rise in blood glucose level which was abolished by antiglucagon serum in the case of 176-191, followed by a later and more sustained rise which is independent of the effects of antiglucagon serum. Peptide 180-191 is inert in both respects.

Although it appears virtually certain that the primary rise in blood glucose is due to the release of glucagon, the secondary effect required further investigation as it could be due to inhibition of activation of pyruvate dehydrogenase (Aylward et al., 1974) and glycogen synthase (Aylward et al., 1975) or lipolysis (Morstyn and Zimmet, in Bornstein et al., 1973) or possibly due to interference with insulin binding to cell receptors. Preliminary data obtained by Miss P. Bacon indicates that these peptides inhibited insulin binding to lipocytes by 33% at a concentration of 3.2 ng/ml.

DISCUSSION

The mechanism and significance of the inhibition and reversal of glyceraldehyde 3-phosphate dehydrogenase

It is seen that whilst HGH 176-191 inhibits GAPD, HGH 180-191 does not, and the inhibition is reversed both by HGH 6-13 and HGH 176-180. This would suggest that binding is due to the Phe-Leu-Arg-Ile part of the structure of HGH 176-191, and the inhibition to the whole or part of the remainder. This data also strongly suggests that the reversal of such inhibition is a non-specific phenomenon based on competition of the Phe-Leu-Arg part of the structure of HGH 176-180 and the Arg-Leu-Phe part of the HGH 6-13 sequence. The fact that Leu-Ser-Arg-Leu-Phe-Asp is inert suggests that there are conformational requirements even in such small peptides if they are to attach to appropriate receptors or binding sites. These findings, together with the fact that reversal of HGH 176-191 inhibition of GAPD is dissociated from the immediate effects of the peptides on glucose tolerance and glucose uptake by the rat diaphragm, further tends to suggest that this reversal is not a physiological phenomenon. The finding that for one sample of the inhibitory peptide significant inhibition of GAPD was observed at 1.33 µg/ml, whereas inhibition of activation of pyruvate dehydrogenase was observed at 40 ng/ml, inhibition of insulin binding to adipocytes at 3.2 ng/ml and an in vivo hyperglycaemic response was seen at 1 ng/g body weight, also tends to eliminate the inhibition of GAPD as a reaction of physiological significance. However, the GAPD assay remains a useful monitoring tool, as so far all peptides which have failed to reverse inhibition by active C-terminal peptides have had no in vivo hypoglycaemic activity, and only those peptides which in the disulphide form inhibit the enzyme have had in vivo hyperglycaemic activity, although the disulphide bond has been shown in preliminary experiments to be not necessary for in vivo action (Miss D. Pakenham, unpublished data).

Further, the fact that inhibition kinetics of GAPD by the appropriate peptides indicate tight binding (M.J. Waters and J.McD. Armstrong, unpublished data) may provide a model of the type of binding needed for the action of the peptides, particularly as the primary and tertiary structures of GAPD have been elucidated (Buehner et al., 1974).

The action of HGH 6-13 and HGH 176-180

From the data presented in this paper and those previously obtained (Ng et al., 1974), it is clear that polypeptides containing the sequence HGH 6-13 in the correct configuration are capable of immediately potentiating the action of insulin both in vitro and in vivo. This early stimulation does not appear to be related to stimulation of insulin release from the β cells, as it was observed in diabetic animals given a standard dose of insulin (Ng et al., 1975) and is also seen in the rat diaphragm experiments. It is tempting to attribute these effects to the observed apparent stimulation of synthesis of insulin receptors in hepatocytes (see Table 6) and similar effects observed in trypsin-treated lipocytes (Bornstein et al., 1975a), as such effects apparently involve de novo protein synthesis being inhibited by cycloheximide and by actinomycin D in trypsin-treated lipocytes. For this hypothesis to be valid, both HGH 6-13 and HGH 176-180 would have to enter the cell and act as derepressors at transcription level, the difference between them in terms of acute and chronic effects being purely one of dose effectiveness. However, this hypothesis does not account for the finding that both peptides on prolonged treatment condition the islets to release more insulin when stimulated by glucose at a concentration of 10×10^{-3} M (see Table 7). Here it is clearly observed that HGH 6-13 is more potent than HGH 176-180, as a single injection of the former peptide is sufficient to produce sensitisation of the β cells whereas repeated treatment with the latter peptide is needed to produce a response. Perhaps the most attractive hypothesis is that such active peptides act to increase the number of effective receptors whilst increasing the sensitivity of the β cells to glucose, and that the low plasma radioimmunoreactive insulin levels observed (Ng et al., 1974) 3 days following the injection of a large dose of HGH 1-15 (6 µg/g) are due to increased hepatic and peripheral usage of the hormone. This hypothesis is being tested at present by estimating portal and peripheral insulin levels under conditions of glucose load following HGH 6-13 and HGH 176-180 treatment. It is difficult at this time to postulate a physiological role for peptides of this type, although a peptide fraction with identical in vivo activity patterns has been isolated from urine. The very high dosage required to produce a measurable in vivo effect (approximately 3 µg/g body weight) is in a pharmacological rather than a physiological range although, in rather curious contradistinction, effects on the receptor capacity of trypsinised lipocytes and preincubated hepatocytes were obtained in the range of 5-1000 ng/ml.

The hyperglycaemic effect of HGH 176-191

It is clearly seen that the minimum active sequence at present known capable of producing an in vivo hyperglycaemic effect is HGH 177-191 — i.e., H-Leu-Arg-Ile-Val-Gln-Cys-Arg-Ser-Val-Glu-Gly-Ser-Cys-Gly-Phe-OH — whereas HGH 180-191 proved inactive both in vivo and in vitro. Whether 178-191 or 179-191 are active is not at present known, but the lack of activity of 180-191 fits in with the fact that the sequence H-Lys-Cys-Arg-Arg-Phe-Gly-Glu-Ala-Ser-Cys-Ala-Phe-OH obtained by cyanogen bromide hydrolysis of ovine growth hormone is also inactive. If these peptides have conformational requirements similar to the N-terminal peptides, then it is quite conceivable that peptides larger than HGH 176-191 may be inactive until a suitable sequence is reached, e.g. HGH 167-191. For example: whereas HGH 167-191 is positively charged, as is HGH 176-191, the C-terminal cyanogen bromide fragment is neutral.

On present data it appears conclusive that the initial rise in blood glucose produced by these peptides is due to stimulation of glucagon release and resulting glycogenolysis, as it is abolished by anti-glucagon serum. Alternatively, depression of insulin release could alter the insulin-glucagon ratio so as to produce such a result. However, a further rise in blood glucose relative to the controls is observed after some 30 minutes when it could be expected that insulin secretion in response to the rising glucose level would tend to reduce that level towards normal. Thus it must be assumed that the initial release of glucagon is followed by the development of insulin resistance. A number of possible mechanisms suggest themselves for this phenomenon: the apparent inhibition of insulin binding at very low concentrations of peptide (see above) or the observed inhibition of the activation of pyruvate dehydrogenase and glycogen synthase also at very low concentrations of peptide (Aylward et al., 1975), this latter acting through resultant mass action phenomena.

The mechanism of action of the lipolysis is not known except that it is not cyclic-AMP-mediated. Competition with insulin for binding sites does not require the peptide to enter the cell, whereas the peptide would need to enter the cell in order to inhibit the dephosphorylation of glycogen synthase and pyruvic dehydrogenase. However, like GAPD, these enzymes could also be models for a receptor mechanism. These phenomena are at present being investigated.

Relationship of these actions to the actions of growth hormone

Firstly, it must be emphasized that neither peptide has any growth or somatomedin activity, and these actions can only be considered in relation to metabolic actions of growth hormone. Clearly, the 'insulin potentiating' action of the HGH 6-13 peptide can be related to the well known hypoglycaemic action of growth hormone and the glucagon-releasing, anti-insulin, lipolytic action of HGH 176-191 to identical actions of growth hormone (Bornstein et al., 1951; Luft and Cerasi, 1967). Although growth hormone is hydrolysed by plasma cell membrane (J. Sowden, in Bornstein et al., 1973), no direct evidence exists that these peptides are indeed the active centres of the whole molecule, particularly as H-Leu-Ser-Arg-Leu-Phe-Asp-Asn-Ala-OH and H-Phe-Leu-Arg-Ile-Val-OH have at least in part identical actions, thus indicating that other sequences giving rise to the correct conformational requirements may exist within the molecule.

Similar considerations apply to the fact that a fraction with similar chromatographic behaviour and apparently identical actions to HGH 6-13 can be isolated from urine, and a fraction with apparently identical actions to HGH 176-191 can be isolated from plasma.

Until such time as more positive identification of the urinary and plasma fragments is possible and their levels fully correlated to physiological phenomena, no positive statement on whether these sequences represent active centres can be made; however, a number of models capable of accounting for the metabolic actions of growth hormone can be developed and experimentally investigated.

REFERENCES

Ashcroft, S.J.H., Weerasinghe, L.C.C., Bassett, J.M. and Randle, P.J. (1972): *Biochem. J., 126*, 525.
Aylward, J.H., Bornstein, J. and Gould, M.K. (1975): *Proc. Aust. biochem. Soc., 8*, 99.
Aylward, J.H., Bornstein, J., Gould, M.K. and Hall, S. (1974): *Biochem. biophys. Res. Commun, 59*, 57.
Bornstein, J. (1972): In: *Growth and Growth Hormone, Proceedings, II International Symposium on Growth Hormone, Milan 1971*, pp. 68-74. Editors: A. Pecile and E.E. Müller. ICS 244, Excerpta Medica, Amsterdam.
Bornstein, J., Armstrong, J.McD., Gould, M.K., Harcourt, J.A. and Jones, M.D. (1969a): *Biochim. biophys. Acta (Amst.), 192*, 265.
Bornstein, J., Armstrong, J.McD. and Jones, M.D. (1968): *Biochim. biophys. Acta (Amst.), 156*, 38.
Bornstein, J., Armstrong, J.McD., Taft, H.P., Ng, F.M. and Gould, M.K. (1973): *Postgrad. med.J., 49*, 219.
Bornstein, J., Bacon, P.A., Welker, C. and Parsons, I.C. (1975a): Submitted for publication.
Bornstein, J., Pakenham, D.E., Browne, C.A. and Ng, F.M. (1975b): Submitted for publication.
Bornstein, J., Reid, E. and Young, F.G. (1951): *Nature (Lond.), 168*, 903.
Bornstein, J., Taylor, W.M., Marshall, L.B., Armstrong, J.McD. and Gould, M.K. (1969b): *Biochim. biophys. Acta (Amst.), 192*, 271.
Buehner, M., Ford, G.C., Moras, D., Olsen, K.W. and Rossmann, M.G. (1974): *J. molec. Biol., 90*, 25.
Howard, R.B., Lee, J.C. and Pesch, LeR.A. (1973): *J. Cell Biol., 57*, 642.
Hunter, W.M. and Greenwood, F.C. (1964): *Biochem. J., 91*, 43.
Kono, T. and Barham, F.W. (1971): *J. biol. Chem., 246*, 6210.
Luft, R. and Cerasi, E. (1967): *Acta endocr. (Kbh.), 56, Suppl. 124*, 9.
Ng, F.M., Bornstein, J., Welker, C., Zimmet, P.Z. and Taft, P. (1974): *Diabetes, 23*, 943.
Ng, F.M., Lawrence, A.S., Welker, C. and Bornstein, J. (1975): Submitted for publication.
Niall, H.D. (1971): *Nature (Lond.), 230*, 90.
Rodbell, M. (1964): *J. biol. Chem., 239*, 375.
Stewart, J.M. and Young, J.D. (1969): *Solid Phase Peptide Synthesis*. W.H. Freeman, San Francisco.

SYNTHESIS OF FRAGMENTS WITH AMINO ACID SEQUENCES OF GROWTH HORMONES*

F. CHILLEMI[1], A. AIELLO[1], A. PECILE[2] and V.R. OLGIATI[2]

[1]Department of Organic Chemistry and [2]Department of Pharmacology, University of Milan, Milan, Italy

The relationship between the structure of growth hormone (GH) and its biological activities has been extensively studied. The entire hormone molecule may not be required to elicit the hormone's characteristic metabolic effects.

Kostyo (1974) comprehensively reviewed the searches for the active core on pituitary growth hormone and stated: 'Knowledge of the structure of the core could have considerable value. Should the core of human growth hormone (HGH) prove to be of moderate size, it might be possible to synthetize it readily and in amounts sufficient to solve the present problem of providing an adequate supply of HGH for therapeutic and investigational needs. Further, knowledge of the structure of the active core should give insight into the molecular basis for species specificity.'

The verification of GH core concept demands studies of: *(a)* the biological activity of enzymatic digests of growth hormones; *(b)* the biological activity of fragments produced by cleavage of various growth hormones with cyanogen bromide; *(c)* synthetic materials sharing amino acid sequences of GH molecular fragments.

Partial digestion with trypsin, chymotrypsin and pepsin cleaves up to 10 peptide bonds of the GH molecule but does not destroy the growth-promoting activity (Li, 1968). Tryptic digestion releases 2 peptides (A I and A II) with biological activity (Yamasaki et al., 1970). The A II consists of 38 amino acid residues, corresponds to the 96-133 sequence of bovine growth hormone (BGH), and is a most interesting small peptide in the weight gain and tibia line test; it also retains in vitro many of the characteristic activities of BGH. In human subjects large doses of A II peptide reduced blood urea nitrogen, urinary nitrogen and creatinine.

Kostyo (1974) determined the nature of the active material obtained through peptide bond cleavage by plasmin. Peptide 1-134 resulted, and acted like HGH when injected intravenously into test animals. The peptide isolated by Sonenberg (A II = BGH residues 96-133) had only moderate activity compared to native growth hormone, so that the almost full activity of peptide 1-134 suggests that additional amino acid residues N-terminal to position 95 are required for full biological activity. Thus, the active core of GH may consist of a substantial portion of the N-terminal two-thirds of the molecule. Li and Gráf (1974), in similar studies on plasmin digests of HGH, found both fragments to be biologically active.

Cyanogen bromide fragments of BGH, porcine growth hormone (PGH), ovine growth hormone (OGH) and HGH have been studied. Fragment A of PGH (residues

---

* This work was supported by the National Research Council (C.N.R.), Italy.

GH fragments

```
             1                        10                          20
HGH    H₂N-Phe-Pro-Thr-Ile-Pro────Leu-Ser-Arg────Leu-Phe-Asp-Asn-Ala────Met-Leu-Arg-Ala-His-Arg-Leu-

OPL    H₂N-Thr-Pro-Val-Cys-Pro(  )Leu────Arg-Asp-Leu-Phe-Asp-Arg-Ala-Val-Met-Val────Ser-His-Tyr-Ile-
            1          5       15              20

                                       30                          40
HGH    -His-Gln-Leu-Ala-Phe-Asp-Thr-Tyr-Gln-Glu-Phe-Glu-Glu-Ala-Tyr-Ile-Pro────Lys-Glu-Gln-

OPL    -His-Asn-Leu-Ser-Ser-Glu-Met-Phe-Asn-Glu-Phe-Asp-Lys-Arg-Tyr-Ala-Gln-Gly-Lys-Gly────
                                  30            40

                            50                          60
HGH    -Lys-Tyr-Ser-Phe-Leu-Gln-Asn-Pro-Gln-Thr-Ser-Leu-Cys-Phe-Ser-Glu-Ser-Ile-Pro-Thr-

OPL    ────────Phe-Ile-Thr-Met-Ala-Leu-Asn-Ser────Cys-His-Thr-Ser-Ser-Leu-Pro-Thr-
                50                              60

                                 70                          80
HGH    -Pro-Ser-Asn-Arg-Glu-Glu-Thr-Gln-Gln-Lys-Ser-Asn-Leu-Gln-Leu-Leu-Arg-Ile-Ser-Leu-

OPL    -Pro-Glu-Asp-Lys-Glu-Gln-Ala-Gln-Gln────Thr-His-His-Glu-Val-Leu────Met-Ser-Leu-
                          70                            80

                                 90                          100
HGH    -Leu-Leu────Ile-Gln-Ser-Trp-Leu-Glu-Pro-Val-Gln-Phe-Leu-Arg-Ser────Val-Phe-Ala-Asn-Ser-

OPL    -Ile-Leu-Gly-Leu-Arg-Ser-Trp-Asn-Asp-Pro-Leu-Tyr-His-Leu-Val-Thr-Glu-Val-Arg-Gly-Met-Lys-
                           90                      100

                                   110                         120
HGH    -Leu-Val-Tyr-Gly-Ala-Ser-Asn-Ser-Asp-Val-Tyr-Asp-Leu-Leu-Lys-Asp-Leu-Glu-Glu-Gly-

OPL    -Gly-Val-Pro-Asp-Ala-Ile-Leu-Ser-Arg-Ala-Ile-Glu-Ile-Glu-Glu-Glu-Asn-Lys-Arg-Leu-
                           110                        120

                             130                          140
HGH    -Ile-Gln-Thr-Leu────Met-Gly-Arg-Leu-Glu-Asp-Gly-Ser-Pro-Arg-Thr-Gly-Gln-Ile-Phe-Lys-

OPL    -Leu-Glu-Gly-Met-Glu-Met-Ile-Phe-Gly-Gln-Val-Ile────Pro-Gly-Ala-Lys-Glu-Thr-Glu-Pro-
                                  130                       140

                                 150                          160
HGH    -Gln-Thr-Tyr────Ser-Lys-Phe-Asp-Thr────Asn-Ser-His-Asn-Asp-Asp-Ala-Leu-Leu-Lys-Asn────Tyr-

OPL    -Tyr-Pro-Val-Trp-Ser-Gly-Leu-Pro-Ser-Leu-Gln-Thr-Lys-Asp-Glu-Asp-Ala-Arg-His-Ser-Ala-Phe-Tyr-
                            150                         160

                                 170                          180
HGH    -Gly-Leu-Leu-Tyr-Cys-Phe-Arg-Lys-Asp-Met-Asp-Lys-Val-Glu-Thr-Phe-Leu-Arg-Ile-Val-

OPL    -Asn-Leu-Leu-His-Cys-Leu-Arg-Arg-Asp-Ser-Ser-Lys-Ile-Asp-Thr-Tyr-Leu-Lys-Leu-Leu-
             170                          180

                                       190
HGH    -Gln-Cys-Arg-Ser-Val────Glu-Gly-Ser-Cys-Gly-Phe-COOH

OPL    -Asn-Cys-Arg-Ile-Ile-Tyr-Asn-Asn-Asn-Cys-COOH
             190                  198
```

Fig. 1. Comparison of human growth hormone and bovine prolactin sequences. Identical residues are indicated by solid lines and highly favored substitutions by dotted lines.

5-125) and fragment A of HGH (residues 15-125) were anabolic in terms of ability to stimulate protein synthesis in muscle and liver, but surprisingly, neither stimulated growth or somatomedin production in hypophysectomized rats (Nutting et al., 1972).

The synthetic approach to the problem of the core concept was initiated in 1971. Chillemi and Pecile (1971) prepared by solid-phase synthesis 2 peptides corresponding to 81-121 and 122-153 of the sequence of Li et al. (1966). Small tibia test activity was found and the synthesis of similar peptides corresponding to the revised sequence was begun. Blake and Li (1973) synthetized a N^{α}-acetyl-95-136 peptide of HGH which showed small but measurable tibia test activity. Niall and Tregear (1973) found the unprotected fragment of HGH 95-134 to be inactive in both in vivo and in vitro tests.

The present study concerns the synthesis of fragments of human and ovine growth hormone molecules which are possibly the active cores of the natural hormone. The criteria for the choice of the synthetic sequences were:

1. The observation of external homologies in the molecule of HGH and of ovine prolactin (OPL) (Chillemi et al., 1972).

External homologies between HGH and OPL are considerable within sequences 1-89, 160-191 of HGH and 1-93, 168-198 of OPL respectively (Fig. 1). Few homologies between the 2 hormones occur in sequences 90-159 of HGH and 94-167 of OPL. Since HGH possesses both growth-promoting and lactogenic activities, and OPL is only lactogenic, the region 90-159 of HGH with minimal homologies by comparison with OPL, is likely to be biologically relevant for growth-promoting actions.

Similar conclusions result from comparisons of the sequences of OGH and of OPL. It is also possible that the structural differences between HGH and OPL do not represent the accumulation of functionally unimportant mutations during evolutionary divergence of human and ovine species, rather they are relevant to their biological activities.

2. Observation of internal homologies in HGH molecule.

Four internal homologous regions occur in HGH molecule: sequences 9-36, 87-116, 124-153 and 157-186 (Fig. 2). These 4 internal homologies suggest the possibility of an evolutionary change of the GH molecule from an ancestral peptide through 2 subsequent replications: the process of the evolutionary origin of GH could be somehow similar to that proposed for immunoglobulins (Hill et al., 1966).

The fact that the primordial peptide has survived long enough for duplication and reduplication may indicate intrinsic biological activity. During peptide evolution such activity could have been modified so that physiological functions complementary to that present in the first peptide could have emerged. Thus, each of the internally homologous regions could represent an active site of HGH molecule.

SOLID-PHASE SYNTHESIS OF PEPTIDES

Peptides were synthetized by stepwise addition of a suitably protected amino acid to terbutyloxycarbonyl amino acid resin. The butyloxycarbonyl (Boc) group or 2-phenylisopropyloxycarbonyl (Ppoc) group was used for α-amino protection. The Ppoc amino acids have been prepared with satisfactory yields both via phenyl ester and via azide (Sieber and Iselin, 1968; Sandberg and Ragnarsson, 1974).

The side chains of amino acids were protected as follows: Asp (β-benzyl ester), Glu (γ-benzyl ester), Ser (benzyl), Thr (benzyl), Tyr (2,6-dichlorobenzyl), Lys (ε-2-chlorobenzyl-oxycarbonyl), Arg (p-toluenesulfonyl), His (2,4-dinitrophenyl).

Coupling was carried out using a 4-fold excess of the appropriate amino acid and N,N'-dicyclohexylcarbodiimide (DCCI) for 3 hours. All the coupling reactions were re-

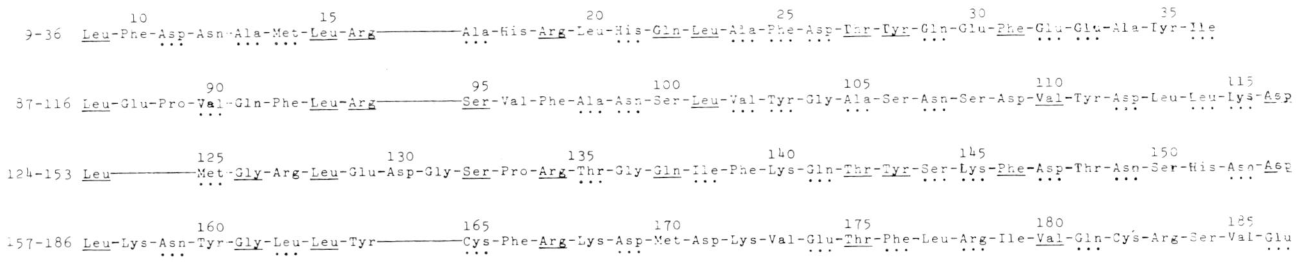

GH fragments

```
  9-36    Leu-Phe-Asp-Asn-Ala-Met-Leu-Arg————Ala-His-Arg-Leu-His-Gln-Leu-Ala-Phe-Asp-Thr-Tyr-Gln-Glu-Phe-Glu-Glu-Ala-Tyr-Ile
          ...     ...     ... ...              ...     ... ... ... ... ...         ... ...     ...         ... ...
 87-116   Leu-Glu-Pro-Val-Gln-Phe-Leu-Arg————Ser-Val-Phe-Ala-Asn-Ser-Leu-Val-Tyr-Gly-Ala-Ser-Asn-Ser-Asp-Val-Tyr-Asp-Leu-Leu-Lys-Asp
          ...     ...     ... ...              ...     ... ... ... ... ...         ... ...     ...         ... ...
124-153   Leu————Met-Gly-Arg-Leu-Glu-Asp-Gly-Ser-Pro-Arg-Thr-Gly-Gln-Ile-Phe-Lys-Gln-Thr-Tyr-Ser-Lys-Phe-Asp-Thr-Asn-Ser-His-Asn-Asp
          ...     ...     ... ...              ...     ... ... ... ... ...         ... ...     ...         ... ...
157-186   Leu-Lys-Asn-Tyr-Gly-Leu-Leu-Tyr————Cys-Phe-Arg-Lys-Asp-Met-Asp-Lys-Val-Glu-Thr-Phe-Leu-Arg-Ile-Val-Gln-Cys-Arg-Ser-Val-Glu
          ...     ...     ... ...              ...     ... ... ... ... ...         ... ...     ...         ... ...
```

Fig. 2. Internal homology in human growth hormone. Identical residues are indicated by solid lines and highly favored substitutions by dotted lines.

TABLE 1

Fragments of HGH and of OGH synthetized

1. By solid-phase method:
   HGH 1-36, HGH 83-130, HGH 88-124, HGH 96-134, HGH 111-134, HGH 125-156, HGH 166-191; OGH 95-133, OGH 111-133, OGH 125-133

2. By conventional methods:
   HGH 139-146, HGH 147-156

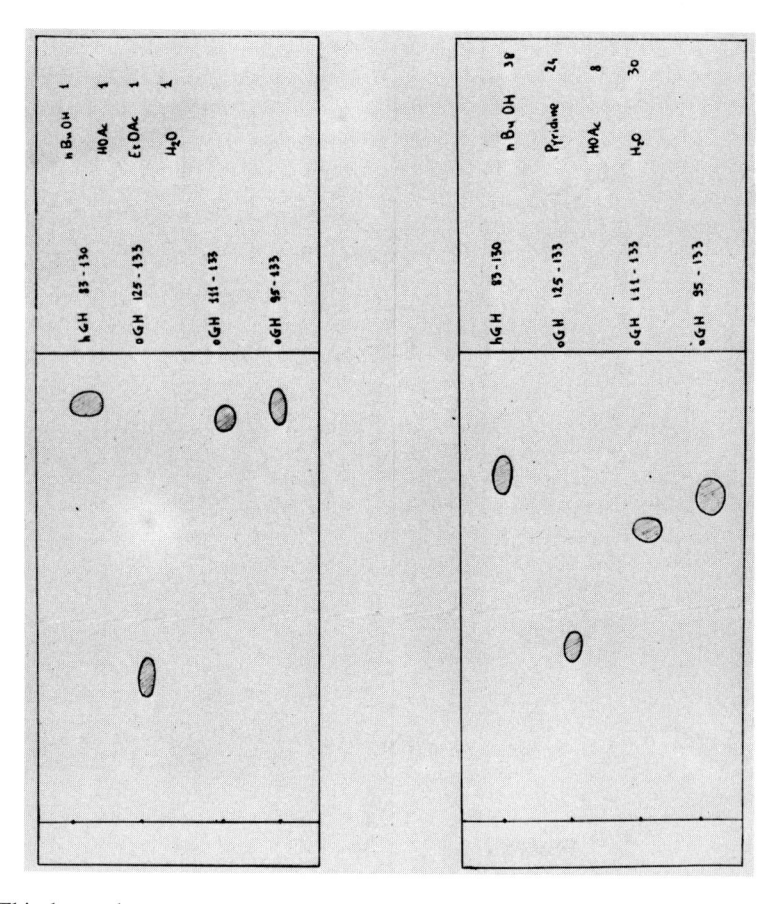

Fig. 3. Thin-layer chromatography of peptides HGH 83-130, OGH 125-133, OGH 111-133, OGH 95-133. Left: n-BuOH, HOAc, EtOAc, H$_2$O (1:1:1:1); right: n-BuOH, pyridine, HOAc, H$_2$O (32:24:8:30). The color was developed with iodine.

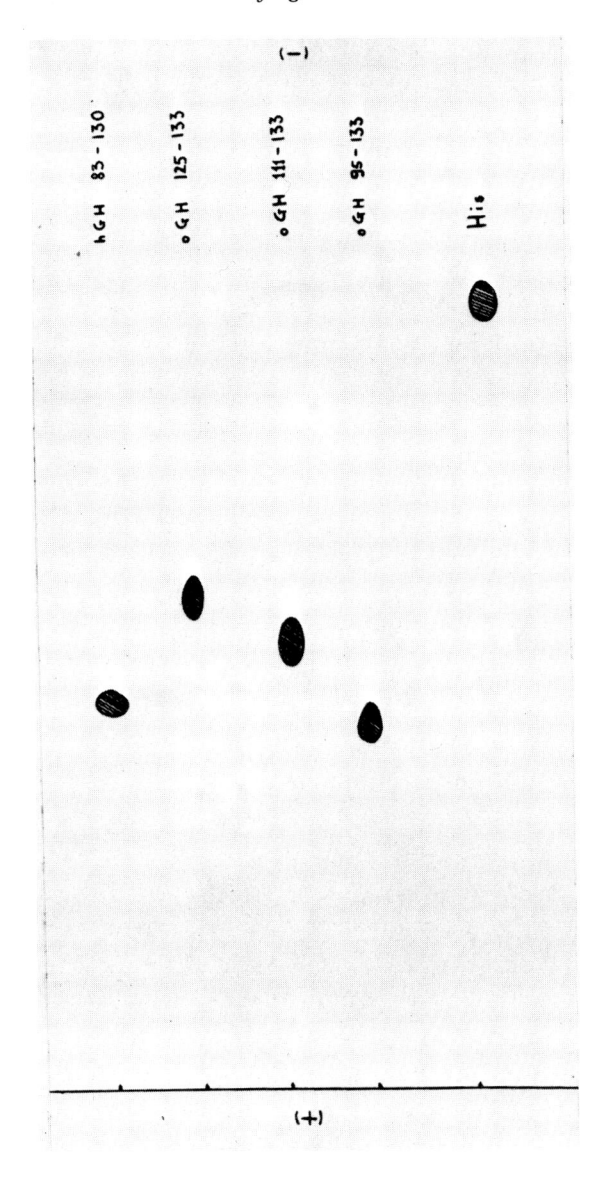

Fig. 4. Paper electrophoresis of peptides HGH 83-130, OGH 125-133, OGH 111-113, OGH 95-133. Buffer: formic acid, acetic acid, water 15:10:75.

TABLE 2
Amino acid analyses*

	HGH 83-130 Experimental	Theoretical	HGH 88-124 Experimental	Theoretical	HGH 125-156 Experimental	Theoretical
Asp	5.75	6	5.35	5	6.23	6
Thr	1.10	1	1.21	1	3.25	3
Ser	4.90	5	3.82	4	2.87	3
Glu	6.80	7	5.23	5	3.27	3
Pro	1.11	1	1.18	1	1.12	1
Gly	3.23	3	2.36	2	3.31	3
Ala	1.97	2	2.17	2	1.00	1
Val	4.07	4	3.83	4	–	–
Met	0.79	1	–	–	0.87	1
Ile	2.12	2	1.00	1	0.91	1
Leu	7.78	8	6.05	6	2.24	2
Tyr	1.85	2	1.92	2	1.03	1
Phe	2.10	2	1.96	2	2.13	2
Lys	1.07	1	0.88	1	1.89	2
His	–	–	–	–	1.05	1
Arg	1.83	2	1.02	1	2.23	2
Trp**	0.92	1	–	–	–	–

* Samples were hydrolysed in 6 N HCl for 24 hours at 110°C and analysed on a 'Technicon' amino acid analyser.
** Determinated by colorimetric method.

TABLE 3
Amino acid analyses*

	OGH 125-133 Experimental	Theoretical	OGH 111-133 Experimental	Theoretical	OGH 95-133 Experimental	Theoretical
Asp	1.04	1	2.05	2	3.93	4
Thr	0.93	1	1.03	1	2.97	3
Ser	–	–	–	–	1.84	2
Glu	2.08	2	4.83	5	4.93	5
Pro	0.84	1	0.84	1	0.95	1
Gly	–	–	1.23	1	2.04	2
Ala	–	–	1.05	1	0.99	1
Val	1.12	1	1.08	1	4.35	4
Met	–	–	0.75	1	0.78	1
Ile	–	–	1.08	1	0.91	1
Leu	1.23	1	4.83	5	6.30	6
Tyr	–	–	–	–	0.89	1
Phe	–	–	–	–	1.76	2
Lys	–	–	2.10	2	1.78	2
Arg	1.77	2	1.94	2	3.67	4

* Samples were hydrolysed in 6 N HCl for 24 hours at 110°C and analysed on a 'Technicon' amino acid analyser.

peated as indicated above. Glutamine and asparagine were added in the form of p-nitrophenyl esters (5-fold excess, reaction time 16 hours).

Boc groups were removed by treating the resin with trifluoroacetic acid in methylene chloride 50%; the Ppoc groups were removed by trifluoroacetic acid in methylene chloride 5%. The final peptides were freed from the protected groups and the solid support by anhydrous hydrogen fluoride for 1 hour at 0°C in the presence of anisole. Synthetized peptides are summarized in Table 1.

All polypeptides were purified by countercurrent distribution in systems 2-butanol-0.1% aqueous dichloroacetic acid (1:1) and 2-butanol-0.0025 N ammonium hydroxide (1:1).

The homogeneity was assessed by thin-layer chromatography (Fig. 3), high-voltage paper electrophoresis at pH 1.3 (Fig. 4) and amino acid analysis of the acid hydrolysates (Tables 2 and 3).

COMMENTS ON PEPTIDES SYNTHESIS

In the solid phase synthesis of some peptides the protective group Ppoc was chosen instead of Boc because it may be more easily removed (trifluoroacetic acid 5%) and produces better yields.

The hydroxyl group of tyrosine was protected by 2,6-dichlorobenzyl (Erickson and Merrifield, 1973a). This protection is better than benzyl in that acid stability is greater, and the intramolecular rearrangement O-benzyltyrosine to 3-benzyltyrosine is partially prevented.

The use of the 2-chlorobenzyloxycarbonyl to protect the ε-amino group of lysine eliminates the formation of the N$^\varepsilon$-branched peptides (Erickson and Merrifield, 1973b).

The guanidine function of arginine was protected by the tosyl group instead of the nitro group to avoid the possible formation of ornithine-containing material during acidolysis (Yamashiro et al., 1972).

It should be noted that during the preparation of peptides HGH 96-134, HGH 111-134, HGH 125-156, a rearrangement of the sequence Asp-Gly is expected with the formation of a succinimido derivative (II) (Fig. 5). However, the natural and the imide synthetic nonapeptides BGH 125-133 gave comparable responses in the tibia test (Wang et al., 1974). The response of the cyclic peptide may have been due to the opening of the imide ring with the formation of a mixture of α- and β-aspartyl peptides. In peptides OGH 95-133, OGH 111-133, OGH 125-133 the sequence Asp-Gly is replaced by Asp-Val and the above-reported difficulty in the synthesis is not present.

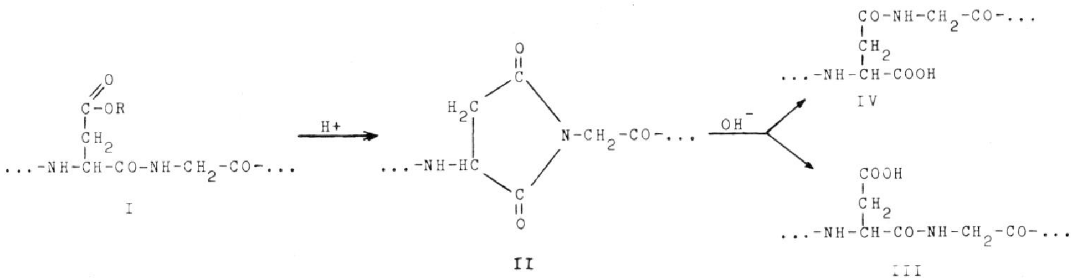

Fig. 5. α-β rearrangement of aspartyl-glycyl sequence.

BIOLOGICAL ACTIVITY OF FRAGMENTS

Tables 4-8 summarize data at present available on the biological activities of the synthetized fragments. The peptides 88-124 and 125-156 have small but measurable effects in the tibia test using intraperitoneal injections into hypophysectomized animals (Greenspan et al., 1949) (Table 4).

TABLE 4

Growth-promoting potency of peptides corresponding to the sequences 88-124 and 125-156 of HGH as assayed by the rat tibia test

Preparation		Daily dose (μg)	Number of rats	Mean response ± S.E. (μm)	P
Saline			7	150.8 ± 3.9	
BGH		18.75	9	194.7 ± 6.1	< 0.001
BGH		75	8	251.8 ± 2.7	< 0.001
HGH	88-124	100	8	186.1 ± 2.9	< 0.001
HGH	88-124	400	6	213.1 ± 7.3	< 0.001
HGH	125-156	100	7	176.4 ± 2.5	< 0.001
HGH	125-156	400	9	202.5 ± 2.8	< 0.001

TABLE 5

Growth-promoting potency of peptides corresponding to the sequences 88-124 and 125-156 of HGH as assayed by the rat tibia test 3 months after the preparation

Preparation		Daily dose (μg)	Number of rats	Mean response ± S.E. (μm)	P
Saline			14	137.4 ± 2.7	
BGH		18.75	15	195.0 ± 4.8	< 0.001
BGH		75	17	253.6 ± 2.7	< 0.001
HGH	88-124	100	14	151.5 ± 3.4	< 0.01
HGH	88-124	400	15	186.5 ± 4.1	< 0.001
HGH	125-156	100	15	139.7 ± 1.1	
HGH	125-156	400	16	188.9 ± 2.9	< 0.001

TABLE 6

Growth-promoting potency of peptides corresponding to the sequences 83-130 and 96-134 of HGH as assayed by the rat tibia test

Preparation		Daily dose (μg)	Number of rats	Mean response ± S.E. (μm)	P
Saline			22	137.4 ± 2.7	
BGH		18.75	21	198.0 ± 4.2	< 0.001
BGH		75	21	247.7 ± 3.9	< 0.001
HGH	83-130	187.5	14	144.7 ± 3.2	
HGH	83-130	750	14	146.3 ± 3.2	
HGH	96-134	100	8	150.0 ± 4.0	< 0.05
HGH	96-134	400	9	152.2 ± 2.5	< 0.01

The specificity of the tibia test has been questioned (Peña et al., 1972) but it nevertheless gives relevant information, particularly if interfering substances (e.g. methylthiocyanate, a by-product in the reaction of cyanogen bromide with the protein) are excluded. Obviously positive tibia test results should be critically evaluated and substantiated with body weight gain tests, tests of metabolic activity in vivo and in vitro, radioreceptor analysis, etc.

It is noteworthy that fragments 88-124 and 125-156 contain one of the 4 internal homologous regions (87-116 and 124-153 respectively). The Sonenberg fragment 96-133 of BGH which is also active in the human, possesses a sequence overlapping the 2 fragments synthetized by us. The same peptides were retested 3 months after the preparation (Table 5). Activity was reduced which is indicative of instability so that assays of the substances at different time intervals after preparation may give different results.

Other fragments of HGH such as 96-134 and 83-130 were assayed on tibia test. Surprisingly, all failed to show a significant biological activity (Table 6). This was rather unexpected particularly since fragment 96-134 represents the region in HGH homologous with Sonenberg's fragment of BGH. Our findings confirm those by Niall and Tregear (1973) who synthetized HGH 96-134 fragments and found no activity either in vitro or in vivo. On the contrary, Blake and Li (1973) who synthetized a very similar peptide (95-136) but protected it by acetylation of the α-amino group, found it able to stimulate tibial width in hypophysectomized rats.

Two fragments of OGH, 95-133 and 111-133, were also assayed by the tibia test (Table 7). The smaller peptide, 111-133, had a measurable effect while the larger was virtually inactive.

TABLE 7

Growth-promoting potency of peptides corresponding to the sequences 95-133 and 111-133 of OGH as assayed by the rat tibia test

Preparation		Daily dose (µg)	Number of rats	Mean response ± S.E. (µm)	P
Saline			15	136.0 ± 1.6	
BGH		18.75	15	190.0 ± 2.5	<0.001
BGH		75	20	225.9 ± 5.3	<0.001
OGH	95-133	187.5	9	145.7 ± 2.6	<0.01
OGH	95-133	750	9	152.8 ± 1.9	<0.001
OGH	111-133	187.5	13	150.3 ± 3.1	<0.001
OGH	111-133	750	13	160.1 ± 2.0	<0.001

TABLE 8

Growth-promoting activity of HGH 83-130 and OGH 95-133 as assayed by the body weight gain

Preparation		Daily dose (µg)	Number of rats	Mean response ± S.E. (g)	P
Saline			10	−1.1 ± 1.5	
BGH		25	10	10.1 ± 0.8	<0.001
BGH		100	10	21.0 ± 1.5	<0.001
HGH	83-130	250	10	1.8 ± 0.6	
OGH	95-133	250	8	9.7 ± 2.2	<0.01

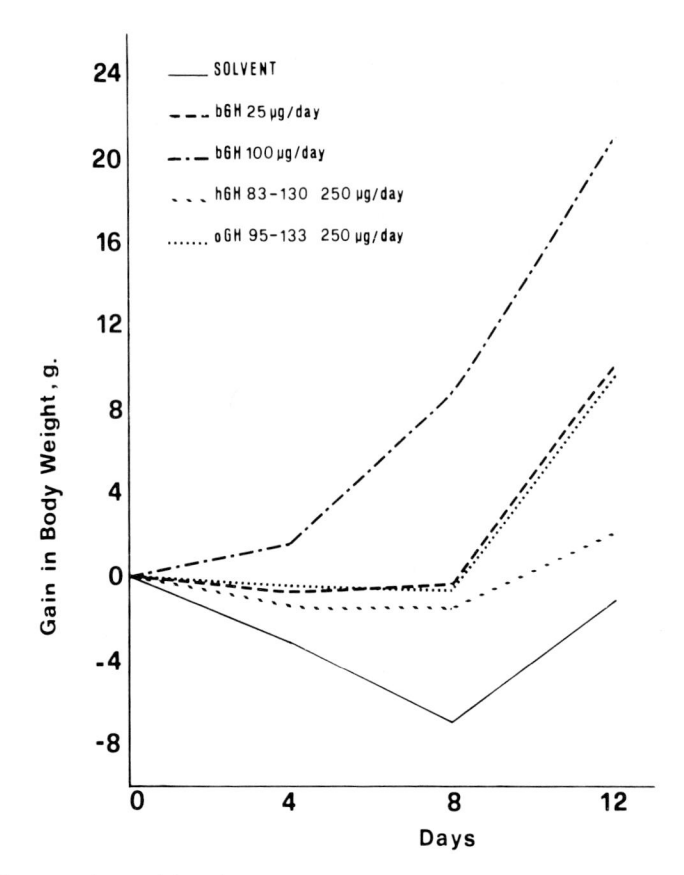

Fig. 6. Growth-promoting activity of HGH 83-130 and OGH 95-133 in immature hypophysectomized female rats as assayed by the body weight gain. Animals operated on at 28 days of age and used 2 weeks postoperatively.

Thus far only 2 peptides have been assayed in a body weight gain test (Marx et al., 1942): the HGH 83-130 and OGH 95-133. Table 8 and Figure 6 show the weight increase of hypophysectomized animals after 12 days of treatment. A positive effect is induced only by the OGH peptide while the relatively large HGH 83-130 fragment produced no effect. Yamasaki et al. (1970) studied a peptide of BGH with a sequence very similar to that of OGH 95-133. The results of Yamasaki and co-workers on the body weight gain test are qualitatively similar.

Figure 7 compares the sequences of OGH, BGH and HGH fragments. Comparison of the structure of OGH fragment with that of HGH fragment reveals that 24 positions have identical amino residues, 8 have highly acceptable replacements and 8 have unacceptable replacements. The fragments of OGH and BGH show a high degree of homology: 37 residues of amino acids are identical, one a highly favored replacement and one an unfavored replacement.

The leucine incorporation into protein test (Martin and Young, 1965) was used only for the HGH 83-130 fragment. A small activity is present at a dose of 200 μg/ml but

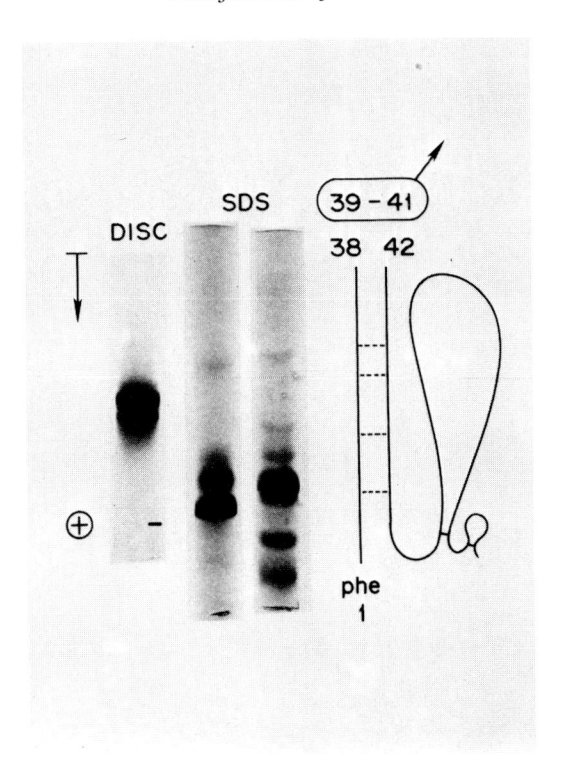

Fig. 3. Disc and SDS electrophoretic patterns of the 20K modification of HGH. The major band (with an R_f value near 0.45) was slightly more basic than HGH at pH 10 and 8% acrylamide. The preparation was not homogeneous in SDS. The major component, however, did not dissociate when reduced. The nature of the minor component that did dissociate when reduced is not known. Residues 39-41 were deleted from 20K but the 1-38 peptide did not separate from the rest of the molecule during purification.

and may actually cause a decrease. The prolactin-like activity of this form, on the other hand, was slightly increased above that seen with intact HGH. Cleavage of HGH at residues 38-39 to form 20K did not decrease the growth activity and definitely potentiated the crop stimulating activity. It will now be most interesting to cleave 20K to produce an α_3-like deletion of residues in the large disulfide loop and determine if this smaller form (approximately 140 residues) has enhanced growth-promoting activity.

It is worth pointing out that the 24K and 20K forms were present in greatest quantities as dimers in pituitary extracts and that chromatography on Sephadex G-100 removed most of these forms from monomeric HGH. The 24K and 20K were indistinguishable from NIH reference HGH when analyzed by radioimmunoassay.

DISULFIDE DIMER

A peak obtained by chromatography of crude HGH on Sephadex G-100 (Singh et al., 1974b) contained rather large quantities of a substance with a molecular weight of 45,000 daltons. This is the 45K band of Figure 1. The substance was isolated and

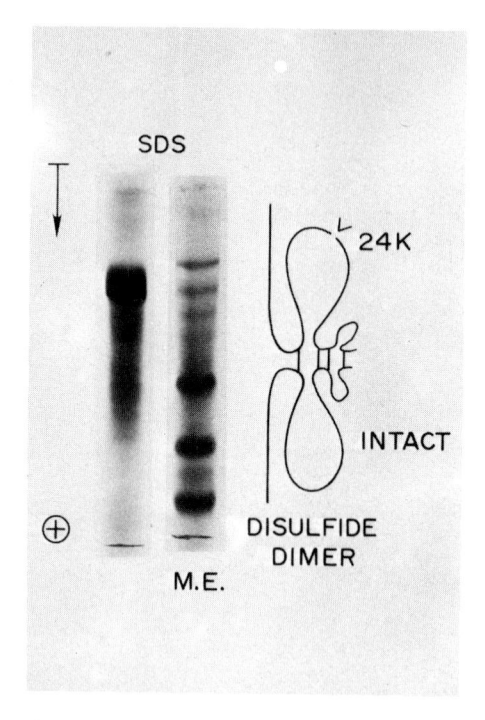

Fig. 4. SDS electrophoretic patterns of the disulfide dimer. Dissociation after reduction with mercaptoethanol (M.E.) is discussed in the text. The sketch indicates that the dimer is a hybrid of HGH and 24K and shows 4 disulfide bridges connecting the 2 monomers. These points are not known with certainty.

found to be another form of HGH. The material behaved as a dimer during chromatography in 6 M urea and during SDS electrophoresis if mercaptoethanol was not used. Upon reduction, however, the substance dissociated into HGH of 22,000 daltons and fragments similar to those formed from 24K (Fig. 4). The results indicated that the urea-SDS stable substance was a disulfide dimer of HGH. It is our feeling that this is probably 'big' HGH and we suggest that the name 'big' be dropped and the term disulfide dimer be considered. Schneider et al. (1975) found a similar form of human placental lactogen and also suggested that this disulfide modification was 'big' placental lactogen. We have observed this type of dimer of human prolactin, and in collaboration with Dr. H.G. Kwa, we isolated a disulfide dimer of both rat and mouse prolactins.

The number of cysteine residues involved in the disulfide bonding of 2 monomeric units is not known, nor is the location of the bridge(s) known. To be determined also is whether the disulfide dimer is a hybrid of intact HGH and a cleaved form, or whether the isolated material is a mixture of dimeric intact HGH and dimeric cleaved forms.

Additional electrophoretic characterization of the disulfide dimer of HGH was made by constructing Ferguson plots (Ferguson, 1964) of the electrophoretic mobility data. These are shown in Figure 5 where it can be seen that the disulfide dimer is a more acidic substance than monomeric, intact HGH. This is also shown in the electrophoretic patterns of Figure 6 where the disulfide dimer is seen to migrate faster than

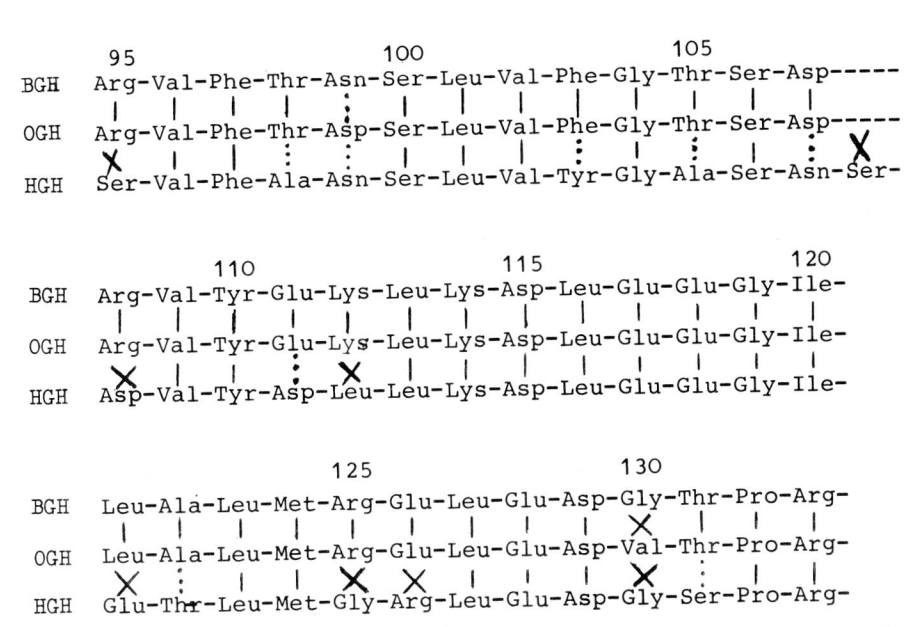

```
              95                  100                           105
BGH   Arg-Val-Phe-Thr-Asn-Ser-Leu-Val-Phe-Gly-Thr-Ser-Asp-----
       |   |   |   |   :   |   |   |   |   |   |   |   |
OGH   Arg-Val-Phe-Thr-Asp-Ser-Leu-Val-Phe-Gly-Thr-Ser-Asp-----
       X   |   |   :   |   |   |   |   :   |   :   |   :   X
HGH   Ser-Val-Phe-Ala-Asn-Ser-Leu-Val-Tyr-Gly-Ala-Ser-Asn-Ser-

                     110                 115                 120
BGH   Arg-Val-Tyr-Glu-Lys-Leu-Lys-Asp-Leu-Glu-Glu-Gly-Ile-
       |   |   |   :   |   |   |   |   |   |   |   |   |
OGH   Arg-Val-Tyr-Glu-Lys-Leu-Lys-Asp-Leu-Glu-Glu-Gly-Ile-
       X   |   |   X   |   |   |   |   |   |   |   |   |
HGH   Asp-Val-Tyr-Asp-Leu-Leu-Lys-Asp-Leu-Glu-Glu-Gly-Ile-

                         125                 130
BGH   Leu-Ala-Leu-Met-Arg-Glu-Leu-Glu-Asp-Gly-Thr-Pro-Arg-
       |   |   |   |   |   |   |   |   |   X   |   |   |
OGH   Leu-Ala-Leu-Met-Arg-Glu-Leu-Glu-Asp-Val-Thr-Pro-Arg-
       X   :   |   |   X   X   |   |   |   X   :   |   |
HGH   Glu-Thr-Leu-Met-Gly-Arg-Leu-Glu-Asp-Gly-Ser-Pro-Arg-
```

Fig. 7. Sequence 95-133 of OGH as compared to that of BGH and HGH. Homology is indicated by: identical pairs (|), highly acceptable replacements (:) and unacceptable replacement (X).

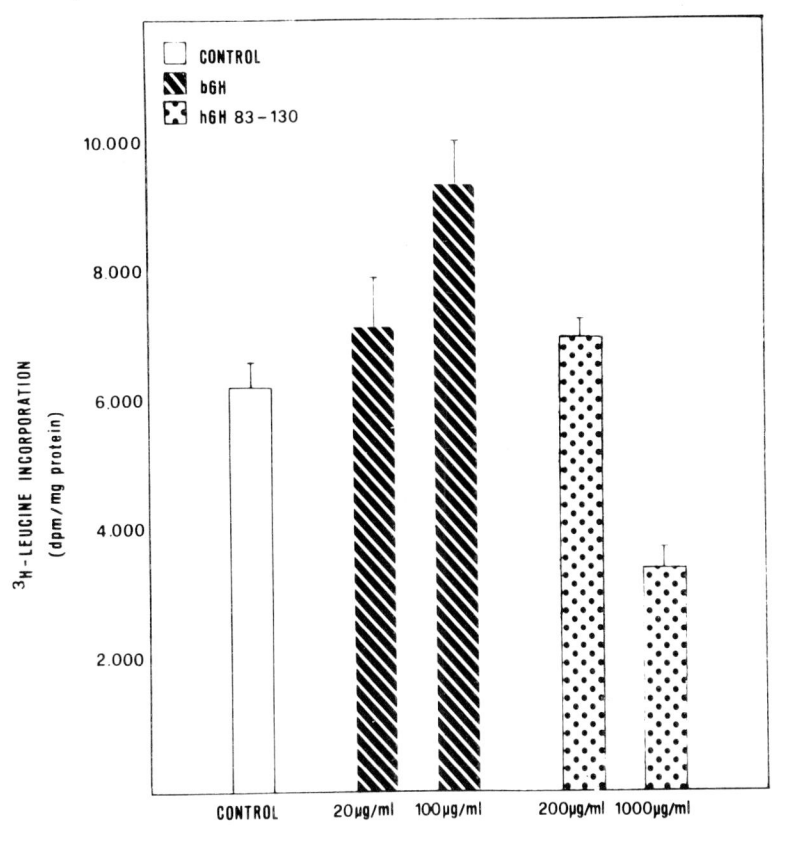

Fig. 8. The in vitro effect of HGH 83-130 on the incorporation of ^{3}H-leucine into proteins extracted from the diaphragms of hypophysectomized rats. The diaphragms were incubated for 2 hours. Bars indicate mean values; the line over the bars represents one standard error of the mean difference.

increasing concentrations, instead of progressively increasing leucine incorporation produced a clear inhibition (Fig. 8).

HGH fragments 1-36 and HGH 166-191 were also assayed in vivo on blood glucose and on the tibia test but as such compounds are insoluble in aqueous media they were used as suspensions in carboxymethyl cellulose 0.5%. In both tests results were negative.

CONCLUSIONS

The present data on the biological activity of various peptides are only preliminary. However, even the low level of activity observed suggests the existence of active cores responsible for selected effects of the growth hormone molecule.

It should be noted that in their synthesis, peptides are exposed to various treatments, including anhydrous hydrogen fluoride; partial inactivation of GH fragments, susceptible under acidic conditions, may occur.

Short fragments exhibit a tendency to aggregate rather more strongly those whole GH molecules, and this could hamper biological evaluation. On the other hand, small peptides may be very susceptible to enzymatic degradation. Even small indications of activity are considered important because the synthesis of natural sequences offers a useful starting material to search minimum structural requirements for GH-like activity. The preparation either of protected peptides which may survive longer in vivo, and larger peptides with the conformational characteristics of the GH molecule necessary for full biological activity is a future step.

ACKNOWLEDGEMENT

Warm thanks are due to Prof. A.E. Wilhelmi for the BGH standard.

REFERENCES

Blake, J. and Li, C.H. (1973): *Int. J. Peptide and Protein Res., 5,* 123.
Chillemi, F. and Pecile, A. (1971): *Experientia (Basel), 27,* 385.
Chillemi, F., Aiello, A. and Pecile, A. (1972): *Nature New Biol., 238,* 243.
Erickson, B.W. and Merrifield, R.B. (1973a): *J. Amer. chem. Soc., 95,* 3750.
Erickson, B.W. and Merrifield, R.B. (1973b): *J. Amer. chem. Soc., 95,* 3757.
Greenspan, F.S., Li, C.H., Simpson, M.E. and Evans, H.M. (1949): *Endocrinology, 45,* 455.
Hill, R.L., Delaney, R., Fellows Jr, R.E. and Lebovitz, H.E. (1966): *Biochemistry, 56,* 1762.
Kostyo, J.L. (1974): *Metabolism, 23,* 885.
Li, C.H. (1968): In: *Growth Hormone,* p. 3. Editors: A. Pecile and E. Müller. ICS 158, Excerpta Medica, Amsterdam.
Li, C.H. and Gráf, L. (1974): *Proc. nat. Acad. Sci. (Wash.), 71,* 1197.
Li, C.H., Liu, W.K. and Dixon, J.S. (1966): *J. Amer. chem. Soc., 88,* 2050.
Martin, T.E. and Young, F.G. (1965): *Nature (Lond.), 208,* 684.
Marx, W., Simpson, M.E. and Evans, H.M. (1942): *Endocrinology, 30,* 1.
Niall, H.D. and Tregear, G.W. (1973): In: *Advances in Human Growth Hormone Research, Baltimore 1973,* p. 394. Editor: S. Raiti. DHEW Publication No. (NIH) 74-612, Government Printing Office, Washington, D.C.
Nutting, D.F., Kostyo, J.L., Mills, J.B. and Wilhelmi, A.E. (1972): *Endocrinology, 90,* 1202.
Peña, C., Hecht, J.P., Santomé, J.A., Dellacha, J.M. and Paladini, A.C. (1972): *FEBS Letters, 27,* 338.

Sandberg, B.E. and Ragnarsson, U. (1974): *Int. J. Peptide and Protein Res., 6,* 111.
Sieber, P. and Iselin, B. (1968): *Helv. chim. Acta, 51,* 622.
Wang, S.S., Yang, C.C., Kulesha, I.D., Sonenberg, M. and Merrifield, R.B. (1974): *Int. J. Peptide and Protein Res., 6,* 103.
Yamasaki, N., Kikutani, M. and Sonenberg, M. (1970): *Biochemistry, 9,* 1107.
Yamashiro, D., Blake, J. and Li, C.H. (1972): *J. Amer. chem. Soc., 94,* 2855.

HUMAN GROWTH HORMONE: A FAMILY OF PROTEINS*

U.J. LEWIS, R.N.P. SINGH, S.M. PETERSON and W.P. VANDERLAAN

Division of Diabetes and Endocrinology, Scripps Clinic and Research Foundation, La Jolla, Calif., U.S.A.

We have known for some years that human growth hormone (HGH) is readily altered by proteinases in pituitary extracts (Lewis and Cheever, 1965). A study of the enzymically modified forms seemed of value, even if these proved to be artifacts, because the proteolysis potentiated the biological potency of the hormone (Lewis, 1966; Singh et al., 1974a). The disc electrophoresis method of Davis (1964) was used in our early work and as newer electrophoretic procedures were developed these were applied to a study of the altered forms of HGH. The topic of this paper will be a cataloging of these modifications. The information gathered from the studies has led us to develop the hypothesis that HGH must be modified to a 2-chain structure during the secretory process to be fully active.

NATURALLY OCCURRING CLEAVED FORMS

α-β series

The isolation and characterization of these forms have been described (Singh et al., 1974a). Briefly, they are more acidic forms best detected by disc electrophoresis at an alkaline pH. The forms noted were designated as α_1, α_2, α_3 and β (Fig. 1A). Insufficient amounts of α_1 were available for characterization but α_2 was found to lack residues 135-140. The α_3 had residues 135-146 deleted and β was a mixture of more highly degraded forms. In the tibial line assay α_2 had a potency similar to purified HGH whereas α_3 was 4-5 times more active than HGH that was free of the altered forms. The pigeon crop sac activity was potentiated in all but the β form; α_3 was the most active. The results indicated that to enhance the growth-promoting activity of HGH, 12 residues of the large disulfide loop had to be deleted. Removal of 6 amino acids, as in the case of α_2, was not sufficient to potentiate the hormone. By radioimmunoassay the α-β forms were indistinguishable from NIH reference HGH.

Yadley and Chrambach (1973) and Yadley et al. (1973) have carried out extensive work on the electrophoretic characterization of altered forms of HGH found in pituitary extracts and those formed by digestion of the hormone with plasmin. They found enhanced biological activities of the same magnitude we noted for the α forms. Struc-

---

* This research was supported by NIH Grants AM-09537, AM-16065 and CA-14025; American Cancer Society Grant BC-104; and awards from the Kroc Foundation, Human Growth Foundation and the Diabetes Association of Southern California. Publication No. 50 from the Department of Clinical Research.

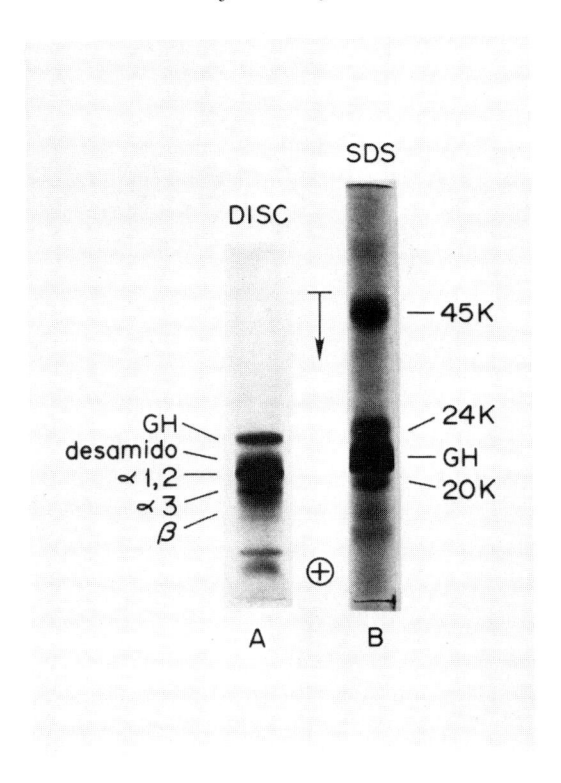

Fig. 1. Electrophoretic patterns of HGH. A: Disc electrophoresis, pH 10, 6.5% acrylamide. B: SDS electrophoresis (K = 1000 daltons).

tural characterization of the modifications studied by Chrambach and his associates will be needed to determine how they are related to the α-β series. Reagan et al. (1975) isolated a plasmin modification of HGH that lacked residues 135-140 which would make it correspond to our α₂. Likewise, Li and Gráf (1974) reported what appears to be the same modification. There was no potentiation in growth-promoting activity by removal of the hexapeptide 135-140.

Molecular size modifications

When we learned to remove the α-β forms from the major component of HGH, we thought we had an essentially homogeneous preparation of the hormone. Electrophoresis in sodium dodecyl sulfate (SDS) by the method of Weber and Osborn (1969) showed this was far from being true. All preparations that we examined contained 2 and many times 3 components in addition to the one thought to be HGH. These are shown in Figure 1B. The major component migrated as a protein with a molecular weight of 22,000 daltons, a value which was in close agreement with the actual molecular weight of the hormone. The 3 contaminating substances had apparent molecular weights of 24,000 (24K), 20,000 (20K) and 45,000 daltons (45K).

The 24K and 20K modifications were isolated (Singh et al., 1974b) and characterized structurally. The 24K form was HGH that had been cleaved in the large disulfide

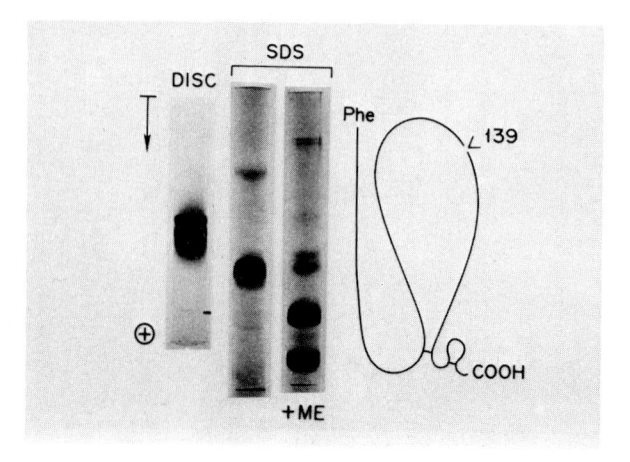

Fig. 2. Disc and SDS electrophoretic patterns of the 24K modification of HGH. The electrophoretic mobilities of the 3 components seen at pH 10 and 8% acrylamide were 0.55, 0.6 and 0.65. Intact HGH had a value near 0.5. The 24K was 1 component in SDS if not reduced but it dissociated into 2 major fragments after reduction (ME = mercaptoethanol). The nature of the 2 minor bands is not known. The 24K has a cleavage of the chain at position 139.

loop but with no deletion of amino acids. By peptide mapping it appeared that the peptide chain had been opened between residues 139 and 140. These data explained the results observed when 24K was analyzed by SDS electrophoresis (Fig. 2). The break in the peptide chain within the large disulfide loop permitted greater unfolding of the molecule, and as a result, 24K migrated as a substance with a larger molecular radius than intact HGH. When treated with mercaptoethanol prior to SDS electrophoresis, 24K dissociated into 2 fragments because the disulfide bridge at 53-165 no longer held the molecule together.

The 20K modification is still somewhat of a puzzle even though we know that the large disulfide loop is intact and that a cleavage has removed residues 39-41. The puzzling fact is that even though a tripeptide was removed from the amino terminal portion of the hormone, the peptide 1-38 did not dissociate from the rest of the molecule. We have been unable to remove the peptide by chromatography on Sephadex in 6 M urea or 10% formic acid. Because of this, the 20K can be considered a 2-chain form held by hydrophobic bonding as shown in Figure 3. Amino terminal analysis indicated both phenylalanine and tyrosine.

Our earlier publication (Singh et al., 1974*b*) reported the crop sac stimulating activities of the 24K and 20K forms to be 2 and 9 IU/mg, respectively. Removal of 20K and 24K from HGH lowered the crop stimulating activity of the intact form to less than 1 IU/mg. This low order of activity is similar to that reported a number of years ago by Chadwick et al. (1961). The results indicate that much of the prolactin-like activity measured by the crop sac assay is a result of enzymatically modified forms in preparations of the hormone. Recently we had sufficient quantities of the 20K and 24K modifications to measure the tibial line activities. The 20K form gave a response identical to that produced by the WHO reference growth hormone (1 IU/mg). The 24K form had only borderline activity since 40 μg produced a width of the tibial line that was not much greater than that noted for uninjected rats. These bioassay results show that the single 'nick' to produce 24K does not increase the growth-promoting activity

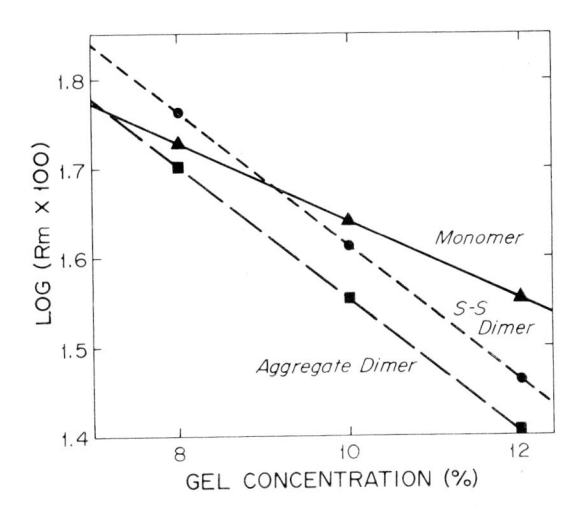

Fig. 5. Plot of electrophoretic mobility (R_m) vs gel concentration for monomeric HGH, the aggregate dimer of HGH and the disulfide dimer.

HGH in an 8% acrylamide gel. With increasing concentrations of acrylamide the dimer was retarded to a greater extent than HGH and migrated behind the monomer (Fig. 5). Figure 6 shows that aggregation of the disulfide dimer could be detected by disc electrophoresis. In SDS electrophoresis, these aggregates dissociated and only the dimeric form was noted. We have found also that highly purified, monomeric intact HGH readily aggregated upon lyophilization and that these higher molecular weight species were easily detected by disc electrophoresis (Fig. 6). In SDS, only the monomeric form was present, again indicating that aggregates were dissociated by the detergent. Estimations of the molecular weights of the aggregates of HGH and of the disul-

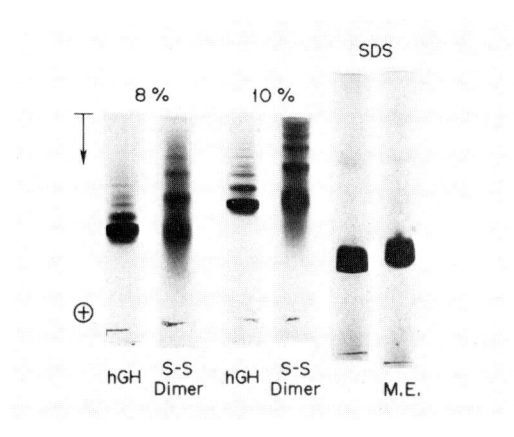

Fig. 6. Disc electrophoretic patterns (pH 10 in 8 and 10% acrylamide) of HGH and the disulfide dimer. Aggregated forms can be seen in both. The SDS patterns are of the HGH preparation analyzed with and without mercaptoethanol (M.E.). The aggregates are not seen by this technique. SDS patterns of the disulfide dimer are shown in Figure 4.

TABLE 1

Aggregates of intact HGH and the disulfide dimer

Form	Slope of R_m vs gel concentration plot	
	HGH	Disulfide dimer
Monomer	4.4	–
Dimer	7.7	7.8
Trimer	9.6	–
Tetramer	11.5	12.0
Pentamer	13.9	–
Hexamer	15.9	16.4
Heptamer	17.7	–
Octamer	19.6	20.4

Data from Cheever and Lewis (1969).

fide dimer were made as described by Cheever and Lewis (1969). These are tabulated in Table 1. As expected, the HGH formed a series of aggregates made up of increasing numbers of the monomer; the disulfide dimer, on the other hand, formed a series of aggregates which corresponded to increasing numbers of dimers.

Often even our best preparations of HGH contained small amounts of the disulfide dimer. Since the dimer was not detected in HGH analyzed directly from a Sephadex column, we feel there must be some disulfide interchange between monomers during workup of the sample.

The tibial line test indicated that the disulfide dimer was essentially inactive as a growth-promoting substance. At 40 µg per rat, no increase in the width of the tibial plate was seen. By the pigeon crop sac assay, however, the material had a potency near 2 IU/mg (Singh et al., 1974b).

Using a radioimmunoassay for HGH which employed NIH HGH (HS1652C) as a reference and commercially available antiserum (Mann) and labeled HGH (Abbott Laboratories), displacement by the dimer was only one-half to one-third that seen by intact HGH. Hopefully, with monovalent antisera to intact HGH and dimer, the 2 forms will be able to be measured individually in serum.

NATURALLY OCCURRING FORMS DETECTED BY ISOELECTRIC FOCUSING

When we learned to prepare HGH that was free of the α-β forms and the molecular size modifications, we thought that finally we had homogeneous HGH. Figure 7 pictures the disc and SDS electrophoresis patterns of our most highly purified HGH. The pattern at pH 10 showed a second faster migrating band which we believe is a desamido form. A single component was seen at pH 4. In SDS the sample was free of 24K, 20K and the disulfide dimer. When examined by isoelectric focusing, however, it was a mixture of at least 3 components (Fig. 7). The center component was the major one and 2 other principal components migrated on each side of it. We have designated them as 1a and 1c, the a denoting an anodal and c a cathodal position relative to the major component. It is interesting that the desamido form showed the same 3 components. Purification of these forms by ion-exchange chromatography was unsuccessful. Using the preparative isoelectric focusing method of Radola (1973) we have been able to obtain only enriched preparations of these forms. Separation achieved by the Radola method is shown in Figure 8 and Figure 9 pictures the analytical focusing pat-

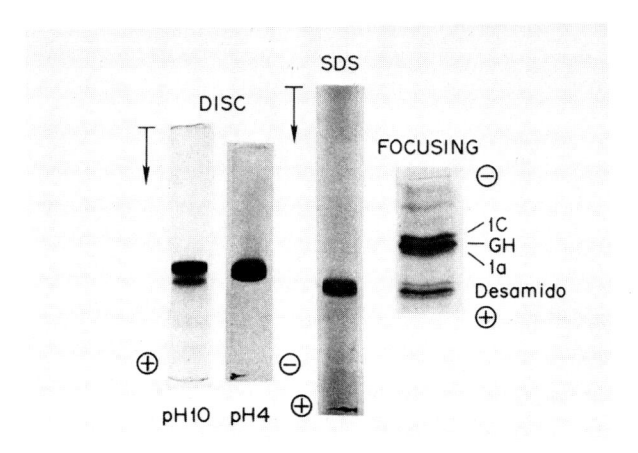

Fig. 7. Disc (8% acrylamide), SDS and isoelectric focusing patterns (pH 4-6) of purified HGH.

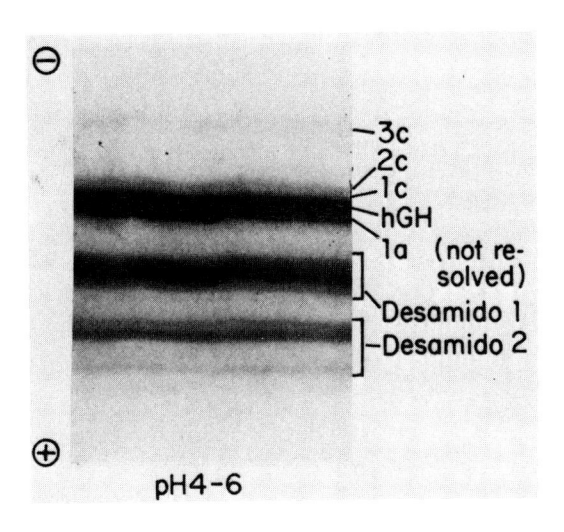

Fig. 8. Print of preparative isoelectric focusing run (pH 4-6) carried out by the procedure of Radola (1973). A partially purified sample of HGH was used.

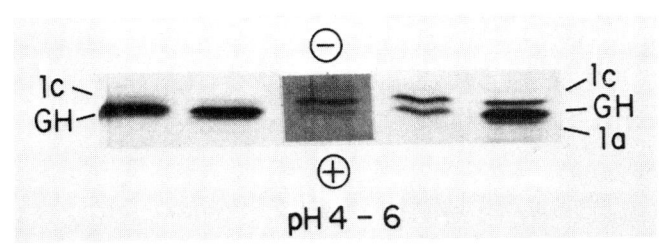

Fig. 9. Analytical isoelectric focusing patterns of fractions prepared by the Radola method. Note enrichment of 1c and samples free of 1a.

terns of material isolated by the procedure. There was definite enrichment of the substances. Because this type of enrichment could be obtained and because the same pattern was seen with 2 different Ampholines (pH ranges 4-6 and 5-7), we feel that these components are not artifacts of the focusing method. They may be allelic modifications where amino acid substitution of neutral residues prevents detection by disc electrophoresis but not by isoelectric focusing. Alternately, the 1a and 1c forms could be true isohormones synthesized in the pituitary gland. The term 'isohormone' is used by Chrambach et al. (1973) to denote enzymatically modified forms. The modifications seen by isoelectric focusing may prove to have differing biological as well as physicochemical properties.

MODIFICATIONS IN FRESH PITUITARY EXTRACTS

We examined extracts of pituitary glands made within 36 hours after death of the individual. We detected no α forms nor were the 24K and 20K modifications seen. In each of the 6 glands examined, however, the 1a and 1c isoelectric components were present along with the major band. These results support the idea that 1a and 1c may be allelic forms or isohormones and that the other modifications detected by disc electrophoresis and SDS electrophoresis are enzymic alterations produced during the isolation procedure.

SUMMARY OF NATURALLY OCCURRING MODIFICATIONS OF HGH

Figure 10 summarizes what is known structurally of the modified forms of HGH. Figure 11 indicates the isolation procedures used in purification of the various forms.

ENHANCEMENT OF GROWTH-PROMOTING ACTIVITY OF HGH BY A BACTERIAL PROTEINASE

Attempts at purifying the enzyme that was responsible for the formation of α_3, the form with enhanced biological activities, were disappointingly unsuccessful. We turned to other known proteinases with the hope that one could be found that would duplicate the formation of α_3-like material. A bacterial proteinase, available commercially from Calbiochem (San Diego, California) looks promising in this respect. The enzyme degraded HGH to 3 well-defined components, each with an estimated molecular weight near 20,000 daltons. The 3 forms were isolated by chromatography on DEAE-cellulose and peptide mapping indicated that the sequences deleted were in the same area as those missing from α_3. The cleavages were not identical, however. The bacterial enzyme removed residues 137-147 and possibly 148 and 149. As shown in Table 2, all 3 forms had enhanced growth-promoting activity when tested in the body weight gain assay. The crop sac stimulating activity was also enhanced when compared to intact HGH. Details of this work wil be published (Lewis et al., 1975).

PROHORMONE HYPOTHESIS FOR INTACT HGH

The above results have led us to formulate the hypothesis that the intact 191 amino acid form of HGH is a prohormone which must be modified enzymically to a 2-chain

Modifications of HGH

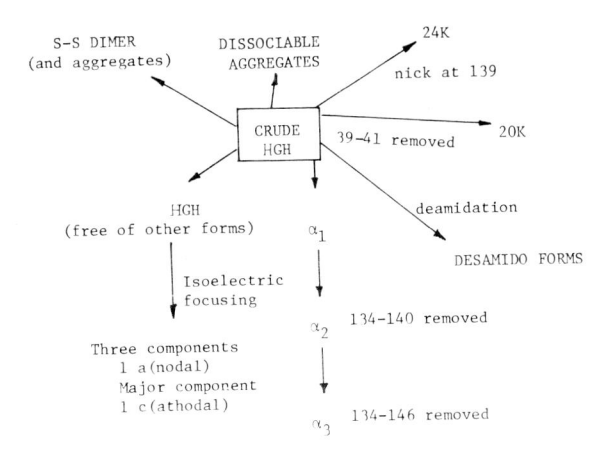

Fig. 10. Summary of the various modifications of HGH described in the text.

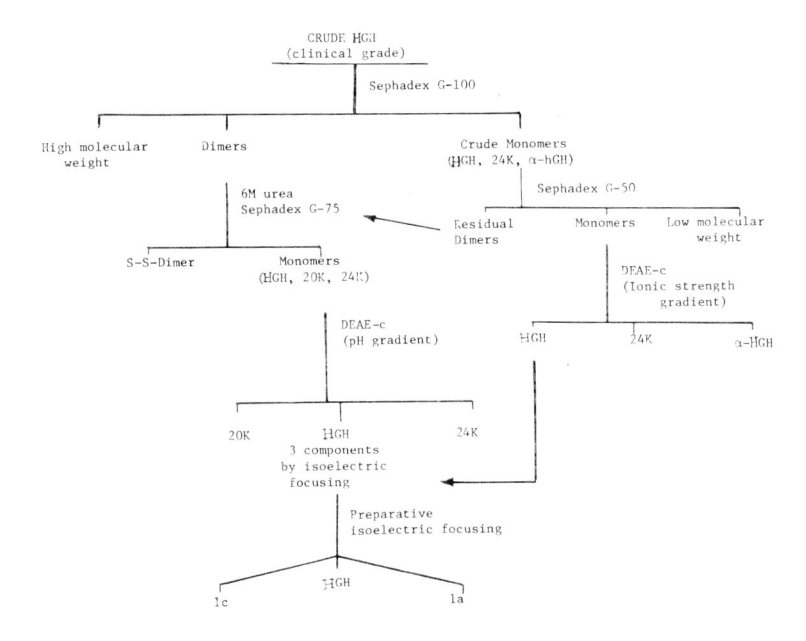

Fig. 11. Summary of the isolation scheme used to purify the modifications of HGH described in the text.

TABLE 2

Bioassays of modifications of HGH produced by digestion with a bacterial proteinase

Sample	Weight gain assay			Crop sac assay		
	Potency (IU/mg)	95% limits (IU/mg)	λ	Potency (IU/mg)	95% limits (IU/mg)	λ
Intact HGH*	0.91	0.67-1.2	0.15	1.6	1.1-2.3	0.17
Form I	3.9	2.9 -5.3	0.15	17.1	7.9-62	0.39
Form II	5.1	3.7 -7.2	0.17	13	6.4-25	0.3
Form III	3.7	2.5 -5.4	0.19	6.7	3 -15.1	0.37

Six rats or pigeons per dosage point. 2 × 2 calculations. Weight gain assay: Parlow et al. (1965). Crop sac assay: Nicoll (1967). * Singh et al. (1947b).

structure for full biological activity. Failure to demonstrate modified forms in fresh pituitary extracts suggests that modification occurs during the secretory process. That activation does not occur systemically is supported by the fact that highly purified intact HGH had a low order of biological activity.

ACKNOWLEDGEMENT

This work was made possible by an award of pituitary glands made by the National Pituitary Agency, NIAMDD. We thank the Medical Research Council, England, for the generous gift of the WHO 1st International Standard for Growth Hormone (Bovine).

REFERENCES

Chadwick, A., Folley, S.J. and Gemzell, C.A. (1961): *Lancet, 2,* 241.
Cheever, E.V. and Lewis, U.J. (1969): *Endocrinology, 85,* 465.
Chrambach, A., Yadley, R.A., Ben-David, M. and Rodbard, D. (1973): *Endocrinology, 93,* 848.
Davis, B.J. (1964): *Ann. N.Y. Acad. Sci., 121,* 404.
Ferguson, K.A. (1964): *Metabolism, 13,* 985.
Lewis, U.J. (1966): In: *Human Pituitary Growth Hormone, Report, Fifty-Fourth Ross Conference on Pediatric Research,* p. 76. Editor: R.M. Blizzard. Ross Laboratories, Columbus, Ohio.
Lewis, U.J. and Cheever, E.V. (1965): *J. biol. Chem., 240,* 247.
Lewis, U.J., Pence, S.J., Singh, R.N.P. and VanderLaan, W.P. (1975): *Biochem. biophys. Res. Commun.,* submitted for publication.
Li, C.H. and Gráf, L. (1974): *Proc. nat. Acad. Sci. (Wash. 1), 71,* 1197.
Nicoll, C.S. (1967): *Endocrinology, 80,* 641.
Parlow, A.F., Wilhelmi, A.E. and Reichert Jr, L.E. (1965): *Endocrinology, 77,* 1126.
Radola, B.J. (1973): *Ann. N.Y. Acad. Sci., 209,* 127.
Reagan, C.R., Mills, J.B., Kostyo, J.L. and Wilhelmi, A.E. (1975): *Endocrinology, 96,* 625.
Schneider, A.B., Kowalski, K. and Sherwood, L.M. (1975): *Biochem. biophys. Res. Commun., 64,* 717.
Singh, R.N.P., Seavey, B.K., Rice, V.P., Lindsey, T.T. and Lewis, U.J. (1974a): *Endocrinology, 94,* 883.
Singh, R.N.P., Seavey, B.K. and Lewis, U.J. (1974b): *Endocrine Res. Commun., 1,* 449.
Weber, K. and Osborn, M. (1969): *J. biol. Chem., 244,* 4406.
Yadley, R.A. and Chrambach, A. (1973): *Endocrinology, 93,* 858.
Yadley, R.A., Rodbard, D. and Chrambach, A. (1973): *Endocrinology, 93,* 866.

STUDIES ON THE NATURE OF PLASMA GROWTH HORMONE

S. ELLIS, R.E. GRINDELAND, T.J. REILLY and S.H. YANG

Biomedical Research Division; Ames Research Center, National Space and Aeronautics Administration, Moffett Field, Calif., U.S.A.

In an earlier report it was noted that the concentrations of growth hormone in human or rat plasma, as determined by the tibial bioassay, were manyfold greater than those obtained by the radioimmunoassay (Ellis and Grindeland, 1974). In contrast the 2 methods showed an excellent correlation in the case of extracts of the whole anterior pituitary gland or its hormone granule fraction. These observations suggested that growth hormone in both human and rat plasma differs significantly in nature from the hormone which can be isolated from the pituitary gland and which is generally used for the preparation of antisera. Further evidence in support of this hypothesis derives from the observation that although antisera against rat pituitary growth hormone can inhibit the growth-promoting action of rat pituitary extracts and purified rat growth hormone in the tibial assay, these antisera cannot inhibit the growth-promoting action of rat plasma. Moreover, concentrates of the growth hormone have been prepared from human plasma which promote the growth of hypophysectomized rats, but exhibit little or no immunoreactivity when tested against antisera to pituitary growth hormone.

The purpose of the present report is to present additional evidence in support of the concept that there are two discrete forms of growth hormone in human plasma. One form is detectable both by radioimmunoassay and bioassay and shall be referred to as the 'immunoreactive' growth hormone. The other form, designated as 'bioactive', can be measured by the tibial bioassay but does not show a significant measure of reactivity with currently available antisera to pituitary growth hormone. In addition, it will be shown that it is possible to achieve a partial purification of the bioactive form from pooled human plasma or the Cohn IV fraction derived therefrom by chemical procedures.

The concentrations of growth hormone in several types of plasmas as determined by the double antibody radioimmunoassay (Schalch and Reichlin, 1966; Schalch and Parker, 1964) and the 4-day tibial bioassay in hypophysectomized rats (Greenspan et al., 1949) are compared in Table 1. It can be seen that the greatest disparity between the bioactive and the immunoactive growth hormone exists in normal human plasma and, to a lesser extent, in normal rat plasma. Much lower ratios of the 2 hormone forms were found in the plasma of an acromegalic subject and in the pooled plasmas of rats implanted with pituitary tumors. In the latter instance, the ratio is close to 1, which is the value generally observed in the case of extracts of the anterior pituitary and of the subcellular hormone granules from the glands of non-stressed rats.

In view of the inability of insulin, cold exposure, or fasting to elevate the plasma concentrations of immunoreactive growth hormone in the plasma of rats coupled with the fact that the bioassayable growth hormone content in the pituitary can be reduced

TABLE 1

Comparison of growth hormone concentrations in human and rat plasmas measured by RIA and bioassay

Plasma	Radioimmunoassay (ng/ml)	Bioassay (ng/ml)	TA*/RIA
Human[1]	1.5	304	202
Acromegalic	40	1090	27
Rat[2]	10.5	494	47
Rat (Furth tumor MtT/W5)	25,100	33,900	1.35

[1] Two pools of human plasma (3 subjects per pool); Factor VIII and fibrinogen-poor plasma. Plasma bioassayed in 40-day old female Simonsen albino rats used 2 weeks posthypophysectomy; 2 + 2 assay design. HGH standard = HS 840 FA (1.7 USP U/mg).
[2] Rat plasma pooled from 50 donor Simonsen albino rats (450 g) anesthetized with ether. Bioassays had 6 rats per group; 2 + 2 assay design.
* TA = tibial assay.

TABLE 2

Plasma biological and immunological growth hormone of insulin-treated rats

	Tibial assay (ng/ml)	RIA (ng/ml)	TA/RIA
Jugular vein plasma			
Saline	1740	8.5	205
Insulin	2830	6.5	435
Cardiac plasma			
Saline	425*	13.7	31
Insulin	475*	7.7	62

Male albino rats (324-354 g) were fasted 16 hr, anesthetized with chloral hydrate, their left jugular veins ligated, and saline or regular insulin (2 U/kg) given i.p. Blood was collected from the right jugular veins by cannula 5-15 minutes after insulin or saline. Cardiac blood was obtained 10 minutes after insulin or saline. Plasma for assay was pooled from groups of 6 or 8 donor rats.
* Potencies of cardiac plasma estimated from one dose level.

by these stimuli, an effort was made to determine whether the plasma concentrations of the bioactive hormone were alterable by insulin administration. In Table 2 are depicted the increases in bioactive hormone which occur in jugular vein plasma following intraperitoneal injection of 2 U/kg of insulin into fasted rats which were anesthetized with chloral hydrate. It can be seen that the concentrations of bioactive hormone were about 4-fold greater in jugular vein plasma than in cardiac plasma, whereas the immunoreactive levels remained unchanged so that the ratio of bioactive hormone to immunoactive hormone was also increased to a similar degree. While insulin did not increase the concentration of immunoreactive hormone in either the jugular vein or cardiac plasma, the concentration of bioactive hormone in the jugular vein plasma was increased by about 1100 ng. These results indicate that bioactive hormone is preferentially released into jugular vein plasma and that the process can be

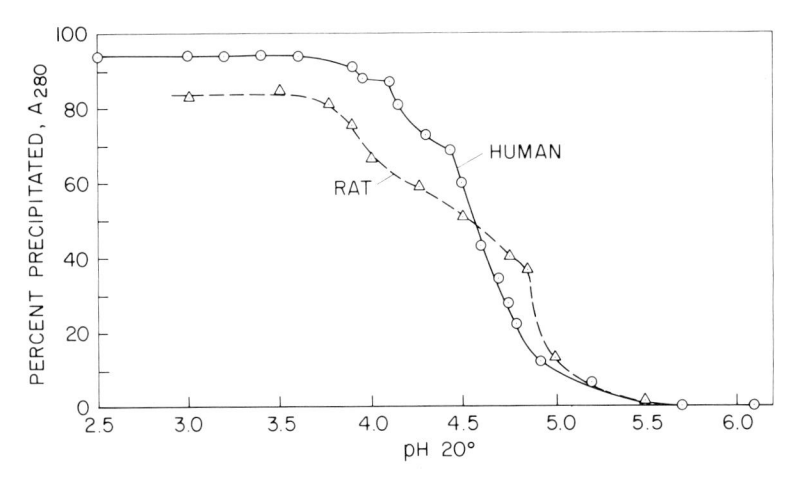

Fig. 1. Precipitation of human and rat plasma proteins (1:1 dilution) in 1% $(NaPO_3)_6$ vs pH.

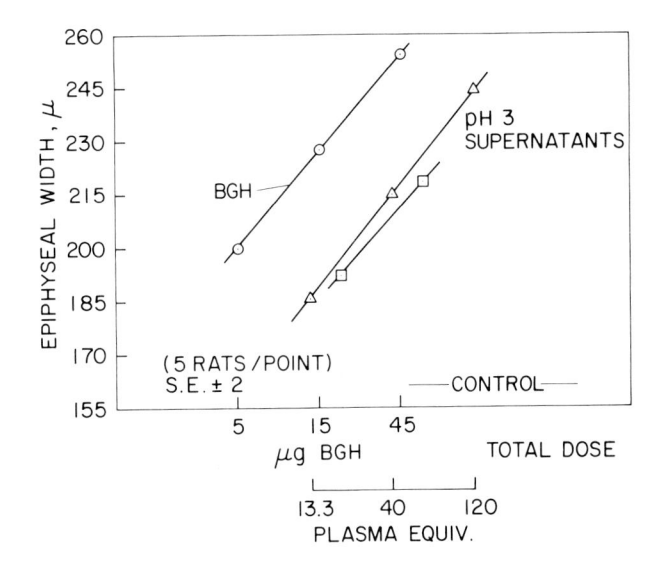

Fig. 2. Tibial assay of human crude plasma growth hormone.

selectively stimulated by insulin, corresponding to the depletion of bioactive hormone from the pituitary gland which has been reported by Müller and Pecile (1968).

Definition of the biological and chemical properties of the bioactive hormone as well as the development of a specific radioimmunoassay would be greatly facilitated by the availability of the pure plasma hormone. Toward this end, studies were undertaken to develop a procedure for purifying the bioactive hormone from pooled human plasma as well as from appropriate plasma fractions which are available as by-products of the commercial fractionation of human plasma.

TABLE 3

Effect of 10-day injection of growth-promoting human plasma fraction on hypox rats

Treatment	Dose (day)	Body weight gain (g)	Tail length (cm)	Epiphyseal width (μ)
Saline	1 ml	0.83 ± 0.95	12.1 ± 0.2³	160.2 ± 1.6
Plasma fraction*	30 p.e. (= 2 µg HGH)/day	3.83 ± 0.60²	12.5 ± 0.1³	213.3 ± 3.5¹
BGH**	10 µg/day	13.5 ± 0.62¹	12.7 ± 0.2³	244.7 ± 3.8¹

[1] $p < 0.001$ vs saline or BGH; [2] $p < 0.05$ vs saline; [3] $p > 0.05$ vs saline; 6 rats per group. Rats: 75 g Simonsen albino; hypox 26 days old; treatment started 17 days p.o.; injected twice daily.
* Human plasma fraction H17-21-8: ≅ 25 ng mg^{-1}, 2.7 mg per plasma equivalent.
**BGH XIV-44-C5, 1.5 USP U/mg.

The initial purification studies were performed by the hexametaphosphate method of Nitschmann et al. (1960) on outdated pools of human plasma containing acid-citrate-dextrose anticoagulant from which much of the fibrinogen was previously removed by cryoprecipitation. The precipitability of human and rat plasma proteins, from 1% solutions of sodium metaphosphate, $(NaPO_3)_6$, as a function of pH is illustrated in Figure 1. About 95% of the protein were precipitated at and below pH 3.5. The resulting supernatants contained the bioactive hormone which yielded log dose-response curves parallel to those given by the bovine growth hormone reference standard (1.5 USP units per mg) (Fig. 2). In order to determine whether the preliminary concentrate of bioactive plasma growth hormone could produce a significant increment in the body weight of hypophysectomized rats, similar to that which occurs following the administration of pituitary growth hormone, the pH 3 supernatant were tested in the 10-day body weight gain assay. Doses of the supernatant protein (80 mg), equivalent to the yield from 30 ml of plasma, were administered daily to 75 g hypophysectomized rats. The effects of the concentrate on the body weight, tail length and epiphyseal cartilage are summarized in Table 3. A small but significant body weight gain was obtained which corresponded to the content of hormone originally determined by the 4-day tibial assay. Though the tail length did not increase significantly, the epiphyseal cartilage showed the expected growth response.

A fractionation method based on the precipitation of non-hormonal proteins at pH 3.0 and retention of bioactive hormone in the supernatant is outlined in Figure 3. The use of $(NH_4)_2SO_4$ fractionation was incorporated to precipitate additional non-hormonal proteins and to separate the bioactive hormone from the high molecular weight metaphosphate ions. The bioactive hormone which was precipitated between 2 and 3.2 M $(NH_4)_2SO_4$ at pH 7 and 20° was obtained in a yield of 3 mg/ml of pooled plasma. The dialyzed freeze-dried protein showed a mean specific activity of 100 ng of human growth hormone per mg of protein referred to standard human growth hormone of 2 USP unit potency per mg. Although the mean yield of activity was 70%, the recoveries varied from 50 to 100% for reasons as yet undetermined. The ratio of bioassayable to immunoassayable activities (TA/RIA) was 200 as compared to 300 for pooled human plasma.

Further purification was performed by submitting the 2 to 3.2 M $(NH_4)_2SO_4$ fraction of the pH 3 supernatant to preparative free-flow electrophoresis and isoelectric

Nature of plasma GH

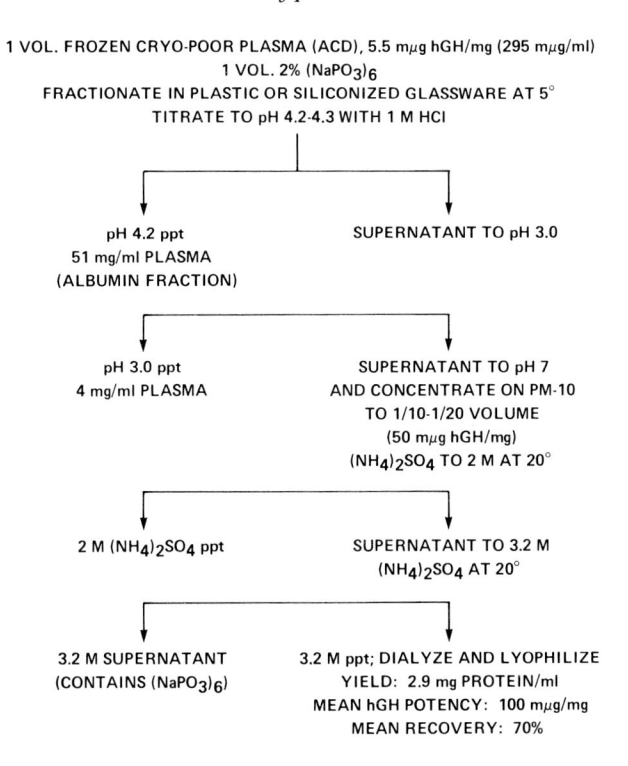

```
1 VOL. FROZEN CRYO-POOR PLASMA (ACD), 5.5 mµg hGH/mg (295 mµg/ml)
              1 VOL. 2% (NaPO₃)₆
FRACTIONATE IN PLASTIC OR SILICONIZED GLASSWARE AT 5°
        TITRATE TO pH 4.2-4.3 WITH 1 M HCl
```

pH 4.2 ppt	SUPERNATANT TO pH 3.0
51 mg/ml PLASMA	
(ALBUMIN FRACTION)	

pH 3.0 ppt	SUPERNATANT TO pH 7
4 mg/ml PLASMA	AND CONCENTRATE ON PM-10
	TO 1/10-1/20 VOLUME
	(50 mµg hGH/mg)
	(NH₄)₂SO₄ TO 2 M AT 20°

2 M (NH₄)₂SO₄ ppt	SUPERNATANT TO 3.2 M
	(NH₄)₂SO₄ AT 20°

3.2 M SUPERNATANT	3.2 M ppt; DIALYZE AND LYOPHILIZE
(CONTAINS (NaPO₃)₆)	YIELD: 2.9 mg PROTEIN/ml
	MEAN hGH POTENCY: 100 mµg/mg
	MEAN RECOVERY: 70%

Fig. 3. Concentration of tibial growth activity from human plasma.

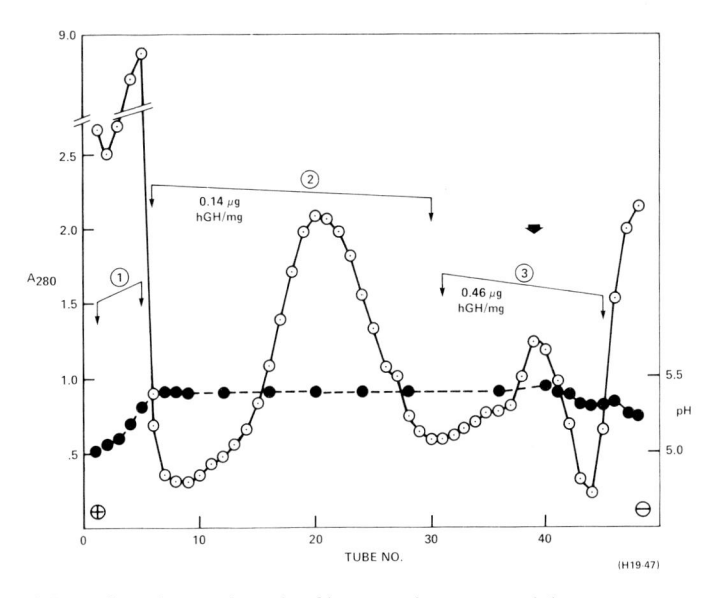

Fig. 4. Free-flow electrophoresis of human plasma growth hormone concentrate. (NaPO₃)₆-(NH₄)₂SO₄ (pyridine acetate, pH 5.4).

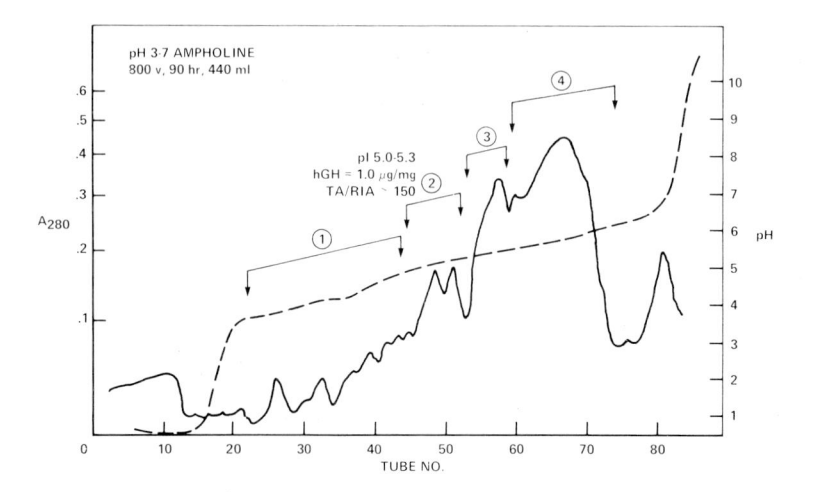

Fig. 5. Isoelectric focusing of human plasma growth hormone concentrate (from free-flow pH 5.4, $(NaPO_3)_6$-$(NH_4)_2SO_4$, ACD plasma).

focusing in Ampholine pH gradients. Preliminary trials with electrofocusing showed that the bioactive hormone had an isoionic point at about pH 5.0. Preparative free-flow electrophoresis was therefore conducted at pH 5.4 in pyridine acetate buffer. The electrophoretic profile of a bioactive hormone concentrate which had been cycled twice through the 2- to 3.2-M $(NH_4)_2SO_4$ step is illustrated in Figure 4. The specific activity of the bioactive hormone was highest in peak 3 where it was concentrated 10-fold to a level of 460 ng equivalents of human growth hormone and with a recovery of 100%.

When the protein derived from peak 3 was electrofocused in Ampholine-sucrose gradients ranging from pH 3 to 7, the protein pattern shown in Figure 5 was obtained. The bioactive hormone was localized between pH 4.9 and 5.3 in the gradient. Radioimmunoassays showed that traces of immunoactive hormone localized in the same pH region as the bioactive hormone. The ratio of bioactive to immunoactive hormone was about 200 and the growth-promoting potency was equivalent to 1 μg of HGH per mg of protein. The overall recovery of bioactive hormone from the starting pool of plasma was 44%.

With a view toward employing plasma fractions prepared commercially by the ethanolic fractionation method of Cohn et al. (1946), the content of bioactive hormone was determined in Cohn fractions I through VI. As can be seen from Table 4, fractions IV and the subfractions IV-1 and IV-4 were found to contain significant amounts of growth hormone of a specific activity which was about equal to that of unfractionated plasma, whereas the specific activity of the immunoactive hormone was less than 0.05 ng HGH per mg. At most, only 38% of the bioactive hormone in plasma could be accounted for by fraction IV. It is presumed that the remainder of the hormone may have been inactivated during preparation of the Cohn fractions or else distributed into fractions I and III. It has not been possible to determine the content of bioactive hormone in fraction I or III because the severe toxicity of high dose levels of these fractions rendered the tibial assay unusable. For these reasons, Cohn fraction IV was considered to be the most practical starting material for the isolation of growth hormone by the metaphosphate method.

TABLE 4

Growth-promoting activity of plasma fractions (Cohn)

Fraction	Yield (mg/ml)	Tibial assay (ng HGH/mg)	HGH (ξng)	Plasma activity recovered (%)
ACD plasma	54	5.5	298	—
IV*	13**	8.7	113	38
IV-1	4.3**	3.8	16	5.4
IV-4	5.8**	10.5	61	21

Fractions I, II, III, V, VI were inactive; I and III were also toxic.
* Somatomedin, a chick cartilage assay (Uthne): maximal content = 0.004 U/mg.
** From Cohn et al., 1946.

The freeze-dried fraction IV was submitted to fractionation according to the scheme outlined in Figure 3, except that the initial protein concentration was 1%. The resulting concentrate of growth hormone contained 0.2 μg of HGH per mg which represents a 20-fold purification. About 10 mg of hormone was estimated to be present in the 54 g of protein which can be derived from 1 kg of Cohn fraction IV. These results indicate that the isolation of highly purified bioactive hormone in the order of milligram quantities lies within the realm of practicability and that this Cohn fraction is a suitable starting material for the isolation of the bioactive hormone.

Other potentially useful methods by which the bioactive hormone might be purified have been tested, but were found to be unsatisfactory because of large losses of activity. For example, the bioactive hormone could be absorbed from the $(NH_4)_2SO_4$ concentrate unto the cation exchange columns of carboxymethyl cellulose, equilibrated with ammonium acetate buffer at pH 4.5, and eluted in association with 20% of the applied protein by means of 0.5 M NH_4HCO_3. However, the expected 5-fold purification was not achieved due to the loss of 60-70% of the biological activity. Losses of similar magnitude occurred when the hormone was precipitated between 15 and 30% ethanol at pH 3. Under these conditions the hormone precipitated due, presumably, to association with certain major protein contaminants such as transferrin, $α_2$-macroglobulin and $α_1$-antitrypsin. Until the reasons for the large measure of inactivation are established, these methods offer little promise of applicability to the isolation of bioactive plasma growth hormone.

The possibility that a transformation of immunoactive growth hormone to the bioactive form may occur within the pituitary gland itself prompted a search for bioactive forms in pituitary subcellular fractions derived from homogenates prepared in 0.25 M sucrose. In a previous report (Ellis and Grindeland, 1974) it was noted that the ratio of bio- to immunoactive hormone was 1.0 in the granular fraction and 1.6 in the cytosol obtained from freshly collected rat pituitary glands. When the cytosol was submitted to gel filtration on Sephadex G-75 or G-100 bioactive hormone, free of immunoassayable activity, was associated with the protein fractions which preceded the elution of the main peak of dimeric rat growth hormone (m.w. = 40-45,000). The distribution of the bioassayable and immunoassayable hormones is illustrated in Figure 6. About 30% of the total bioactive hormone content in the cytosol was present in fractions 1 and 2, whereas fraction 3 emerged at a K_{av} characteristic of pituitary growth hormone, yielded a bio- to immunoactive ratio of 1.0, and accounted for 70% of the applied bioactive hormone.

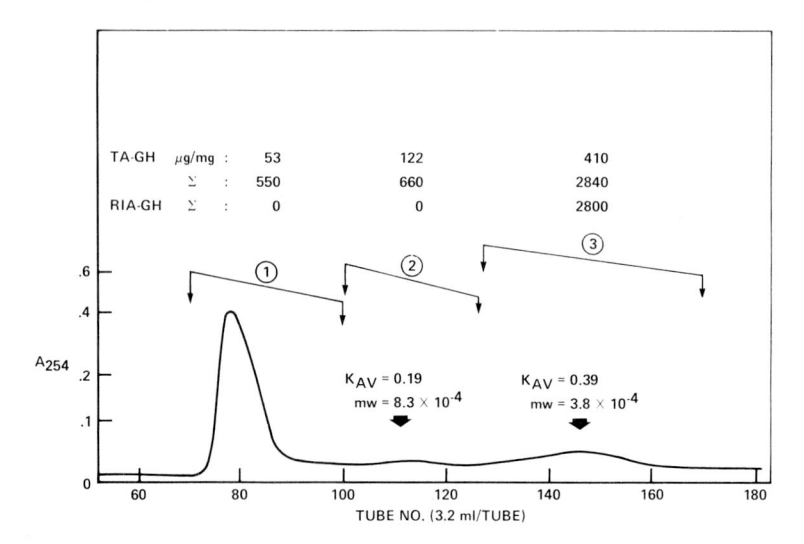

Fig. 6. Gel filtration of rat pituitary cytosol. 5 ml cytosol in M/4 sucrose from 580 mg anterior pituitary (G-100, 3.2×95 cm in M/10 NH$_4$HCO$_3$).

Since approximately 20% of the growth hormone in the rat pituitary gland is localized in the cytosol (McShan, 1971) and, as shown above, 30% is in the bioactive, non-immunoreactive form, there should be about 1 to 2 mg of this hormone per gram of rat anterior pituitary, calculating on the basis of a content of 25 mg of growth hormone per gram (wet weight) of gland. Thus, the rat pituitary tissue may also be employed as a starting material for the isolation of the bioactive form of pituitary growth hormone.

In summary, it has been shown that (1) bioactive growth hormone, which is by far the preponderant form of the hormone in human plasma, can be concentrated about 200-fold from pooled outdated human plasma with a recovery of 40%; (2) the bioactive hormone was detectable by tibial assays in Cohn fractions IV, IV-1 and IV-4, and can be concentrated about 40-fold by fractionation with (NaPO$_3$)$_6$ and (NH$_4$)$_2$SO$_4$; (3) the ratio of the bio- to immunoactivity in the final products remains similar to that found in the starting plasmas, namely, 200 to 300:1; (4) the bioactive hormone purified from whole plasma is isoionic at pH 4.9 to 5.3 as determined by isoelectric focusing in Ampholine pH gradients; and (5) bioactive growth hormone which has no immunoactivity could also be found in rat pituitary cytosol to the extent of 6% of the total growth content of the gland.

ACKNOWLEDGMENTS

We wish to express our indebtedness to the Cutter Laboratories, Berkeley, California, for generously providing Cohn fractions of human plasma, to Dr. W.P. Vander Laan of the Scripps Clinic, La Jolla, California, for supplying the acromegalic plasma, and to Mr. O.K. Chee for rendering technical assistance.

REFERENCES

Cohn, E.J. et al. (1946): *J. Amer. chem. Soc., 68,* 459.
Ellis, S. and Grindeland, R.E. (1974): In: *Advances in Human Growth Hormone Research,* pp. 409-433. Editor: S. Raiti. U.S. Government Printing Office, Washington, D.C.
Greenspan, F.S., Li, C.H., Simpson, M.E. and Evans, H.M. (1949): *Endocrinology, 45,* 355.
McShan, W.H. (1971): In: *Subcellular Organization and Function in Endocrine Tissues,* pp. 161-184. Editors: H. Heller and K. Lederis. Cambridge University Press, London.
Müller, E.E. and Pecile, A. (1968): In: *Proceedings, I International Symposium on Growth Hormone, Milan, 1967,* pp. 253-266. Editors: A. Pecile and E.E. Müller. ICS 158, Excerpta Medica, Amsterdam.
Nitschmann, H., Rickli, E. and Kistler, P. (1960): *Vox Sang. (Basel), 5,* 232.
Schalch, D.S. and Parker, M.L. (1964): *Nature (Lond.), 203,* 1141.
Schalch, D.S. and Reichlin, S. (1966): *Endocrinology, 79,* 275.

STUDIES OF GROWTH HORMONE SYNTHESIS IN CULTURED RAT PITUITARY CELLS AND IN CELL-FREE SYSTEMS*

F. CARTER BANCROFT, PHYLLIS M. SUSSMAN and ROBERT J. TUSHINSKI

Department of Biological Sciences, Columbia University, New York, N.Y., U.S.A.

A complete understanding of the mechanisms which regulate the production of growth hormone (GH) by the anterior pituitary will ultimately require detailed information about the synthesis, processing, and secretion of GH molecules by pituitary cells. The regulation of the secretion of GH by the pituitary has been the subject of intensive study by a number of investigators. However, to date there have been relatively few studies reported concerning the synthesis and intracellular processing of GH. The relative paucity of such studies is probably due to the fact that the assay presently employed in most biochemical studies of GH metabolism, the radioimmunoassay, cannot be used to perform direct measurements of the synthesis of a protein (Bancroft, 1973a).

During the past several years, my laboratory has been engaged in the development of techniques for the measurement of GH synthesis, and in investigating the location and mode of the synthesis of GH by pituitary cells. For our studies we have employed a clonal strain of rat pituitary tumor cells (GH$_3$), which produce both GH and prolactin (Tashjian et al., 1970). We have chosen this system because we believe that the advantages of studying a cloned population of animal cells under the simple and well-defined conditions of tissue culture outweighs the disadvantages inherent in using cells which are both neoplastic and in a somewhat unnatural environment. The present report describes what we have learned to date about GH synthesis in unstimulated cells. We hope to report at a later time on current studies of the mechanisms whereby various external stimuli regulate GH synthesis.

EXPERIMENTAL SYSTEM AND TECHNIQUES

The investigations to be described here were carried out using a clonal strain of rat pituitary tumor cells, designated GC, in which GH represents about 8% of the total protein synthesis (Bancroft, 1973a). The GC cells are a subclone of the GH$_3$ strain whose origin has been described (Bancroft and Tashjian, 1970). The GH$_3$ cells produce both GH and prolactin (Tashjian et al., 1970) while the GC cells appear to produce little or no prolactin (Sussman and Bancroft, in preparation). To measure GH synthesis by the GC cells, an immunoprecipitation technique was developed. This technique, which has been described in detail (Bancroft, 1973a), involves the use of baboon anti-GH antiserum plus carrier GH to specifically precipitate newly synthe-

---

* This work was supported by grants from the National Science Foundation, the National Institutes of Health, the American Cancer Society, and the Human Growth Foundation.

Studies of GH synthesis

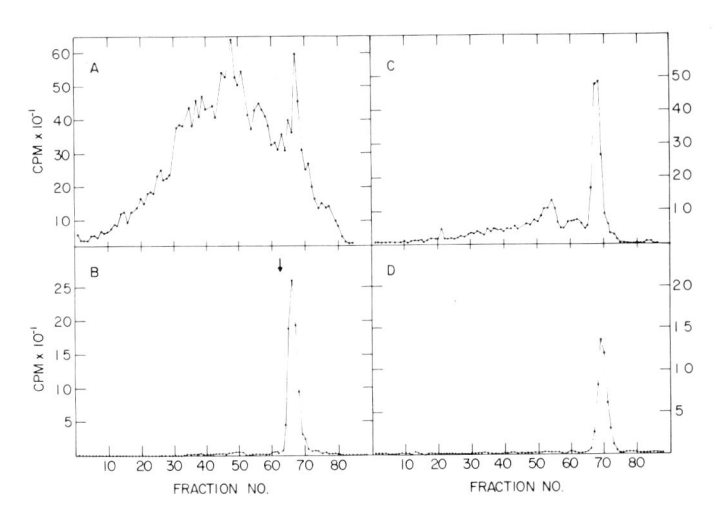

Fig. 1. Specificity of the precipitation of GH by immune serum. GC cells were incubated in the presence of [^{14}C]amino acids. After 12 minutes, cell cytoplasm was prepared. After 90 minutes, labeled medium was prepared. Aliquots of cytoplasm and medium were precipitated with immune serum. Cytoplasm, medium, and immune precipitates of each were subjected to electrophoresis on 5%, 20 cm continuous SDS-polyacrylamide gels. A. Cytoplasm. B. Immune precipitate of cytoplasm. C. Medium. D. Immune precipitate of medium. The arrow in B represents the position of authentic GH on a fifth gel. (Reproduced from Bancroft, *Endocrinology* (1973a); by courtesy of Charles C Thomas, Publisher.)

sized, radioactive GH molecules. The specificity of this technique for GH has been demonstrated in a number of ways, most notably by solubilization of the immune precipitate in the presence of sodium dodecyl sulfate (SDS) plus mercaptoethanol, followed by analysis by SDS-polyacrylamide gel electrophoresis. The result of such an experiment is shown in Figure 1. Figure 1A demonstrates the range of polypeptide chains in GC cytoplasm that are labeled during a brief exposure to [^{14}C]amino acids. Figure 1B demonstrates that when this cytoplasm is precipitated with immune serum, only a single peak of radioactivity with the same mobility as GH is precipitated. Figures 1C and 1D demonstrate the specificity of the immune serum for the precipitation of radioactive GH from the growth medium of the GC cells.

INTRACELLULAR LOCATION OF GH AND GH MESSENGER RNA

Utilizing the immune precipitation technique described above, we have investigated a number of aspects of the synthesis of GH. The first concerns the location in pituitary cells of the apparatus for the synthesis of GH. GH is of course a secretory protein. Thus its synthesis might reasonably be expected to have certain features in common with that of other secretory proteins. The current model for proteins of this type involves synthesis on ribosomes attached to the outside of the rough endoplasmic reticulum, with the nascent polypeptide chain being transferred into the inside of the rough endoplasmic reticulum. Thus the model predicts that newly synthesized GH will be found within membrane-enclosed structures at very early times after its synthesis. The experiment depicted in Table 1 was performed to test this prediction. GC cells

TABLE 1

Radioactivity in TCA-insoluble material or in GH, in various cell fractions

Cellular fraction	(a) TCA-insoluble c.p.m. ($\times 10^{-3}$)	(b) GH c.p.m. ($\times 10^{-3}$)	(b)/(a) (%)	$\dfrac{\text{Membrane GH}}{\text{Membrane GH + free GH}}$ (%)
Cytoplasm	189	9.55	5.1	
Free	117	1.13	1.0	86
Membrane	58.2	7.12	12	

Conditions as described in the text. TCA, trichloroacetic acid. (Reprinted from Bancroft (1973b), by courtesy of the Editors of *Experimental Cell Research*.)

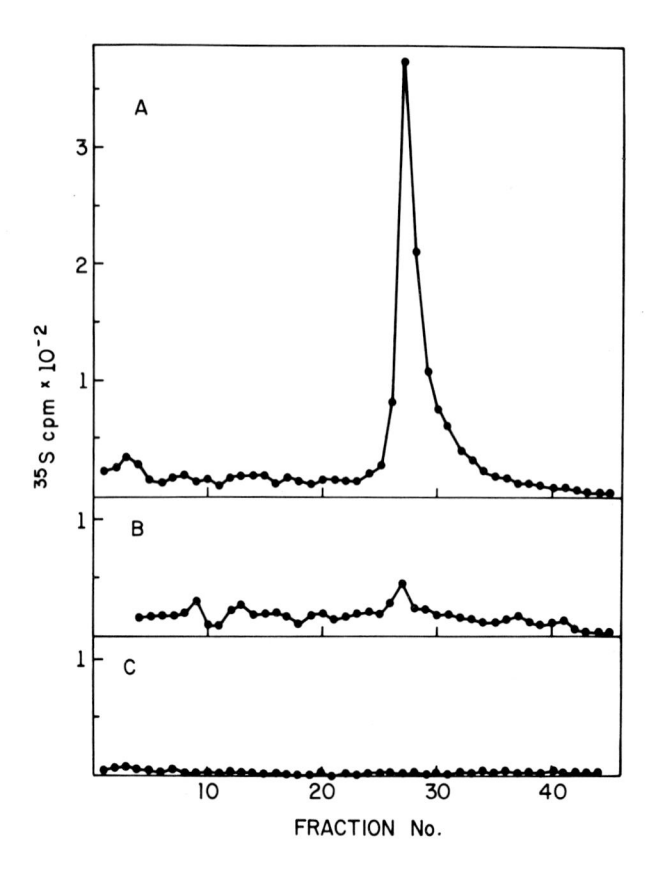

Fig. 2. SDS-polyacrylamide gel electrophoresis of material synthesized in the Krebs ascites cell-free system and precipitated with anti-GH antiserum. Immune precipitates were analyzed on 10%, 10 cm continuous SDS-polyacrylamide gels. The reaction mixtures received RNA isolated from GC cell cytoplasm fractions as follows: A. 255 µg/ml of membrane-associated RNA; B. 240 µg/ml of post-membrane RNA; C. no exogenous RNA. (Reproduced from Bancroft et al. (1973); by courtesy of the Editors of the *Proceedings of the National Academy of Sciences of the U.S.A.)*

were pulsed with [³H]leucine for 2 minutes, then incubated with excess cold leucine for 2.5 minutes to chase nascent proteins off of the polysomes. Separation of GC cytoplasm into membrane and post-membrane ('free') fractions by centrifugation (21,500 × g for 10 minutes), followed by measurement of radioactive GH in the two fractions indicated that the membrane fraction is greatly enriched for GH compared to the free fraction (Table 1, column 3), and that most (86%) of the newly synthesized GH is membrane-associated (Table 1, column 4). This result clearly agrees with the prediction stated above. It should also be noted in this connection that immunocytochemical studies have confirmed the localization of GH within the rough endoplasmic reticulum of individual GC cells (Masur et al., 1973).

Next a series of studies were undertaken which involved the translation of messenger RNA (mRNA) for GH in various cell-free systems. The purpose of these studies was both to continue to investigate the intracellular location of the GH-synthesizing apparatus, and also, as described in detail below, to investigate the question of the synthesis of GH in the form of a precursor molecule.

We first asked where in the cell GH mRNA is located. GC cell cytoplasm was separated by centrifugation into membrane and post-membrane fractions. RNA was then isolated from either fraction, and incubated in extracts of mouse Krebs ascites tumor cells, in the presence of [³⁵S]methionine. After removal of an aliquot for measurement of total acid-insoluble radioactivity, the reaction mixture was immunoprecipitated with anti-GH antiserum. The precipitate was then solubilized and analyzed by SDS-

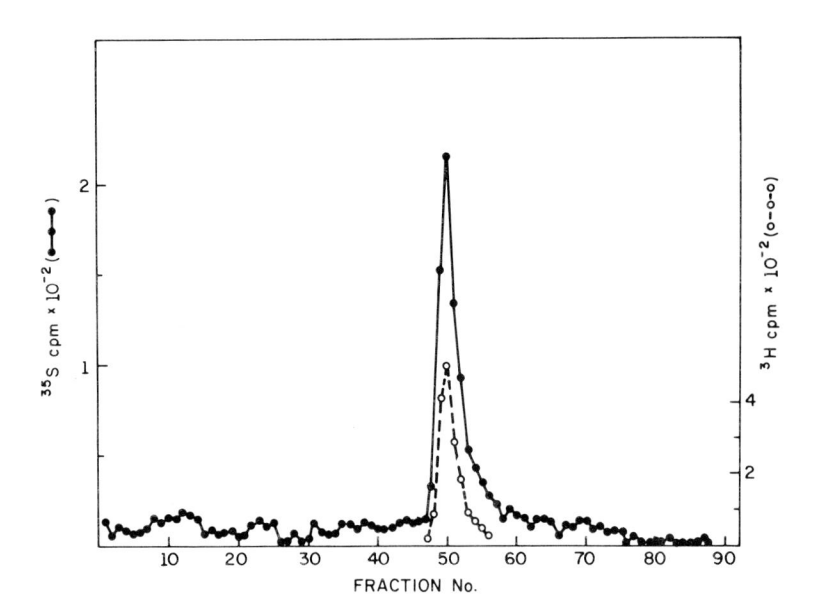

Fig. 3. Relative mobilities during SDS-polyacrylamide gel electrophoresis of anti-GH antiserum-precipitated material synthesized in vitro and in vivo. The antiserum precipitate of [³⁵S]-labeled material synthesized in the Krebs ascites cell-free system in the presence of GC cell membrane-associated RNA was combined with an antiserum precipitate containing [³H]-labeled GH synthesized and secreted by the GC cells. The sample was then analyzed on a 10%, 20 cm continuous SDS-polyacrylamide gel. (•—•), ³⁵S; (o----o), ³H. (Reproduced from Bancroft et al. (1973); by courtesy of the Editors of the *Proceedings of the National Academy of Sciences of the U.S.A.*)

polyacrylamide gel electrophoresis. (It is interesting that about one-third of the total cytoplasmic RNA was found in the membrane fraction of the GC cells, since this agrees well with the observation that about one-third of the total newly synthesized protein in these cells is membrane-associated (Table 1, column 1).) When equal concentrations of membrane-associated or post-membrane RNA were incubated in the Krebs ascites cell-free system, they were found to yield equal stimulation of total protein synthesis (Bancroft et al., 1973), indicating that the percentage of the two fractions which was functional mRNA was about the same. However, the amount of GH synthesized in response to RNA from the two fractions was vastly different. It is seen in Figure 2 that virtually all (>95%) of the GH mRNA activity was found in the membrane fraction of the cells. This result demonstrates that, as expected, GH mRNA is membrane-associated, presumably in the form of polysomes bound to the outside of the rough endoplasmic reticulum. Further evidence that the peak in Figure 2A actually represents GH, as well as a determination of the size of the Krebs ascites cell-free product, came from a double label experiment in which [^{3}H]-labeled GH which had been synthesized and secreted by the GC cells was analyzed together with the [^{35}S]-labeled Krebs ascites cell-free product by SDS-polyacrylamide gel electrophoresis. It was observed that the cell-free product comigrated exactly with secreted GH (Fig. 3), a strong indication that the two products have the same molecular weight.

IS THERE A GH PRECURSOR?

The synthesis of peptide hormones in the form of precursor molecules appears to be a quite common phenomenon (Tager and Steiner, 1974). The concept that GH may also be synthesized as a precursor has been repeatedly considered (Stachura and Frohman, 1973, 1974; Tager and Steiner, 1974; Zanini et al., 1974). However, to date there has been no hard evidence to support this concept. The 'large growth hormone' which has been observed in rat pituitary organ cultures represents GH which is apparently non-covalently bound to another protein (Stachura and Frohman, 1973) and/or bound to RNA (Stachura and Frohman, 1974). Thus 'large growth hormone' is not likely to represent a GH precursor in the usual sense, i.e., a single polypeptide chain in which a specific peptide bond(s) is cleaved to yield the mature form of the protein.

We have looked for a GH precursor in the intact pituitary cells by pulsing the cells with [^{35}S]methionine for a very short time (2 minutes). Cytoplasm was then prepared as described (Bancroft, 1973a), the ribosomes removed by high speed centrifugation (100,000 × g for 1 hour), and immunoprecipitation and gel analysis performed as described (Sussman et al., 1976). The only discrete radioactive peak observed in the immune precipitate had a mobility identical to that of GH synthesized and secreted by the GC cells (Fig. 4), suggesting that if GH is synthesized as a larger precursor, it must be cleaved very soon thereafter to its mature form. Similar results have been observed in rat pituitary organ cultures (Zanini et al., 1974).

This result, plus the observation that the product of the translation of GH mRNA in the mouse Krebs ascites cell-free system comigrated with GH on an SDS-polyacrylamide gel (Fig. 3), appeared to constitute strong evidence against the existence of a GH precursor. However, it seemed possible that the latter observation could be explained by the presence in mouse tumor cells of an enzyme(s) capable of processing a rat secretory protein precursor to its mature form. We therefore initiated studies of the translation of membrane fraction RNA from the GC cells in extracts of cells evolutionarily far removed from the rat; i.e., wheat germ. Immune precipitation and gel analysis of a reaction mixture containing [^{35}S]methionine yielded the results shown

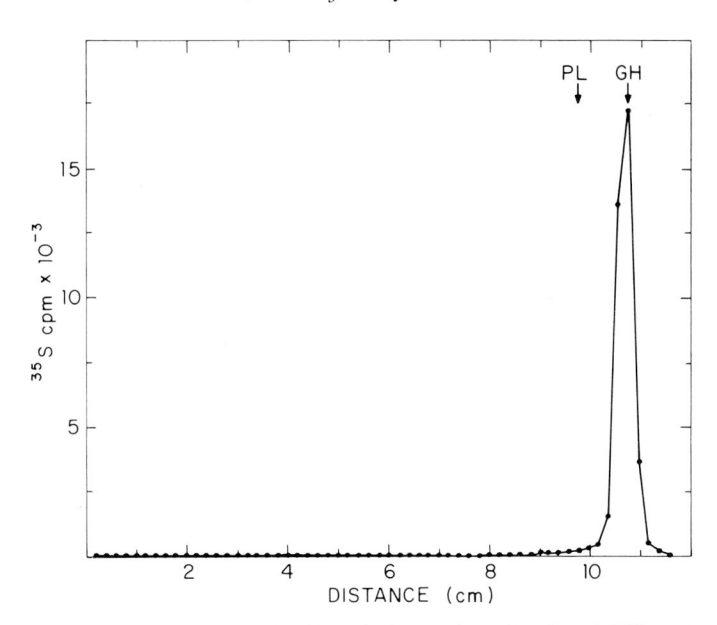

Fig. 4. Mobility during SDS-polyacrylamide gel electrophoresis of anti-GH antiserum-precipitated material synthesized and released from polysomes during a short pulse of GC cells with [^{35}S]methionine. Labeling (2 minutes) and cell fractionation were as described in the text. The immune precipitate was analyzed on a 15%, 13 cm discontinuous ('stacking') SDS-polyacrylamide gel. The positions of GH and prolactin (PL) indicated by the arrows represent [^{3}H]-labeled GH and prolactin markers applied to the same gel.

in Figure 5. It is seen that the major immunoprecipitated peak migrated considerably more slowly than GH (19,500 daltons), and slightly more slowly than prolactin (22,500 daltons) (Fig. 5A). The observations that the major peak is precipitated by anti-GH antiserum (Fig. 5A), that it is not observed when GC cell membrane fraction RNA is omitted from the cell-free reaction mixture (Fig. 5A), and that only a non-specific amount of the major peak is coprecipitated with a heterologous albumin-anti-albumin immune precipitate (Fig. 5B), suggests strongly that this peak is related to GH. Even stronger evidence for a structural relationship between this peak and GH comes from a comparison of their methionine-containing tryptic peptides (Fig. 6). It is seen that the major immunoprecipitated product of the wheat germ cell-free system yielded four major methionine-containing peptides, each of which comigrated with a corresponding peptide from GH. This result demonstrates that the major immunoprecipitated cell-free product and GH have amino acid sequences in common. Treatment of the cell-free product with KOH or RNase did not alter its mobility on gels (Sussman et al., 1976), showing that its large size relative to GH is not due to attached RNA. Finally, by comparison with the mobilities on gels of eight proteins of known molecular weights, the major immunoprecipitated wheat germ cell-free product and GH were estimated to have molecular weights of 24,000 and 19,500, respectively (Sussman et al., 1976).

These results suggest that translation of GH mRNA in wheat germ extracts yields a precursor of GH ('preGH') not hitherto observed. The fact that wheat germ extracts have been found to translate faithfully a number of animal and animal virus mRNA's

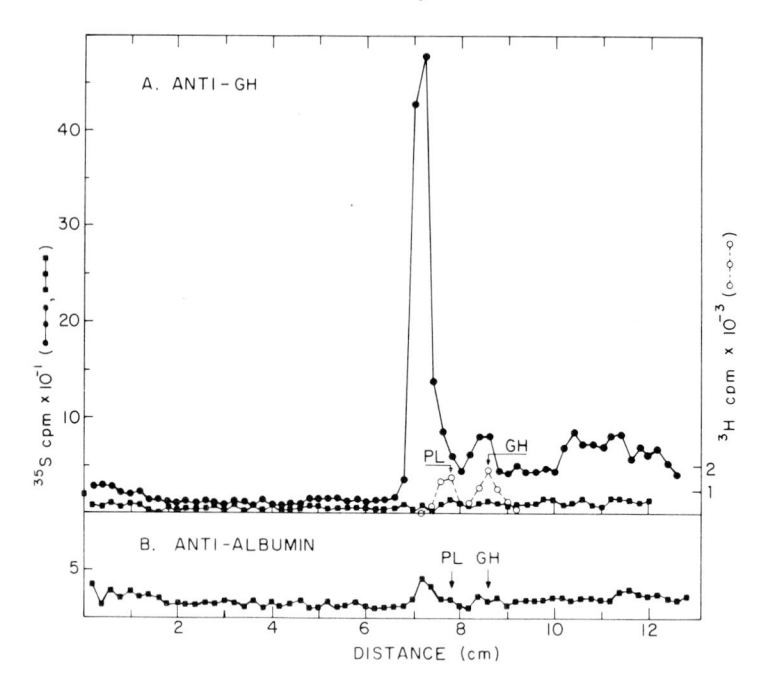

Fig. 5. SDS-polyacrylamide gel electrophoresis of material synthesized in the wheat germ cell-free system. Electrophoresis was as in Figure 4. A. Anti-GH antiserum precipitates of [^{35}S]-methionine-labeled material synthesized in the presence (•—•) or absence (■—■) of GC cell membrane fraction RNA. B. Anti-rat serum albumin precipitate of [^{35}S]methionine-labeled material synthesized in the presence of GC cell membrane fraction RNA. Internal [^{3}H]-labeled GH and prolactin (PL) markers (o----o) were included in all 3 gels; for clarity they are plotted only in A. (Reproduced from Sussman et al. (1976), by courtesy of the Editors of the *Proceedings of the National Academy of Sciences of the U.S.A.*)

(references in Sussman et al., 1976), and have apparently also revealed the existence of a large precursor of parathyroid hormone (Kemper et al., 1974), suggests that preGH is not simply an artifact of the translation system employed. However, convincing evidence that preGH is a physiological precursor of GH would require a demonstration of the synthesis of preGH by intact pituitary cells. The fact that no preGH is observed among the completed proteins in GC cells after a very short pulse of radioactive methionine (Fig. 4) suggests that preGH may in fact be normally cleaved while it is still undergoing synthesis. Hence we undertook experiments involving the use of protease inhibitors to attempt to inhibit the conversion by GC cells of preGH to GH. To date we have had the greatest success with 1-chloro-4-phenyl-3-tosylamido-2-butanone (TPCK), an inhibitor of chymotrypsin (Shaw, 1970). Analysis by SDS-polyacrylamide gel electrophoresis of the anti-GH antiserum-precipitated proteins synthesized by GC cells in the presence of TPCK is shown in Figure 7. It is seen that although the major peak obtained from the TPCK-treated cells (Fig. 7B) has the same mobility as GH obtained from the untreated cells (Fig. 7A), a distinct peak (8.4 cm) was observed in the TPCK-treated cells, but not in the untreated cells. The peak at 8.4 cm in Fig. 7B migrated two fractions behind the prolactin marker, and thus had a mobility identical to that of the preGH observed upon translation of GH mRNA in wheat germ extracts

Studies of GH synthesis

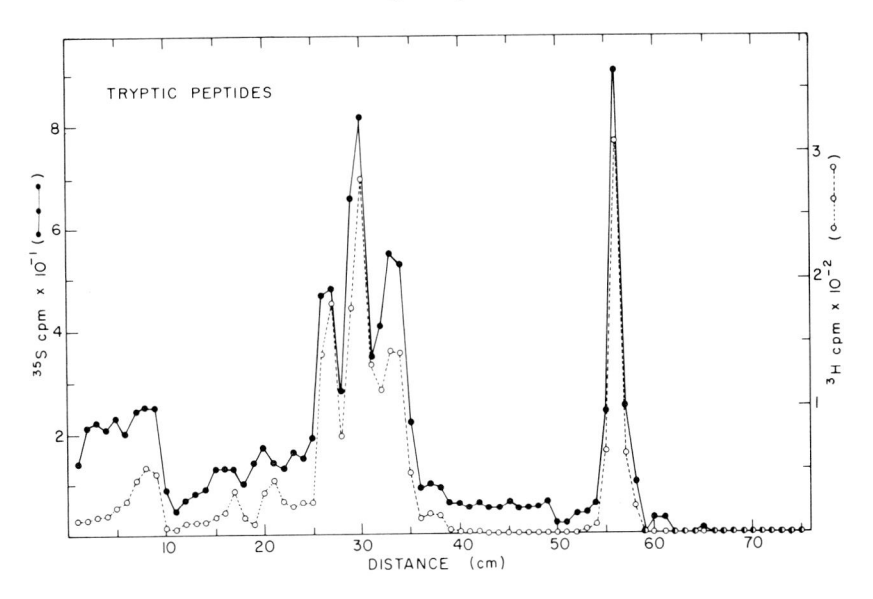

Fig. 6. Methionine-containing tryptic peptides of the major immunoprecipitated product of the wheat germ system, and of GH. The major [^{35}S]methionine-labeled peak in Figure 5A, and [^{3}H]methionine-labeled GH synthesized and secreted by GC cells were isolated from separate gels, combined, oxidized with performic acid, digested with trypsin, and subjected to high voltage paper electrophoresis at pH 3.5. (●—●), [^{35}S]methionine; (o----o), [^{3}H]methionine. (Reproduced from Sussman et al. (1976), by courtesy of the Editors of the *Proceedings of the National Academy of Sciences of the U.S.A.*)

(Fig. 5A). The relatively small amount of material in the peak at 8.4 cm in Figure 7B suggests that under the conditions presently employed, TPCK only partially inhibits the cleavage of preGH to GH. We are currently attempting to find conditions which will inhibit the cleavage more completely.

It is thus seen that experiments involving either translation of GH mRNA in wheat germ extracts (Fig. 5A), or treatment of intact GC cells with a protease inhibitor (Fig. 7) demonstrate the existence of a GH-related molecule which is about 23% larger than GH. Although either type of experiment alone would only suggest that this molecule (preGH) is a true physiological precursor of GH, the observation of preGH in two entirely different sorts of experiments, one performed in vitro and the other in vivo, constitutes strong evidence in favor of this concept.

DISCUSSION

In this report we have described techniques for the study of GH synthesis by a clonal strain of rat pituitary tumor cells (GC). These techniques involve the use of baboon anti-GH antiserum to precipitate newly synthesized, radioactive GH, analysis of immune precipitates by SDS-polyacrylamide gel electrophoresis, and the use of cell-free systems for the translation of GH mRNA. Using these techniques, we have found that both GH mRNA and newly synthesized GH are membrane-associated, in agreement with expectations for a secretory protein. In addition, experiments involving either

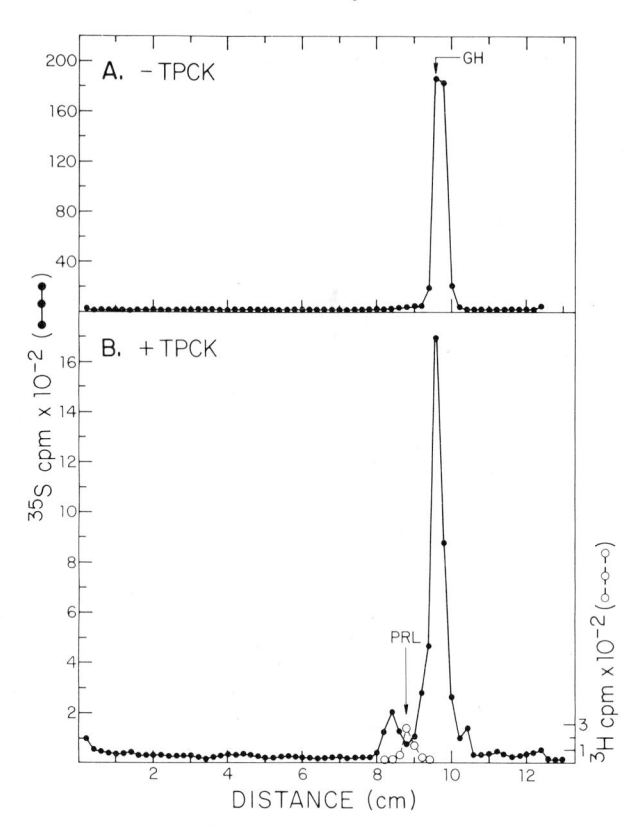

Fig. 7. SDS-polyacrylamide gel electrophoresis of anti-GH antiserum-precipitated material synthesized by GC cells in the precence of TPCK. GC cells were either untreated (A) or exposed to 100 µg/ml TPCK (B) for 2 minutes. [^{35}S]methionine was then added, and incubation continued for 30 minutes. Cytoplasm was then prepared and immune-precipitated, and electrophoresis of the immune precipitates was performed as in Figure 4. An internal [^{3}H]-labeled prolactin (PRL) marker was included. (●—●), [^{35}S], (o----o), [^{3}H].

translation of GH mRNA in wheat germ extracts or treatment of GC cells with a protease inhibitor have provided strong evidence that GH is synthesized in the form of a precursor molecule, which we have termed pregrowth hormone (preGH). PreGH appears to have a quite transient existence in pituitary cells, and may even be normally converted to GH while it is still being synthesized.

At this point we can only speculate on the intracellular function of preGH. The first polypeptide hormone for which a biosynthetic precursor was demonstrated was of course insulin (Steiner et al., 1967). It was observed that after reduction of disulfide bonds, proinsulin renatured more rapidly than did insulin (Steiner and Clark, 1968). This result suggested that the connecting peptide in proinsulin insures the proper folding necessary for the formation of the proper disulfide bridges between the two chains of insulin. Human GH has two intrachain disulfide bonds (Li et al., 1969), and presumably rat GH does also, although the latter molecule has been only partially sequenced (see Wallis and Davies, *This Volume*, p. 1). However, it seems somewhat unlikely that preGH would serve a function analogous to proinsulin, since the folding of the single GH polypeptide chain would be expected to insure the formation of the proper intrachain disulfide bonds.

There appear at present to be two types of precursors of secretory proteins. The first type, of which proinsulin and the more recently discovered proparathyroid hormone (Kemper et al., 1972) are examples, are relatively stable and can be detected in the intact cell. The second type, which includes a precursor of gamma-globulin L chain (termed P) (Milstein et al., 1972), pre-proparathyroid hormone (Kemper et al., 1974), and now preGH have very transient intracellular existences and have not been observed in normal intact cells. It has been suggested that precursors of the second class may function in the binding of polysomes engaged in the synthesis of secretory proteins to the endoplasmic reticulum (Milstein et al., 1972). Our observations that both GH mRNA and newly synthesized GH are membrane-associated are certainly consistent with the concept that preGH serves a similar function in pituitary cells. It is also possible that preGH serves some function in the passage of newly synthesized GH to the inside of the endoplasmic reticulum, or that the conversion of preGH to GH serves as a point for the control of GH synthesis by pituitary cells.

ACKNOWLEDGEMENTS

We thank Aileen Feldman, Claire Nisonger, and Roberta Sugar for expert technical assistance during various stages of these studies. Rat growth hormone and prolactin were supplied by the NIAMDD Rat Pituitary Hormone Distribution Program.

REFERENCES

Bancroft, F.C. (1973a): *Endocrinology, 92*, 1014.
Bancroft, F.C. (1973b): *Exp. Cell Res., 79*, 275.
Bancroft, F.C. and Tashjian Jr, A.H. (1970): *In Vitro, 6*, 180.
Bancroft, F.C., Wu, G. and Zubay, G. (1973): *Proc. nat. Acad. Sci. (Wash.), 70*, 3646.
Kemper, B., Habener, J.R., Mulligan, R.C., Potts Jr, J.T. and Rich, A. (1974): *Proc. nat. Acad. Sci. (Wash.), 71*, 3731.
Kemper, B., Habener, J.R., Potts Jr, J.T. and Rich, A. (1972): *Proc. nat. Acad. Sci. (Wash.), 69*, 643.
Li, C.H., Dixon, J.S. and Liu, W.K. (1969): *Arch. Biochem., 133*, 70.
Masur, S.K., Holtzman, E. and Bancroft, F.C. (1973): *J. Histochem. Cytochem., 22*, 385.
Milstein, C., Brownlee, G.G., Harrison, T.M. and Mathews, M.B. (1972): *Nature New Biol., 239*, 117.
Shaw, E. (1970): *Physiol. Rev., 50*, 244.
Stachura, M.E. and Frohman, L.A. (1973): *Endocrinology, 92*, 1708.
Stachura, M.E. and Frohman, L.A. (1974): *Endocrinology, 94*, 701.
Steiner, D.R., Cunningham, D., Spiegelman, L. and Aten, B. (1967): *Science, 157*, 697.
Steiner, D.F. and Clark, J.L. (1968): *Proc. nat. Acad. Sci. (Wash.), 60*, 622.
Sussman, P.M., Tushinski, R.J. and Bancroft, F.C. (1976): *Proc. nat. Acad. Sci. (Wash.), 73*, 24.
Tager, H.S. and Steiner, D.F. (1974): *Ann. Rev. Biochem., 43*, 509.
Tashjian Jr, A.H., Bancroft, R.C. and Levine, L. (1970): *J. Cell Biol., 47*, 61.
Zanini, A., Giannattasio, G. and Meldolesi, J. (1974): *Endocrinology, 94*, 104.

II. Growth hormones and growth factors

CELLULAR MECHANISMS OF THE ACUTE STIMULATORY EFFECT OF GROWTH HORMONE*

K. AHRÉN, K. ALBERTSSON-WIKLAND, O. ISAKSSON and J.L. KOSTYO

Department of Physiology, University of Göteborg, Göteborg, Sweden, and Department of Physiology, Emory University, Atlanta, Ga, U.S.A.

It has been known for a number of years that the pituitary growth hormone (GH) can stimulate amino acid incorporation into protein and membrane transport of amino acids and sugars into skeletal muscle tissue, both in vivo and in vitro, when the hypophysectomized rat is used as the experimental animal (see Kostyo and Nutting, 1974). The cellular mechanisms involved in these acute effects of GH are to a large extent unknown. Moreover, the relationship between these 'experimental' effects of GH and the physiological important anabolic action of the hormone in the normal animal has still to be clarified. This review summarizes observations made by our two laboratories during recent years on the cellular events taking place in the rat diaphragm after the initial interaction between GH and the muscle cell. An attempt will also be made to place these observations into physiological perspective.

TIME COURSE OF THE STIMULATORY EFFECT OF GH

The stimulatory effect of GH on skeletal muscle tissue, as well as that on the heart, liver and adipose tissue, is usually defined (as in the title of this review) as 'acute'. This gives the impression that the effect develops rapidly. In fact, a lag period of 20-30 minutes exists both under in vivo and in vitro conditions following the addition of the hormone, before a significant effect can be seen. For example, Hjalmarson et al. (1969) reported a lag period of more than 15 minutes following the addition of GH to medium perfusing the rat heart before a significant effect on amino acid transport was seen. Moreover, Goodman (1968) was unable to detect an effect of GH on glucose oxidation to CO_2 by isolated adipose tissue earlier than 15 minutes after the addition of the hormone to the incubation medium. A more extensive examination of this inductive phase of the in vitro effect of GH on amino acid transport and protein synthesis was reported by Rillema and Kostyo (1971). A lag phase of more than 20 minutes was seen after the addition of GH to the incubation medium before an effect could be detected on the uptake of the non-utilizable amino acid, α-aminoisobutyric acid (AIB) or leucine incorporation into protein by the diaphragm of the hypophysectomized rat. This inductive phase could not be eliminated by prolonged incubation of the diaphragms at 2°C.

---

* This work was supported by grants from the Swedish Medical Research Council (B74-14X-4250-01, B76-14X-00027-12), the Medical Faculty, University of Göteborg, and the U.S. National Institutes of Health (AM 12782).

Certain evidence suggests that the acute stimulation of protein synthesis by GH may be one of the more primary events mediating the stimulatory effect of GH on amino acid transport. As noted above, the effect of the hormone on protein synthesis has an early time course which is similar to that of the transport effect. It will also occur in the absence of an effect of the hormone on amino acid transport, e.g. when the transport effect is blocked by replacing Na$^+$ in the incubation medium with choline (Kostyo, 1964). Particularly pertinent is the finding that pre-incubation of the diaphragm with inhibitors of protein synthesis such as puromycin and cycloheximide prior to exposure of the tissue to GH will severely inhibit or abolish the effect of the hormone on AIB transport (Kostyo, 1968). Thus it is possible that GH stimulates the synthesis of protein(s) which mediates the activation of the amino acid transport mechanism. Alternatively, the inhibitor experiments could be interpreted to mean that a protein with a high rate of turnover is involved in the effect of GH on amino acid transport, and that during the pre-incubation phase of the experiments described above, the cellular supply of this protein was depleted.

Another important aspect of the time course of the stimulatory action of GH is its duration. In 1967 it was reported from one of our laboratories (Ahrén and Hjalmarson, 1968) that the stimulation of AIB transport and sugar transport in diaphragms of hypophysectomized rats lasts only 2-3 hours after in vivo or in vitro administration of the hormone. Then the rate of AIB and sugar transport returns to the basal level. Furthermore, diaphragms exposed to GH in vivo or in vitro for 3 or more hours, which then have gone through the stimulatory phase of transport, are no longer responsive to further stimulation by GH in vitro. This phase of refractoriness, which in the original publications by Hjalmarson and Ahrén (1967a,b) was called the 'late inhibitory effect' of GH, lasts for 24-48 hours following in vivo administration of the hormone. In contrast, the stimulatory effect of GH on protein synthesis, measured as amino acid incorporation into protein, does not have the same short duration. Addition of GH to the isolated rat diaphragm stimulated incorporation of amino acids for at least 3-4 hours after the initial exposure to GH, when an effect on AIB or sugar transport can no longer be detected (Hjalmarson, 1968a).

The stimulatory phase of the action of GH on amino acid and sugar transport in the rat diaphragm can be prolonged if the stimulated tissue is subsequently incubated in the presence of actinomycin D or puromycin (Hjalmarson, 1968a,b). That the stimulatory effect of the hormone can be prolonged by exposure of the muscle to poisons of RNA and protein synthesis suggests that the latter processes are involved in turning off the stimulatory effect on the hormone on the transport mechanisms, i.e. that a protein(s) is synthesized during the initial action of the hormone which, in some way, deactivates the transport mechanism and prevents it from being reactivated for some period to time. Such a deactivator or 'refractory protein' has not yet been isolated, however.

CHANGES IN THE SENSITIVITY OF THE DIAPHRAGM MUSCLE TO THE STIMULATORY ACTION OF GH

A finding of considerable interest is that GH shows consistent acute stimulatory effects on isolated tissues only when they are taken from hypophysectomized animals. All of the experiments described above, for example, were performed on tissues of hypophysectomized rats for this reason. Several years ago we (Ahrén, Arvill, Hjalmarson and Isaksson, unpublished data) performed a series of experiments on isolated diaphragms from normal, very young rats (15-20 days old). Addition of GH in vitro

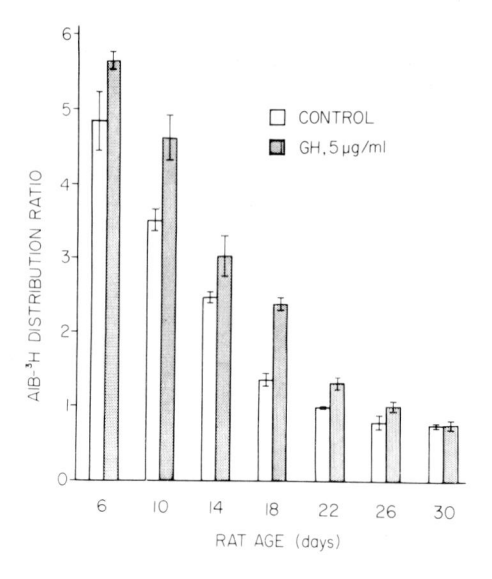

Fig. 1. Effect of GH on AIB-^{3}H uptake by intact hemidiaphragms from fed rats of different ages. The diaphragms were incubated for 90 minutes in Krebs bicarbonate buffer containing glucose (2.5 mg/ml) and 0.1 mM AIB-^{3}H with and without bovine GH (NIH-B-17; 5 µg/ml). There were 6 diaphragms in each group and the standard error is indicated at the top of each bar. GH significantly (paired t-test) increased the uptake in diaphragms from 6-, 10-, 14-, 18- and 22-day-old rats. (From Albertsson-Wikland and Isaksson, 1976. By courtesy of the Editors of *Metabolism*.)

markedly stimulated the rate of AIB uptake in *some* of these experiments, while the hormone had no effect on diaphragms from older rats (30-35 days old) under the same experimental conditions. A clear response was only seen in approximately 50% of the small diaphragms; the remainder gave little or no response. Therefore, it was decided that a systematic analysis was required to establish the reason for this variability in response; such an analysis has been in progress for the past 2 years (Albertsson-Wikland and Isaksson, 1975, 1976).

Figure 1 shows AIB uptake by diaphragms from 6- to 30-day-old rats. A slight stimulatory effect of GH on the uptake of AIB was seen in diaphragms from 6-day-old rats. The stimulatory effect then increased progressively with the age of the animal, reaching a maximum in diaphragms from 18-day-old rats. Responsiveness then declined rapidly, and GH had no effect at all on diaphragms from 30-day-old animals. A similar pattern of sensitivity was found for the effect of GH on the uptake of monosaccharides. GH increased the uptake of the non-metabolizable sugar 3-O-methylglucose by diaphragms from 17- to 22-day-old rats but had no effect on muscles from 32-day-old animals.

When studying the effect of a hormone that influences metabolism, the nutritional state of the animal at the time of the experiment is a crucial factor. The results shown in Figure 1 were obtained from rats that had free access to food. The time between the last meal and the experiment could have varied considerably for the animals in the various age groups, however. The only nutrients available to the small rats (6-10 days old) came from the milk of their mothers. The 18-day-old animals received some milk from their mothers but in addition ate the pelleted diet. The older rats, which were weaned at 21 days of age, had access only to the pelleted diet. A series of experiments were therefore performed with rats that were fasted for 10 or 20 hours before being sacrificed. Fasting did not change the age-dependent pattern of sensitivity of GH, but it increased the magnitude of the in vitro effect and the effect was also more consistent. Table 1 shows results obtained on diaphragms from 18-day-old rats that were fed ad

TABLE 1

Effect of fasting upon the stimulatory effect of GH on the accumulation of AIB-^{14}C in diaphragms from 18-day-old rats

Experimental condition	AIB-^{14}C distribution ratio	
	Control	GH, 5 µg/ml
Fed	1.51 ± 0.12 (7)	1.89 ± 0.13 (7)
Fasting (10 hr)	1.04 ± 0.09 (6)*	1.93 ± 0.13 (6)
Fasting (20 hr)	0.97 ± 0.07 (7)*	1.95 ± 0.13 (7)

Intact hemidiaphragms were incubated for 90 minutes in Krebs bicarbonate buffer containing glucose (2.5 mg/ml) and 0.1 mM AIB-^{14}C with and without bovine GH (NIH-B-17; 5 µg/ml). Values are means ± S.E.M. Number of diaphragms in each group is indicated in parentheses.
* Significantly different from the control value of muscles from fed animals (p < 0.05; analysis of variance).

libitum or fasted. Fasting reduced the rate of AIB uptake in control diaphragms, but the uptake of AIB by diaphragms exposed to GH was the same whether they were obtained from fed or fasted animals. Possibly, fasting reduces some factor(s) in the animal (insulin? GH?) which stimulates amino acid transport in the fed animal, thus accounting for lower AIB uptake by the control tissues. It was thought therefore that diaphragms from fasted animals might be a better preparation on which to test in vitro sensitivity to GH. The results of an experiment in which diaphragms from fasted rats have been used to test the effects of various concentrations of GH on AIB uptake is shown in Figure 2. A stimulatory effect is seen with a hormone concentration as low

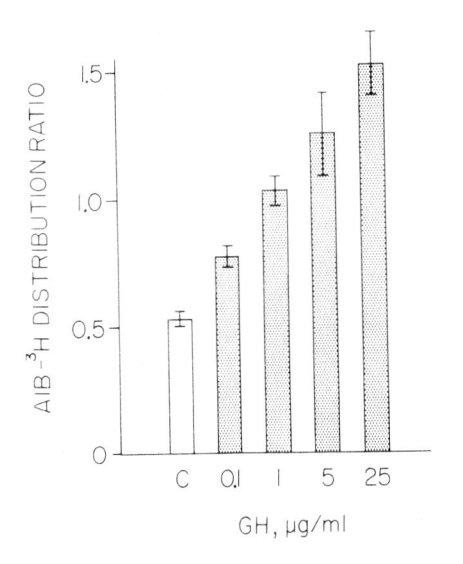

Fig. 2. Effect of various concentrations of bovine GH (NIH-B-17) or AIB-^{3}H uptake by intact hemidiaphragms from fasted (20 hr) 18-day-old rats. Incubation conditions as in Figure 1. The effect of GH is significant (analysis of variance) at all concentrations tested. (From Albertsson-Wikland and Isaksson, 1976. By courtesy of the Editors of *Metabolism*.)

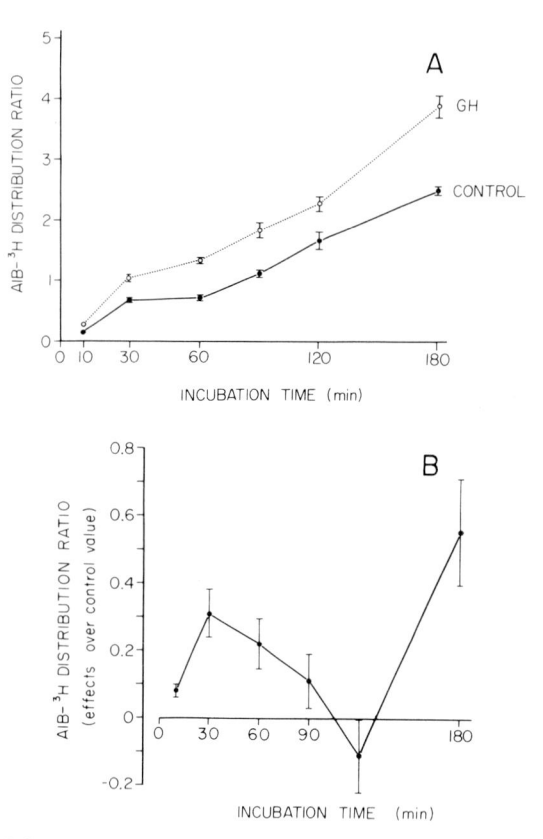

Fig. 3. Time course of the stimulatory effect of bovine GH on the accumulation of AIB-^{3}H in diaphragms from fasted (20 hr) 18-day-old rats. Incubation conditions as in Figure 1. A represents the absolute distribution ratios of AIB-^{3}H in the absence or presence of GH with time. In B the difference in distribution ratios between muscles incubated with and without bovine GH has been plotted during the consecutive phases of the incubation. Values presented are means ± S.E.M. of 12-18 observations. Data were pooled from 4 individual experiments. The stimulatory effect of bovine GH was significant (paired t-test) during 0-10, 10-30, 30-60 and 120-180 minutes of incubation. (From Albertsson-Wikland and Isaksson, 1976. By courtesy of the Editors of *Metabolism*.)

as 0.1 µg/ml of medium, which is well within the range of the concentration of GH circulating in the rat (Martin et al., 1974; Tannenbaum and Martin, 1975). Recently, Nutting (1975) has also found, in an independent study, that rats of various ages show a sensitivity pattern nearly identical to the one described above, as well as the same influence of fasting.

It was mentioned above that the stimulatory effect of GH on amino acid transport in diaphragms from small rats has a duration of 2-3 hours followed by a phase of refractoriness, when the muscle can no longer be stimulated by the hormone. Figure 3 shows that the time course of the stimulatory effect of GH on diaphragms of small rats is different. The initial lag period appears shorter and a marked effect is already apparent after 30 minutes of incubation with the hormone. In the period between 60 and 120 minutes, no further effect of GH is seen. However, a clear effect is once again evident in the period between 120 and 180 minutes. One possible interpretation of these results is that the diaphragm of small rats develops refractoriness within 60 minutes of exposure to GH, but that this phase of refractoriness has a duration of only 2-3 hours. Further experiments will be required to verify that the time course of the action of GH is different in the small rat and to determine if the phase of refractoriness can be altered by inhibitors of RNA and protein synthesis.

THE POSSIBLE INVOLVEMENT OF CYCLIC AMP IN THE ACTION OF GH

A crucial question in the analysis of the stimulatory effect of GH is the nature of the metabolic events that occur during the lag period, i.e. during the 30 minutes that precede the stimulatory effects of GH on protein synthesis and membrane transport. Recent work in our laboratories indicates that one of these events may involve an alteration in cyclic nucleotide metabolism. Theophylline, papaverine and quinine blocked the in vitro effects of GH on protein synthesis and on AIB and 3-O-methylglucose transport into the isolated diaphragm of the hypophysectomized rat (Payne and Kostyo, 1970; Rillema et al., 1973). The extent to which the action of GH was blocked correlated directly with the degree of glycogenolysis produced by the phosphodiesterase inhibitor. These findings suggest that reduction in the concentration of cyclic AMP or cyclic GMP in the cells might be one of the early events mediating the effects of GH on protein synthesis and membrane transport. Efforts (Isaksson et al., 1974) to demonstrate a reduction in the total cyclic AMP content of the isolated diaphragm in response to GH in vitro were, however, unsuccessful. This is perhaps not surprising considering that much of the cyclic AMP present in the diaphragm in the resting state is probably bound to intracellular structures and hence not subject to metabolic regulation.

Attempts were also made to determine if GH in vitro could influence the total amount of cyclic GMP in the rat diaphragm. Diaphragms of hypophysectomized rats were incubated for 5, 30 or 120 minutes with GH and then processed for the measurement of cyclic GMP. It can be seen from Table 2 that the hormone had no consistent effect on the total amount of cyclic GMP in the diaphragm under these conditions.

It is often easier to demonstrate a hormone-induced reduction in the cellular level of cyclic AMP if the cyclic nucleotide content of the tissue is increased by some agent prior or subsequent to exposure of the tissue to the hormone in question. Therefore, this approach was also used in an attempt to determine if GH has an effect on the amount of cyclic AMP in the diaphragm. Isolated diaphragms of hypophysectomized rats were incubated in the presence or absence of GH (NIH-B-17; 5 μg/ml) for 10 or 20 minutes. Then epinephrine was added to the incubation medium to a final concentration of either 0.01 or 10 μg/ml, and incubation was continued for an additional 2 minutes. The muscles were then processed for the measurement of cyclic AMP. Although the addition of epinephrine to the incubation medium markedly increased the amount of cyclic AMP in the muscles, no effect of GH was apparent. Using this same experimental design, it has been possible to demonstrate a significant reduction in the epinephrine-raised level of cyclic AMP with the hormone insulin (Isaksson et al., 1974).

Although the above experiments did not reveal an effect of GH on the total amount of cyclic AMP or cyclic GMP in the diaphragm, they did not rule out the possibility that the hormone might cause a reduction in the level of one or both of these nucleotides in a small free pool in the cells. The hypothesis that GH affects the amount of free cyclic AMP in the diaphragm was tested using a modification (Kostyo et al., 1975) of the method of Kuo and De Renzo (1969) to estimate changes in the amount of free

TABLE 2

*Effect of GH on the cyclic GMP content of diaphragms of hypophysectomized rats**

Incubation time (min)	Cyclic GMP content (pmoles/mg protein)	
	Control	BGH
5	0.52 ± 0.06 (7)	0.50 ± 0.05 (7)
30	0.78 ± 0.20 (8)	0.47 ± 0.06 (8)
120	0.36 ± 0.05 (8)	0.30 ± 0.02 (8)

* Intact hemidiaphragms were washed for 20 minutes in Krebs bicarbonate buffer containing glucose (2.5 mg/ml) and then incubated for the times indicated with or without bovine growth hormone (GH; 5 μg/ml). At the end of incubation, the muscles were frozen between blocks of aluminium that had been cooled to the temperature of liquid nitrogen. The frozen tissue was homogenized in 5% trichloroacetic acid at 0°C. Aliquots of the homogenate were taken for protein in measurement and the remainder was centrifuged to obtain a protein-free supernatant, which was used for the determination of cyclic GMP by radioimmunoassay. Values are means ± S.E.M.; (n) indicates the number of muscles used.

cyclic AMP in the cells. Diaphragms of hypophysectomized rats were first incubated for 1 hour with ^{3}H-adenine to label tissue nucleotides. Then the muscles were washed for 30 minutes to remove unincorporated adenine. After the washing step, the tissues were exposed to GH (BGH; 5 μg/ml) or various drugs for appropriate periods of time. Then they were placed in medium containing carrier cyclic AMP and the polyene filipin (to increase membrane permeability to intracellular nucleotides) and incubated for a final period to collect the tritiated cyclic AMP released from the cells. To determine the amount of cyclic AMP released, ^{14}C-cyclic AMP was added at the end of incubation as a recovery tracer, and then the media were subjected to Dowex-50 chromatography and repeated precipitation with barium hydroxide and zinc sulfate to remove labeled substances other than cyclic AMP before radioactivity measurements were made.

When labeled and washed diaphragms were preincubated with GH for periods ranging from 5 to 45 minutes and then placed in incubation medium containing filipin to collect the labeled cyclic AMP that was released from the cells, it was found that 15 to 30 minutes of exposure of the muscles to GH caused a decrease in the release of tritiated cyclic nucleotide from the cells, presumably reflecting a reduction in the amount of free cyclic AMP in the cells. Since this reduction occurred some time between 15 and 30 minutes of exposure of the cells to GH, it precedes or at most coincides with the stimulatory effects produced by the hormone on protein synthesis and membrane transport. Thus, these findings are consistent with the hypothesis that a reduction in the amount of free cyclic AMP is one of the events mediating the effects of GH on protein synthesis and membrane transport.

The reduction in cyclic AMP release produced by GH is only transitory. When the preincubation period with GH was extended (45-120 minutes) the amount of radioactive cyclic AMP released during the final incubation period was actually increased. Thus it would appear that after 30-45 minutes' exposure of the cells to GH, metabolic events occur that lead to an increase in the amount of radioactive cyclic AMP that is available for release from the cells.

The secondary rise in free cyclic AMP in the diaphragm appears to be dependent upon protein synthesis. In the experiment shown in Figure 4 labeled and washed

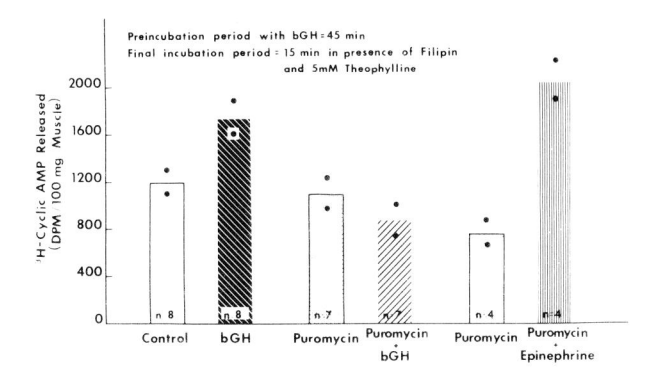

Fig. 4. The ability of puromycin to block the GH-induced secondary rise in ^{3}H-cyclic AMP release from the isolated rat diaphragm. Intact hemidiaphragms from hypophysectomized rats were incubated for 1 hour with ^{3}H-adenine (20 μCi/ml), washed for 30 minutes in Krebs buffer and then exposed to puromycin (200 μg/ml) for 15 minutes. The tissues were then incubated with or without bovine GH (5 μg/ml) in the presence of puromycin for 45 minutes. They were then transferred to medium containing filipin (0.07 mM), theophylline (5 mM), cyclic AMP (1 mM), puromycin, and GH, where indicated, and incubated for a final 15 minutes to collect the ^{3}H-cyclic AMP released from the tissue. In the experiment shown by the pair of bars on the right, the muscles were not exposed to GH. Instead, epinephrine (10 μg/ml) was added to the medium during the final 15-minute incubation period. The dots represent the standard error.

diaphragms were first exposed to puromycin for 15 minutes to inhibit protein synthesis and then preincubated with bovine GH and puromycin for 45 minutes. Labeled cyclic AMP release was then monitored during a final 15-minute incubation period. The first pair of bars shows the expected rise in tritiated cyclic AMP release due to GH in diaphragms not exposed to puromycin. The second pair of bars shows the results obtained with muscles exposed to puromycin. The secondary rise in cyclic AMP release did not occur. If anything, the results suggest that by inhibiting protein synthesis the early depressive effect of GH on cyclic AMP release is preserved. To be sure that the diaphragms were still viable under these circumstances the control experiment shown by the last pair of bars was performed. Labeled and washed diaphragms were incubated with puromycin for 75 minutes and then incubated for the final 15 minutes in the presence of epinephrine. It can be seen that the puromycin-treated muscles responded to epinephrine with a marked release of cyclic AMP.

From the foregoing results it would appear that GH has a biphasic effect on the metabolism of cyclic AMP in the isolated rat diaphragm. There is a relatively rapid effect occurring within 15-30 minutes of exposure of the tissue to GH, which if the interpretation of the findings is correct, involves a reduction in the amount of free cyclic AMP in the cells. Such a reduction in free cyclic AMP could result from a variety of mechanisms. The hormone might inhibit the adenylate cyclase reaction or affect the availability of substrate for that reaction, increase the intracellular binding of cyclic AMP or its destruction by phosphodiesterase, or less likely increase its rate of release from the cells. Whether or not any of these mechanisms are involved remains to be established. The presumed reduction in tissue cyclic AMP is soon followed by a secondary rise in the level of the nucleotide. This rise appears to be prevented by inhibiting protein synthesis. As noted earlier, protein synthesis inhibitors prolong the stimulatory phase of GH's action on membrane transport. Thus, the secondary rise in free cyclic AMP in response to GH may be involved in some manner in restoring the rates of sugar and amino acid transport to basal levels and the development of refractoriness.

CONCLUSIONS

Although much remains to be learned about the cellular events that produce the stimulatory effects of GH on protein synthesis and membrane transport in muscle, a working hypothesis can be constructed from our present knowledge, which outlines the mechanism involved in the physiological regulation of these processes by GH. Available evidence suggests that GH is secreted by the pituitary in episodic bursts, resulting in transitory but dramatic increases in the concentration of GH in the blood. We propose that once a young animal develops its initial sensitivity to GH these peaks or spikes in the concentration of the hormone in the blood trigger the cellular events that lead to the stimulation of protein synthesis and membrane transport in muscle cells. Certainly, the levels of GH that have been found in the blood during the bursts of hormone secretion are quite similar to the concentrations of GH required to influence metabolic processes in isolated tissues. In response to the rise in GH secretion, sufficient receptors on the muscle cell interact with GH to trigger an initial metabolic event. This then leads to a series of metabolic reactions that eventually culminate in the stimulation of protein synthesis. Whether or not a reduction in cyclic AMP is one of these events remains to be established. In any event, a series of metabolic events appears to be involved, since there is a considerable lag period after a cell is exposed to GH before protein synthesis is stimulated. The change produced in the protein synthetic machinery is a fairly stable change, since, once stimulated, the rate of protein synthesis remains elevated for a number of hours. Whether or not the rate would eventually decline to a basal level (like that found in tissues of the hypophysectomized animal) would depend upon the frequency of the bursts of GH secretion. Conceivably, if the bursts are sufficiently frequent, the rate of protein synthesis could remain elevated constantly. This may be the reason that the isolated diaphragm of the older growing rat shows little, if any, change in protein synthesis in response to the addition of GH to the incubation medium.

One consequence of the stimulation of protein synthesis by GH would be the formation of the protein that makes the amino acid and sugar transport processes refractory to the insulin-like action of GH. If the episodic bursts of GH are frequent enough to maintain the protein synthetic machinery in a permanently stimulated state, then the amounts of 'refractory protein' would remain elevated in the muscle cell and, as a consequence, the rates of amino acid and sugar transport would remain at the basal level, even during bursts of GH secretion. In contrast, in the hypophysectomized animal that has been deprived of GH for some period of time, the supply of 'refractory protein' would be depleted, hence explaining why sugar and amino acid transport can be stimulated readily in the muscle cells of this animal either in vivo or in vitro. From our observations, it would appear that the half-life of the 'refractory protein' is quite short in the very young organism and that it becomes longer as the animal ages. Thus, when the diaphragm is removed from very young normal rats and placed in vitro, and hence isolated from the oscillations in circulating GH level, its membrane transport processes either are or soon become sensitive to the stimulatory effect of GH. In the very young animal, depending upon the frequency of bursts of GH secretion and hence the amount of 'refractory protein' in the cells, the rate of amino acid and sugar transport could increase in response to a burst of GH secretion. Consequently, in the very young animal, GH may be quite important as a physiological regulator of plasma amino acid and glucose levels, particularly during the period prior to weaning when there is an almost continual input of these metabolites into the blood. After weaning, when the organism undergoes periods of feeding and fasting, it would be particularly vulnerable to an insulin-like agent that is secreted in a pattern not intimately tied to the feeding-fasting cycle, as GH is not. From a physiological point of view, it is quite fascinating that in the rat the development of virtual permanent refractoriness of the transport mechanisms to GH occurs at or shortly after weaning.

Thus, if the above hypothesis is correct, the frequency of the bursts of GH secretion and the rate of protein turnover will determine whether or not the rates of protein synthesis and membrane transport in the muscles of normal growing animals will oscillate and whether or not they will respond to exogenous GH. Clearly, much further work will be necessary to establish or refute this hypothesis.

REFERENCES

Ahrén, K. and Hjalmarson, Å. (1968): In: *Growth Hormone,* p. 143. Editors: A. Pecile and E.E. Müller. ICS 158, Excerpta Medica, Amsterdam.
Albertsson-Wikland, K. and Isaksson, O. (1975): *Acta physiol. scand.,* in press.
Albertsson-Wikland, K. and Isaksson, O. (1976): *Metabolism,* in press.
Goodman, H.M. (1968): In: *Growth Hormone,* p. 153. Editors: A. Pecile and E.E. Muller. ICS 158, Excerpta Medica, Amsterdam.
Hjalmarson, Å. (1968a): *Acta endocr. (Kbh.), 57, Suppl. 126.*
Hjalmarson, Å. (1968b): *Effects of Growth Hormone on the Metabolism of the Isolated Rat Diaphragm.* Thesis. Orstadius Boktryckeri AB, Göteborg.
Hjalmarson, Å. and Ahrén, K. (1967a): *Life Sci., 6,* 809.
Hjalmarson, Å. and Ahrén, K. (1967b): *Acta endocr. (Kbh.), 56,* 347.
Hjalmarson, Å., Isaksson, O. and Ahrén, K. (1969): *Amer. J. Physiol., 217,* 1795.
Isaksson, O., Gimpel, L.P., Ahrén, K. and Kostyo, J.L. (1974): *Acta endocr. (Kbh.), Suppl. 191, 73.*
Kostyo, J.L. (1964): *Endocrinology, 75,* 113.
Kostyo, J.L. (1968): In: *Growth Hormone,* p. 175. Editors: A. Pecile and E.E. Müller. ICS 158, Excerpta Medica, Amsterdam.
Kostyo, J.L., Gimpel, L.P. and Isaksson, O. (1975): *Advanc. metab. Disorders, 8,* 529.
Kostyo, J.L. and Nutting, D.F. (1974): In: *Handbook of Physiology, Sect. 7: Endocrinology, Vol. IV,* Part 2, p. 187. Editors: R.O. Greep and E.B. Astwood. American Physiological Society, Washington, D.C.
Kuo, J.F. and De Renzo, E.C. (1969): *J. biol. Chem., 244,* 2252.
Martin, J.B., Renaud, L.P. and Brazeau Jr, P. (1974): *Science, 186,* 538.
Nutting, D.F. (1975): *Fed. Proc., 34,* Abstract No. 699, 343.
Payne, S.G. and Kostyo, J.L. (1970): *Endocrinology, 87,* 1186.
Rillema, J.A. and Kostyo, J.L. (1971): *Endocrinology, 88,* 240.
Rillema, J.A., Kostyo, J.L. and Gimpel, L.P. (1973): *Biochim. biophys. Acta (Amst.), 297,* 527.
Tannenbaum, G.S. and Martin, J.B. (1975): *Fed. Proc., 34,* 273
Walker, D.G., Simpson, M.E., Asling, J.W. and Evans, H.M. (1950): *Anat. Rec., 106,* 539.

GROWTH HORMONE ACTION ON THYMUS AND LYMPHOID CELLS*

G.P. TALWAR, S.N.S. HANJAN, Z. KIDWAI, P.D. GUPTA, N.N. MEHROTRA, R. SAXENA and Q. BHATTARAI

Department of Biochemistry, All India Institute of Medical Sciences, Ansari Nagar, New Delhi, India

Growth hormone influences many tissue and metabolic processes. This presentation is concerned with an interesting, recently discovered, action of growth hormone on the lymphoid cells. Other aspects of growth hormone action have been reviewed (Talwar, 1972; Talwar et al., 1975a,b).

PITUITARY CONTROL OF THYMUS

Evidence for pituitary regulation of thymic functions has come from a variety of observations. Snell Bagg dwarf mice with hypopituitarism show poor thymic development and the animals die with deficient thymic function (Baroni, 1967; Duquesnoy et al., 1970; Pierpaoli et al., 1970). The thymus progressively atrophies after hypophysectomy (Smith, 1930; Feldman, 1951; Talwar et al., 1975a), and administration of pituitary or growth hormone antibodies reduces thymus size (Pierpaoli and Sorkin, 1967, 1968). When antibodies against growth hormone are injected for a period, the capacity of rats to produce plaque-forming cells against sheep erythrocytes is reduced as is the incorporation of ^{3}H-thymidine into DNA in secondary lymphoid organs such as the spleen where clonal expansion of cells takes place upon antigenic stimulus (Pandian and Talwar, 1971).

ACTION OF GROWTH HORMONE ON THYMUS

The marked reductions in plaque-forming cells, haemagglutination titres and DNA synthesis in thymocytes and splenocytes, seen 15 weeks after hypophysectomy, are ameliorated by growth hormone (Table 1).

Growth hormone stimulates metabolic activities in both cortical and medullary regions of thymus as gauged by thymidine and ^{35}S-sulphate incorporation into biopolymers. The former precursor is incorporated at a high rate in the cortex, where active cellular replication predominates while the incorporation of ^{35}S-sulphate is essentially confined to the medulla.

---

* This work was supported by research grants from The Indian Council of Medical Research, The Population Council Inc., New York, The World Health Organization (Immunology Division) and a PL-480 grant No. N 00014-70-C-0179, NR 202-028 from the office of Naval Research.

TABLE 1
Effect of growth hormone on the immune system of rats

Animals used	PFC/10^6 spleen cells***	^{3}H-thymidine incorporated into DNA c.p.m./mg DNA	
		Thymus	Spleen
Normal	328 ± 57	113 ± 1.1	295 ± 4.9
Hypox.*	31 ± 15	53 ± 4.1	176 ± 2.4
Hypox. + GH**	236 ± 26	171 ± 4.7	601 ± 8.6
Normal + GH	393 ± 70	201 ± 2.9	475 ± 4.8

* Hypophysectomized 15 weeks before experiments.
** 2.1 mg BGH given subcutaneously in 2 doses for 5 days.
***2 × 10^8 SRBC injected 102 hours before assay. PFC determined as described (Pandian and Talwar, 1971). ^{3}H-thymidine (25 μCi/100 g body weight) was given intraperitoneally 30 minutes before sacrifice.

TABLE 2
Effect of pituitary hormones on 4-week-old rat thymocytes

Addition	Concentration	^{3}H-uridine incorporated into RNA (% of control ± SEM)
None	—	100.0 ± 1.0 (7)
Rat GH	50 μg/ml	127.6 ± 1.8 (5)
Bovine GH	50 μg/ml	130.6 ± 0.9 (7)
Ovine GH	50 μg/ml	124.1 ± 3.1 (5)
Human GH	50 μg/ml	129.6 ± 2.1 (6)
Fish GH	50 μg/ml	98.7 ± 3.4 (6)
Boiled bovine GH*	50 μg/ml	100.5 ± 0.9 (5)
Prolactin	50 μg/ml	124.0 ± 1.2 (5)
Prolactin + GH	50 μg/ml each	128.1 ± 2.9 (5)
Ovine LH	50 μg/ml	100.1 ± 3.2 (6)
Ovine FSH	50 μg/ml	98.1 ± 2.7 (6)
TSH	50 μg/ml	100.6 ± 0.9 (5)
ACTH	0.2 IU/ml	97.5 ± 3.2 (6)
Oxytocin	0.5 IU/ml	98.2 ± 2.9 (4)

Thymocytes were incubated for 3 hours at 37°C in a medium containing 5.0 mM tris-HCl buffer, pH 7.2, 120 mM NaCl, 5.0 mM Na_2HPO_4, 5.0 mM KCl, 1.0 mM $MgSO_4$, 0.8 mM $CaCl_2$ and 5.5 mM glucose, with or without the hormone. ^{3}H-uridine was added at a concentration of 0.5 μCi/ml at 2 hours. Incorporated radioactivity in 60 minutes was measured as described (Saxena and Talwar, 1974).

* Hormone heated at 100°C for 15 minutes before use.

Direct action of growth hormone on some metabolic activities is demonstrable in vitro. The hormone promotes the uptake and incorporation of radioactive uridine into RNA in isolated teased-out cells. The effect disappears if the hormone is inactivated by heating. Other pituitary hormones such as TSH, FSH, LH, ACTH have no effects (Table 2). Prolactin, however, has a similar effect to growth hormone suggesting that they share molecular features necessary for actions on thymocytes (Talwar et al., 1975b). The two hormones are known to have significant homologies in amino acid sequences (Li, 1972; Niall et al., 1973).

Rat thymocytes respond to a variety of growth hormones including human, bovine, and ovine. However, fish growth hormone elicits no response. The effect of growth hormone is manifest in a variety of media. Foetal calf serum is not necessary, but there is an obligatory requirement for calcium (Pandian and Talwar, 1971).

The growth hormone-stimulated incorporation of uridine into RNA has been proposed as a simple, rapid and sensitive method for bioassay of growth hormone (Saxena and Talwar, 1974).

LOCUS OF ACTION OF GROWTH HORMONE

The mechanisms by which growth hormone exercises its many effects in the body are not fully understood (Fain and Saperstein, 1970; Tanner, 1972; Talwar, 1972; Talwar et al., 1975a). The facilitated transport of glucose and amino acids in adipose tissue (Goodman, 1968; Pandian et al., 1971) and diaphragm (Kostyo et al., 1959), and of calcium in thymocytes (Hanjan and Talwar, unpublished data) suggest that one locus of action may be the cell membrane.

The hormonal stimulation of the uptake and incorporation of uridine could result from interactions of the hormone with membrane sites. This is borne out by the fact that growth hormone conjugated to Sepharose-4B beads, larger than the cells, stimulates uridine incorporation in intact thymocytes (Table 3).

Localization of the hormone

The binding of immunologically reactive determinants of the growth hormone to thymocytic membranes has been recently demonstrated by high-resolution electron micrographs of preparations incubated sequentially with rabbit anti-growth hormone globulins and monospecific sheep anti-rabbit globulins tagged with horse-radish peroxidase (Pandian et al., 1975). Electron-dense deposits (corresponding to the bound hormone) are present in patches along the plasma membrane of the thymocytes (Fig. 1A). The cells incubated with heat denatured OGH or without the hormone have no such electron-dense material (Fig. 1B).

TABLE 3

Biological activity of Sepharose-tagged growth hormone

Addition	^{3}H-uridine incorporated into RNA (c.p.m./10^6 cells)
None	4862 ± 163 (8)
Sepharose	5069 ± 92 (8)
Rat growth hormone (RGH)	6397 ± 161 (7)
Sepharose-RGH	6209 ± 166 (8)

Rat thymocytes (3 × 10^6/ml) were incubated for 2 hours at 37°C in medium containing 5 mM tris-HCl buffer pH 7.2, 120 mM NaCl, 5 mM KCl, 5 mM Na$_2$HPO$_4$, 1 mM MgSO$_4$, 0.8 mM CaCl$_2$ and 5.5 mM glucose and tritiated uridine (2.5 μCi/ml, 2 μM) ± the agent tested. Values in parentheses denote the number of observations.

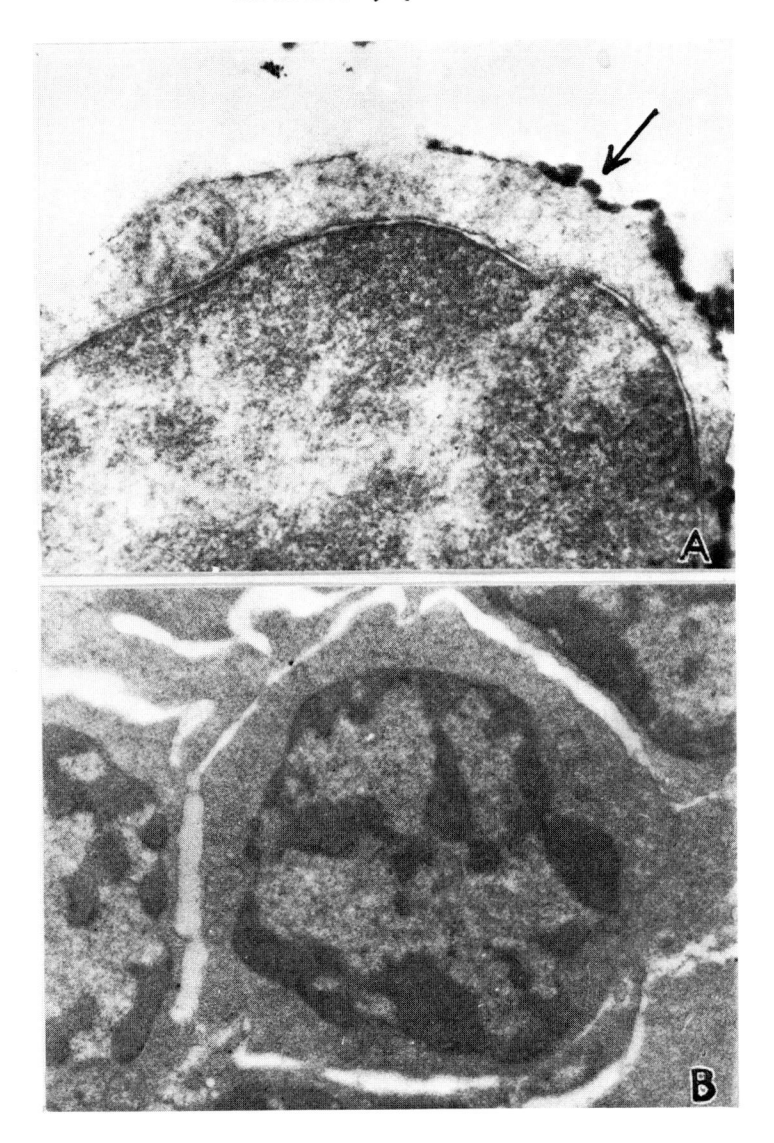

Fig. 1. Electron micrograph of rat thymocyte showing (A) electron-dense deposits at irregular intervals (arrows) on plasma membrane. Control preparations (incubation with growth hormone) do not show such deposits (B). The experimental procedure was the same as described (Pandian et al., 1975). Sections are stained with uranyl acetate only, scanned at 80 kV (A: × 79,500, B: × 15,000).

DECLINE IN RESPONSIVENESS OF THYMUS TO GROWTH HORMONE WITH AGE

The thymus involutes with age and this impinges upon the lowering of immunological capacity of the body to resist disease (Greenberg and Yunis, 1972). The yield of thymocytes from aged rats is greatly reduced (Talwar et al., 1975b). Furthermore the hormone-induced uptake and incorporation of uridine into these cells in vitro progressively decreases with age (Table 4).

TABLE 4

Effect of growth hormone on the uptake and incorporation of ^{3}H-uridine in thymocytes derived from rats of different ages

Age of the animal (months)	c.p.m. of ^{3}H-uridine incorporated per 10×10^6 cells		Per cent stimulation
	− Control	+ Bovine GH (20 µg/ml)	
1	317 ± 24.1 (4)	431 ± 44.2 (4)	24.7
2	216 ± 13.1 (4)	283.75 ± 20.5 (4)	31.2
12	108 ± 7.8 (3)	104.66 ± 10.5 (3)	—
24	127 ± 17.0 (3)	123 ± 14 (3)	—

The experimental procedure was the same as described for Table 3.

QUANTIFICATION OF 'RECEPTORS' ON THYMOCYTES FROM YOUNG AND AGED RATS

To investigate the possible reasons for the decline in the end organ responsiveness, the receptors for the hormone were quantified by measuring the antibody peroxidase complex in conditions in which the activity was proportional to the amount of the enzyme. Monospecific antibodies to growth hormone were tagged to peroxidase in equimolar proportions. Thymocytes obtained from young rats bound 3 to 9 times more growth hormone than an equal number of cells obtained from thymuses of old rats (Talwar et al., 1974).

These studies do not, however, reveal whether all cells are sensitive to growth hormone. The decreased response to growth hormone could result from either a reduced proportion of hormone-sensitive cells in the thymus or from a decline in the 'density' of hormone receptors on the cells. A biophysical approach, enabling the analysis and quantification of cells interacting with growth hormone, was thus employed.

HORMONE-SENSITIVE THYMOCYTE SUB-POPULATIONS

Figure 2A shows the profile of the electrophoretic mobility (EPM) of thymocytes from 5-week-old rats. The EPM is spread over a wide range. However, the distribution of mobilities of cells around the mean appeared homogenous. This was confirmed by the linear relation obtained on transfer of the EPM data as an integrated curve on a sum-probability graph (Fig. 3).

After incubation with growth hormone for 45 minutes at 37°C, and then washed, the cells displayed an altered mobility and surface charge. The average mobility of the

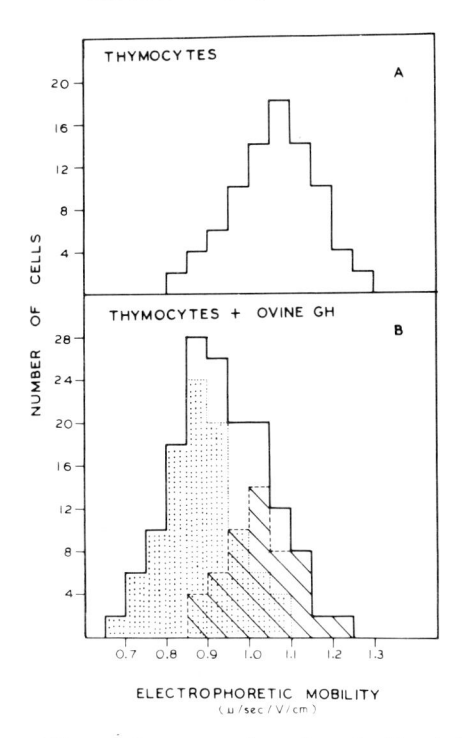

Fig. 2. Electrophoretic mobility of thymocytes from 5-week-old rats before (A) and after (B) incubation with growth hormone. Thymocytes from 4 rats were pooled. Aliquots of 10×10^6 cells were incubated in 1 ml of minimum Eagle's medium with 50 µg of GH for 45 minutes at 37°C. The cells were washed and EPM measured as described. Untreated thymocytes show a Gaussian distribution as assessed by sum-probability analysis (see Fig. 3.). Growth hormone treated thymocytes (Fig. 2B) being heterogenous are further split by stripping technique into 2 Gaussian populations. Population (▦) represents cells whose EPM is decreased by GH and population (▧) represents the unaffected cells. (Data from Hanjan and Talwar, 1975b.)

cells fell from 1.02 ± 0.08 to 0.93 ± 0.09 µ/sec/V/cm. The EPM profile of these cells was heterogenous (Fig. 2B). The histogram of the EPM was stripped (Ruhenstroth-Bauer and Lucke-Huhle, 1968) and 2 sub-populations of cells each with a homogenous distribution were obtained. The quantitative preponderance of these populations and their mean EPM was calculated (Hanjan and Talwar, 1975a).

In 5-week-old rats, growth hormone decreased the EPM of 65% of the isolated thymocytes, with 35% of the cells being unaffected.

DECLINE OF THE HORMONE-SENSITIVE SUB-POPULATION WITH AGE

Evidence for the presence of growth hormone receptors on thymocyte membranes has been presented above. The EPM of a cell is largely a function of the net surface change. Alterations in this parameter are indicative of the changes in the charged groups present on the cell surface within 10 Å of the hydrodynamic plane of shear (Wallach and Esandi, 1964). The results presented suggest that the decrease in surface charge of the thymocytes after their incubation with growth hormone results from

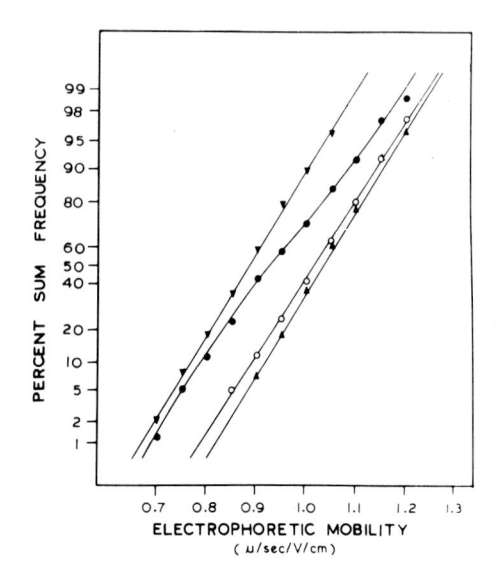

Fig. 3. Sum-probability distribution of the electrophoretic mobilities of thymocytes from 5-week-old rats before (o----o) and after (●----●) incubation with growth hormone. The growth hormone-treated population being heterogenous, as indicated by the non-linear curve, was resolved into 2 sub-populations by stripping technique; ▼----▼ (population affected by growth hormone) and ▲----▲ (population unaffected by hormone treatment). (Data from Hanjan and Talwar, 1975*b*.)

binding of the hormone with plasma membrane receptors. The change in the surface charge is neither due to new protein synthesized in the cells under the influence of the hormone nor to hormone-modulated cellular metabolism (Hanjan and Talwar, 1975*b*).

Experiments were performed to quantify the hormone-sensitive sub-populations in thymocytes from rats of different ages. Not all cells in the thymus interact with growth

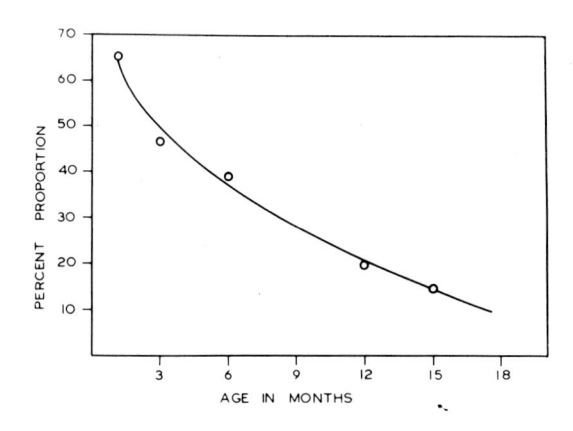

Fig. 4. The GH-sensitive population of thymocytes derived from rats of different ages. The sub-population was calculated as described in Figure 3.

TABLE 5

*Effect of growth hormone on the electrophoretic mobility (EPM)
of thymocytes from 5-week- and 1-year-old rats*

Age	Treatment	Sub-population	Proportion	Mean EPM ± SD	Charge density
5 weeks	None	All cells	100	1.02 ± 0.08	3424
	GH	Affected	65	0.88 ± 0.09	2953
		Unaffected	35	1.03 ± 0.08	3457
1 year	None	All cells	100	1.035 ± 0.124	3474
	GH	Affected	19	0.869 ± 0.68	2916
		Unaffected	81	1.039 ± 0.08	3457

hormone, and reactive cells diminish in number with age (Fig. 4). Actual counting of cells bearing growth hormone receptors in thymocyte preparations from 4-week- and 1-year-old rats confirmed this view (Talwar et al., 1974).

Furthermore, the extent of the change in EPM of the hormone-sensitive sub-population of cells at various ages is of the same order (Table 5) even though their numbers decrease with age. The growth hormone receptors on the cells on which they are present, thus remain quantitatively similar.

SITUATION IN WHICH THE LYMPHOID CELLS HAVE GROWTH HORMONE RECEPTORS

Two further questions arise: (1) Do human lymphocytes carry receptors for growth hormone? (2) Are there fluctuations in the number of cells bearing receptors for the hormone? If receptors are present, under what circumstances do they occur? Human peripheral blood lymphocytes, purified on Ficoll-Visotrast gradients, were investigated.

BINDING OF GROWTH HORMONE TO PERIPHERAL BLOOD LYMPHOCYTES

Figure 5A shows the EPM profile of the circulating lymphocytes from a healthy human subject. The histogram can be stripped into 3 sub-populations of cells. Their mean mobilities and other properties are given in Table 6.

On incubation of these cells with human growth hormone (100 μg/ml) at 37°C for 45 minutes, their mean mobility was altered slightly (Fig. 5B). The EPM profile of the altered cells on stripping revealed that the affected cells were part of the T_5 min sub-population. The remaining T cells as well as the B cells did not seem to bind growth hormone.

Lesniak et al. (1973) have reported the presence of receptors for growth hormone in human lymphocyte cell lines. We have observed only a small percentage of the cells with such receptors in the circulating lymphocytes of healthy human subjects. The circulating lymphocytes are mostly resting cells. It was, therefore, of interest to investigate the interaction of growth hormone with the human lymphocytes triggered to multiply by phytomitogens.

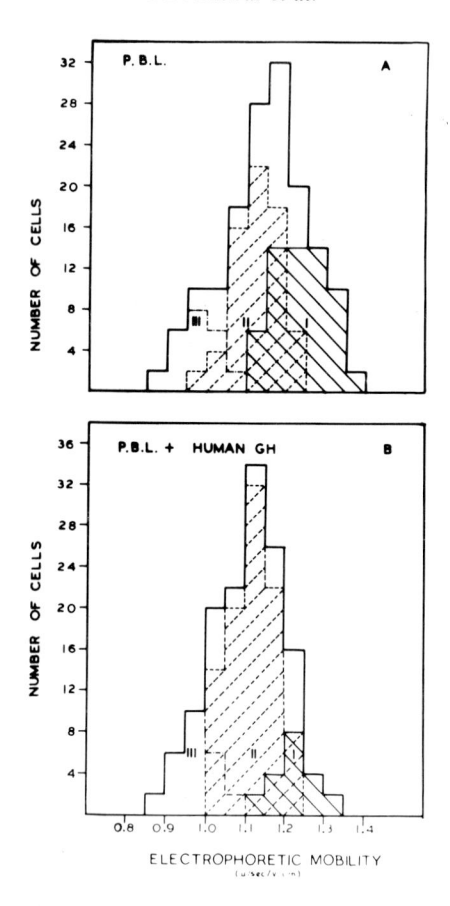

Fig. 5. Electrophoretic mobility (EPM) of normal human peripheral blood lymphocytes before (A) and after (B) incubation with human growth hormone. Sub-population III corresponds to the number of cells forming EAC rosettes and hence represents B cells. Sub-population I corresponds to the number of T cells forming spontaneous rosettes with sheep erythrocytes in 5 minutes and sub-population II corresponds to the number of cells forming spontaneous rosettes in 24 hours. Upon incubation of these cells with human growth hormone (10 μg/ml) at 37°C for 45 minutes, the average mobilities of the 3 populations described above were unaffected. Only a part of the sub-population I was affected.

OBSERVATIONS ON LYMPHOCYTES CULTIVATED WITH PHYTOMITOGENS

Phytohaemagglutinin (PHA) induces the blast transformation of a large number of peripheral blood T (or thymus-derived) lymphocytes (Krishnaraj and Talwar, 1973). Healthy human peripheral lymphocytes were cultured with PHA and examined after 96 hours, when the rate of transformation of these cells is optimal (Mehra et al., 1972; Jha et al., 1975).

The mean EPM of the PHA transformed lymphocytes cultured for 96 hours was 1.05 ± 0.11, whereas the cells cultured in the absence of PHA was 0.89 ± 0.128 μ/sec/

TABLE 6
Sub-populations of lymphocytes in human peripheral blood

Sub-population	Mean EPM ± SD (μ/sec/V/cm)	Per cent proportion quantificated from EPM data	Per cent proportion quantificated from rosette data	Corresponding nomenclature	Functional properties
I	1.266 ± 0.029	40.624 ± 8.29	39.032 ± 11.77	T_5 min	Cell-mediated immunity and helper function in antibody formation
II	1.108 ± 0.019	43.03 ± 6.81	36.31 ± 9.37	T_{24} hr	
III	0.939 ± 0.034	16.35 ± 3.59	14.63 ± 5.32	B	Precursors of antibody-forming cells

V/cm; when incubated with human growth hormone (10 μg/ml) their mobility was 0.84 ± 0.128 and 0.82 ± 0.11 μ/sec/V/cm respectively. These results suggest that the number of cells interacting with growth hormone is markedly increased in the presence of PHA.

SUMMARY

The pituitary appears to have a thymotropic role. Somatotropin is one of the pituitary secretions influencing lymphoid organs. Growth hormone (GH) stimulates the incorporation of radioactive precursors into both medullary and cortical portions of the thymus. The antibody response of the hypophysectomized rats to sheep erythrocytes is improved by treatment with growth hormone.

GH acts directly on isolated thymocytes. The uptake and incorporation of radioactive uridine in these cells is promoted by the hormone. With the exception of prolactin, other known pituitary hormones are inactive in this assay. The stimulation is also lost by thermal inactivation of the hormone.

The receptors for this hormone appear to be on the cell membrane. Not all cells from the rat thymus interact with the hormone. In young, 5-week-old, rats, a subpopulation consisting of 65% of cells interacts with the hormone. This sub-population progressively decreases with age, as does the hormonal stimulation of uridine incorporation. These observations are relevant to the age involution of the organ.

Studies on human peripheral blood lymphocytes indicate that few resting lymphocytes interact with growth hormone. On the other hand, lymphocytes cultured with phytohaemagglutinin on a peak day of thymidine incorporation show markedly enhanced reactivity with growth hormone, judged from shifts in the electrophoretic mobility and surface charge of hormone-treated cells.

REFERENCES

Baroni, C. (1967): *Experientia (Basel), 23,* 282.
Duquesnoy, R.J., Kalpaktsoglon, P.K. and Good, R.A. (1970): *Proc. Soc. exp. Biol. (N.Y.), 133,* 201.
Fain, J.N. and Saperstein, R. (1970): *Hormone metab. Res., 2,* Suppl. 2, 20.
Feldman, J.D. (1951): *Anat. Rec., 110,* 17.
Goodman, H.M. (1968): *Ann. N.Y. Acad. Sci., 148,* 419.
Greenberg, L.J. and Yunis, E.J. (1972): *Gerontologia, 18,* 247.
Hanjan, S.N.S. and Talwar, G.P. (1975a): *J. Immunol., 114,* 55.
Hanjan, S.N.S. and Talwar, G.P. (1975b): *Molec. cell. Endocr., 3,* 185.
Jha, P., Talwar, G.P. and Hingorani, V. (1975): *Amer. J. Obstet. Gynaec., 122,* 965.
Kostyo, J.L., Hotchkiss, J. and Knobil, E. (1959): *Science, 130,* 1653.
Krishnaraj, R. and Talwar, G.P. (1973): *J. Immunol., 111,* 1010.
Lesniak, M.A., Roth, J., Gorden, P. and Gavin, J.R. (1973): *Nature New Biol., 241,* 20.
Li, C.H. (1972): In: *Lactogenic Hormones,* p. 7. Editors: G.E.W. Wolstenholme and J. Knight. Churchill-Livingstone, London.
Mehra, V.L., Talwar, G.P., Balakrishnan, K. and Bhutani, L.K. (1972): *Clin. exp. Immunol., 12,* 205.
Niall, H.D., Hagan, M.L., Tregear, G.W., Segre, G.V., Hwang, P. and Friesen, H. (1973): *Recent Progr. Hormone Res., 29,* 387.
Pandian, M.R., Gupta, S.L. and Talwar, G.P. (1971): *Endocrinology, 88,* 928.
Pandian, M.R. and Talwar, G.P. (1971): *J. exp. Med., 134,* 1095.
Pandian, M.R., Gupta, P.D., Talwar, G.P. and Avrameas, S. (1975): *Acta endocr. (Kbh.), 78,* 781.

Pierpaoli, W. and Sorkin, E. (1967): *Nature (Lond.), 215,* 834.
Pierpaoli, W. and Sorkin, E. (1968): *J. Immunol., 101,* 1036.
Pierpaoli, W., Fabris, N. and Sorkin, E. (1970): In: *Hormones and Immune Response,* p. 126. Editors: G.E.W. Wolstenholme and J. Knight. Churchill-Livingstone, London.
Ruhenstroth-Bauer, G. and Luke-Huhle, C. (1968): *J. Cell Biol., 37,* 196.
Saxena, R.K. and Talwar, G.P. (1974): *Acta endocr. (Kbh.), 78,* 248.
Smith, P.E. (1930): *Anat. Rec., 47,* 119.
Talwar, G.P. (1972): *Int. J. Biochem., 3,* 39.
Talwar, G.P., Kumar, N., Pandian, M.R. and Gupta, P.D. (1974): *Molec. cell. Endocr., 1,* 209.
Talwar, G.P., Pandian, M.R., Nirbhay Kumar, Hanjan, S.N.S., Saxena, R.K., Krishnaraj, R. and Gupta, S.L. (1975a): *Recent Progr. Hormone Res., 31,* 141.
Talwar, G.P., Hanjan, S.N.S., Saxena, R.K., Pandian, M.R., Gupta, P.D. and Bhattarai, Q.B. (1975b): In: *Regulation of Growth and Differentiated Functions in Eukaryotic Cells,* p. 271. Editor: G.P. Talwar. Raven Press, New York.
Tanner, J.M. (1972): *Nature (Lond.), 237,* 433.
Wallach, D.F.H. and Esandi, M.V.P. (1964): *Biochim. biophys. Acta (Amst.), 83,* 363.

GROWTH HORMONE AND INSULIN SECRETION*

LESLIE L. BENNETT and DONALD L. CURRY

Department of Physiology and Hormone Research Laboratory, University of California, San Francisco, and Department of Physiological Sciences, University of California, Davis, Calif., U.S.A.

In the early 30's almost simultaneously Houssay et al. (1932) and Evans et al. (1932) demonstrated that the injection of crude anterior hypophyseal extracts into normal dogs had a diabetogenic effect producing hyperglycemia and glycosuria, i.e., diabetes mellitus. Surprisingly, one of the dogs injected by Evans et al. (1932) continued its diabetic state following cessation of treatment by anterior lobe extract. This persistent diabetes mellitus could only be explained by assuming there must be some impairment of the ability of the pancreas to secrete insulin which persisted following treatment. Young (1937) also reported that the treatment of normal dogs with crude anterior lobe extracts would produce hyperglycemia and glycosuria and that if the treatment was continued for a sufficient period of time the diabetic state would reproducibly persist following cessation of treatment. These observations were confirmed and extended by Houssay et al. (1942) who not only produced a permanent state of diabetes (metahypophyseal diabetes) but also demonstrated that the pancreas in dogs with metahypophyseal diabetes had reduced ability or no ability to secrete insulin. Following the isolation and purification of growth hormone (GH) Cotes et al. (1949) demonstrated its diabetogenic action in intact cats, Houssay and Anderson (1949) demonstrated its diabetogenic effect in partially depancreatized dogs and Campbell et al. (1950) produced permanent diabetes in dogs. These observations all supported the earlier hypothesis that GH was the active component in the crude extracts.

Injection of crude pituitary extracts into normal rats never reproduced the effect seen in normal adult dogs, i.e., hyperglycemia, glycosuria and the state of metahypophyseal diabetes. Young (1941) pointed out that GH was diabetogenic only in the adult cat or dog, not in the growing kitten or pup. The growing young animal may have the ability to increase its insulin-producing tissue whereas the adult does not. The difference in response of dogs and rats might reside in the fact that the rat could increase its number of islets whereas the dog could not. Such an increase in islet number was reported over 40 years ago by Anselmino et al. (1933). Following the isolation of GH, Bennett and Li (1947) reported the administration of purified bovine growth hormone (BGH) to rats with alloxan diabetes was inconsistent in producing an increase in glycosuria. In fact, the time course of the glycosuria showed many animals had only an initial increase followd by a decline in glycosuria which persisted throughout the period of treatment. Upon cessation of treatment the glycosuria again increased to

---

* These studies were supported in part by NIH grant AM 17668, the American Diabetes Association and the American Cancer Society.

essentially pretreatment levels. Bennett (1948) reported that BGH did not increase the glycosuria of hypophysectomized diabetic rats (alloxan diabetes). Subsequently, Batts et al. (1956) using partially depancreatized rats demonstrated that this decline in glycosuria upon treatment with BGH was associated with cytological changes in the beta cells of the remaining pancreatic islets. These cytological changes took the form of enlargement of beta cells, disappearance of pyknotic nuclei and the appearance of new clusters of beta cells that seem to be regenerating cells. These authors concluded that an effect of BGH was to enable the pancreas of the rat to respond by secreting more insulin in response to hyperglycemia than it was able to secrete prior to the BGH treatment. Following the development of immunochemical assay for insulin and the ready availability of highly purified GH many investigators again approached the general question of whether GH in particular might exert what could be called pancreatropic effect on the islets of Langerhans. Typical types of experiments are those of Campbell and Rastogi (1966) who demonstrated that normal dogs treated with BGH and then given a standard challenge by glucose developed higher immunoassayable blood insulin levels than did control dogs. Comparable experiments were done in humans by Luft and Cerasi (1964) who reported studies in 3 patients with panhypopituitarism whom they treated with human growth hormone (HGH). They report both higher immunoassayable blood insulin levels while receiving HGH treatment and an exaggerated insulin response to a glucose challenge. The ability of GH to alter insulin secretion by the rat pancreas has been studied by many investigators in in vitro systems utilizing both isolated pancreatic islets and fragments of pancreas. For example, Martin and Gagliardino (1967) working with isolated pancreatic islets demonstrated that islets from hypophysectomized rats release less insulin in response to glucose stimulus than islets from normal rats. They also demonstrated that pretreatment with GH would partially restore the ability of isolated islets to secrete insulin. The same investigators also demonstrated that islets from hypophysectomized rats had reduced ability to incorporate radioactively labeled leucine into newly synthesized insulin and that this was fully corrected by pretreatment with GH. Similar results were reported by Malaisse et al. (1968) using pancreas fragments. In our laboratory we have primarily utilized the isolated perfused rat pancreas as the technique for assessing the ability of various hormonal manipulations to alter insulin secretion. This technique has previously been described in detail by Curry et al. (1968) as has the perfusion medium. All the data to be reported subsequently in this paper were obtained by the use of the isolated perfused rat or hamster pancreas preparation.

EFFECTS OF HYPOPHYSECTOMY, GROWTH HORMONE AND RELATED HORMONES

Typically, as demonstrated by Curry et al. (1968), the isolated perfused rat pancreas responds to a constant glucose stimulus in a diphasic pattern. This diphasic response consists of an early transient rapid stage of insulin release (first phase) succeeded by a fall in rate and then a slowly rising secondary release phase (second phase). Figure 1 illustrates the typical diphasic secretory response by the pancreases of 6 normal rats. Figure 2 shows the time course of insulin release by the pancreases of 5 hypophysectomized rats 5 weeks postoperative in response to the same glucose stimulus. Particularly striking is the reduction of the first phase of secretion, approximately 70% ($p < 0.001$). The second phase is also attenuated but only by about 35% ($p < 0.01$). Since the first phase occurs with such rapidity and as shown by Curry et al. (1968) is not abolished or reduced by inhibitors of protein synthesis, it is generally thought to repre-

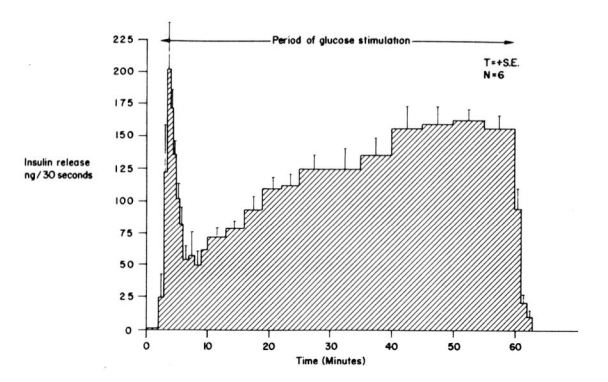

Fig. 1. Time course of insulin release by normal rat pancreases in response to 300 mg/dl glucose stimulus. (Reproduced from Curry and Bennett (1973), by courtesy of the Editors of *Endocrinology*.)

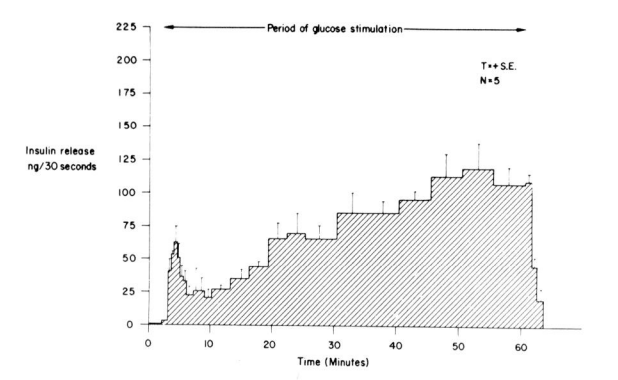

Fig. 2. Time course of insulin release by the pancreases of 5 weeks postoperative hypophysectomized rats in response to a 300 mg/dl glucose stimulus. (Reproduced from Curry and Bennett (1973), by courtesy of the Editors of *Endocrinology*.)

sent release from beta cells of granules stored adjacent to the cell membrane. The reduction of the first phase response following hypophysectomy could result either from a reduction in the numbers of granules stored adjacent to the beta cell membrane or result from some alteration of the secretory process per se. In Figure 3 is shown an electronphotomicrograph of an islet from a hypophysectomized rat pancreas. Clearly seen are numerous storage granules adjacent to the beta cell membrane which are not reduced in number when compared to those seen by Lee et al. (1970) in the normal rat pancreas. Hence, the first hypothesis above, namely that the reduction of the first phase results from a reduction in the numbers of granules stored adjacent to the beta cell membrane, cannot be the correct hypothesis, and the second is the more likely.

Treatment of hypophysectomized rats for 15 days with 500 µg of BGH, as shown by Curry and Bennett (1973), completely restores the second phase of secretion but is without effect on the first phase.

It has been demonstrated by Curry et al. (1968) that the second phase of insulin

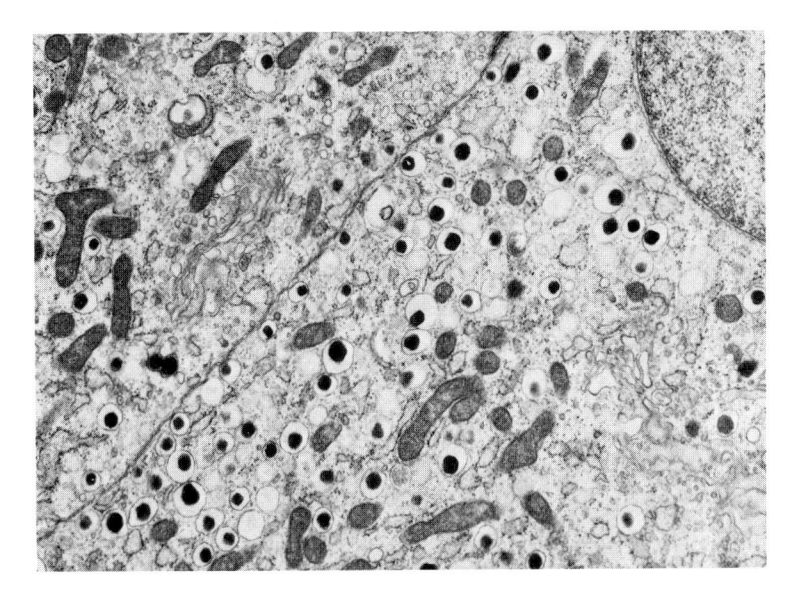

Fig. 3. An electronphotomicrograph (× 2850) of a beta cell from a fasted hypophysectomized rat pancreas. A small portion of the beta cell nucleus is shown in the upper right hand corner and numerous storage granules are seen adjacent to the beta cell membrane.

release is reduced by addition to the perfusion medium of puromycin or other inhibitors of protein synthesis. Thus, the second phase of insulin release is dependent upon protein synthetic mechanisms but this does not necessarily mean that newly synthesized insulin must be appearing as a component of the secreted insulin, as shown by Sando and Grodsky (1973). Since BGH has a striking effect in restoring the second phase of insulin secretion by the pancreas of hypophysectomized rats one is led to the conclusion that GH improves the ability of the pancreas to synthesize insulin in response to glucose stimulus. These findings and conclusions are in accord with the results of Martin and Gagliardino (1967) which were previously mentioned.

It is important to note that all of these experiments involved the use of *fasting* rats. Curry and Bennett (unpublished observations) have not been able to demonstrate comparable effects of hypophysectomy or of GH treatment using *unfasted* rats.

We have also reported experiments in which GH was administered not to the hypophysectomized animal but was introduced only into the perfusion medium. In these experiments by Curry et al. (1974) a perfusion medium containing 7.5 μg/ml of BGH was recycled through the hamster pancreas preparations for 1 hour prior to stimulation by glucose. The total insulin released by the control pancreases during the first phase (minutes 2-10) was 1.08 ± 0.19 μg (n = 5) as compared to 2.48 ± 0.56 μg (n = 5) from the pancreases through which the BGH was recycled (p < 0.05). For the second phase (minutes 30-60) the total insulin release was 2.99 ± 0.64 μg and 6.77 ± 0.86 μg respectively for the control pancreases and the pancreases exposed to BGH (p < 0.01).

In Table 1 are shown comparable data of insulin secretion by pancreases from normal rats which were exposed to BGH at a concentration of 10 μg/ml for a 1-hour recycling period prior to glucose stimulation. In these experiments 2 different levels

TABLE 1

Total insulin release in µg by the isolated perfused rat pancreas through which BGH had been recycled for 1 hour

	First phase (min 2-6)	p	Second phase (min 25-40)	p
Controls* (5)[+]	0.090 ± 0.02[++]		0.69 ± 0.24	
Recycled* (5)	0.12 ± 0.02	NS	1.26 ± 0.19	NS
Controls** (5)	0.22 ± 0.03		2.33 ± 0.11	
Recycled ** (5)	0.27 ± 0.03	NS	3.48 ± 0.31	< 0.01

[+] number of animals in the group; * glucose 200 mg/dl; ** glucose 300 mg/dl; [++] µg of insulin ± standard error.

of glucose stimulation were used, 200 mg/dl and 300 mg/dl. It is quite apparent that the pancreases exposed to BGH during the 1 hour of recycling released more insulin both during the first and second phases at both glucose concentrations. The magnitude of increase in insulin secretion during the first phase by the pancreases exposed to BGH is relatively small and at neither glucose concentration is significant. This is in agreement with the data reported above regarding the lack of effect of BGH upon the first phase secretion by pancreases of hypophysectomized rats. The magnitude of increase in insulin secretion during the second phase is much greater and at the higher glucose concentration it is statistically significant to the < 0.01 level. One can only speculate on the mechanism of action of BGH in these experiments. It is possible that growth hormone increases the ability of the beta cells to synthesize insulin in response to glucose stimulus or it may enhance transport to the cell periphery from which it is secreted. Another mechanism which may be operating perhaps in combination with one or both of the above is that GH could increase the number of glucose receptor sites on beta cell membranes. Such receptor sites have been postulated by Matschinsky et al. (1971). Regardless of the mechanism of action it is clear that it cannot be through conversion by liver or kidney to somatomedin or sulfation factors.

The ability of human growth hormone (HGH), ovine prolactin (OP), human chorionic somatomammotropin (HCS) and plasmin modified HGH (PL-HGH) have been compared in their abilities to alter the secretory response by the hypophysectomized rat pancreas. These hormones were all made available through the generosity of Dr. C.H. Li and were prepared in this laboratory by the following methods: HCS by the method of Li (1970), HGH by the method of Li et al. (1962), OP by the method of Li et al. (1970) and PL-HGH by the method of Li and Gráf (1974). The final product of PL-HGH was then further purified by exclusion chromatography on Sephadex G-100. In all of these experiments hypophysectomized rats were roughly 4 weeks postoperative at the time of perfusion and the 4 hormones were given at doses of 200 µg/day for the final 6 days with the last injection being the morning of perfusion, i.e., during the approximate middle of the 20- to 24-hour fast.

These experiments were of somewhat shorter duration than the experiments shown in Figures 1 and 2 and the data are summarized in Table 2. These data clearly show that HGH, as is true for BGH, increases the second phase of insulin secretion ($p < 0.05$). Although there is an increase in the first phase it is not statistically significant. PL-HGH likewise produces a significant increase in the second phase of secretion and an increase in the first phase which although numerically greater than that produced by HGH still is not significantly different from the first phase secretion by the

TABLE 2

Total insulin release in µg by the isolated perfused pancreas of hypx rats treated with HGH and related hormones

	First phase (min 2-6)	p	Second phase (min 25-40)	p
Hypx controls (7)[+]	0.13 ± 0.03**		1.86 ± 0.17	
HGH* (7)	0.23 ± 0.04	NS	2.62 ± 0.29	<0.05
PL-HGH* (4)	0.32 ± 0.08	NS	3.31 ± 0.44	<0.02
HCS* (8)	0.26 ± 0.05	<0.05	3.38 ± 0.35	<0.01
OP* (4)	0.17 ± 0.03	NS	2.41 ± 0.45	NS

[+] number of animals in the group; * all hormones at 200 µg/day for 6 days; ** µg of insulin ± standard error.

hypophysectomized controls. In a comparable fashion HCS increases in the second phase of insulin secretion significantly ($p < 0.01$); in contrast prolactin has no significant effect on either the first or second phases of secretion.

The animals treated with HGH and PL-HGH gained approximately 5 g of body weight/day of treatment while the animals treated with HCS and OP lost approximately 1 g/day. Thus, the growth-promoting activity of the hormones does not correlate with their ability to produce additional insulin secretion. On the other hand, the amino acid sequence reported by Li (1972) for HGH (a good growth promoter) and the sequence reported by Li et al. (1973) for HCS (a poor growth promoter) show 96% homology. Both HCS and OP are potent lactogenic hormones but only the one with close homology to HGH, i.e., HCS, promotes insulin secretion.

EFFECTS OF SOMATOSTATIN AND CALCIUM ION (Ca^{2+})

Somatostatin has been shown by Alberti et al. (1973) to inhibit both the first and second phases of insulin secretion. We have carried out dose-response relationship studies for somatostatin for these 2 phases and have observed that there is a marked difference in the sensitivity of these 2 phases to somatostatin inhibition. The somatostatin used was the synthetic cyclic compound synthesized by Yamashiro and Li (1973) by the solid state method. When we investigated the dose-response relationship for somatostatin inhibition of the first phase we did experiments involving double square wave pulses of glucose, to produce a glucose concentration of 300 mg/dl perfusate, separated by a rest period of 10 minutes. The duration of glucose pulse used was 3 minutes and when somatostatin was present in the infusion medium its presence preceded the beginning of the glucose pulse by 2 minutes and continued for 1 minute beyond the glucose pulse. For each concentration of somatostatin there were 4 preparations and each preparation served as its own control, having one glucose pulse with somatostatin present and one glucose pulse without somatostatin being present. Successive stimulations by short glucose pulses do not necessarily produce identical amounts of insulin release as shown by Curry (1971) but rather the amount of insulin release may decline with successive pulses. The sequence of glucose alone first, or glucose combined with somatostatin first, was altered from one preparation to the next to randomize this potential source of variation. When we investigated the dose-response characteristics of somatostatin inhibition in the second phase of insulin

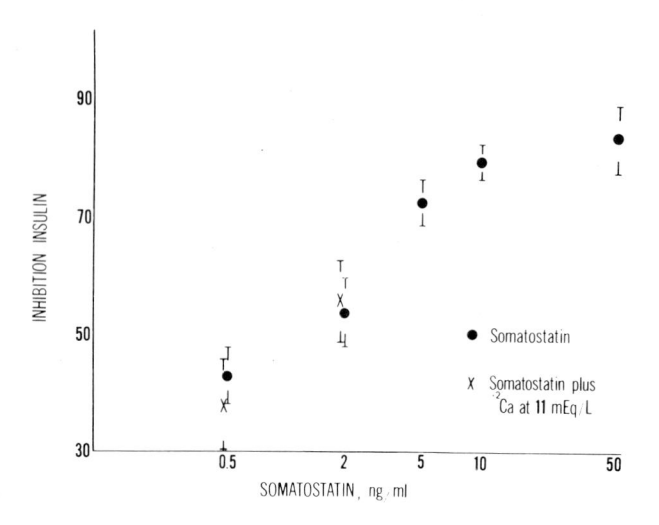

Fig. 4. Log somatostatin concentration vs. percent inhibition of first phase insulin release. Ca^{2+} elevation to 11 mEq/l at the 2 lowest somatostatin concentrations shows no reversal of inhibition.

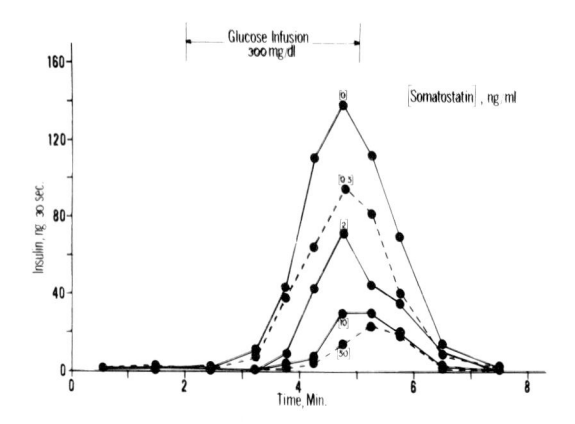

Fig. 5. Time course of insulin release during somatostatin inhibition of the first phase secretion. The control curve is the average of 24 preparations; 4 preparations were used at each somatostatin concentration.

secretion, we superimposed an infusion of somatostatin in the middle of a long stimulus by glucose. In Figure 4 is shown the dose-response curve for somatostatin inhibition of the first phase of insulin release. These data are presented as percent inhibition of insulin release during the first phase by somatostatin concentrations ranging from 50 to 0.5 ng/ml. One should particularly note that 2 ng/ml produces approximately 55% inhibition of the first phase and that 5 ng/ml produces approximately 75% inhibition.

In Figure 5 is shown the time course of insulin release for the data given in Figure 4. In addition this graph shows the mean insulin release in response to glucose alone for

GH and insulin

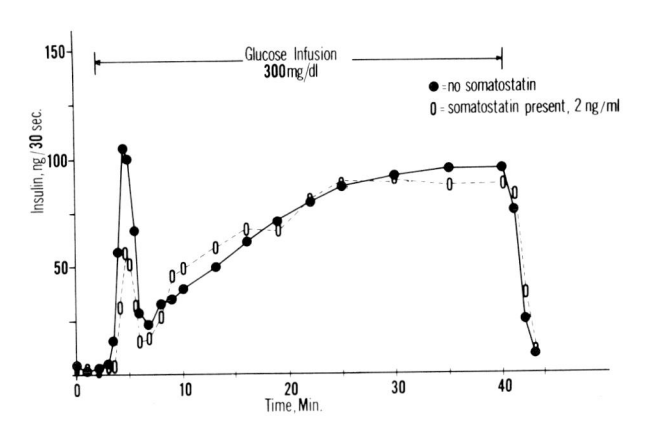

Fig. 6. Time course of insulin release with somatostatin present in the medium both before and during the entire period of stimulation. There were 3 preparations in each group.

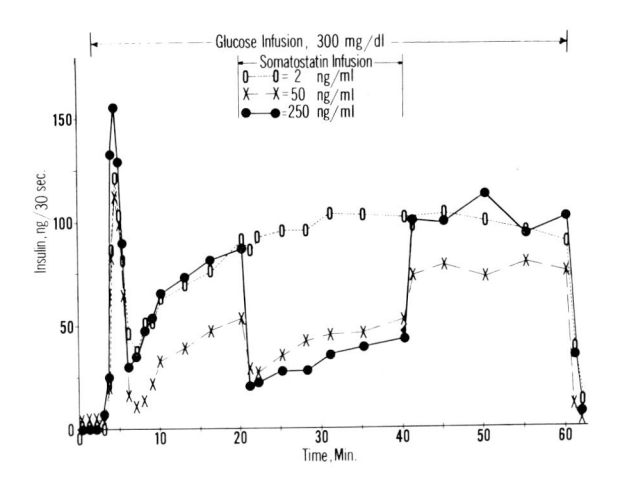

Fig. 7. Time course of insulin release with somatostatin infused during the middle period of a 60-minute glucose stimulus. Each curve represents the average insulin release by 3 different pancreas preparations.

all control perfusions (n = 24). These data show the dose-response effect as in Figure 4, but more importantly they demonstrate that with increasing concentrations of somatostatin not only is the initial release of insulin delayed but also the peak release occurs later in time. Since these experiments were done by having somatostatin present prior to the introduction of glucose, the results are consistent with the hypothesis that somatostatin and glucose are competing for the same membrane receptor site from which somatostatin is displaced by glucose before the secretory response can occur.

Figure 6 shows data from experiments in which somatostatin was present in the perfusate both during the equilibration period and throughout the remainder of the experiment. Striking is the fact that 2 ng/ml of somatostatin in the perfusate

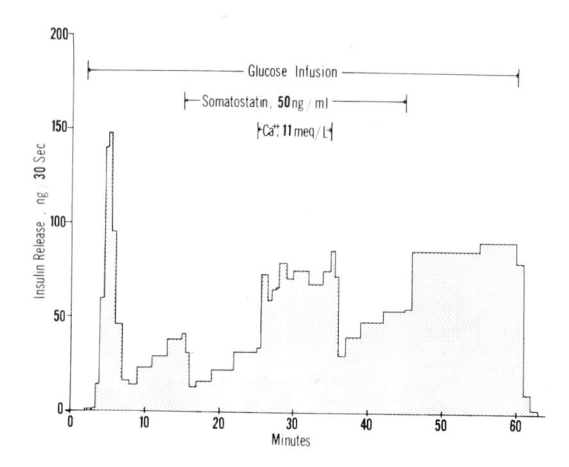

Fig. 8. Time course of insulin release by 2 rat pancreas preparations in response to a constant glucose infusion showing somatostatin inhibition and reversal of somatostatin inhibition by elevation of Ca^{2+} to 11 mEq/l during the middle of the somatostatin infusion. (Reproduced from Curry and Bennett (1974), by courtesy of the Editors of *Biochemical and Biophysical Research Communications*.)

produces attenuation of the first phase that is quite comparable to that shown in Figures 4 and 5. On the other hand, somatostatin at 2 ng/ml is without effect upon insulin release during the second phase of secretion. Data from experiments in which somatostatin was introduced at concentrations of 2, 50 and 250 ng/ml in the middle of a sustained glucose stimulation period are given in Figure 7. It again is clear that somatostatin at 2 ng/ml does not inhibit the second phase of insulin release, and that somatostatin at 250 ng/ml produces about 75% inhibition, a degree of inhibition comparable to that produced by 5 ng/ml during the first phase. Inhibition here is calculated by comparing the amount of insulin released during somatostatin with the average release of the 10-minute periods immediately before and after. Thus the first phase insulin release is much more sensitive to somatostatin inhibition than is the second phase and the order of magnitude of this differential sensitivity is approximately 25- to 50-fold.

In Figure 8 is shown the effect of elevating calcium ion (Ca^{2+}) in the middle of a period of somatostatin inhibition. Clearly, elevation of Ca^{2+} accentuates insulin release and in a sense can be said to 'reverse' the somatostatin inhibition. It should be pointed out that if Ca^{2+} is elevated to the same extent in the middle of a glucose stimulation without somatostatin an increase in insulin release of comparable magnitude is seen, thus it can really be questioned whether this represents in any true sense antagonism of somatostatin inhibition per se or simply a potentiation of insulin secretion in response to glucose stimulation. On the other hand, if Ca^{2+} is elevated simultaneously with somatostatin infusion during the experiments in which the ability of somatostatin to inhibit the first phase was investigated no effect on insulin release was observed. These data also are shown in Figure 5 and demonstrate that the elevation of Ca^{2+} in the two lowest somatostatin concentrations does not in any way alter the degree of inhibition produced by the somatostatin.

The fact that there is a marked difference in sensitivity to somatostatin inhibition for the first and second phase with the first phase being more sensitive to somatostatin inhibition; and the fact that elevating Ca^{2+} is without effect upon insulin relase during somatostatin inhibition of the first phase whereas it 'reverses' the somatostatin inhibition of the second phase lead almost inescapably to the conclusion that somatostatin inhibits insulin secretion through 2 different mechanisms. These 2 mechanisms could well be: first, a somatostatin effect on the cell membrane involving the release process itself. This mechanism is remarkably sensitive to somatostatin and is not affected by elevating the Ca^{2+} concentration and perhaps involves competition between somatostatin and glucose for some receptor binding site. A second mechanism could be an effect of somatostatin upon the intracellular transport of granules to the cell surface for release. This mechanism is undoubtedly dependent upon intracellular Ca^{2+}. It is much less sensitive to somatostatin inhibition and this somatostatin inhibition can be 'reversed' (glucose stimulated insulin release potentiated?) by elevating Ca^{2+} above that normally present in the perfusate. Either of these possible mechanisms of action could also involve inhibition of some metabolic process involving or triggered by glucose.

SUMMARY

1. BGH in vivo restores the ability of the perfused hypophysectomized rat pancreas to secrete insulin (particularly the second phase of secretion) in response to a glucose stimulus. BGH recycled in vitro enhances the ability of perfused pancreases of normal rats and hamsters to secrete insulin in response to a glucose stimulus.
2. HGH and related compounds with a high degree of homology in amino acid sequences (HCS and PL-HGH) have comparable effects.
3. Somatostatin inhibits both the first and second phases of insulin secretion. The first phase is 25-50 times more sensitive to somatostatin inhibition than is the second phase.
4. Elevating Ca^{2+} does not antagonize somatostatin inhibition of the first phase but does antagonize somatostatin inhibition of the second phase (potentiates insulin secretion?). These observations (3) and (4) lead to the hypothesis that somatostatin has 2 mechanisms of action.

REFERENCES

Alberti, K.G.M., Christensen, N.J., Christensen, S.E., Hansen, A.P., Iversen, J.I., Lundbaek, K., Seyer-Hansen, K. and Orskov, H. (1973): *Lancet, 66,* 1299.
Anselmino, K.J., Herold, L. and Hoffmann, F. (1933): *Klin. Wschr., 12,* 1245.
Batts, A.A., Bennett, L.L., Ellis, S. and George, R.A. (1956): *Endocrinology, 59,* 620.
Bennett, L.L. (1948): *Amer. J. Physiol., 155,* 24.
Bennett, L.L. and Li, C.H. (1947): *Amer. J. Physiol., 150,* 400.
Campbell, J. and Rastogi, K.S. (1966): *Diabetes, 15,* 30.
Campbell, J., Davidson, I.W.F. and Lei, H.P. (1950): *Endocrinology, 46,* 488.
Cotes, P.M., Reid, E. and Young, F.G. (1949): *Nature (Lond.), 164,* 992.
Curry, D.L. (1971): *Amer. J. Physiol., 220,* 319.
Curry, D.L. and Bennett, L.L. (1973): *Endocrinology, 93,* 602.
Curry, D.L. and Bennett, L.L. (1974): *Biochem. biophys. Res. Commun., 60,* 1015.
Curry, D.L., Bennett, L.L. and Grodsky, G.M. (1968): *Endocrinology, 83,* 572.
Curry, D.L., Bennett, L.L. and Li, C.H. (1974): *Biochem. biophys. Res. Commun., 58,* 885.
Evans, H.M., Meyer, K., Simpson, M.E. and Reichert, F.L. (1932): *Proc. Soc. exp. Biol. (N.Y.), 29,* 857.

Houssay, B.A. and Anderson, E. (1949): *Endocrinology, 45,* 627.
Houssay, B.A., Biasotti, A. and Rietti, C.T. (1932): *Rev. Soc. argent. Biol., 8,* 469.
Houssay, B.A., Foglia, V.G., Smyth, F.S., Rietti, C.T. and Houssay, A.B. (1942): *J. exp. Med., 75,* 547.
Lee, J.C., Grodsky, G.M., Bennett, L.L., Smith-Kyle, D.F. and Craw, L. (1970): *Diabetologia, 6,* 542.
Li, C.H. (1970): *Ann. Sclavo, 12,* 651.
Li, C.H. (1972): *Proc. Amer. philos. Ass., 116,* 365.
Li, C.H., Dixon, J.S. and Chung, D. (1973): *Arch. Biochem., 155,* 95.
Li, C.H., Dixon, J.S., Lo, T-B., Schmidt, K.D. and Pankov, Y.A. (1970): *Arch. Biochem., 141,* 705.
Li, C.H. and Gráf, L. (1974): *Proc. nat. Acad. Sci. (Wash.), 71,* 1197.
Li, C.H., Liu, W-K. and Dixon, J.S. (1962): *Arch. Biochem., Suppl., 1,* 327.
Luft, R. and Cerasi, E. (1964): *Lancet, 2,* 124.
Malaisse, W.J., Malaisse-Lagae, F., King, S. and Wright, P.H. (1968): *Amer. J. Physiol., 215,* 423.
Martin, J.M. and Gagliardino, J.J. (1967): *Nature (Lond.), 213,* 630.
Matschinsky, F.M., Ellerman, J.E., Krsanowski, J., Kotter-Brajbury, J., Landgroft, R. and Fertel, R. (1971): *J. biol. Chem., 243,* 1007.
Sando, H. and Grodsky, G.M. (1973): *Diabetes, 22,* 354.
Yamashiro, D. and Li, C.H. (1973): *Biochem. biophys. Res. Commun., 54,* 882.
Young, F.G. (1937): *Lancet, 2,* 372.
Young, F.G. (1941): *Brit. med. J., 2,* 897.

RADIORECEPTOR ASSAY OF PLASMA NSILA-s IN MAN: BASAL AND STIMULATED LEVELS IN NORMAL AND PATHOLOGIC STATES

KLARA MEGYESI*, C. RONALD KAHN, JESSE ROTH, PHILLIP GORDEN and DAVID M. NEVILLE Jr**

Diabetes Branch, National Institute of Arthritis, Metabolism, and Digestive Diseases, National Institutes of Health, Bethesda, Md., U.S.A.

INSULIN-LIKE ACTIVITY OF PLASMA

The total insulin-like activity of plasma in the post-absorbative state is about 200 μU/ml as measured by in vitro bioassays. Only about 10% of this insulin-like activity is due to insulin and proinsulin; these components are measured as immunoreactive insulin and their bioactivities are *suppressed* by anti-insulin antibodies. Most of the insulin-like activity of plasma *is not suppressed* by anti-insulin antibodies and is referred to as non-suppressible insulin-like activity, or NSILA. NSILA itself consists of at least 2 components: NSILA-s which is soluble in acid ethanol, and NSILA-p which is precipitated in acid ethanol. Different authors used different terminology for insulin-like materials in blood (Table 1).

We studied NSILA-s which by definition is soluble in acid ethanol and does not react with anti-insulin antibodies; it has a molecular weight of 7400, has no interchain disulfide bonds and its spectrum of biological activity is similar to insulin:

In vitro: It stimulates glucose utilization (less potent), stimulates cell growth (more potent). It acts as a growth factor by stimulating DNA synthesis in mouse, chick embryo and human fibroblasts and ^{35}S incorporation into rat and chicken cartilage.

In vivo: NSILA-s produces hypoglycemia (longer duration than by insulin). These studies together with the purification of NSILA-s from human plasma and the characterization of chemical properties and biological actions of this peptide were done by Froesch et al. (1967). Their generosity by giving us some NSILA-s helped us to contribute a little to this puzzling and exciting field of small peptides with growth-promoting activity.

TABLE 1
Terminology

	SILA	NSILA
Froesch		
Samaan	Typical insulin	Atypical insulin
Poffenbarger	Inter α	β-globulin associated
Antoniades	└── Free ──────────┘ └── Bound ──────┘

* Guest Worker on leave from the Semmelweis Medical University, Budapest, Hungary.
** Section of Biophysical Chemistry, Laboratory of Neurochemistry, National Institute of Mental Health, National Institutes of Health, Bethesda, Md.

INSULIN RECEPTORS AND NSILA-s RECEPTORS

The first step in the action of polypeptide hormones is binding of the hormone to specific receptor sites on the cell. In previous studies (Freychet et al., 1971) we have shown that ^{125}I-insulin binding to its receptors was inhibited by unlabeled insulin analogues (Fig. 1). Porcine insulin, porcine proinsulin and guinea-pig insulin inhibit the binding of ^{125}I-insulin in direct proportion to their bioactivities. Likewise a preparation of NSILA-s which has 1/300th of the biological potency of insulin was 1/300th as active as insulin.

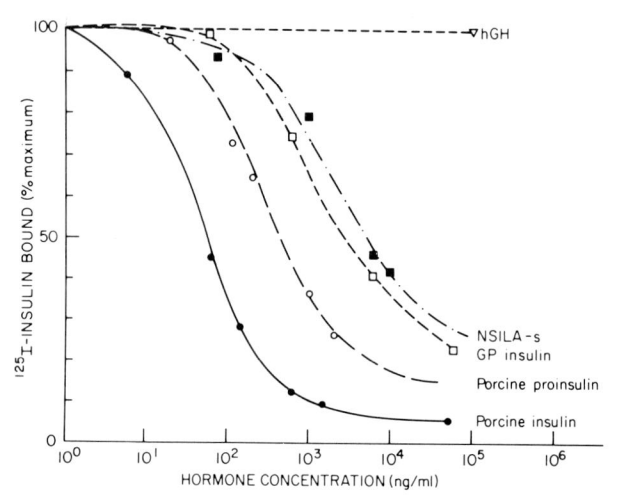

Fig. 1. ^{125}I-insulin binding to liver. The ^{125}I-insulin bound to purified liver plasma membranes, expressed as percent of maximum, is plotted as a function of unlabeled hormone concentration in a nanogram per milliliter (Freychet et al., 1971).

In 1974 we reported a radioreceptor assay for NSILA-s (Megyesi et al., 1974a). We iodinated NSILA-s with ^{125}I using the modification of chloramine T method (Roth, 1973), and bound the labeled hormone to purified rat liver plasma membrane (Neville, 1968), in the absence or presence of increasing amounts of unlabeled NSILA-s. Figure 2 shows the binding of ^{125}I-NSILA-s to *its* receptors on liver. In contrast to the pre-

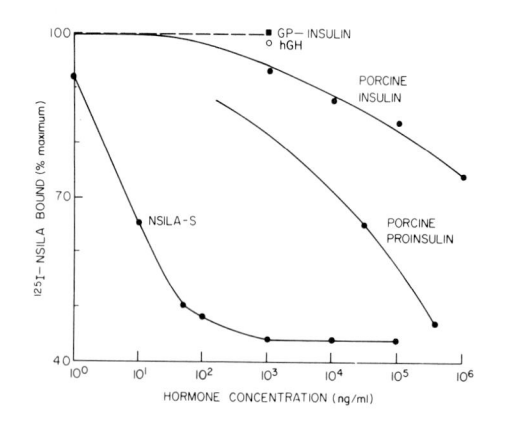

Fig. 2. ^{125}I-NSILA-s binding to liver. ^{125}I-NSILA-s was incubated with highly purified rat liver plasma membranes with or without unlabeled peptide in a total volume of 150 μl. After 90 minutes at 20° receptor-bound radioactivity was sedimented (microfuge, 1 minute) and counted (Megyesi et al., 1974b).

vious experiment, unlabeled NSILA-s is very potent in displacing the tracer, with some displacement even at 1 ng/ml. Proinsulin and insulin are only 1/1000 and 1/300,000 as potent as NSILA-s in displacing the labeled hormone, while guinea-pig insulin and growth hormone, at concentrations up to 1 μg/ml are unreactive. These data suggest that in liver insulin and NSILA-s bind to 2 types of receptors but with markedly different affinities. The insulin receptor has a high affinity for insulin, and low affinity for NSILA-s, while the NSILA-s receptor has a high affinity for NSILA-s and low affinity for insulin. Similar findings had been reported previously by Hintz et al. (1972) for somatomedin.

To further characterize the specificity of the NSILA-s receptor, 7 preparations of NSILA-s which varied over 70-fold in biological potency were tested (Fig. 3). These preparations inhibited ^{125}I-NSILA-s binding to liver membranes in order of their biological potencies measured by fat pad bioassay. Thus, the 70 mU/mg preparation of NSILA-s produced 50% displacement of the specifically bound ^{125}I-NSILA-s at a concentration of 3-5 ng/ml, while the 31 mU/mg preparation required 10 ng/ml, the 21 mU/mg preparation 100 ng/ml and so forth. NSILA-s which had been biologically inactivated by aminoaethyletion produced only minimal displacement of the tracer even at concentrations of 10 μg/ml (Megyesi et al., 1975a).

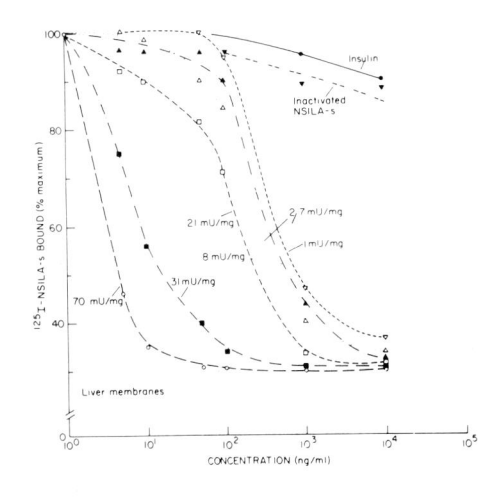

Fig. 3. Effect of NSILA-s preparations on ^{125}I-NSILA-s binding to its receptor on liver. ^{125}I-NSILA-s was incubated with rat liver membrane in the presence of the indicated concentrations of various NSILA-s preparations. The percent ^{125}I-NSILA-s bound was determined and the data plotted as percent of maximal binding as a function of unlabeled hormone concentration.

BINDING OF ^{125}I-NSILA-s TO PLASMA COMPONENTS

In an attempt to develop a radioreceptor assay for NSILA-s in plasma, we studied the behavior of ^{125}I-NSILA-s on gel filtration. When ^{125}I-NSILA-s was filtered on Sephadex G-50 at pH 7.5 in the absence of plasma the majority of radioactivity was recovered as a peak that corresponded to an apparent molecular weight of 7400. When ^{125}I-NSILA-s was added to plasma and filtered at pH 7.5 the radioactivity was recovered in 1 or 2 peaks: one in its normal elution volume and one in the region of the void volume. This suggested that plasma contained substances of high molecular weight that bound NSILA-s at neutral pH (Megyesi et al., 1974b). When ^{125}I-NSILA-s in plasma was brought to pH 2.3 and filtered at that pH, the effluent radioactivity was restored to its original position and clearly separated from the majority of plasma proteins (Fig. 4, bottom panel).

TABLE 2

Tumors associated with fasting hypoglycemia

Tumors	Hypoglycemia	Inappropriate elevation of plasma insulin	Elevation of plasma NSILA-s
Islet cell	Yes	Yes	No

Non-islet cell
  Mesenchymal
  Hepatoma
  Adrenocortical
    carcinoma
  Other
	Yes	No	Yes

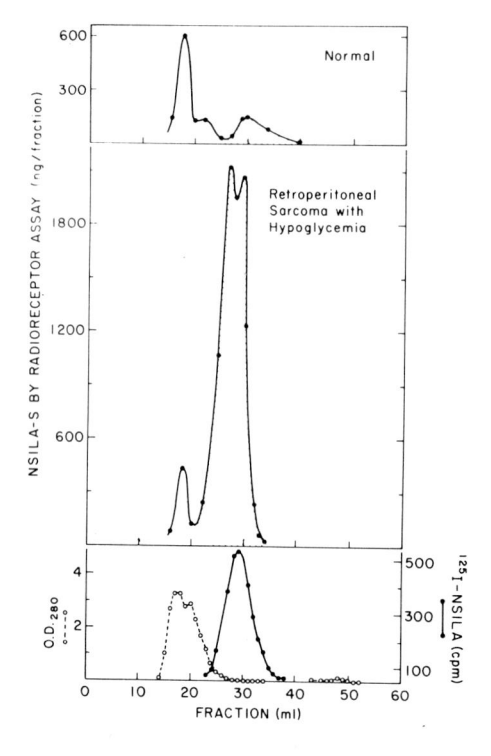

Fig. 4. Gel filtration pattern of NSILA-s in plasma. At acid pH from a normal subject (upper panel) and from a retroperitoneal sarcoma patient with hypoglycemia (middle panel). The hatched area denotes the region of NSILA-s radioreceptor activity used to calculate the concentration of NSILA-s in plasma. Elution position of ^{125}I-NSILA-s marker and optical density of plasma at 280 mµ seen on the bottom panel (Megyesi et al., 1974b).

HYPOGLYCEMIA WITH NON-ISLET CELL TUMORS

Table 2 shows certain features of tumors associated with fasting hypoglycemia. The etiology of the hypoglycemia associated with non-islet cell tumors has been obscure. Previous studies have suggested excessive glucose utilization by the tumor or secretion of some insulin-like substance as measured in non-specific bioassay. Using our specific radioreceptor assay we demonstrated that the hypoglycemia in some patients with non-islet cell tumors is related to an elevation in plasma level of NSILA-s (Table 2).

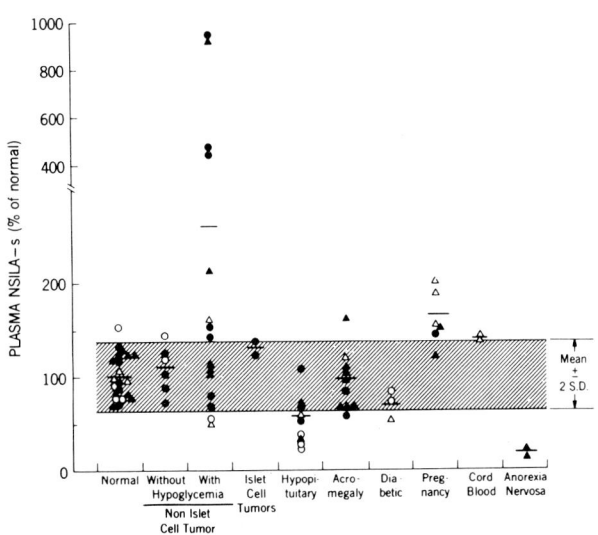

Fig. 5. Plasma concentrations of NSILA-s in basal state. On the vertical axis is plotted the plasma NSILA-s concentration expressed as percent of normal. The shaded area represents the mean ± 2 standard deviation for the normal adults. The short horizontal lines represent the arithmetic mean for each group of patients.

Figure 4 shows the gel filtration pattern *in acid* of NSILA-s in plasma from a normal subject and from a patient with retroperitoneal sarcoma and hypoglycemia. When the plasmas were filtered, 1 ml fractions collected and neutralized and assayed by radioreceptor assay for NSILA-s, the activity eluted in 2 peaks, one of which corresponds to the void volume and the other, which coelutes with the NSILA-s marker (Fig. 4, upper panel). The radioreceptor activity in the void did not dilute in parallel with the NSILA-s standard and might represent binding proteins, NSILA-p or other substances. The activity which coeluted with the NSILA-s marker, shown by the shaded area, was considered to represent plasma NSILA-s and diluted in parallel with the standard over a 50-fold range. Note on the middle panel that the plasma from the patient with retroperitoneal sarcoma and hypoglycemia had 9-fold more NSILA-s than did the normal subject. We tested 76 plasmas after gel filtration in acid in the NSILA-s radioreceptor assay. The results are shown in Figure 5 (Megyesi et al., 1975*b*).

The third patient group has the extrapancreatic tumor patients *with* hypoglycemia. Four of 15 had markedly (4- to 9-fold) elevated NSILA-s concentration, 4 were just above the upper limit of normal, while the remainder were at or below normal. That suggests that the hypoglycemia with this syndrome has multiple etiologies, of which NSILA-s excess is only one. Patients with islet cell tumors had normal levels (4th group), so the increase in NSILA-s was not due to the hypoglycemia per se. Patients in the second group had extrapancreatic tumors of the same tissue type but *no* hypoglycemia and NSILA-s levels were normal.

In this study the only other group with clearly elevated NSILA-s levels were the pregnant patients. It is also interesting that the concentration of this peptide in the maternal circulation was somewhat higher than that in the cord blood. The source of the increased NSILA-s in pregnancy is unknown, as is its source in the normal subjects. We think that the quantity or quality of the binding proteins might play a

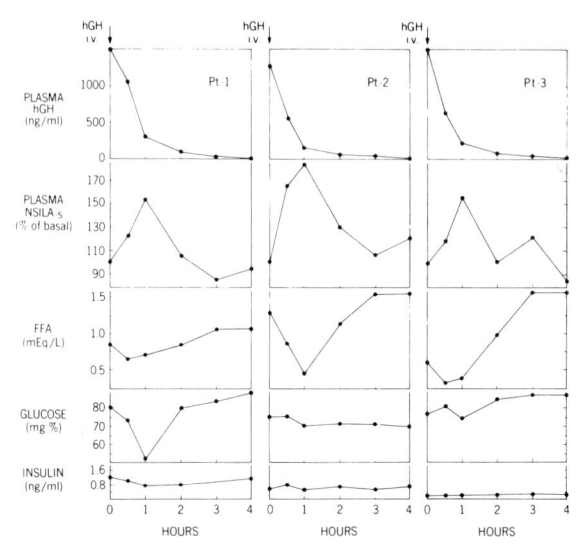

Fig. 6. Effects of intravenous growth hormone in three hypopituitary patients. Human growth hormone were given (4-8 mg) i.v. over a period of 5 minutes to 3 hypopituitary patients. Blood samples were drawn prior to and at time intervals following HGH administration for measurements of plasma HGH and insulin by radioimmunoassay, NSILA-s by radioreceptor assay, free fatty acids and glucose. The plasma NSILA-s is expressed as percent of basal. All plasmas from a single experiment were studied in a single assay.

role in the high levels of NSILA-s in pregnancy.

Hypopituitarism was associated with decreased levels of plasma NSILA-s averaging 55% of normal; this is similar to data obtained by Froesch et al. (1974) using a bioassay for NSILA-s and by Marshall et al. (1974) using a radioreceptor assay for somatomedin C.

In 3 hypopituitary patients with subnormal basal plasma NSILA-s values, intravenous human growth hormone (HGH) administration was followed by a rise in NSILA-s to normal levels at 1 hour (Fig. 6). In all 3 of our patients the rise in NSILA-s was coincident with the fall in blood glucose and/or free fatty acids, which suggests the possibility that NSILA-s is the factor responsible for the early insulin-like effects of HGH (Zahnd et al., 1960), i.e. antipolytic and hypoglycemic without insulin secretion. The patient with the significant fall in blood glucose also had pronounced adrenal glucocorticoid deficiency, which might have rendered him more 'NSILA-s sensitive'.

In our study out of 10 acromegalic patients only 1 had plasma NSILA-s above the normal range. That is in contrast to results of Marshall et al. (1974) who found elevated levels of somatomedin C in acromegaly.

A relationship of NSILA-s to the somatomedins has been suggested by a number of similarities (Table 3). Observations on point 9 by Frankel and Jenkins (1975) and by Phillips and Young (1975) respectively. The studies on somatomedin C were mostly done by Van Wyk et al. (1971) on somatomedin A, and B by Hall and Uthne (1972). Besides these similarities some possible differences exist between these peptides, like the rise after HGH which comes after 3 hours in somatomedin A, the high levels of somatomedins A and C in acromegaly and the lack of it in our study. This may suggest that the somatomedins are closely related but discrete peptides. The tendency to low

TABLE 3

Similarities between NSILA-s and somatomedin C

1. Source and steps of purification
2. Molecular weight
3. Isoelectric point (basic)
4. Insulin-like activity (metabolic and growth promoting)
5. Partial growth hormone dependency
6. Relation to other growth factors
7. Action on the insulin receptor
8. At neutral pH both are bound to larger molecules
9. In clinical studies both peptides were found low in anorexia nervosa. In acute diabetes of male rats the somatomedin levels were decreased

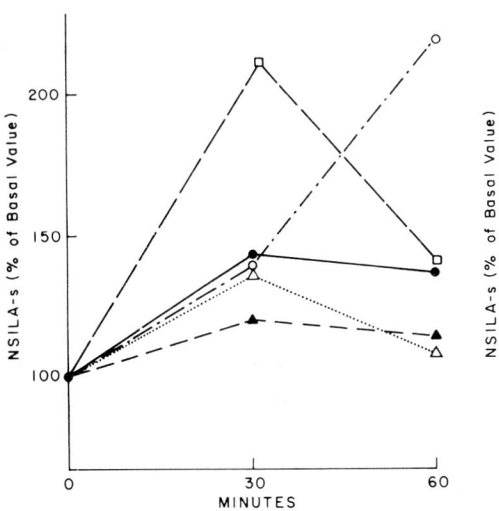

Fig. 7. Effect of oral glucose on plasma NSILA-s. 100 g glucose was administered orally. Blood was drawn before and at intervals following glucose administration. The plasma NSILA-s concentrations after oral glucose are expressed as percentage of the initial value.

Fig. 8. Effect of intravenous insulin on plasma NSILA-s. Regular pork insulin (0.1 U/kg body weight) was administered intravenously. Values are expressed as a percentage of the basal values for each patient.

values in diabetics and high values in umbilical cord blood that is shown on Figure 5 needs further investigation. The extremely low levels in anorexia nervosa (18% of normal) is very interesting. Both patients had high growth hormone levels and one had hypoglycemia at the time the plasmas were obtained. After weight gain and psychological improvement the hypoglycemia disappeared and NSILA-s level was 4 times the earlier value.

The oral administration of glucose resulted in a rise of plasma NSILA-s in all 5 normal subjects studied (Fig. 7). Insulin-induced hypoglycemia tended to lower NSILA-s levels (Fig. 8) but was observed only in 3 patients out of 6.

All of the studies of plasma NSILA-s described above in Figure 5 were done with gel filtration at pH 2.3. To study the state of endogenous NSILA-s at physiological pH, a

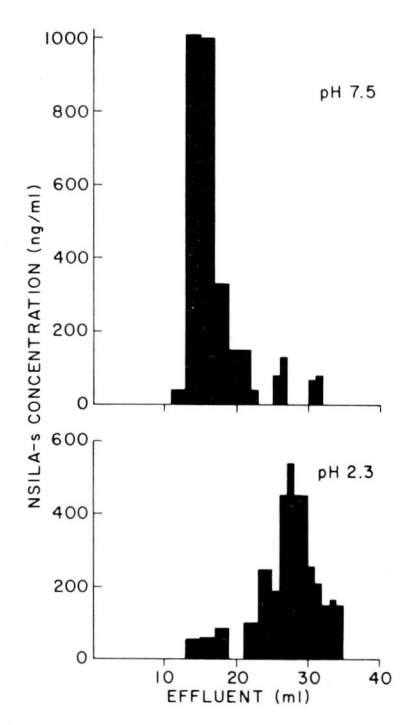

Fig. 9. Gel filtration of plasma at neutral and at acid pH. *Upper panel:* 0.5 ml of plasma was applied to a Sephadex G-50 column equilibrated in Krebs-Ringer phosphate buffer, pH 7.5, 1 ml fractions were collected and assayed directly in the radioreceptor assay. *Lower panel:* 0.5 ml of the same plasma was acidified and applied to a Sephadex G-50 column equilibrated in 1 M acetic acid. The neutralized effluent fractions were assayed. The NSILA-s radioreceptor activity is plotted as a function of the effluent volume of the columns.

single plasma was filtered at both pH 7.5 and pH 2.3 and assayed under identical conditions (Fig. 9). At acid pH, most of the radioreceptor activity was recovered in the region of 7400 daltons (bottom panel). In contrast, at neutral pH about 89% of the radioreceptor activity was recovered in the fractions that corresponded to the void volumes (upper panel). The remainder was in the region that corresponded to molecules of 5000-10,000 daltons. When the fractions that depicted the void volume of the neutral column were pooled, concentrated and filtered again at pH 7.5, 90% of the receptor activity was recovered in the region of the void volume and 10% in the low molecular weight region. This suggests that NSILA-s binds *reversibly* to a high molecular weight component of plasma and that the receptor activity in the void volume from the neutral column is *'bound NSILA-s'* while the activity in the low molecular weight region is *'free NSILA-s'*. The sum of bound and free NSILA-s from the neutral column nearly equaled the total NSILA-s receptor activity that coeluted with the ^{125}I-NSILA-s marker of the acid column. That is in keeping with our previous observations on the bound form of NSILA-s at neutral pH (Megyesi et al., 1974*b*) and that of Hintz et al. (1974) on the state of somatomedin C in human plasma.

When plasmas from a normal patient and a patient with fibrosarcoma associated with hypoglycemia were filtered on Sephadex G-50 at neutral pH (Fig. 10), similar results were obtained. In both cases 93% of the radioreceptor activity was recovered as

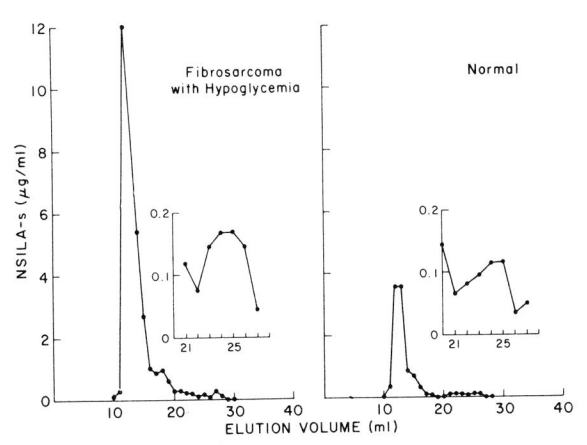

Fig. 10. Gel filtration of plasma at neutral pH. At left are results of a plasma from a patient with fibrosarcoma and hypoglycemia. At the right are the results of a normal subject's plasma. The results of repeat filtrations are shown in the insets of the respective figures.

bound NSILA-s. When the effluent fractions that corresponded to low molecular weight NSILA-s were pooled, lyophilized, and refiltered at neutral pH, most of the NSILA-s was recovered as a single distinct peak that coeluted with the ^{125}I-NSILA-s marker (insets). Since the fraction of NSILA-s bound (93%) and free (7%) was the same in both patients, the free NSILA-s and the total NSILA-s concentrations in the plasma of the patient with tumor hypoglycemia exceeded that of the normal by the same ratio.

It had been suggested that under normal circumstances NSILA does not play a major role in glucose metabolisms in vivo since pancreatectomy or injection of insulin antiserum can produce marked hyperglycemia in laboratory animals and since patients with diabetic ketoacidosis have normal levels of plasma NSILA. However, this argument is complicated by our observations that about 90% of NSILA-s appears to be bound to large molecules in plasma which may be a biologically inactive form. Recently Zapf et al. (1975) published data on ^{125}I-NSILA-s binding to plasma and found also that the majority of NSILA-s is bound to larger molecules at neutral pH. Their interesting experiments indicated that the large molecules are specific binding proteins, with a molecular weight of about 50-70,000 daltons.

Based on our NSILA-s standard (70 mU/mg) total plasma NSILA-s after gel filtration in acid in normal adults is equivalent to about 100 μU/ml of insulin (70 μU/μg × 1.5 μg/ml) (Table 4). At neutral pH only 10% or less NSILA-s is free. In normal persons, the amount of insulin-like bioactivity due to free NSILA-s approximated that due to insulin. In diabetics this small amount of endogenous insulin is insufficient to prevent hyperglycemia or ketoacidosis. On the other hand in patients with extra-pancreatic tumors associated with hypoglycemia and high total NSILA-s levels, the concentration of free NSILA-s *is sufficient to produce hypoglycemia, since it has insulin-like activity similar to the circulating insulin levels in patients with islet cell tumors* (Table 4). All of these calculations assume that the free NSILA-s from plasma is fully active on the insulin receptor.

This assumption was validated by measuring the insulin-like activity of these plasmas after gel filtration in the 2 insulin-receptor assays. In each case, the 'measured' free insulin-like activity correlated well with the 'estimated' free insulin-like activity calculated by summing the free NSILA-s and immunoreactive insulin.

TABLE 4

The insulin-like activity of plasma

Patient type	NSILA-s radioreceptor activity Total μU/ml	NSILA-s radioreceptor activity Free μU/ml	Immuno-reactive insulin μU/ml	Total free insulin-like activity Estimated μU/ml	Total free insulin-like activity Insulin radioreceptor assay Liver μU/ml	Total free insulin-like activity Insulin radioreceptor assay IM-9 cells μU/ml
Normal	100	10	8	18	<10	27
Insulinoma	120	5	128	133	65	48
Tumor hypoglycemia	970	116	15	131	150	150

Note that the insulinoma plasma showed higher result in the estimated free insulin-like activity than measured by insulin radioreceptor assays. That is in keeping with the high amount of proinsulin in this plasma which is about as active as insulin in the radioimmunoassay, but being biologically less active, produces lower results in the radioreceptor assays.

We also believe that the 'bound NSILA-s' is biologically ineffective. At least 2 possible mechanisms exist which could account for this result. First, the binding to plasma components may cover the part(s) of the molecule necessary for interaction with either the NSILA-s receptor, the insulin receptor, or both. Alternatively, the binding of NSILA-s to high molecular weight plasma components may restrict its space of distribution and only the 'free NSILA-s' might reach some of the target tissues. This question is now under study.

It has long been known, that multiplication of animal cells in vitro requires certain macromolecular factors which are usually found in serum. Gey and Thalhimer (1924) observed that insulin at very high concentrations could partly replace the serum requirement for HeLa cell growth. Later on the isolation of a variety of insulin-like peptides with growth-promoting activity have been described.

Table 5 gives a summary of the growth factors and compares those by certain characteristics. Insulin as a growth factor is much less potent than are the others. The striking similarities among somatomedin A, somatomedin C, NSILA-s and MSA (multiplication stimulating activity) are quite clear. Some of these growth factors had been tested in human cell cultures and proved to be effective. The insulin-like property of some has not been described or investigated yet.

When some of the growth factors were tested in our NSILA-s radioreceptor assay (Fig. 11) the most purified preparation of NSILA-s and MSA exhibited almost identical competition inhibition curves for labeled NSILA-s binding: nerve growth factor (NGF) was weakly reactive, producing about a 25% inhibition of specific ^{125}I-NSILA-s binding at a concentration of 1 μg/ml, while fibroblast growth factor (FGF), epidermal growth factor (EGF) and somatomedin B (SMB) were even less effective.

When we iodinated MSA itself and tested its binding to liver membranes in a manner identical to NSILA-s (Fig. 12), competition inhibition with NSILA-s, MSA, NGF and insulin were almost identical to those obtained with ^{125}I-NSILA-s. NGF, FGF, EGF, and SMB exert their effects through other receptors. Thus far we have not been able to obtain any somatomedin A or C to test in this assay. NSILA-s and MSA appear to interact with the same receptor in rat liver, which made it simple for us to establish a radioreceptor assay for MSA simply by substituting MSA for NSILA-s in our NSILA-s radioreceptor assay.

TABLE 5
Growth factors

Name	MW	Source	Insulin-like activity	Isoelectric point	Promotes growth by stimulating	Assay	Reference
1. Insulin	6000	Pancreatic β cells (beef, pork human)	+	pH 5-6	DNA synthesis in chick embryo fibroblasts	Bioassay, radioimmunoassay, radioreceptor assay	Schwartz and Amos (1968)
2. Somatomedin A	7000	Human plasma	+	pH 7.1-7.5	^{35}S incorporation into chick cartilage	Bioassay, radioreceptor assay	Uthne (1973)
3. Somatomedin B	5000	Human plasma	+	pH 5	DNA synthesis in human glia cells	Bioassay, radioimmunoassay	Uthne (1973)
4. Somatomedin C	7000	Human plasma	+	pH 8.4	^{35}S incorporation into rat cartilage, DNA synthesis in human fibroblasts	Bioassay, radioreceptor assay	Van Wyk (1974)
5. NSILA-s	7400	Human plasma	+	Basic	DNA synthesis in mouse, chicken embryo and human fibroblasts	Bioassay, radioreceptor assay	Froesch et al. (1974); Megyesi et al. (1974a); Rechler et al. (1974)
6. MSA	6000 10000	Fetal calf serum Rat liver cell conditioned medium	+	pH 6.2-7	DNA synthesis in chick embryo fibroblasts	Bioassay, radioreceptor assay	Pierson and Temin (1972); Dulak and Temin (1973a, b)
7. Nerve growth factor	13,259	Submaxillary glands of adult male mice	+	?	DNA synthesis in the mature sympathetic neuron of chick embryo	Radioreceptor assay	Bocchini and Angeletti (1969); Frazier et al. (1972)
8. Fibroblast growth factor	13,400	Bovine pituitary gland and brain	?	?	DNA synthesis in BALB/c 3T3 fibroblasts		Gospodarowitz (1975)
9. Epidermal growth factor	6500	Submaxillary gland of adult mice	?	pH 4.6	RNA synthesis in chick embryo epidermal cells	Bioassay (precocious opening of the eyelids of newborn mice), radioimmunoassay	Taylor et al. (1970)
10. Ovarian growth factor	13,400	Bovine pituitary	?	pH 8.5	Growth of an ovarian cell line	Bioassay	Gospodarowitz et al. (1974)
11. Platelet growth factor	20-35,000	Human platelet	?	?	Mitogenic activity for 3T3 mouse fibroblasts		Kohler and Lipton (1974)
12. Erythropoietin	20-60,000	Sheep plasma Anaemic rat kidney	?	?	DNA synthesis in rat bone marrow cells		Dukes (1968); Contrera and Gordon (1968); Goldwasser and Kung (1968)

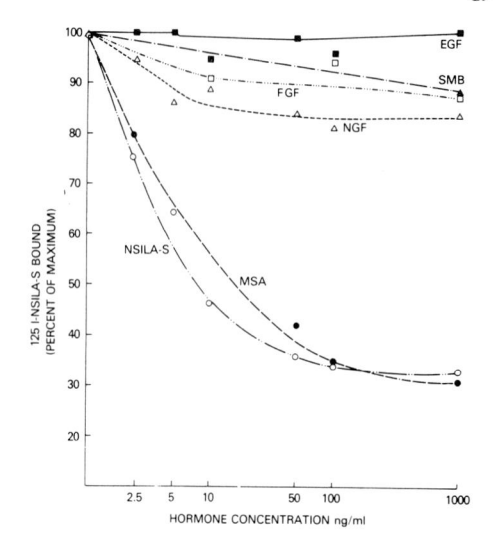

Fig. 11. Effect of various growth factors on NSILA-s binding to liver membranes. The ^{125}I-NSILA-s bound (percent of maximum) is plotted as a function of unlabeled peptide concentration.

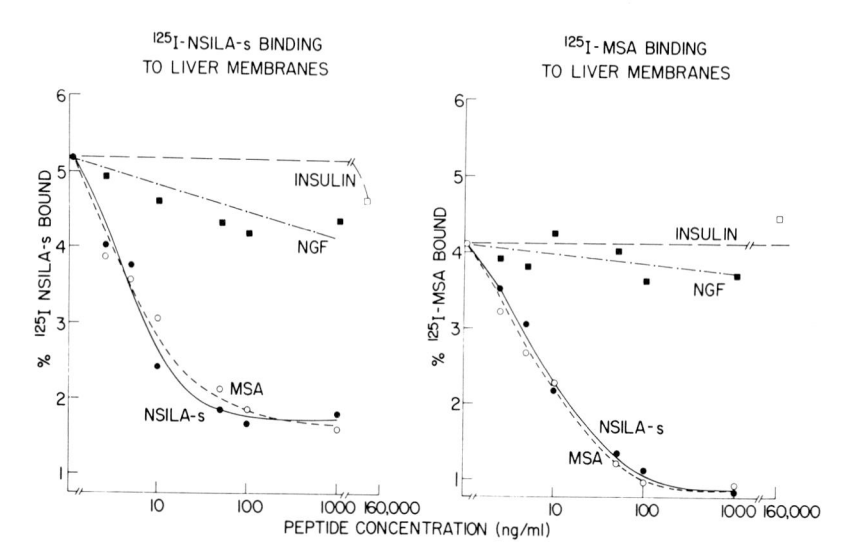

Fig. 12. ^{125}I-NSILA-s and ^{125}I-MSA binding to rat liver membranes. In **parallel** experiments ^{125}I-NSILA-s and ^{125}I-MSA were incubated with liver membranes for 90 minutes at 20° with the indicated concentrations of unlabeled NSILA-s, MSA, NGF and porcine insulin. Receptor bound ^{125}I-peptide was determined as indicated in legend to Figure 2.

SUMMARY

In 1974 we developed a radioreceptor assay for NSILA-s in human plasma. The normal basal plasma concentration of this peptide is about 100 μU/ml. Based on our study, more than 90% of that material is bound to larger molecules, less than 10% is free. We found high plasma NSILA-s levels in some patients with extrapancreatic tumors and hypoglycemia and presented evidence that the free NSILA-s levels in these patients are sufficient to cause hypoglycemia. The plasma levels in pregnancy were also high. We observed in hypopituitarism subnormal levels of this peptide which rise following intravenous HGH administration, indicating some growth hormone dependency. The lack of elevated levels in our acromegalic patients might question that dependency or simply indicate that chronic stimuli (acromegaly) might not have the same effect as acute HGH administration. The very low NSILA-s levels in anorexia nervosa can be of interest in the studies of that disease.

Comparing several growth factors we show that multiplication stimulating activity is as fully potent as NSILA-s on the NSILA-s receptor of rat liver.

The most important question has not yet been settled (and we might say the same about the other growth factors) on whether NSILA-s can produce growth in vivo? To investigate this basic problem and to study the possible physiological importance of NSILA-s, large amounts of pure peptide would be needed.

Note added in proof

Since the time of submission of this paper we tested somatomedin A and C (supplied generously by Dr. R.L. Hintz) in our radioreceptor assay. Interestingly, somatomedin A seemed to act on the same receptor, somatomedin C was at most 1% as effective as NSILA-s in competing for ^{125}I-NSILA-s binding.

REFERENCES

Bocchini, V. and Angeletti, P.U. (1969): *Proc. nat. Acad. Sci. (Wash.), 64,* 787.
Contrera, J.F. and Gordon, A. (1968): *Ann. N.Y. Acad. Sci., 149,* 114.
Dukes, P.P. (1968): *Ann. N.Y. Acad. Sci., 149,* 437.
Dulak, N.C. and Temin, H.M. (1973a): *J. Cell Physiol., 81,* 153.
Dulak, N.C. and Temin, H.M. (1973b): *J. Cell Physiol., 81,* 161.
Frankel, R.J. and Jenkins, J.S. (1975): *Acta endocr. (Kbh.), 78,* 209.
Frazier, W.A., Andeletti, R.H. and Brandshaw, R.A. (1972): *Science, 176,* 482.
Freychet, P., Roth, J. and Neville Jr, D.M. (1971): *Proc. nat. Acad. Sci. (Wash.), 68,* 1833.
Froesch, E.R., Bürgi, H., Müller, W.A., Humbel, R.E., Jakob, A. and Labhart, A. (1967): *Recent Progr. Hormone Res., 23,* 565.
Froesch, E.R., Morell, B., Zapf, J., Zingg, A.E., Meuli, C., Schlumpf, U., Heimann, R., Eigenmann, E. and Humbel, R.E. (1974): *Acta endocr. (Kbh.), Suppl. 184,* 183.
Gey, G.O. and Thalhimer, W. (1924): *J. Amer. med. Ass., 82,* 1609.
Goldwasser, E. and Kung, Ch.K.H. (1968): *Ann. N.Y. Acad. Sci., 149,* 49.
Gospodarowitz, D. (1975): *J. biol. Chem., 250,* 2515.
Gospodarowitz, D., Johnes, K.L. and Sato, G. (1974): *Proc. nat. Acad. Sci. (Wash.), 71,* 2295.
Hall, K. and Uthne, K. (1972): *Acta med. scand., 190,* 137.
Hintz, R.L., Clemmons, D.R., Underwood, L.E. and Van Wyk, J.J. (1972): *Proc. nat. Acad. Sci. (Wash.), 69,* 2351.
Hintz, R.L., Orsini, E.M. and Van Camp, M.G. (1974): In: *Program, 56th Annual Meeting of the Endocrine Society, Atlanta, Georgia.* (Abstract.)
Kohler, N. and Lipton, A. (1974): *Exp. Cell Res., 87,* 297.
Marshall, R.N., Underwood, L.E., Voina, S.J., Foushee, D.B. and Van Wyk, J.J. (1974): *J. clin. Endocr., 39,* 283.
Megyesi, K., Kahn, C.R., Roth, J., Froesch, E.R., Humbel, R.E., Zapf, J. and Neville Jr, D.M. (1974a): *Biochem. biophys. Res. Commun., 57,* 307.
Megyesi, K., Kahn, C.R., Roth, J. and Gorden, P. (1974b): *J. clin. Endocr., 38,* 931.
Megyesi, K., Kahn, C.R., Roth, J., Neville Jr, D.M., Nissley, P.S., Humbel, R.E. and Froesch, E.R. (1975a): *J. biol. Chem., 250,* 8990.
Megyesi, K., Kahn, C.R., Roth, J. and Gorden, P. (1975b): *J. clin. Endocr., 41,* 475.

Neville Jr, D.M. (1968): *Biochim. biophys. Acta (Amst.), 154,* 540.
Phillips, L.S. and Young, H.S. (1975): In: *35th Annual Meeting of the American Diabetes Association, New York,* Abstract 57.
Pierson Jr, R.W. and Temin, H.M. (1972): *J. Cell Physiol., 79,* 319.
Rechler, M.M., Podskalny, J.M., Goldfine, I.D. and Wells, C.A. (1974): *J. clin. Endocr., 39,* 512.
Roth, J. (1973): *Metabolism, 22,* 1059.
Schwartz, A.G. and Amos, J. (1968): *Nature (Lond.), 219,* 1366.
Taylor, J.M., Cohen, S. and Mitchell, W.M. (1970): *Proc. nat. Acad. Sci. (Wash.), 67,* 164.
Uthne, K. (1973): *Acta endocr. (Kbh.), Suppl. 73,* 175.
Van Wyk, J.J., Hall, K., Van den Brande, L. and Weaver, R.P. (1971): *J. clin. Endocr., 32,* 389.
Van Wyk, J.J. (1974): *Recent Progr. Hormone Res., 30,* 259.
Zahnd, G.R., Steinke, J. and Renold, A.E. (1960): *Proc. Soc. exp. Biol. (N.Y.), 105,* 455.
Zapf, J., Waldvogel, M. and Froesch, E.R. (1975): *Arch. Biochem., 168,* 638.

MITOGENIC FACTORS FROM THE BRAIN AND THE PITUITARY: PHYSIOLOGICAL SIGNIFICANCE*

D. GOSPODAROWICZ, J.S. MORAN and H. BIALECKI

The Salk Institute for Biological Studies, San Diego, Calif., U.S.A.

The multiplication of animal cells in vitro requires mitogenic macromolecular factors that are usually found in serum. Due to the complexity of serum, attempts to isolate these factors have been only partially successful. Since any mitogenic factor present in serum must be synthesized in some tissue, we have explored the possibility that these factors may be more easily purified from specific organs than from serum.

Our first encounter with mitogenic agents distinct from known hormones came with the establishment of the first cell line which was supposed to be dependent on a trophic hormone for its growth. This cell line was derived from ovarian tumors obtained by the implantation of luteal phase ovaries into the spleens of ovariectomized rats (Clark et al., 1972). The ovarian tumors which developed were assumed to be dependent on gonadotropins for their growth. However, while we demonstrated that the ovarian cell line (31A) derived from one of these tumors was dependent on a partially-purified preparation of luteinizing hormone (NIH-LH) for its proliferation, highly purified preparations of LH did not have any effect (Clark et al., 1972). This led us to conclude that the mitogenic agent present in NIH-LH was not LH but was a contaminant. Further work resulted in the purification of a growth factor from pituitary gland, ovarian growth factor (OGF) (Gospodarowicz et al., 1974; Jones and Gospodarowicz, 1974). OGF was shown to be distinct from known pituitary hormones (Gospodarowicz et al., 1974).

Reports of mitogenic activity for various cell types in various pituitary hormone preparations led us to investigate whether the pituitary could be a major site of synthesis of mitogenic agents (for a general review, see Gospodarowicz and Moran, 1976b). A comparison of the mitogenic activity of crude extracts from various organs has led us to conclude that the pituitary and the brain contain 10 to 100 times as much mitogenic activity for mammalian mesodermal-derived cells as do other organs. We have recently reported the purification from pituitary of a mitogenic factor named fibroblast growth factor (FGF) which promotes the growth of fibroblasts, among other cell types (Gospodarowicz, 1974, 1975). In this communication we describe the purification from brain of a mitogenic factor similar to FGF, and the identification of a neutral peptide from brains and pituitary, which promotes the clonal growth of myoblasts in tissue culture (Gospodarowicz et al., 1975b). We also report the effect of brain and pituitary FGF on the proliferation of a variety of cell types. We suggest that FGF may be involved in wound healing in vivo.

---

* Supported by National Institutes of Health grant No. HD 07651 and the American Cancer Society (BC 152).

BRAIN FGF

Purification

Whole cow brains were used as starting material, and the initial steps of purification were similar to those described for the pituitary FGF with the exception that the HPO_3 step was omitted. Brains were homogenized in the presence of 0.15 M $(NH_4)_2SO_4$. The suspension was adjusted to pH 4.5 with HCl, and then centrifuged. The supernatant was adjusted to pH 6.5-7.0 by the addition of 1 M NaOH. Addition of $(NH_4)_2SO_4$ (220 g/l) produced a precipitate that was inactive. More $(NH_4)_2SO_4$ (250 g/l, final molarity 3.56) was added to the supernatant, precipitating the FGF activity. The precipitate was dissolved in cold distilled water and dialysed against water for at least 24 hours before being lyophilized.

Further purification was achieved by dissolving the lyophilized fraction at a concentration of 20 mg of protein per ml in 0.1 M sodium phosphate, pH 6.0, applying this solution to a column of CM-Sephadex-C50 and chromatographing the proteins as described for the purification of FGF from the pituitary (Gospodarowicz, 1974, 1975).

80% of the proteins passed through the column unadsorbed. They did not contain more than 10% of the initial mitogenic activity. Elution with 0.1 M sodium phosphate with 0.15 M NaCl removed a fraction with little FGF activity. Most of the FGF activity was eluted with O.1 M sodium phosphate with 0.6 M NaCl at pH 6.0. That fraction contained 2 to 3% of the protein and 80% of the mitogenic activity applied to the column.

FGF was further purified by gel filtration on Sephadex-G75. A typical pattern is shown in Figure 1A. The mitogenic activity for 3T3 was contained in fractions 32 to 38 and eluted just before the cytochrome C marker. After lyophilization, the partially purified preparation of FGF was dissolved in 0.2 M NH_4COOH, pH 6.0, applied to a column of carboxymethyl cellulose (Whatman CM 52) and eluted with a linear gradient (0.2 M to 0.5 M NH_4COOH) (Fig. 1B). The active fraction was recovered after the main peak. After lyophilization, it was reapplied to a small column of carboxymethyl cellulose (CM 52) and eluted with a linear gradient from 0.3 M to 0.45 M NH_4COOH (pH 6.0). The proteins emerge as a single peak, and the specific mitogenic activity was similar for all the fractions. When submitted to gel filtration on Sephadex-G 75, a single peak was observed (Fig. 1C).

The homogeneity of the final fraction was determined by polyacrylamide gel electrophoresis at pH 4.5 and at pH 8.5. A single band was observed at pH 4.5. At pH 8.5 the proteins did not run. Isoelectric focusing gave an isoelectric point of 9.6 (Fig. 1D).

AMINO ACID, CARBOHYDRATE AND N-TERMINAL DETERMINATION

The amino acid composition of the brain FGF reflects its basic nature. It is rich in lysine and arginine (Table 1), and differs from the FGF purified from the pituitary gland (Gospodarowicz, 1975) by containing one methionine and no cystine (Table 1). Since no carbohydrates such as glucosamine or galactosamine can be detected, the brain FGF is not a glycoprotein. No N-terminal product could be detected after dansylation of the peptide indicating that the N-terminal was either blocked or was lysine. No lipid analysis has been conducted. The apparent molecular weight of the FGF determined by SDS polyacrylamide gel electrophoresis was $13,000 \pm 1,200$, similar to that observed for the pituitary FGF (Gospodarowicz, 1975). As was the case for the pituitary FGF, both heat and acid treatment destroyed the mitogenic activity of

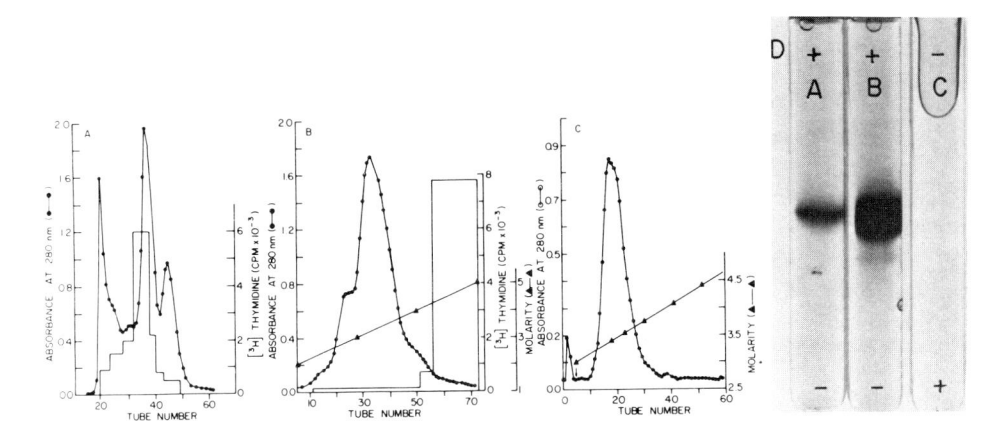

Fig. 1. A. 400 mg of proteins was dissolved in 3 ml of 0.1 M ammonium carbonate (pH 8.5) and applied to a column (2.5 × 90 cm) of Sephadex-G75 equilibrated in the same buffer. Fractions (5.8 ml each) were collected. The FGF activity was found between fractions 32 and 38. The histogram shows the relative activity of the different fractions on the initiation of DNA synthesis in sparse, resting populations of 3T3 cells. Fractions were tested at 10 ng/ml.
B. Gradient elution of FGF on carboxymethyl cellulose (Whatman CM 52). 320 mg of partially purified FGF (Sephadex-G75 fraction, Fig. 1A) was dissolved in 20 ml of 0.2 M ammonium formate (pH 6.0) and applied to a column (1.5 × 20 cm) of carboxymethyl cellulose (CM 52) equilibrated in the same buffer. The column was eluted with a linear gradient (0.2 M to 0.5 M ammonium formate) (•—•) (total volume of 500 ml). The flow rate was adjusted to 20 ml per hour, and 6-ml fractions were collected. The FGF activity was recovered between fractions 51 and 78. The histogram shows the relative activity of the fractions on the initiation of DNA synthesis in sparse, resting populations of 3T3 cells. Fractions were tested at 2.5 ng/ml.
C. Rerun of the active fraction on carboxymethyl cellulose (CM 52). Material was applied in 0.3 M ammonium formate, pH 6.0, and the gradient was run from 0.3 M to 0.5 M ammonium formate, pH 6.0. All fractions had the same specific activity when tested at 1 ng/ml for the initiation of DNA synthesis in sparse, resting populations of 3T3 cells.
D. Polyacrylamide gel electrophoresis of FGF at pH 4.5 and 8.5. FGF obtained from the carboxymethyl cellulose (CM 52) gradient (Fig. 1C) was submitted to polyacrylamide gel electrophoresis at pH 4.5 (A: 10 µg; B: 50 µg) and at pH 8.5 (C: 100 µg). The electrophoresis at pH 4.5 was performed at 7 mAmp per tube for 45 minutes. Electrophoresis at pH 8.5 was performed at 4 mAmp per tube for 3 hours. Staining was done with 1% amido Schwarz — 7% acetic acid. Marked problems have been encountered with the stability of brain FGF when stored at room temperature after lyophilization. Because of deamination, losses of biological activity are encountered. This results in the formation of a double band on polyacrylamide gel electrophoresis at pH 4.5. Storage at −20°C greatly slows the degradation.

brain FGF. When heated at 60°C at pH 7 for 5 minutes, 80% of the mitogenic activity was lost. Treatment with 1 N acetic acid or 1% formic acid destroyed 90-95% of the biological activity. In the presence of 0.1 N acetic acid or 0.1% formic acid (final pH: 3.5), the FGF activity was reduced by 20 and 45% respectively. This indicates that FGF was not stable at acid pH's and, thus, differs significantly from MSA, NSILA-S, EGF and the somatomedins, which are stable under similar conditions.

The brain FGF was devoid of nerve growth factor activity in the chick dorsal root ganglion fiber outgrowth assay; however, it was 60% as potent as the nerve growth factor when the attachment of neurons was used as an assay (S. Norr, private commu-

TABLE 1

Comparison of the amino acid composition of the brain and pituitary FGF

Amino acid (or sugar)	Brain	Pituitary
	nmoles of amino acid*	nmoles of amino acid*
Lysine	10	13
Histidine	6	4
Arginine	8	6
Aspartic acid	7	6
Threonine	5	5
Serine	8	6
Glutamic acid	8	8
Proline	13	10
Glycine	8	10
Alanine	8	9
Valine	3	3
Methionine	1	ND***
Half cystine**	ND	2
Isoleucine	2	3
Leucine	6	4
Tyrosine	3	2
Phenylalanine	3	2
Tryptophan**	–	ND
Glucosamine	ND	ND
Galactosamine	ND	ND

* nmoles of amino acid per nmole of FGF (apparent molecular weight to be 13,000); **p-toluenesulfonic acid hydrolysis; *** not detectable.

nication). It did not have any significant somatomedin activity (J.J. van Wyk, private communication) nor EGF activity (S. Cohen, private communication).

No esterase activity was detected, even at FGF concentrations as high as 1 mg/ml with either denatured bovine serum albumin or N α-benzoyl-L-arginine ethyl ester as substrates. This indicates that if any trypsin-like activity is present in FGF, it is less than 0.005%. Treatment with proteolytic enzymes such as trypsin chymotrypsin, pepsin, or proteases destroyed 90% of the FGF activity.

COMPARISON OF THE BIOLOGICAL ACTIVITY OF THE BRAIN AND THE PITUITARY FGF ON THE INITIATION OF DNA SYNTHESIS

The mitogenic activity of the brain FGF has been tested on the Balb/c 3T3 cell line (Fig. 2). It was active at concentrations as low as 0.01 ng/ml, and between 0.5 and 1 ng/ml, a maximal effect was achieved. Its potency was similar to that of the pituitary FGF in the lower concentration range (0.01 to 1 ng). However, it did not fully replace serum for the initiation of DNA synthesis. As was the case with the FGF purified from the pituitary gland, the presence of glucocorticoids was required in order for brain FGF to mimic the full effect of serum on the initiation of DNA synthesis (Gospodarowicz, 1974, 1975).

To confirm that the brain FGF was a true mitogenic agent, capable not only of

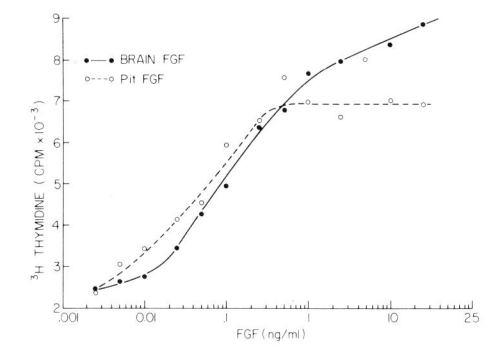

Fig. 2. Comparison of the stimulation of DNA synthesis in 3T3 cells in response to various concentrations of brain and pituitary FGF. 3T3 cells, 2×10^4 cells per 35×15 mm dish, were plated. The determination of ^{3}H-thymidine incorporation into DNA was done as described by Gospodarowicz (1974, 1975). ^{3}H-thymidine in corporation was 110 c.p.m. per dish in controls, and 15,200 c.p.m. when 10% calf serum was added.

provoking the initiation of DNA synthesis, but also of provoking mitosis, the effect of brain FGF on the rate of division of 3T3 cells has been analyzed. When 3T3 cells were maintained resting in the presence of plasma (which contains little growth-promoting activity compared to serum for 3T3 cells), the addition of FGF and glucocorticoids makes the cells proliferate (Gospodarowicz et al., 1975*a*). When cells were maintained in 2.5% plasma, the final cell density obtained when FGF and dexamethasone was added was either equal, or 50% higher than that observed with 10% serum (Gospodarowicz et al., 1975*a*). The dependency of the mitogenic effect of FGF on the serum or plasma concentration is probably due to the presence in serum of conditioning factors, distinct from mitogenic agents. This problem has been reviewed by Gospodarowicz and Moran (1975*a*).

EFFECT OF FGF ON THE PROLIFERATION OF MAMMALIAN CELLS IN TISSUE CULTURE

FGF purified from either the brain or the pituitary has been shown to be a mitogen for mesodermal-derived cell types. In the presence of glucocorticoids, it stimulates the proliferation of sparse, resting populations of Balb/c 3T3 cells (Gospodarowicz and Moran, 1974*a,b*). Its effect is more pronounced when the cells are maintained in the presence of plasma than in serum, and it has been shown to be equivalent to the platelet factor(s) (Gospodarowicz et al., 1975*a*). When the cells are density-inhibited, the addition of FGF, or FGF plus glucocorticoids to the medium provokes a resumption of growth (Gospodarowicz and Moran, 1974*a,b*) (Fig. 3). FGF has also been shown to control the proliferation of thermosensitive mutants of polyoma transformed 3T3 cells at the unpermissive temperature but not at the permissive temperature (Rudland et al., 1974*a*). FGF has been shown to be a mitogenic agent not only for established polyploid cell lines such as 3T3, but also for diploid cells in their early passage, such as human foreskin fibroblasts. With human foreskin fibroblasts in low serum (0.2%) or no serum at all, FGF induced one cycle of division while in high serum (1-10%), FGF had an effect additive over serum. It reduced the division time from 2 days to 1 day, and the final density was 4- to 5-fold greater than that observed with optimal serum concentrations (Gospodarowicz and Moran, 1975*b*) (Fig. 4). A similar effect was observed with secondary cultures of bovine endothelial cells (Fig. 5). With Y1 adrenal cells, FGF replaced serum, and its mitogenic effect was counteracted by ACTH (Gospodarowicz and Handley, 1975) (Fig. 6). With bovine myoblasts seeded at high

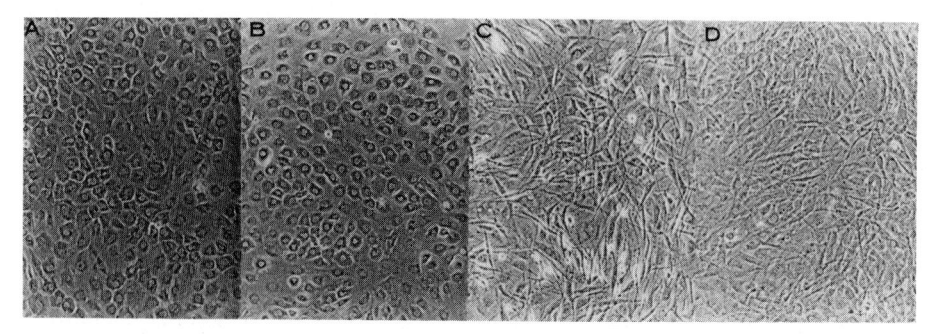

Fig. 3. Effect of FGF and glucocorticoid on Balb/c 3T3 at 'density inhibition' in the presence of 10% serum. Balb/c 3T3 cells were grown in Dulbecco's modified Eagle's medium with 10% calf serum. Two days after reaching confluency (A), either 10% serum (B), 25 ng/ml of FGF (C), 25 ng/ml of FGF plus 1 µg/ml of dexamethasone (D), were added to the cells. FGF or FGF plus dexamethasone were added daily; serum was added only once. Four days later, photographs were taken at a magnification of 40 × with phase contrast optics. While the addition of serum did not change the morphology of the cells, the addition of FGF alone to cultures grown in 10% serum caused the cells to become elongated, density inhibition was overcome, and the cells lost their orientation and grew in multiple layers. When FGF plus dexamethasone was added, the cells were flatter but grew to an even higher density than with FGF alone (Gospodarowicz and Moran, 1974a; Gospodarowicz et al., 1974). When the cells were trypsinized, replated, and grown in the presence of serum, they resumed their usual morphology, thus demonstrating that the transformation observed in the presence of FGF is reversible (Gospodarowicz and Moran, 1974a,b).

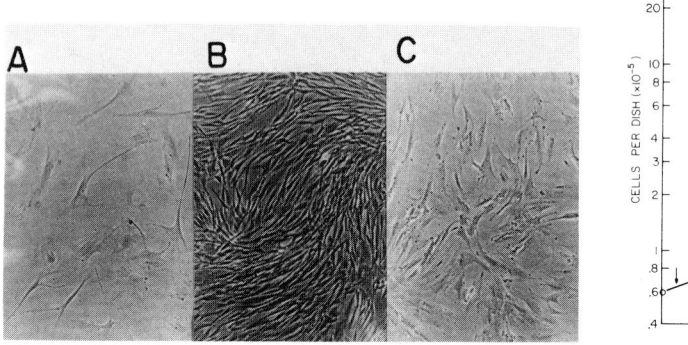

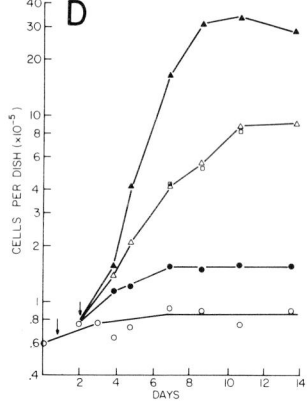

Fig. 4. A, B, C. Appearance of human foreskin fibroblast cells maintained in 0.2% calf serum, 10% calf serum, or 10% calf serum plus FGF. Cells were plated and grown as described for the growth curve (Fig. 4D). Photos were taken at 100 × magnification under phase contrast after 6 days of growth, 2 days before the cultures reached their maximal density. A. 0.2% calf serum; B. 10% calf serum plus FGF (25 ng/ml added daily); C. 10% calf serum. As for the 3T3 cells, these cells were morphologically transformed when grown in the presence of FGF; this transformation was reversed when FGF was removed from the medium.
D. Growth of human foreskin fibroblasts in the presence and in the absence of FGF. 6×10^4 cells were plated per 6 cm dish in 2.5% calf serum (day 0). On day 1 (first arrow), the medium was changed to 0.2% calf serum. On day 2 (second arrow), either 20% fetal calf serum (□—□), 10% calf serum (△—△), FGF (●—●), or 10% calf serum plus FGF (▲—▲) were added. FGF was added daily (25 ng/ml). Duplicate cultures were counted in a Coulter counter after trypsinization. Control (●—●). Cells maintained in the presence of 0.2% serum went through only one cycle of division when FGF was added; however, in the presence of high serum concentrations (10 or 20%) the addition of FGF reduced the division cycle from 2 days to 1 day, and final cell density was increased 4-fold. (Reproduced from Gospodarowicz and Moran, 1975b, by courtesy of the Editors of the *Journal of Cell Biology*.)

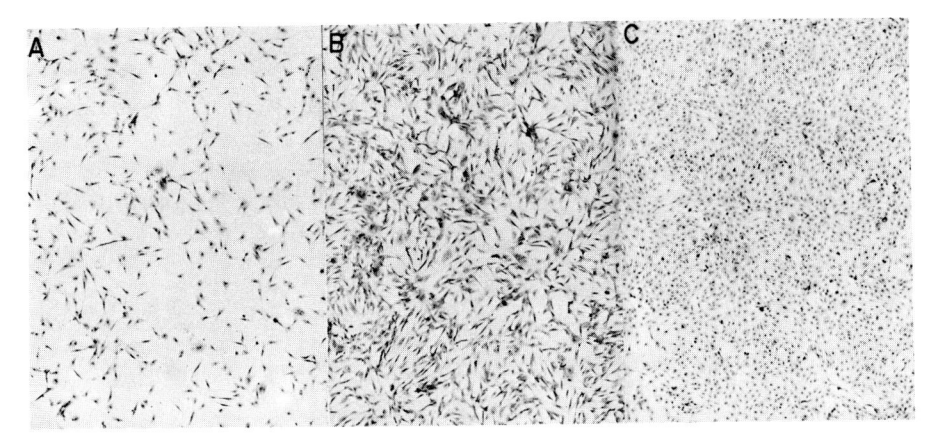

Fig. 5. Effect of FGF on the proliferation of bovine endothelial cells. Endothelial cells were obtained by scraping the interior surface of a bovine umbilical cord blood vessel. The cells thus obtained were cultivated for 7 days with daily changes of medium containing 50% fetal calf serum. These conditions resulted in the development of a homogeneous culture of endothelial cells. The cells were then trypsinized and seeded at 1500 cells/cm² in 35-mm dishes in 2 ml of Dulbecco's modified Eagle's medium containing either 10% fetal calf serum (A), 50% fetal calf serum (B), or 10% fetal calf serum + 50 ng/ml pituitary FGF (C). Four days later the plates were fixed with formalin and stained with Giemsa and photographed at 79×. The addition of FGF resulted in rapid cell proliferation leading to the development of a well-differentiated monolayer.

density, FGF induced proliferation and delayed the fusion of myoblasts into myotubes (Gospodarowicz et al., 1976) (Fig. 7).

It has also been shown to promote the clonal growth of rabbit ear chondrocytes (Fig. 8), of amniotic fluid-derived fibroblasts (Moran et al., 1976) (Fig. 9), and of normal adrenal cells in their early passage (Fig. 10). In all cases, extremely low concentrations of FGF are needed to induce the cells to proliferate.

Of the cell types investigated, fibroblasts, chondrocytes and myoblasts belong to the same family since chondrocytes and myoblasts are considered to be fibroblast-derived and come from the mesoderm. Adrenal cells and endothelial cells are also mesoderm-derived.

FGF is not a mitogen for epithelial cells that are derived from the ectoderm (A. Freeman, private communication), nor is it a mitogen for cells which are derived from the endoderm such as liver cells (Leffert, 1974) or pancreatic cells (R. Pictet, private communication). It has also been shown not to be a mitogen for avian cells, either fibroblasts or myoblasts (Gospodarowicz et al., 1976).

INDUCTION OF THE PLEIOTYPIC RESPONSE BY FGF

The addition of FGF to cultures of fibroblasts induces the pleiotypic response which accompanies the transition from a resting to a growing state. The sequential changes under FGF control are the alterations in cyclic nucleotide levels (Rudland et al., 1974b), stimulation of cellular transport systems, polyribosome formation, protein synthesis, ribosomal RNA and tRNA synthesis followed by the induction of DNA synthesis (Rudland et al., 1974c). The effect of FGF and hydrocortisone on the pleioty-

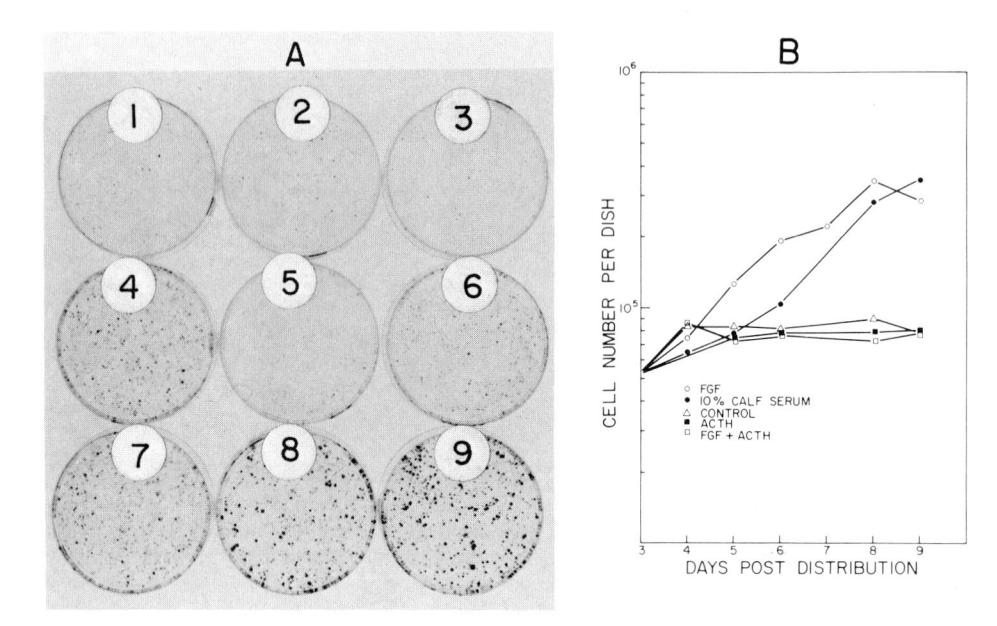

Fig. 6. Effect of FGF on the growth of Y1 adrenal cells. A. Effect of various agents on the final cell density of Y1 adrenal cells. 2×10^4 cells in 5 ml of Ham's F10 medium with 5% calf serum were seeded in 6-cm dishes. One day later the serum concentration was changed to 0.2% calf serum. On day 3, the following additions were made: (1) control, (2) 1 μg/ml dexamethasone, (3) 500 ng/ml insulin, (4) 50 ng/ml FGF, (5) 50 ng/ml FGF plus 1 μg/ml dexamethasone, (6) 50 ng/ml FGF plus 1 μg/ml dexamethasone plus 500 ng/ml insulin, (7) 50 ng/ml FGF plus 500 ng/ml insulin, (8) 10% calf serum, and (9) 12.5% horse serum plus 2.5% fetal calf serum. All agents were added every other day for 9 days, except serum, which was added only on day 3. At day 7, the plates were fixed with ethanol-methanol (1:1) and stained with 1% crystal violet. Of all the agents added, only FGF was mitogenic. Insulin had no effect, and dexamethasone was slightly inhibitory. With the Y1, FGF replaced serum in contrast to other cell lines, where FGF has an additive effect. (Reproduced from Gospodarowicz and Handley, 1975, by courtesy of the Editors of *Endocrinology*.)

B. Effect of ACTH on the growth rate of Y1 cells maintained in the presence of serum or FGF. Cells were cultured as described in A. FGF was added at a final concentration of 5 ng/ml and ACTH at 0.75 IU/ml. While with FGF the cells proliferated at the same rate as in the presence of 10% calf serum, as soon as ACTH was added to cultures maintained in the presence of FGF, the cells ceased proliferating. (Reproduced from Gospodarowicz and Handley, 1975, by courtesy of the Editors of *Endocrinology*.)

pic response has been described in detail elsewhere (Gospodarowicz and Moran, 1976*a*).

NEUTRAL FACTOR OR MYOBLAST GROWTH FACTOR

In the course of investigating the effect of FGF on the proliferation of myoblasts, we noticed that while either FGF or crude extract of brain prepared at pH 4.5 were mitogens for myoblasts at high density (126 cells/mm^2), at low density (10^{-5} cells/mm^2) they had little effect. However, extraction of brain or pituitary at pH 8.5 (a pH at which little or no FGF activity can be extracted) yields an agent mitogenic for myo-

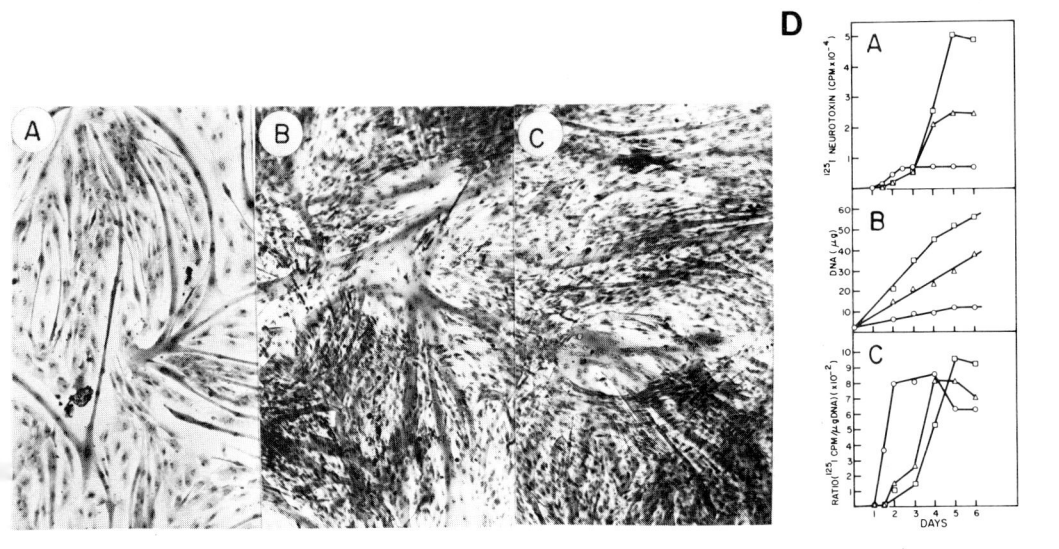

Fig. 7. A, B, C. Appearance of myoblast cultures maintained in the presence or absence of FGF after 5 days. Myoblasts were obtained by trypsinization of hind leg muscles of 3-month bovine fetuses. They were seeded on gelatinized dishes at a density of 8000 cells/cm² in medium 199: Dulbecco's modified Eagle's medium:horse serum (8:2:1). FGF, at a final concentration of 100 ng/ml (B), or 1 µg/ml (C), was added every other day. Controls (A) did not receive any. At day 5, the plates were washed and fixed for 1 hour with 10% formalin and stained with 1% crystal violet.

D. Comparison of the binding of neurotoxin and the synthesis of DNA as a function of time in cultures of bovine myoblasts. (A) Determination of the binding of [^{125}I]neurotoxin as a function of time in myoblast cultures maintained in the presence of FGF (0.1 µg/ml △——△, 1 µg/ml □——□), or in its absence (○——○). (B) Determination of DNA content as a function of time in myoblast cultures maintained in either the presence of FGF (0.1 µg/ml △——△ and 1 µg/ml □——□) or in its absence (·○——○). (C) Ratio of the binding of neurotoxin to the DNA content of the myoblast cultures maintained in either the presence of FGF (0.1 µg/ml △——△ and 1 µg/ml □——□) or in its absence (○——○). In controls, the neurotoxin specific binding was noticeable at 24 hours and increased until 60 hours when it reached a plateau. With 0.1 µg or 1 µg of brain or pituitary FGF per ml, the amount of binding during the first 72 hours was lower than in controls, reflecting a lower rate of fusion in the presence of FGF. At 72 hours, the cultures in the presence of 0.1 µg of FGF bound as much neurotoxin as did controls. Binding increased to a maximum between days 4 and 5. Maximal binding to cultures maintained in the presence of 0.1 µg per ml of FGF was 5-fold higher than in controls; with 1 µg per ml, it was 10-fold higher.

The final specific binding per culture under each condition was proportional to the final DNA content per culture (Fig. 7D-B). As shown in Fig. 7D-C, the number of binding sites per cell (as determined by c.p.m. of toxin bound/µg DNA) was similar at the end of the experiment. This demonstrates that the increased cell proliferation seen with FGF was due to an increased proliferation of myoblasts, and that the percentage of non-myoblasts, if any, was similar under all experimental conditions. Although the amounts of toxin bound per cell were nearly the same at the end of the experiment for all cultures, the time course of appearance of toxin binding sites was quite different. Whereas the ratio of toxin bound per cell reached a maximum on day 2 in controls, it reached a maximum on day 4 with 0.1 µg/ml FGF and on day 5 with 1 µg/ml FGF. In all cases, the maximum toxin binding per cell reached a maximum shortly after the rate of DNA synthesis per culture (determined by ³H-thymidine pulse labeling, dropped from its maximum) (Gospodarowicz et al., 1976).

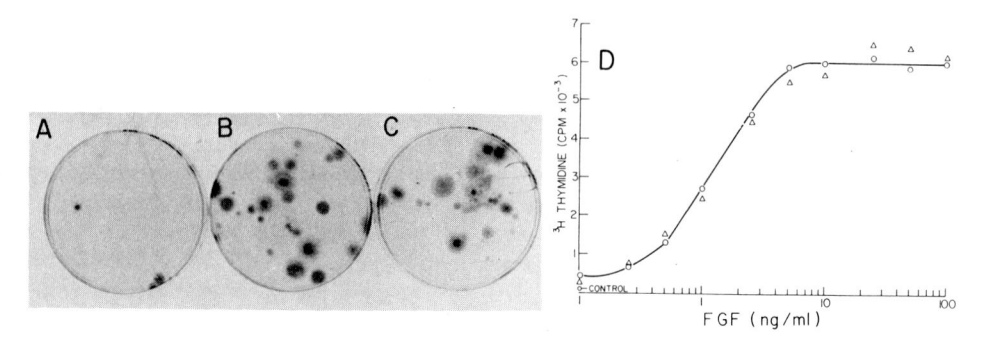

Fig. 8. A, B, C. Comparison of the effect of brain and pituitary FGF on the clonal growth of chondrocytes plated at low density. Chrondrocytes obtained by collagenase dissociation of rabbit ear cartilage were plated at a concentration of 50 cells per 6-cm dish in the presence of 5 ml of F12 medium with 2.5% fetal calf serum. The following additions were made: A. none; B. pituitary FGF (100 ng/ml); C. brain FGF (100 ng/ml). Media were renewed once every 5 days. FGF was added. After 12 days, the plates were washed and fixed with 10% formalin pH 7.2, and the cells were stained with Alcian green to stain the cartilage in green and counter-stained with metanil yellow which stains the chondrocytes in yellow — differentiated colonies appear as yellow with a green spot in the center (Ham, 1972) (on the figure, colonies are gray with dark spots indicating deep green staining).

D. Effect of various concentrations of brain and pituitary FGF on the initiation of DNA synthesis in rabbit chondrocytes. The experiment was done as described by Gospodarowicz (1974, 1975) using rabbit ear chondrocytes instead of 3T3 cells with 4×10^4 cells per 35-mm dish. Each point represents the average of 3 plates. Brain FGF (o———o); pituitary FGF (△———△).

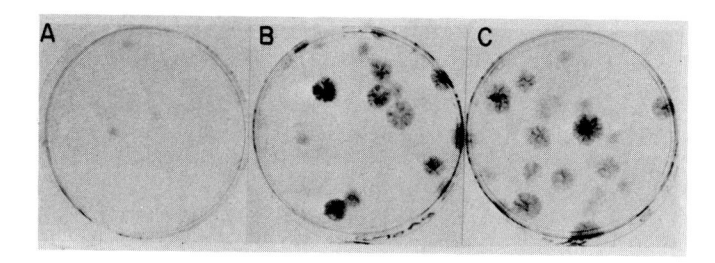

Fig. 9. Human amniotic fluid-derived fibroblastic cells grown for 15 days in the presence of Dulbecco's modified Eagle's medium containing 20% fetal calf serum without FGF (A), or with pituitary FGF. 10 ng/ml (B), or brain FGF, 10 ng/ml (C). Cells were plated at an initial concentration of 100 cells/6-cm dish. On day 1, FGF was added, and the medium was renewed 2 days later and every other day thereafter. Dishes were fixed with 10% formalin pH 7.2 on day 15 and stained with 1% crystal violet.

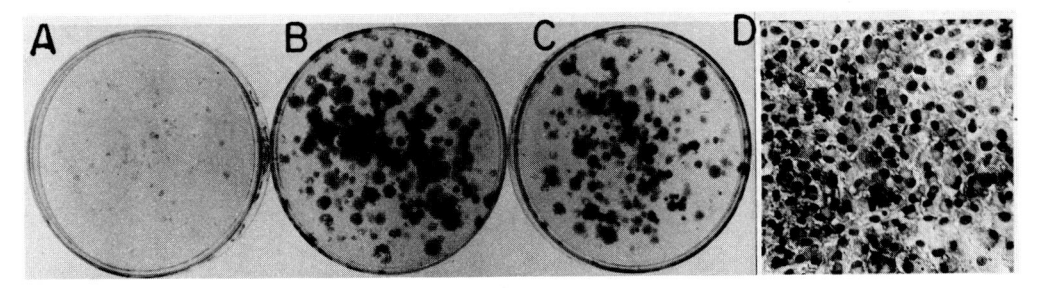

Fig. 10. A, B, C. Effect of FGF on the proliferation of normal adrenal cells. Normal adrenal cells were obtained by collagenase dissociation of mature bovine adrenal glands. 1000 cells were plated per 6-cm dish in the presence of 5 ml of F12 medium with 10% horse serum. One day later, pituitary FGF, 100 ng/ml (B); brain FGF, 100 ng/ml (C), was added to the dishes. Control (A) did not receive any addition. The media were changed every other day, and FGF was added with the media change. After 10 days, the plates were fixed with 10% formalin pH 7.2 and stained with 1% crystal violet. D. Higher magnification (207 ×) of the colony seen in (B). Epithelial cell types can be clearly seen.

blasts seeded at very low density (10^{-5} cells/mm^2), and, not only does it promote proliferation, but it also promoted fusion at extremely low density (Gospodarowicz et al., 1975c) (Fig. 11). This points out that the proliferation of myoblasts in tissue culture can be under the control of different agents depending on the cell density in the culture dish. The factor extracted at pH 8.5, but not yet purified, has been identified as a polypeptide since it is trypsin and heat sensitive. It has a neutral isoelectric point and a higher apparent molecular weight (24,000) than does FGF. The use of a partially purified preparation containing these factors has already been useful in obtaining clones of normal cells, a process which normally requires considerable time-consuming labor.

DISCUSSION

The observation that FGF can be mitogenic for a wide variety of cell types which are mesoderm-derived leads us to wonder what its physiological role may be.

Growth factors of vertebrates can be grouped in 3 categories. (1) The growth factors involved in controlling normal post-natal development. These are represented by growth hormone and thyroxine. Regardless of whether the action of growth hormone is direct or is mediated by somatomedins, it is, nevertheless, the leading growth factor in vivo. In amphibians, thyroxine is involved in metamorphosis which is a good example of precisely controlled cell proliferation. (2) The growth factors responsible for regulating the proliferation of cell types with a rapid turnover. The best characterized example in this category is erythropoietin. It is quite possible that other cell types are regulated by similar mechanisms. (3) The factors involved in wound healing and regeneration. These factors have not yet been identified, but it appears that growth factors may be released from platelets as blood clots to enhance the repair of superficial wounds. In lower vertebrates, regeneration of limbs can proceed under the influence of growth factors released from neural tissue.

Since FGF can be found in neural tissue, and since the role of nerves is well established in regeneration (Singer, 1974), our first hypothesis was that FGF could be a neurotrophic factor involved in the regeneration of the limbs of lower vertebrates.

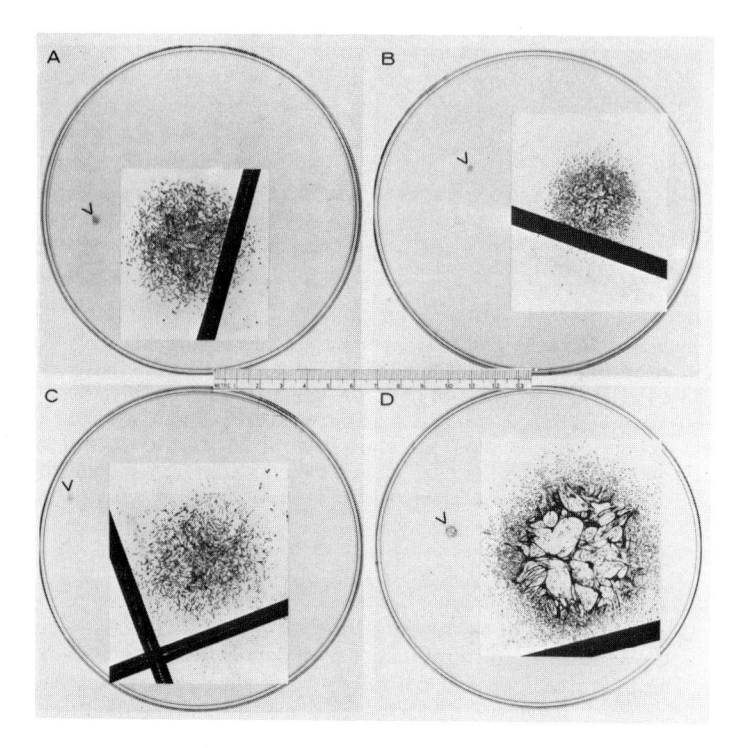

Fig. 11. Effect of the neutral peptide on the proliferation of bovine myoblasts. Five (± 2) myoblasts obtained by trypsinization of the hind leg muscles of a bovine fetus (Gospodarowicz et al., 1975c) were plated in 15-cm dishes in 30 ml of 1:4 Dulbecco's modified Eagle's medium 199 with 10% horse serum. 24 hours later, 100 µg/ml of a pH 8.5 crude extract of brain (Gospodarowicz et al., 1975c) was added to the dishes. In each instance, a single clone of myoblasts became macroscopically visible by 7 days. In controls, which did not receive an addition of the pH 8.5 crude extract of brain, no cells were visible after 7 days. A, B, and C show representative experimental dishes after 7 days. The clones are shown at 11 × magnification in the inserts. In B, fusion has started; D was fixed and stained at day 9, fusion has become extensive, and proliferating myoblasts can be seen at the periphery of the clone. Dishes were fixed and stained as described for Fig. 7.

We have shown in recent studies that FGF can induce the proliferation of undifferentiated cells resembling those of a regeneration blastema in the stumps of amputated frog limbs (Gospodarowicz et al., 1975b), thus inducing the first step of regeneration — blastema formation. Partial redifferentiation was observed but no morphogenetic event occurred, indicating that, in addition to FGF, other agents (neurotrophic or non-neurotrophic) are needed to obtain full differentiation of the limbs. Proof that FGF is the neurotrophic factor involved in regeneration can only be obtained when the factor or factors necessary for morphogenesis are identified.

Since blastema cells are believed to be derived from muscle and/or chondrocytes (Hay, 1962; Thornton, 1968), we have investigated the effect of FGF on the rate of proliferation and differentiation of primary cultures of myoblasts. This cell type was selected because the differentiation of myoblasts into myotubes can be easily followed in vitro by observing the morphology of the culture as well as by measuring the binding of neurotoxin to acetylcholine receptor sites which appear when myoblasts differentiate into myotubes. Thus, with myoblast cultures, one can follow simultaneously the effects of FGF on proliferation and differentiation.

Our conclusion is that FGF cannot block the differentiation of myoblasts, which always occurs when cells are crowded. FGF can, however, delay the fusion of myoblasts to a great extent, thus retarding the process of differentiation of myoblasts into myotubes (Gospodarowicz et al., 1976).

The mitogenic effect of FGF on chondrocytes was also marked, and while it appears that FGF does not belong to the somatomedins (since it lacks sulfation-factor activity) (Jones and Allison, 1975), its effect on the proliferation of chondrocytes in tissue culture is quite remarkable, not only because the concentration of FGF needed (1 to 10 ng/ml) is very low, but also because these cells will eventually differentiate to form cartilage. If the role of FGF is not found in the normal metabolic processes which control the proliferation of articular chondrocytes during the growth process, it could be found in the bone repair mechanism or in regeneration.

The effect of FGF on cell proliferation in tissue culture seems to be similar to that expected of a wound-healing hormone. FGF has been shown to be a potent mitogen for both established cell lines (Gospodarowicz, 1974; Gospodarowicz and Moran, 1974a,b) and diploid fibroblasts or other normal cells in early passage cultures (Gospodarowicz, 1975; Gospodarowicz and Handley, 1975; Gospodarowicz and Moran, 1975a,b; Gospodarowicz et al., 1975a,b,c). It has also been shown to be a mitogenic agent for endothelial cells as well as smooth muscle. Those cell types (endothelial cells and fibroblasts) are known to play an important role in the formation of the granulation tissue which is the first step of wound healing. The effect of FGF on endothelial cells and smooth muscle of the arterial wall also calls our attention to its role in the formation of new capillaries which are formed in granulation tissue by endothelial cells.

We have also shown that FGF is similar to the mitogenic factor(s) present in platelets since the addition of FGF or platelet extract to 3T3 cells maintained in the presence of plasma promotes the proliferation of the cells to the same extent as does serum (Gospodarowicz and Moran, 1975b).

For adrenal cells, FGF has been shown to be mitogenic for both early passage bovine adrenal cells and for the highly differentiated Y1 adrenal cell line which has ACTH receptor sites. Its role in adrenal regeneration should be tested, especially in view of the fact that ACTH, the trophic hormone for the adrenal gland, has been shown to inhibit the proliferation of adrenal cells in vitro (Masui and Garren, 1971; Ramachadran and Suyama, 1975).

While it is relatively easy to study the action of growth factors in vitro, the extension of this work in vivo may prove to be difficult. The main reason is that every organ contains growth-promoting activity; even if some, such as the brain and pituitary, contain much more than others (for a review, see Gospodarowicz and Moran, 1976b). To deprive an animal of specific growth factors by removing the producing organ(s) may be an impossible task. Another approach could be the use of specific antibodies to lower the concentration of growth factors present in the blood stream. However, even this approach may be limited. While one can neutralize one class of growth factors (such as FGF), cells have been shown to be responsive to more than one class of growth factors. A typical example is the human foreskin fibroblast which responds to FGF (Gospodarowicz and Moran, 1975b) as well as to EGF (Cohen and Carpenter, 1975), not to mention growth factors yet to be identified.

However, since it has been observed in tissue culture that with some cell types, such as the endothelial cells, human foreskin fibroblasts, and myoblasts, the effect of FGF is additive over serum, we hope that, at least in the case of the wound-healing process, FGF can speed up the processes, and, thus, provide an adequate model to study its role in vivo.

ACKNOWLEDGEMENTS

We thank D. Braun, J. Weseman, G. Greene, H. Bialecki, and D. Goldminz for excellent technical assistance. Since this paper was limited in length, we have presented mostly our work; contributions by others to the field of growth factors are acknowledged in the references of the publications to which the reader is referred.

REFERENCES

Clark, J., Jones, K.L., Gospodarowicz, D. and Sato, G. (1972): *Nature New Biol., 236,* 180.
Cohen, S. and Carpenter, G. (1975): *Proc. nat. Acad. Sci. (Wash.), 72,* 1317.
Gospodarowicz, D. (1974): *Nature (Lond.), 249,* 123.
Gospodarowicz, D. (1975): *J. biol. Chem., 250,* 2515.
Gospodarowicz, D., Greene, G. and Moran, J. (1975a): *Biochem. biophys. Res. Commun., 65,* 779.
Gospodarowicz, D. and Handley, H.H. (1975): *Endocrinology, 97,* 102.
Gospodarowicz, D., Jones, K.L. and Sato, G. (1974): *Proc. nat. Acad. Sci. (Wash.), 71,* 2295.
Gospodarowicz, D. and Moran, J. (1974a): *Proc. nat. Acad. Sci. (Wash.), 71,* 4648.
Gospodarowicz, D. and Moran, J. (1974b): *Proc. nat. Acad. Sci. (Wash.), 71,* 4584.
Gospodarowicz, D. and Moran, J. (1975a): *Exp. Cell Res., 90,* 279.
Gospodarowicz, D. and Moran, J. (1975b): *J. Cell Biol., 60,* 451.
Gospodarowicz, D. and Moran, J. (1976a): In: *Cell Culture and Its Application.* Editor: R. Acton. Plenum Press, New York. In press.
Gospodarowicz, D. and Moran, J. (1976b): *Ann. Rev. Biochem., 45,* in press.
Gospodarowicz, D., Rudland, P., Lindstrom, J. and Benirschke, K. (1975b): *Advanc. metab. Disorders, 8,* 301.
Gospodarowicz, D., Weseman, J. and Moran, J. (1975c): *Nature (Lond.), 256,* 216.
Gospodarowicz, D., Weseman, J., Moran, J. and Lindstrom, J. (1976): *J. Cell Biol.,* in press.
Ham, R.G. (1972): *Methods Cell Physiol., 5,* 37.
Hay, E.D. (1962): In: *Regeneration, p. 117.* Editor: R. Ross. Academic Press, New York.
Jones, K.L. and Allison, J. (1975): *Endocrinology, 97,* 359.
Jones, K.L. and Gospodarowicz, D. (1974): *Proc. nat. Acad. Sci. (Wash.), 71,* 3372.
Leffert, H. (1974): *J. Cell Biol., 62,* 792.
Masui, H. and Garren, L.D. (1971): *Proc. nat. Acad. Sci. (Wash.), 68,* 3206.
Moran, J., Gospodarowicz, D. and Owashi, N. (1976): submitted for publication.
Rudland, P., Eckardt, W., Gospodarowicz, D. and Seifert, W. (1974a): *Nature New Biol., 250,* 337.
Rudland, P., Gospodarowicz, D. and Seifert, W. (1974b): *Nature New Biol., 250,* 741.
Rudland, P., Seifert, W. and Gospodarowicz, D. (1974c): *Proc. nat. Acad. Sci. (Wash.), 71,* 2600.
Ramachadran, J. and Suyama, A.T. (1975): *Proc. nat. Acad. Sci. (Wash.), 71,* 113.
Singer, M. (1974): *Ann. N.Y. Acad. Sci., 228,* 308.
Thornton, C.S. (1968): *Advanc. Morphogenes., 7,* 205.

ADDENDUM

The experiments on endothelial cells from the umbilical cord (Fig. 5) have been repeated and extended using endothelial cells from the fetal and adult bovine aorta. These cells grow approxi-

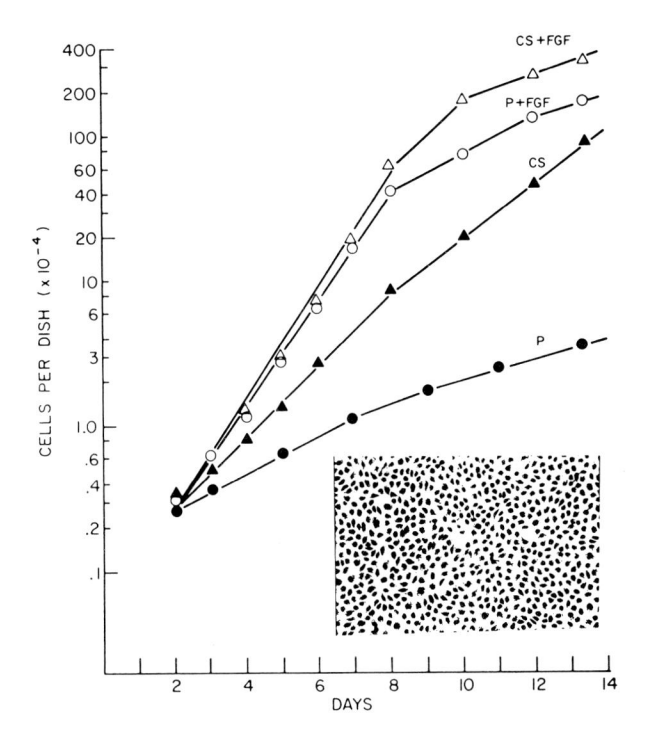

Fig. 1. Proliferation of endothelial cells maintained in plasma or serum with and without FGF. Endothelial cells obtained bij collagenase digestion from the fetal bovine aortic arch were seeded at low density in 10% calf serum in 6-cm dishes. After 2 days, the medium was removed and replaced with either 10% calf serum (CS) or 10% bovine plasma serum (P) with or without 10 ng/ml fibroblast growth factor (FGF). The media were changed every other day. In the presence of FGF, the cells proliferate as monolayers of small, densely-packed cells typical of endothelium (inset).

mately twice as fast in serum (or 4 times as fast in plasma serum) supplemented with FGF than they do in the absence of FGF (Fig. 1). When fetal aortic endothelial cells are seeded at low density in 10% serum the slowly growing colonies consist of large, flat, irregular and sometimes vacuolated cells. In contrast, under the same conditions except with 10-100 ng/ml FGF present, the fast growing colonies consist of homogeneous monolayers of small cells with a typical endothelial morphology (Fig. 1). The possibility that FGF may be related to the endotheliotrophic factor in platelets is currently under investigation. In addition, the mitogenic effect of FGF on endothelial cells is being examined to determine whether or not FGF may be a tumor angiogenesis factor.

In the field of limb regeneration we have been able to demonstrate that FGF is mitogenic for the mesenchymal cells of denervated blastemas of the newt, *Triturus viridescens*. The mitotic index of the denervated limb blastema into which FGF has been injected is 85-90% of the contralateral, undenervated blastema (A. Mescher and D. Gospodarowicz, in preparation).

III. Somatomedin

SOMATOMEDINS A AND B.
Isolation, chemistry and in vivo effects*

L. FRYKLUND[1], A. SKOTTNER[1], H. SIEVERTSSON[1] and K. HALL[2]

[1]The Recip Polypeptide Laboratory, AB Kabi, Stockholm and [2]Department of Endocrinology and Metabolism, Karolinska Hospital, Stockholm, Sweden

Many reviews have been written in recent years on the role of somatomedins and the postulated relationship to growth hormone (Van Wyk et al., 1973; Hall and Luft, 1974; Van Wyk and Underwood, 1975; Hall et al., 1975). There has been a great deal of discussion on the choice of the name *somatomedins* by Daughaday et al. (1972) and the substances to be included. In 1974 we presented evidence (Fryklund et al., 1974*a,b*; Sievertsson et al., 1975) of the isolation and characterization of somatomedins A and B as distinct and unique chemical entities, still retaining the ability to stimulate sulphate and thymidine uptake in vitro which had been used to identify them during isolation. Recent data from assays which utilize the displacement of these pure substances would also indicate that the serum levels of somatomedin A (Hall et al., 1974) and somatomedin B (Yalow et al., 1975) change in conditions where growth hormone is known to vary. Our preliminary in vivo studies, presented here, which examine leucine and sulphate incorporation in rats indicate that these somatomedin A and B preparations in vivo mimic certain known effects of growth hormone. For the first time direct evidence is obtained that somatomedins may carry a growth hormone-like activity, and a possibility of defining more clearly that the respective physiological role is apparent.

ISOLATION

The starting material is an acid-ethanol extract from Cohn's plasma Fraction IV as described by Uthne (1973). The isolation scheme for the preparation of somatomedins A and B is basically that described earlier (Fryklund et al., 1974*b*). Table 1 shows the yields and activity/mg protein at the various stages. The activity in the acid-ethanol extract is excluded here since anomalous results were obtained. By adding the pure iodinated substance to a small batch of the Cohn Fraction IV we have been able to study the distribution in the various precipitation stages, and several places where losses occur have been observed. The particular clay-like constitution of Fraction IV and the resulting precipitates, however, means that efficient extraction is hindered mechanically. The ideal extraction procedure should probably involve a specific adsorption step using for example antibodies. The subsequent stages of gel filtration and column electrophoresis on cellulose are easy to run and good yields of product and

---

* Research was supported by financial assistance from AB Kabi and grants from the Swedish Medical Research Council.

TABLE 1

Yields and activity data for a typical fractionation

Stage	System	Somatomedin A weight	Somatomedin A U/mg*	Somatomedin A μg standard/μg**	Somatomedin B weight	Somatomedin B μg standard/μg†
1	Acid-ethanol extract from Cohn Fraction IV equivalent to 1 ton starting plasma	100 g		—	100 g	—
2	Sephadex G-50, 1% HCOOH (10 × 100 cm)	200 mg	55	0.18	83 mg	0.5
3	Cellulose column electrophoresis. pH 7.5, 0.05 M N-ethyl morpholine-acetate buffer (2 × 100 cm)	80 mg	100	1	20 mg	0.7
4	Cellulose column electrophoresis, pH 5.0, 0.05 M pyridine-acetate buffer (1 × 100 cm)	10 mg	200	1.9	2.5 mg	1
5	Sephadex G-50 fine, 0.02 N HCl (1.6 × 93 cm)	5 mg	400	5.8	2 mg	1.5
6	Cellulose column electrophoresis, pH 2.0, 7.8% HOAc, 2.5 % HCOOH (0.75 × 50)	0.6 mg	3600	12		

* chick bioassay; ** radioreceptor assay; † radioimmunoassay.
It should be noted that all the various amidated forms of somatomedin B are approximately equal in antigenicity and that the somatomedin A standard used in the receptor assay is not pure.

activity are obtained. We can obviously use the electrophoresis systems in a flexible manner, samples from the pH 5.0 stage can be re-run at this pH or any other. Recently we have started using a smaller column, 0.75 × 50 cm, at either pH 5.0 or pH 2.0 which is proving useful for small quantities of material and also has a short running time.

AMINO-ACID COMPOSITIONS

The amino-acid compositions of somatomedins A and B have been presented earlier (Fryklund et al., 1974a,b) and are shown in Table 2. We have also presented evidence for 2 forms of somatomedin A and described multiple forms of somatomedin B which presumably result from deamidation. It is evident from Table 2 that A and B are quite different from each other, and also distinct from both insulin and growth hormone. End-terminal analysis by the Edman procedure and carboxypeptidase A as described by Fryklund and Eaker (1973) shows that residues 1 and 2 in somatomedin B are Asp-Gln and residues 43 and 44 are Val-Thr.

TABLE 2

Amino-acid compositions for somatomedins

	Somatomedin A		Somatomedin B	
Asp	4.75	5	5.23	5
Thr	1.85	2	3.96	4
Ser	3.50	3-4	3.09	3
Glu	4.65	5	8.12	8
Pro	7.30	7	0.77	1
Gly	7.10	7	2.07	2
Ala	7.00	7	1.06	1
Cys	1.00	1	8.00	8
Val	4.90	5	1.83	2
Met	1.00	1	—	
Ile	0.81	1	—	
Leu	3.30	3	0.95	1
Tyr	0.97	1	2.90	3
Phe	2.20	1	0.99	1
His	3.75	4	—	
Lys	2.50	2-3	3.83	4
Arg	4.65	5	1.27	1
Trp		—		—
No. of residues		60		44
N-terminal	Asn		Asp	

HETEROGENEITY OF SOMATOMEDIN A

During the last year, we have encountered several preparations of somatomedin A which have shown varying degrees of activity, and although the amino-acid compositions have had the same basic pattern, the actual mole ratios have not been identical.

Somatomedins A and B

This cannot be ascribed simply to impure preparations. We began to wonder whether we were not in fact obtaining a spectrum of peptides as a result of some form of degradation, containing a core with ragged edges. Tryptic digests of all these preparations were therefore performed under identical conditions (pH 7.5, 1:50 enzyme to substrate ratio, incubation at 40° for 2 hours). The digests were examined by running thin-layer chromatography on cellulose plates in 2 different systems. As shown in Figure 1 we can see 4 peptides which appear in the same amounts in all fractions, whereas the other peptides vary. These results would appear to support the core hypothesis, and structure analysis of these peptides should provide some information as to the nature of the radioreceptor binding site (Hall et al., 1974) as well as the requirements for biological activity.

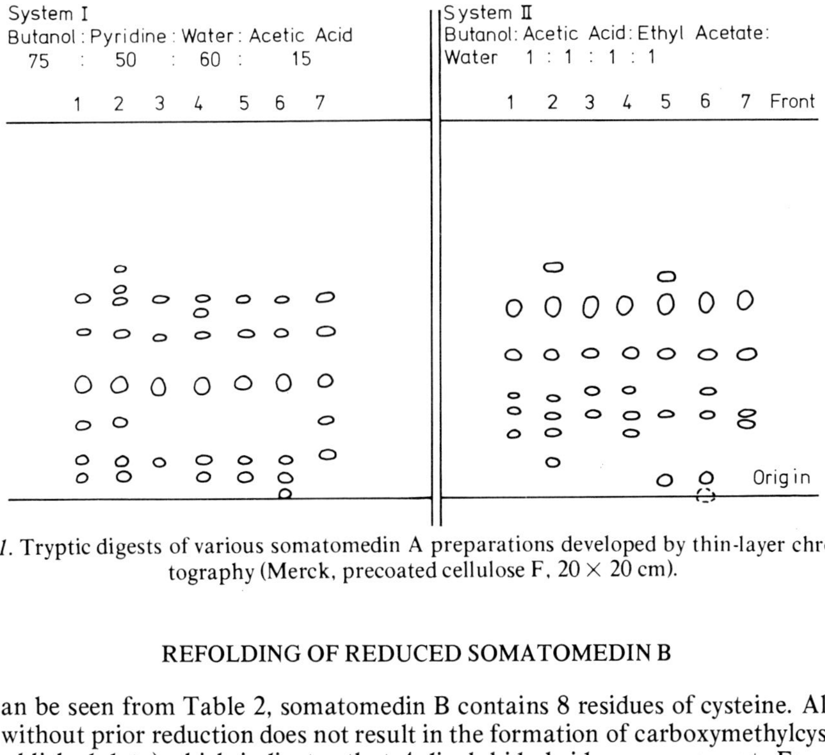

Fig. 1. Tryptic digests of various somatomedin A preparations developed by thin-layer chromatography (Merck, precoated cellulose F, 20 × 20 cm).

REFOLDING OF REDUCED SOMATOMEDIN B

As can be seen from Table 2, somatomedin B contains 8 residues of cysteine. Alkylation without prior reduction does not result in the formation of carboxymethylcysteine (unpublished data) which indicates that 4 disulphide bridges are present. From the point of view of synthesis it is important to determine whether the reduced molecule can refold into the correct conformation with the right disulphide pairing. A sample of somatomedin B was reduced under nitrogen by the standard procedure (Crestfield et al., 1963) for 4 hours using β-mercaptoethanol. An aliquot was removed for alkylation and the remaining solution gel-filtered to remove excess reagents.

The reduced material was pooled, a sample was removed for radioimmunoassay (RIA) and the remaining solution diluted. Compressed air was bubbled through the solution, which was allowed to stand at room temperature with stirring. Samples were removed for alkylation to determine the extent of reoxidation, and RIA, at 3, 6.5 and 24 hours. The 24-hour sample was then gel-filtered on a calibrated Sephadex G-50 fine column (1.6 × 93 cm in 0.02 N HCl). As seen from the gel-filtration pattern shown in

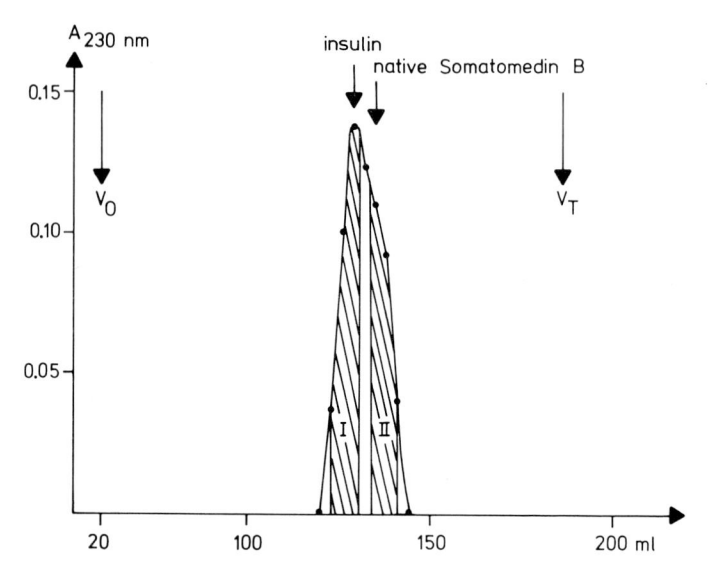

Fig. 2. Gel filtration of the 24-hour sample of reoxidized somatomedin B on Sephadex G-50 fine, 1.6 × 93 cm, in 0.02 N HCl. Fractions of 3.0 ml were collected. V_o and V_t indicate the void and total volume of the column, respectively.

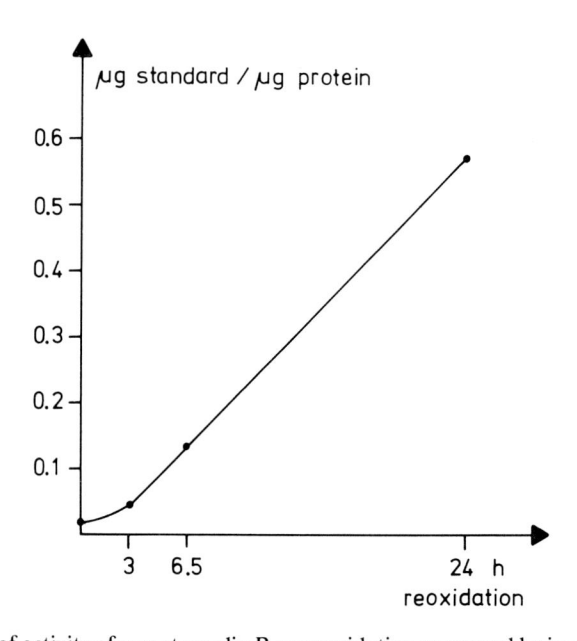

Fig. 3. Recovery of activity of somatomedin B on reoxidation measured by immunoassay.

TABLE 3
Reoxidation of somatomedin B

	Radioimmunoassay (μg standard/μg protein)	CM-Cys* content
Starting material	1.18	0
Fully reduced	0.010	8
3-hour reoxidation	0.043	2.5
6.5-hour reoxidation	0.132	1.4
24-hour reoxidation	0.57	0
24-hour Fraction I	0.01	–
24-hour Fraction II	1.00	–

* CM-Cys = carboxymethylcysteine.

Figure 2, an asymmetrical peak was obtained, which was divided into Fractions I and II as indicated. Fraction I eluted approximately as insulin, and Fraction II as native somatomedin B. Both fractions were gel-filtered again separately on the same column and aliquots taken for RIA. The results of the alkylation and RIA during reoxidation are shown in Table 3. The fully reduced sample has lost virtually all antigenic activity but recovery of antigenicity is seen with increasing reoxidation time (Fig. 3). Fraction II, after gel filtration, contained all the antigenic activity of the mixture, and presumably represents the native configuration. Fraction I is presumably refolded incorrectly — the shift in elution position indicates this. This experiment would therefore indicate that, using the criterion of RIA, somatomedin B can be refolded again correctly. When the experiment was performed at pH 7.5 in 1 M urea no immunoactivity was recovered.

IN VITRO ASSAYS

Somatomedin A is analysed routinely in our laboratory by the chick assay (Hall, 1970) and the radioreceptor assay using placental membranes (Hall et al., 1974; Takano et al., 1975). The displacement curve for somatomedin A is shown in Figure 4. Typical

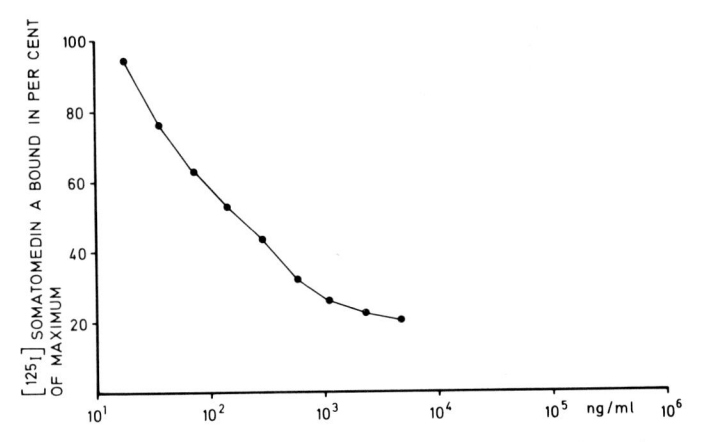

Fig. 4. Radioreceptor assay for somatomedin A using human placental membrane. Displacement of the ^{125}I-labelled somatomedin A by unlabelled material.

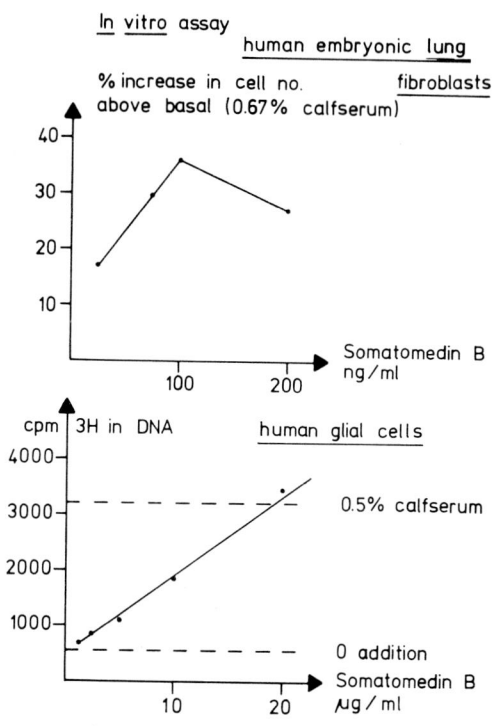

Fig. 5. Dose-response curves for somatomedin B measured in human embryonic lung fibroblasts and human glial cells.

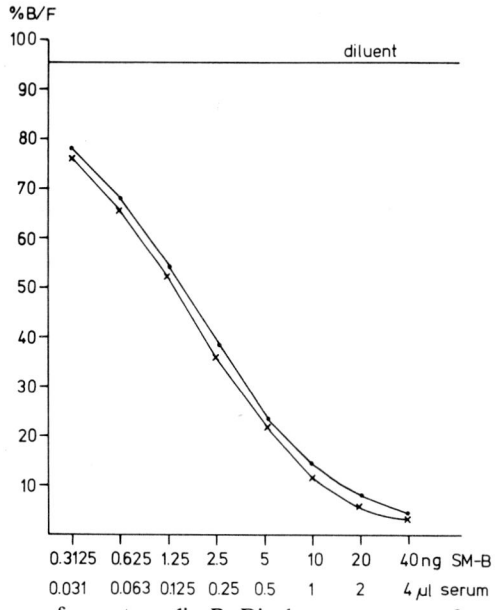

Fig. 6. Radioimmunoassay of somatomedin B. Displacement curves for pure somatomedin B ●—●, standard serum ×—×.

chick cartilage assay data have been presented earlier (Sievertsson et al., 1975). Somatomedin B is analysed by lung fibroblasts, and glial cells as described previously (Fryklund et al., 1974b). The dose-response curves are shown in Figure 5. A RIA was originally developed for somatomedin B by Yalow et al. (1975) and we have produced at Kabi our own antisera towards pure somatomedin B in the rabbit by adsorbing the substance to polyvinylpyrrolidone. The displacement curve is shown in Figure 6.

IN VIVO EXPERIMENTS

These were designed to study the effects of our purified preparations. Two series of preliminary experiments have been run using hypophysectomized female rats (Sprague-Dawley) weighing between 60 and 70 g. These rats were first used 14 days after operation to allow decrease of endogenous levels of rat growth hormone (RGH). Any animals showing a weight change of ± 5 g were rejected. Human growth hormone (HGH) (Crescormon®, AB Kabi) was used as a positive control in these experiments. The somatomedin A preparation used contained 78 U/mg as determined by chick cartilage assay and corresponded to the neutral fraction at pH 7.5 (Table 1, step 3). The somatomedin B preparation corresponded to a mixture of the 4 components described by Fryklund et al. (1974b), i.e., the pH 5.0 electrophoresis had been omitted (Table 1, step 4). RIA data indicated approximately 1 µg/µg pure standard.

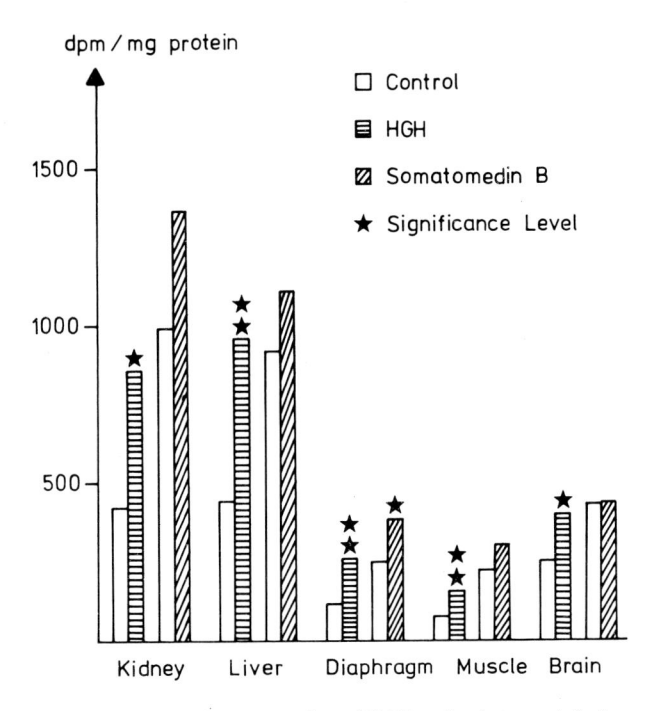

Fig. 7. Staple diagram illustrating incorporation of ^{14}C-leucine into protein in various organs in the hypophysectomized rat in vivo, using growth hormone and somatomedin B. (See Table 4 for the definition of significance levels.)

TABLE 4
Protein synthesis in vivo (d.p.m./mg protein)

	Rat no.	Kidney	Liver	Diaphragm	Muscle	Brain
Buffer	1	347.0	314.4	90.8	69.0	92.3
	2	149.3	163.1	74.6	53.7	117.8
	3	724.1	731.1	115.0	71.6	374.0
	4	301.7	313.0	95.7	39.6	251.7
	5	400.8	654.7	177.6	89.6	378.3
	6	593.0	492.0	126.6	123.4	285.1
	mean	419.3	444.7	113.4	74.5	249.9
	SD	207.5	219.9	36.4	29.4	122.8
	P	—	—	—	—	—
HGH	7	909.3	805.8	276.6	128.6	339.3
	8	1070.8	961.4	297.0	234.1	505.0
	9	407.1	854.7	282.8	171.4	421.6
	10	1167.0	1088.3	349.1	183.8	452.3
	11	563.6	535.0	148.0	88.4	244.1
	12	1024.7	1157.8	197.8	132.4	410.7
	mean	857.1	900.5	258.6	156.5	395.5
	SD	303.7	223.5	72.0	50.9	91.9
	P	*	**	**	**	*
Buffer	13	1209.7	1472.9	228.9	296.0	620.1
	14	794.3	827.6	263.7	192.0	462.1
	15	1259.1	776.7	333.9	285.0	398.1
	16	1253.0	1132.4	286.4	221.4	446.9
	17	464.4	395.0	109.7	111.9	220.0
	mean	996.1	920.9	244.5	221.3	429.4
	SD	355.0	404.8	84.5	74.9	143.7
	P	—	—	—	—	—
Somato-medin B	18	1299.3	1729.0	327.0	227.0	372.1
	19	1521.7	1101.1	450.7	325.0	331.4
	20	1005.6	911.3	339.3	247.7	315.7
	21	1457.1	1312.7	454.4	302.7	607.1
	22	1549.0	1041.4	334.9	332.7	544.7
	mean	1366.5	1219.1	381.3	299.0	434.2
	SD	223.8	319.8	65.2	43.7	132.8
	P	*	N.S.	*		N.S.

*P < 0.05, ** P < 0.01.

Protein synthesis in vivo

The procedure followed was as described by Kostyo and Nutting (1973). The rats were injected into one caudal vein, with 0.2 ml respectively of buffer, HGH (0.2 mg) and somatomedin B (2 mg). Thirty minutes later they received 0.2 ml ^{14}C-leucine (7.5 μCi). After a further 15 minutes the rats were killed by ether and a number of organs (liver, diaphragm, kidney, muscle and brain) were removed and frozen. Sections of each organ were extracted and the protein precipitated by a slightly modified procedure to that described by Kostyo and Knobil (1959). The weighed filter papers were then combusted and the carbon dioxide measured by liquid scintillation.

The results from scintillation were calculated as d.p.m./mg dry weight protein and analysed by Student's t-test for SD and P as shown in Table 4 and Figure 7. In HGH-treated rats the ^{14}C-leucine incorporation in all examined organs and tissues (except serum) differed significantly from that of the control rats. In the rats treated with somatomedin B only diaphragm differed significantly; the p values for kidney and muscle were $0.1 > p > 0.05$. These results indicate that this method might be useful for measuring the biological activity of somatomedin B. Obviously, further experiments must be carried out to determine dose-response and time curves and also the effect of somatomedin A in this system. We also made the same observation as Kostyo (personal communication) that rats larger than 70 g tend to show too great a variation for good statistical calculations, although the same tendencies described in Figure 7 can be observed.

Incorporation of ^{35}S-sulphate into cartilage and skin

The method of Collins and Baker (1960) and Collins et al. (1961) was employed. The rats, weighing about 70 g were injected intraperitoneally each morning and afternoon (7- to 8-hour interval) with 0.5 ml containing 2.5 μCi ^{35}S (as Na$_2$SO$_4$) and respectively HGH (25 μg), somatomedin A (0.5 mg protein), somatomedin B (0.3 mg protein), and buffer as control, for 4 days. On the morning of the 5th day the rats were weighed, and killed by ether. Pieces of skin, from the same place on the left side of the animals, and

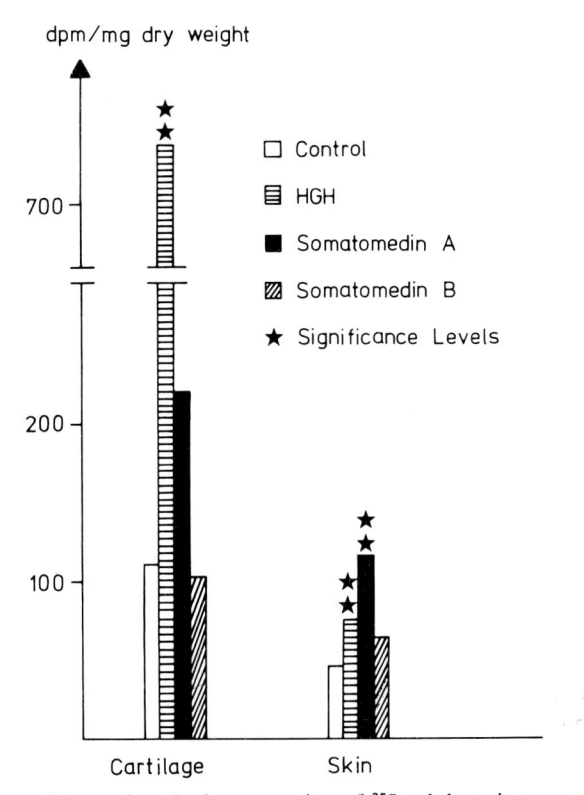

Fig. 8. Staple diagram illustrating the incorporation of ^{35}S-sulphate into costal cartilage and skin of the hypophysectomized rat in vivo, using growth hormone and somatomedins A and B. (See Table 4 for the definition of significance levels.)

TABLE 5
^{35}S-Incorporation in cartilage and skin (d.p.m./mg dry weight)

	Rat no.	Weight change (g)	Cartilage	Skin
Buffer	1	−11.5	71.1	40.9
	2	− 8.5	165.9	74.9
	3	− 3.5	68.6	41.3
	4	− 1.5	138.6	29.7
	mean	− 6.25	111.1	46.7
	SD		48.9	19.5
	P		−	−
HGH	5	+11	665.9	81.4
	6	+15	−	69.3
	7	+14.5	638.9	69.9
	8	+14	744.7	69.1
	9	+13.5	805.0	73.4
	10	+10.5	839.3	87.1
	mean	+13.1	738.8	75.0
	SD		86.3	7.5
	P		**	**
Somato-medin A	11	−12	312.6	102.6
	12	− 9.5	304.9	150.7
	13	− 6.5	172.3	107.4
	14	− 1.5	140.3	76.6
	15	− 6	172.7	146.0
	mean	− 7.1	220.6	116.7
	SD		81.6	31.3
	P			**
Somato-medin B	16	− 2	104.3	32.0
	17	− 2	61.7	45.4
	18	− 3	86.9	65.6
	19	+ 2	109.3	66.3
	20	− 4	138.1	65.7
	21	− 1	119.1	110.1
	mean	− 1.7	103.2	64.2
	SD		26.4	26.5
	P		N.S.	N.S.

* $P < 0.05$, ** $P < 0.01$.

also pieces of cartilage from the 6th and 7th ribs were removed, rinsed and frozen individually. Small pieces from each sample were treated in a similar manner to that described by Hall (1970) for chick pelvic leaflets. The weighed samples were then digested for 2 hours at 60° in 0.5 ml 2% papain solution before liquid scintillation counting. The results were corrected to give d.p.m. ^{35}S/mg cartilage and analysed by Student's t-test as shown in Table 5 and Figure 8. Both the HGH- and somatomedin A-treated rats showed an increased incorporation of sulphate into both cartilage and skin, compared to the controls. The HGH-treated rats had gained between 10 and 15 g during the experiments and had an obese appearance. The somatomedin A-treated rats had, on the other hand, lost about 7 g in weight. This indicates, with regards to somatomedin A, that the elevated incorporation of isotope does not result in total growth of the rat. However, no skeletal comparisons have as yet been made. The rats treated with somatomedin B remained more or less constant in weight, unlike the controls which also showed a decrease. Somatomedin B did not increase the level of sulphate incorporation above that of the controls.

These results suggest that we can use this model for examining the in vivo effects of somatomedin A. Further work is necessary, first, by using purer preparations and second, to determine dose-response curves.

CONCLUSIONS

Our isolation scheme, though not very efficient in terms of total yield from the starting material, has allowed us to obtain sufficient material for chemical characterization and some structural and biological studies. Many of our problems encountered during the isolation of somatomedin A, such as low yields, loss of biological activity, and spreading of activity into several fractions can possibly be accounted for by degradation of the parent molecule. With this in mind it should be pointed out that the activity data in Table 1 from step 4 onwards do not describe all the fractions obtained. Lack of parallelism in the chick cartilage assay is not surprising either. Perhaps the most important factor we have seen here is that somatomedin B can refold into a correct conformation after reduction. We have also shown that the pure substances we have isolated can be used in specific displacement assays to distinguish between various syndromes resulting from malfunctioning of growth-hormone metabolism as was discussed by Dr. Kerstin Hall (*This Volume*, p. 178). Our in vivo studies indicate, even though the results are preliminary, that our preparations can mimic certain in vivo effects of growth hormone.

The serum levels of these substances appear to be approximately 1000-fold that of HGH (Tall and Takano, unpublished observations; see also Fig. 6) which would in fact be expected if an amplification effect was involved. In our in vivo experiments we can see effects at a dose 10-fold higher than that of HGH, though, as we have stated earlier, we are not yet certain if this is optimal. However, as shown by Hall et al. (*This Volume*, p. 178) and Yalow et al. (1975) both somatomedins A and B are present in serum as high molecular weight forms which are perhaps not physiologically active. Time will show whether this is the explanation of the serum levels, which for polypeptide hormones are unusually high. A corollary to this is of course that we could probably never have isolated these substances from serum if they had been present at a concentration of ng/ml since we did not observe any insulin during our isolation work.

SUMMARY

The isolation and chemical characterization of somatomedins A and B are discussed with special emphasis on heterogeneity and refolding and behaviour in in vitro assay systems. Preliminary in vivo data where effects of growth hormone can be mimicked are also presented.

ACKNOWLEDGEMENTS

We are much indebted to Professor Bertil Åberg, Director of Research and Development, AB Kabi, and Professor Rolf Luft, Department of Endocrinology and Metabolism, Karolinska Hospital, Stockholm, for their continuous support. We also thank Dr. Bengt Westermark, Wallenberg Laboratory, Uppsala, for performing the glial cell assays.

REFERENCES

Collins, E.J. and Baker, V.F. (1960): *Metabolism, 9,* 556.
Collins, E.J., Lyster, S.C. and Carpenter, O.S. (1961): *Acta endocr. (Kbh.), 36,* 51.
Crestfield, A.M., Moore, S. and Stein, W.H. (1963): *J. biol. Chem., 238,* 622.
Daughaday, W.H., Hall, K., Raben, M.S., Salmon Jr, W.D., Van den Brande, J.L. and Van Wyk, J.J. (1972): *Nature (Lond.), 236,* 107.
Fryklund, L. and Eaker, D. (1973): *Biochemistry, 12,* 661.
Fryklund, L., Uthne, K. and Sievertsson, H. (1974a): *Biochem. biophys. Res. Commun., 61,* 957.
Fryklund, L., Uthne, K., Sievertsson, H. and Westermark, B. (1974b): *Biochem. biophys. Res. Commun., 61,* 950.
Hall, K. (1970): *Acta endocr. (Kbh.), 63,* 338.
Hall, K. and Luft, R. (1974): *Advanc. metab. Disorders, 7,* 1.
Hall, K., Takano, K., Fryklund, L. and Sievertsson, H. (1974): *J. clin. Endocr., 39,* 973.
Hall, K., Takano, K., Fryklund, L. and Sievertsson, H. (1975): *Advanc. metab. Disorders, 8,* 19.
Kostyo, J.L. and Knobil, E. (1959): *Endocrinology, 65,* 395.
Kostyo, J.L. and Nutting, D.F. (1973): *Hormone metab. Res., 5,* 167.
Sievertsson, H., Fryklund, L., Uthne, K. and Hall, K. (1975): *Advanc. metab. Disorders, 8,* 47.
Takano, K., Hall, K., Fryklund, L., Holmgren, A., Sievertsson, H. and Uthne, K. (1975): *Acta endocr. (Kbh.),* in press.
Uthne, K. (1973): *Acta endocr. (Kbh.), Suppl. 175.*
Van Wyk, J.J., Underwood, L.E., Lister, R.C. and Marshall, R.N. (1973): *Amer. J. Dis. Child., 126,* 705.
Van Wyk, J.J. and Underwood, L.E. (1975): *Ann. Rev. Med., 26,* 427.
Yalow, R., Hall, K. and Luft, R. (1975): *J. clin. Invest., 55,* 127.

REGULATION OF SOMATOMEDIN GENERATION*

WILLIAM H. DAUGHADAY, LAWRENCE S. PHILLIPS[1] and ADRIAN C. HERINGTON[2]

Department of Medicine, Metabolism Division, Washington University School of Medicine, St. Louis, Mo., U.S.A.

The hormonal factors regulating the concentration of somatomedin in plasma remain incompletely defined. It was already shown at an earlier time that growth hormone (GH) treatment of hypophysectomized rats and human beings resulted in a restoration of somatomedin activity of plasma. The effects of GH treatment in rats has recently been restudied by Phillips and Herington in my laboratory (Phillips et al., 1973). While a clear effect of GH treatment was confirmed, the dose of GH required in this relatively short experiment was considerably higher than that required to induce normal growth and sulfate uptake in the recipients' cartilage. These experiments suggested to us that newly synthesized somatomedin was exerting its target effects before accumulating in the plasma. Under conditions of submaximal doses of GH cortisone appeared to inhibit restoration of plasma somatomedin in hypox rat plasma. Wiedemann and Schwartz (1972) have presented evidence in man that estrogens also can suppress the effects of GH in stimulating the release of somatomedin.

These in vivo experiments have certain inherent limitations which makes unequivocal interpretation of results difficult. First of all, differences in plasma binding, tissue utilization or degradation of somatomedin may influence plasma levels independent of secretion. Second, hormonal agents may produce effects in vivo indirectly by altering the level of other hormones. It is indeed difficult to change the concentration of GH without inducing changes in insulin secretion. Lastly, hormone treatment may lead to nonspecific changes in nutrition or metabolism which could influence somatomedin generation. For these reasons we wished to examine the effects of hormonal agents directly on the perfused liver.

We were led to study the perfused rat liver because of the reports of McConaghey and Sledge (1970) and McConaghey (1972) that perfused liver from normal rats released little somatomedin activity unless GH was added to the perfusion medium. In their experiments, after the addition of bovine growth hormone (BGH) 10 μg/ml, a remarkable increase in activity resulted.

Further evidence that the liver may be important in the release of somatomedin was provided by Uthne and Uthne (1971) who showed that partial hepatectomy led to a rapid fall in plasma somatomedin activity with a restoration of somatomedin activity

---

* This study was supported in part by Research Grant Nr. AM01526 from the National Institute of Arthritis, Metabolism, and Digestive Diseases, Bethesda, Maryland.
[1] Current address: Division of Endocrinology, Northwestern University Medical School, Chicago, Ill. 60611, U.S.A.
[2] Current address: Medical Research Centre, Prince Henry's Hospital, Melbourne, Australia 3004.

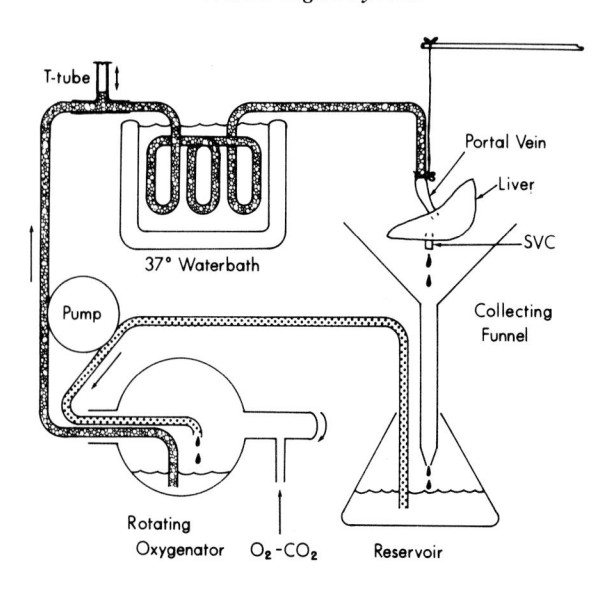

Fig. 1. Diagram of recirculating perfusion apparatus for rat livers. (From Phillips et al., 1976, by courtesy of the Editors of *Endocrinology*.)

with hepatic regeneration. Furthermore, the addition of growth hormone to liver slices (McConaghey, 1972) and microsomes (Hall and Uthne, 1971) has been reported to lead to increased formation of cartilage-stimulating substances. While experiments with perfusion of the kidney and with kidney slices by McConaghey and Dehnel (1972) also suggest that GH can stimulate the release of cartilage-stimulating material, the significance of renal formation of somatomedin remains in doubt.

With the above considerations in mind we undertook a systematic study of the hormonal regulation of somatomedin release in the isolated perfused rat liver. We have utilized a recirculating perfusion system which permitted efficient gas exchange from a rotary oxygenator and a high rate of flow of 15-16 ml/minute (Fig. 1). The perfusion system permitted the virtual complete change of medium at 30-minute intervals.

The medium which we selected for these studies was similar to that used by McConaghey, i.e. Waymouth's medium, MB-752/1 with 1% albumin. Under our conditions of perfusion, oxidative metabolism was well maintained as shown by the lactate and pyruvate levels in perfusates of livers from rats weighing less than 115 grams. Livers of larger rats showed variable increases in lactate pyruvate ratios (Table 1). With the extensive flushing of the liver before perfusion began gross contamination of the internal perfusates with plasma was negligible.

Somatomedin activity was determined by measuring the sulfate uptake by hypophysectomized rat cartilage in an incubation medium which contained 80% unperfused or perfused medium. A single hypophysectomized rat provided 60 segments of cartilage which were distributed into 12 treatment groups permitting the measurement of sulfate uptake for each perfusion period with the appropriate controls of unperfused medium. The results were calculated as sulfate uptake per mg of hypox cartilage and expressed as percent stimulation of sulfate uptake greater than observed in the unperfused medium. The conditions of incubation permit recognition of changes in somatomedin activity without a rigorous calculation of potency.

TABLE 1
Viability of perfused livers

Rat	Weight (g)	Perfusion volume (ml)	Time (final medium change) (min)	Lactate/ pyruvate ratio
Hypox	90	80	20 – 140	11.7
Hypox	110	80	20 – 140	9.3
Hypox	110	80	20 – 140	10.8
Normal	80	40	150 – 180	7.8
Normal	105	40	150 – 180	9.8
Normal	170	40	150 – 180	8.4
Normal	170	80	90 – 180	14.1
Normal	185	80	90 – 180	11.9
Normal	200	80	90 – 180	15.9

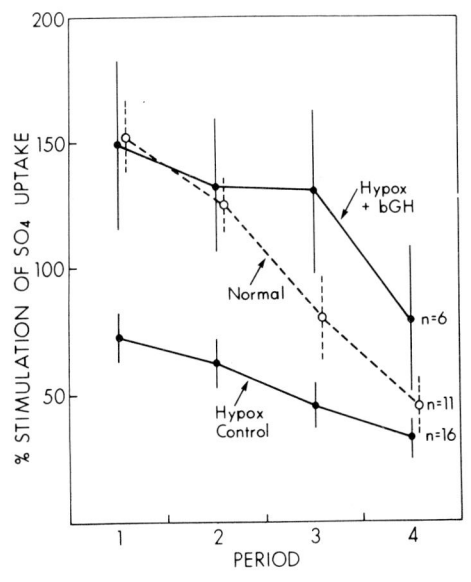

Fig. 2. Somatomedin activity of perfusates of normal, hypox and hypox rats pretreated with BGH. The ordinate is the percent stimulation of sulfate uptake as compared to unperfused medium. Time periods were each 20 minutes. (From Phillips et al., 1976, by courtesy of the Editors of *Endocrinology*.)

We first compared the release of somatomedin by normal and hypox rat livers (Fig. 2). It was observed that the degree of stimulation of sulfate uptake by perfusates of normal rat livers was significantly greater than that released by hypox livers for the first 4 incubation periods. In the perfusates of both the normal and hypox livers there was a progressive fall in somatomedin activity. These results differ substantially from those reported by McConaghey (1972) who found little evidence of somatomedin release by normal livers in the absence of GH. We attribute this difference to the fact that our medium was recirculated extensively permitting concentration of somatomedin, while this apparently was not done by McConaghey.

When hypophysectomized rats were treated with 500 μg of GH at 53, 29 and 6 hours prior to perfusion of their liver, the subsequent release of somatomedin during perfusion was identical to that of normal rats (Fig. 2).

When normal rats were fasted for 72 hours prior to liver perfusion the release of somatomedin was similar to that of hypox rat liver (Fig. 3). These experiments showed that normal and hypox rats differed in their release of somatomedin activity even when hormones were not added to the medium. GH treatment of hypox rats and starvation of normal rats produced the expected change in somatomedin activity. Despite the well maintained oxygenation of our perfusion and the use of a highly enriched medium, somatomedin release was not maintained. This decline in release may indicate attenuation of hormonal influences acquired in vivo or physiologic subtle deterioration of liver function.

We have studied the effects of 2 concentrations of growth hormone and insulin on the release of somatomedin from the perfused liver. The concentrations of hormone which we added to the perfusion medium during the second to sixth periods did not in themselves cause significant stimulation of sulfate uptake in our test cartilage (Table 2). Any slight stimulatory effect which was observed was neutralized by comparing the uptake of sulfate induced by perfusates with the unperfused medium containing the added hormone.

When normal rat livers were perfused with a medium containing 25 μg/ml BGH little or no increase in somatomedin activity was observed. The difference between perfusates with and without BGH was barely significant. Perhaps a greater number of experiments would have established a larger difference.

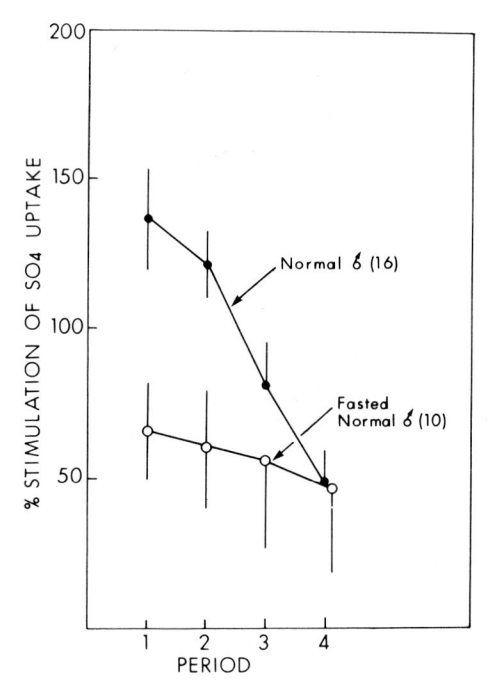

Fig. 3. The effect of fasting of male rats for 72 hours on the somatomedin activity of liver perfusates. (From Phillips et al., 1976, by courtesy of the Editors of *Endocrinology*.)

TABLE 2

Effect of hormone addition to buffer on uptake of sulfate by hypox rat cartilage

Addition	N	% Stimulation (mean ± SE)
BGH 25 µg/ml	10	16.6 ± 10
BGH 0.25 µg/ml	11	3.6 ± 4.5
Insulin 1000 µU/ml	10	12.0 ± 8.3
Insulin 100 µU/ml	10	5.7 ± 8.7
Insulin 1000 µU/ml + BGH 25 µg/ml	10	4.8 ± 7
Insulin 100 µU/ml + BGH 0.25 µg/ml	10	−5.7 ± 5.4

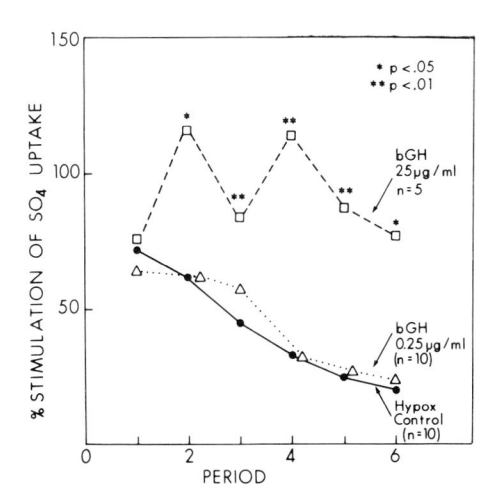

Fig. 4. Effect of BGH when added to perfusion medium in a concentration of 25 µg/ml and 0.25 µg/ml on somatomedin activity of perfusates of hypox rat liver. (From Phillips et al., 1976, by courtesy of the Editors of *Endocrinology*.)

When hypox rat livers were perfused with a medium containing BGH a relative stimulation of somatomedin activity was obtained when perfusates contained 25 µg/ml of BGH but no significant effect was evident with 0.25 µg/ml. The absolute effect of the higher concentration of BGH was to prevent the expected decline in somatomedin activity of the perfusion rather than to increase greatly the rate of somatomedin release (Fig. 4).

After Francis and Hill (1975) reported that addition of ovine prolactin in concentrations as low as 50 ng/ml caused the release of somatomedin from perfused normal male rat livers, we tested the effect of ovine prolactin, 500 ng/ml, in our perfusion system. In the 5 perfusions which we have completed there was no increase in apparent somatomedin activity of the perfusate.

We have been troubled by the apparent discrepancies of our results with those reported by McConaghey (1972) and by Francis and Hill (1975). We found evidence of continued release of somatomedin activity even in the absence of added GH in the medium. In our experiments significant responses to GH were observed only with hypox rats and the responses were far less dramatic than reported by McConaghey. A few differences in perfusion technique exist. We have used smaller rats (80 to 115 grams as compared to 220 to 260 grams). We continuously recirculate both during the control periods and after addition of BGH. Lastly, our high rate of perfusion (15-16 ml/min versus 4-5 ml/min) maintains normal lactate pyruvate ratios in our perfusate.

Another methodologic concern we had was whether during recirculation there was any significant expansion of the medium sulfate pool. The results of measurements of inorganic sulfate indicate that there is a modest expansion of the sulfate pool in perfusates of hypox livers (Fig. 5). The results of measurements of medium inorganic sulfate before and after incubation in 2 experiments suggest that insulin may further increase the sulfate pool. The effect of such an expansion would be to mask the total uptake of sulfate to decrease calculated hormonal effects. Our results have not been corrected for these minor increases in sulfate pool. The effect of such a pool change would be to decrease the apparent effect of insulin.

A number of clinical observations suggest that plasma somatomedin concentrations do not correlate with growth hormone concentrations in certain conditions of abnormal nutrition. In simple obesity spontaneous GH secretion and GH secretion in response to provocative stimuli is depressed yet somatomedin concentrations are normal or elevated (Van den Brande and Du Caju, 1973). Conversely, in protein calorie deficiency as present in Kwashiorkor growth hormone secretion is markedly elevated yet somatomedin activity of plasma has been reported to be depressed. These conditions suggest that nutritional factors, perhaps mediated by alterations in serum insulin concentration are also determinants of somatomedin activity.

We have also been interested in the bizarre growth patterns of certain patients with craniopharyngioma who before operation are short but postoperatively exhibit normal or 'catch-up' growth at a time that spontaneous or induced GH secretion is markedly depressed. Normal levels of somatomedin have been reported in such children. The participation of insulin in this phenomenon is suggested by the rapid weight gain that most of these children exhibit and the fact that plasma insulin levels are much higher than would be present in simple hypopituitarism (Costin et al., 1976). We hypothesize that subnormal concentrations of GH can sustain growth if hypernutrition and hyperinsulinism are present.

The isolated perfused liver has permitted us to examine the effects of growth hormone and insulin independently of one another. This is difficult or impossible in the whole animal.

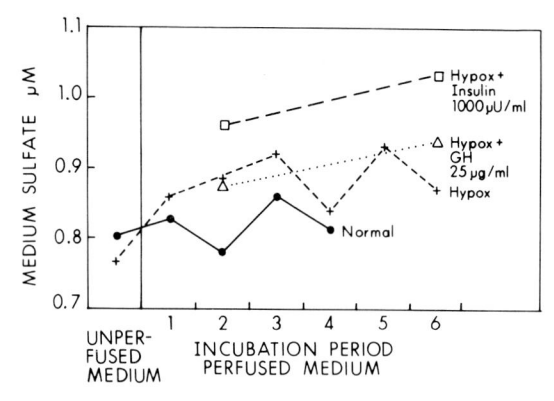

Fig. 5. Medium inorganic sulfate before and after liver perfusion with and without hormone additions.

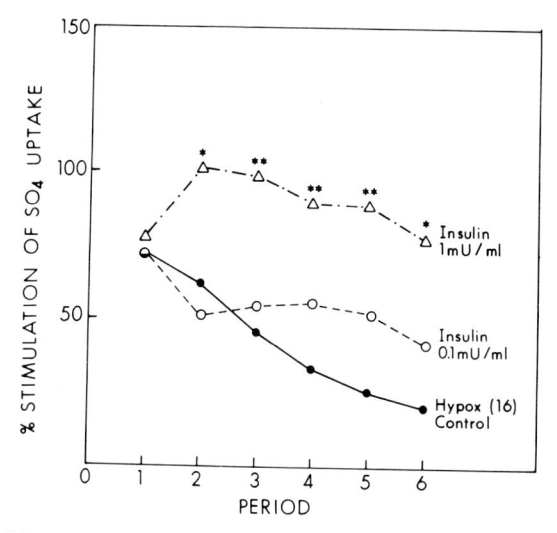

Fig. 6. The effect of 2 concentrations of insulin on the somatomedin activity of perfusates of hypox livers. *p <0.05, **p <0.01.

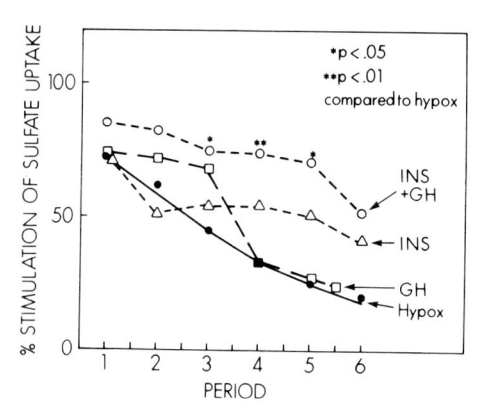

Fig. 7. Effect of addition of 250 ng/ml of BGH and 100 μU/ml insulin alone and in combination on somatomedin activity of perfusates of hypox rat liver.

We have perfused hypox livers with 2 concentrations of insulin, 1000 μU/ml and 100 μU/ml and the effects on perfusate somatomedin activity determined. The higher concentration of insulin induced a significant increase in the relative somatomedin activity while perfusion with 100 μU/ml had no significant effect (Fig. 6).

We have examined possible interactions of GH and insulin in the release of somatomedin by liver. For these experiments we selected a concentration of BGH (250 ng/ml) which is ineffective in our system. When this concentration of BGH was added to a perfusion medium containing 1000 μU/ml of insulin no increase in somatomedin activity release was observed. However, when this concentration of BGH was added to

100 µU/ml of insulin the relative somatomedin response was significantly greater than in the absence of hormone (Fig. 7). Neither concentration by itself was able to increase relative somatomedin activity. The data do not permit us to distinguish additive and synergistic effects.

We were interested in how much insulin remained in the perfusates after 30 minutes of recirculation. This was measured by radioimmunoassay (Fig. 8). With an initial concentration of 1000 µU/ml the perfusates contained 32% of the initial insulin during period 2 and this rose to 83% for period 6. When the original medium concentration was 100 µU/ml the residual insulin was 23% in period 2 and 32% in period 6. It is evident that our perfused livers remove much of the insulin originally added. At higher concentrations of insulin the clearance falls during the period of perfusion.

These experiments with insulin provide experimental support for our hypothesis that it may be a physiologic regulation of somatomedin release by the liver. Such an action of insulin would be consistent with the known effect of insulin in stimulating the release of other plasma proteins by liver. In the careful experiments of Griffin and Miller (1974) livers of hypophysectomized rats were perfused for 12-24 hours. It was observed that a hormone mixture containing growth hormone, insulin and triiodothyronine stimulated the release of a number of plasma proteins. Omission of either insulin or growth hormone led to a marked decrease in the release of fibrinogen, haptoglobin and α_2 (acute phase) globulin; both hormones were required for sustained plasma protein release. The experiments of Griffin and Miller (1974) differ from those reported here in that hormonal induction of plasma protein required 4-8 hours to become evident and 12-24 hours to reach a maximum. These time relationships are similar to the restoration of somatomedin in plasma after GH treatment of rats and human beings. It is possible that with more physiologic conditions of perfusion continued for a longer time that the full effects of hormonal induction would be more evident. The fact that somatomedin activity of perfusates declines progressively in both normal and hypox livers in the absence of hormones indicates to us that a deterioration of liver function was occurring and that our conditions were suboptimal.

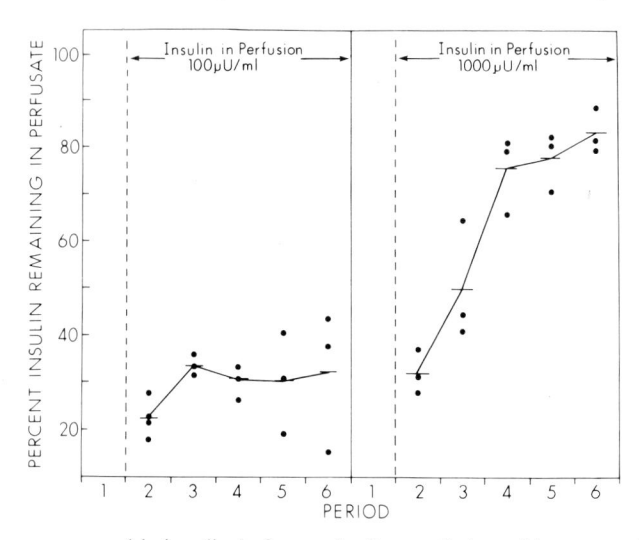

Fig. 8. Radioimmunoassayable insulin before and after perfusion of hypox rat liver. Left: results when initial insulin was 100 µU/ml; right: when the initial insulin was 1000 µU/ml. (From Daughaday et al., 1976, by courtesy of the Editors of *Endocrinology*.)

The liver perfusion experiments which have been conducted thus far do not permit us to distinguish simple release of preformed somatomedin from synthesis of new somatomedin. As indicated earlier we do not believe that contamination of medium with plasma was significant. Our bioassay did not permit us to compare the total somatomedin released into the perfusates as compared to the initial content in the liver.

Another unresolved question is whether the somatomedin activity released into the medium under the various hormonal situations is the same or different than normal rat plasma somatomedin. We have attempted to characterize the molecular size of the somatomedin by Sephadex G-75 filtration and to measure the ability of the fractions to compete with ^{125}I-insulin for binding on human placental membranes. A high molecular weight component similar to that found in rat plasma cross-reacted with the insulin binding site. These preliminary experiments suggest that the somatomedin activity was being released in bound form.

In summary our results with liver perfusion support the role of this organ in the release of somatomedin activity. We have shown that normal liver releases more somatomedin activity than hypox livers. GH stimulates the release of somatomedin activity from hypox livers but has little effect on normal liver. Insulin also can stimulate the release of somatomedin from hypox rat livers. The combination of GH and insulin may be required for normal hepatic release of somatomedin.

REFERENCES

Costin, G., Kogut, M.D., Phillips, L.S. and Daughaday, W.H. (1975): *J. clin. Endocr.*, in press.
Daughaday, W.H., Phillips, L.S. and Mueller, M.C. (1976): *Endocrinology, May issue.*
Francis, M.J.O. and Hill, D.J. (1975): *Nature (Lond.)*, 255, 167.
Griffin, E.E. and Miller, L.E. (1974): *J. biol. Chem.*, 249, 5062.
Hall, K. and Uthne, K. (1971): *Acta med. scand.*, 190, 137.
McConaghey, P. (1972): *J. Endocr.*, 52, 1.
McConaghey, P. and Dehnel, J. (1972): *J. Endocr.*, 52, 587.
McConaghey, P. and Sledge, C.B. (1970): *Nature (Lond.)*, 225, 1249.
Phillips, L.S., Herington, A.C. and Daughaday, W.H. (1973): In: *Advances in Human Growth Hormone Research*, p. 50. Editor: S. Raiti. DHEW (NIH) Nr. 74-612. U.S. Government, Washington, D.C.
Phillips, L.S., Herington, A.C., Karl, I.E. and Daughaday, W.H. (1976): *Endocrinology, April issue.*
Uthne, K.O. and Uthne, T. (1971): *Acta endocr. (Kbh.)*, Suppl. 155, 228.
Van den Brande, J.L. and Du Caju, M.V.L. (1973): In: *Advances in Human Growth Hormone Research*, p. 98. Editor: S. Raiti. DHEW (NIH) Nr. 74-612. U.S. Government, Washington, D.C.
Wiedemann, E. and Schwartz, E. (1972): *J. clin. Endocr.*, 34, 51.

STUDIES ON THE REGULATION OF SOMATOMEDINS A AND B*

KERSTIN HALL[1], KAZUE TAKANO[1], GÖSTA ENBERG[1] and LINDA FRYKLUND[2]

[1]Department of Endocrinology and Metabolism, Karolinska Hospital, and [2]The Recip Polypeptide Laboratory, AB Kabi, Stockholm, Sweden

Although growth hormone (GH) is generally regarded as the principal hormone regulating skeletal growth, evidence has accumulated that growth hormone by itself does not stimulate linear growth directly but rather induces the formation of secondary growth-promoting factors. Since the discovery by Salmon and Daughaday in 1957, that serum contained a GH-dependent activity, several such factors have been purified. Different purification procedures and bioassays have led to at least 4 different peptides being purified from human plasma; somatomedins A, B, and C and non-suppressible insulin-like activity (NSILA-S). Somatomedin A designates the factor which stimulates the incorporation of sulphate into chick cartilage, somatomedin B the factor which increases the DNA synthesis in glial cells, and somatomedin C the factor enhancing both sulphate and thymidine uptake into rat cartilage (Hall, 1972; Uthne, 1973; Fryklund et al., 1974a,b; Van Wyk et al., 1974, 1975). Somatomedin A, somatomedin C, and NSILA-S have been shown to have many target organs in common. All exert an insulin-like action on epididymal fat from rats and stimulate sulphate uptake into cartilage (Hall and Uthne, 1971; Underwood et al., 1972; Zingg and Froesch, 1973; Clemmons et al., 1974; Werner et al., 1974).

All bioassays respond to both stimulatory and inhibitory factors in serum. It is therefore often difficult to decide whether a decrease in the biological activity of serum is due to a decrease in stimulatory factors or an increase in inhibitory factors. The availability of purified somatomedin A and somatomedin B has led to the development of more specific and precise methods for their determination, such as a radioreceptor assay for somatomedin A (Hall et al., 1974; Takano et al., 1975) and a radioimmunoassay for somatomedin B (Yalow et al., 1975a).

RADIOLIGAND ASSAYS FOR SOMATOMEDIN A AND B

The somatomedin preparations used for labelling and standards have been purified at the Recip Polypeptide Laboratory, AB Kabi, Stockholm. In most of the present studies the somatomedin A used was identical with peptide 2 described by Fryklund et al. (*This Volume*, p. 156). This peptide, with a molecular weight of 7000 was homogeneous by electrophoresis and gel chromatography. The biological activity was 3,600 U/mg protein when determined by chick bioassay according to Hall (1970). The somatomedin B used for labelling and standards was identical with the peptide 3 earlier

---

* Research was supported by grants from the Swedish Medical Research Council (0424-03A).

characterized by Fryklund et al. (1974b). The iodination was performed according to the peroxidase method (Thorell and Johansson, 1971) with some modification (Takano et al., 1975). Antibodies against somatomedin B were originally raised by Rosalyn Yalow, who also developed a sensitive radioimmunoassay for somatomedin B (Yalow et al., 1975a). The antisera used in the present study have been produced at AB Kabi and in a final dilution of 1/1500 gave a linear dose-response curve between the ratio B/F and the logarithm of dosages between 0.5 and 50 ng/ml. No cross-reaction was found with somatomedin A, somatomedin C, insulin, or HGH as also shown by Yalow et al. (1975a).

The radioreceptor assays are based on the finding that the primary step in the action of polypeptide hormones is the interaction between hormone and its recognition sites on plasma membranes (Lefkowitz et al., 1970). Human placental membranes were shown to be rich in binding sites not only for insulin but also for somatomedins A and C (Posner, 1972, 1974; Haour and Bertrand, 1974; Marshall et al., 1974; Hall et al., 1974; Takano et al., 1975). Utilizing particulate placental membrane a radioreceptor assay has been developed for somatomedin A (Hall et al., 1974; Takano et al., 1975).

The bound labelled somatomedin A was readily displaced by somatomedin A. The dose-response curve between the displacement and the logarithm of dosages between 0.4 ng and 400 ng/ml (0.0014 U and 1.44 U/ml) was linear (Fig. 1). The Scatchard plot revealed more than one binding site. Assuming 2 binding sites, the apparent calcu-

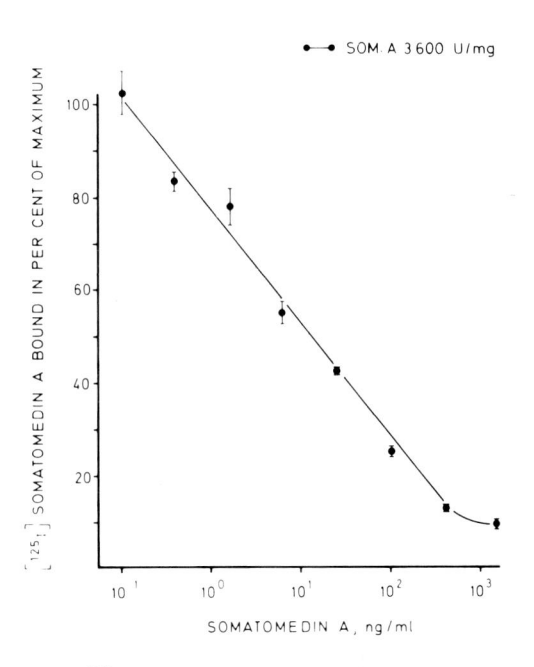

Fig. 1. The displacement of ^{125}I-somatomedin A bound to placental membrane by increasing amounts of somatomedin A. The per cent bound ^{125}I-somatomedin A is expressed in per cent of maximum binding which was 9% of total radioactivity. The bars indicate SEM and the calculated regression line is shown. The somatomedin A contained a biological activity of 3600 U/mg protein according to the chick bioassay. Note, that non-specific binding has not been subtracted.

TABLE 1

Specificity studies of the placental somatomedin A binding sites

Hormone	Concentration (ng/ml)	% of initial binding
Competitive:		
Somatomedin A, 3600 U/mg	15	50
Somatomedin C	40	50
Porcine insulin	10×10^3	50
Porcine proinsulin	25×10^3	50
Nerve growth factor	100×10^3	50
Porcine calcitonin	250×10^3	50
Non-competitive:		
Somatostatin	3×10^3	100
Somatomedin B	10×10^3	100
Fibroblast growth factor	11×10^3	100
HGH	14×10^3	100
Bovine ACTH	700×10^3	100

lated association constant for the high-affinity binding of somatomedin A was $3.6 \times 10^8 \, M^{-1}$ with 1.1×10^{-9} binding sites per mg protein of placental membrane (Takano, 1975). Of other hormones and factors somatomedin C, porcine insulin, porcine proinsulin, nerve growth factor, and porcine calcitonin could compete with somatomedin A for its binding sites. Somatomedin C was nearly as potent as somatomedin A in displacing bound labelled somatomedin A whereas insulin and other peptides with structural similarities to insulin required a dosage more than 1000 times higher (Table 1). Somatomedin B, somatostatin, HGH, ACTH, and fibroblast growth factor had no effect on the displacement of labelled somatomedin A.

THE LEVELS OF SOMATOMEDIN A AND B IN SERUM

Normal human serum in concentrations of 0.2-100 µl/ml caused a dose-dependent displacement of bound labelled somatomedin A and the dose-response curve of serum was congruent with that of somatomedin A. Furthermore, somatomedin A added was recovered to 100% in the radioreceptor assay (Takano et al., 1975). The limited availability of pure somatomedin A has led to the use of a normal serum instead of somatomedin A as standard in the determinations of somatomedin A levels in serum. This arbitrary reference serum was identical with that used in the chick bioassay and contained 2.06 (1.47-3.11) µg of somatomedin A per ml.

Serum from various animals also caused a dose-dependent displacement of bound labelled somatomedin A from its binding sites in human placental membranes (Takano, 1975). The level of somatomedin A decreased with increasing size of the animal species (Fig. 2).

Serum from patients with acromegaly was 6 times more potent than serum from patients with growth-hormone deficiency in displacing labelled somatomedin A from its binding sites. This finding indicates that the purified peptide somatomedin A really is a GH-dependent factor. However, no patient has been found to completely lack somatomedin A. The lowest levels of somatomedin A were found in 2 patients with Laron dwarfism (0.8 and 0.12 U/ml). The distribution of somatomedin A levels in

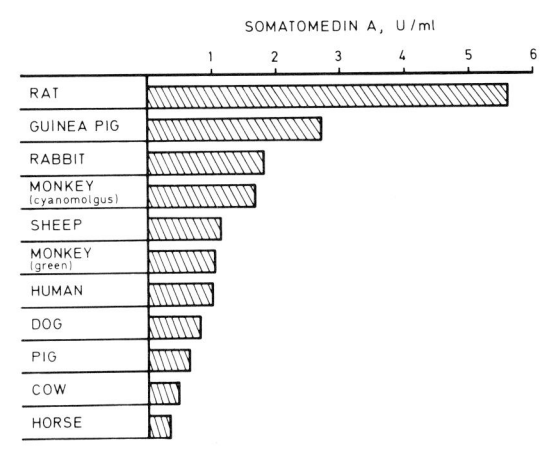

Fig. 2. Level of somatomedin A in serum from different animals in comparison with that of human serum (From Takano, 1975, by courtesy of the Author.)

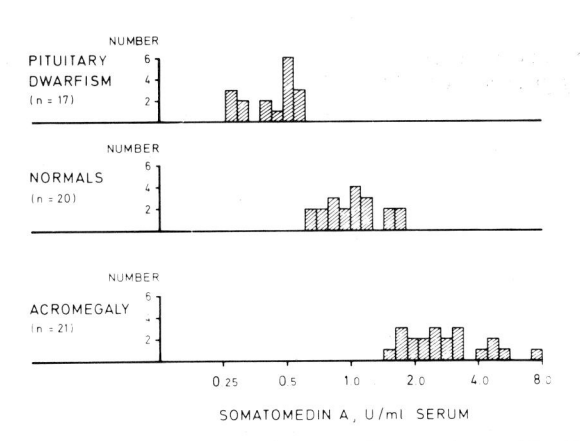

Fig. 3. Distribution of somatomedin A values in serum from patients with pituitary dwarsfism, normal subjects, and patients with acromegaly (From Takano et al., 1976*b*, by courtesy of the Editors of *Acta Endocrinologica*.)

normal adults and patients with growth-hormone deficiency and acromegaly are shown in Figure 3. The mean serum levels of somatomedin A in patients with growth-hormone deficiency and with acromegaly were 0.44 ± 0.03 U/ml and 3.02 ± 0.38 U/ml, respectively, and these values differed significantly from that in normals (Takano et al., 1976*b*).

Dilution curves of human serum and plasma were superimposable on the standard curve of somatomedin B. The arbitrary reference serum contained 10.3 µg of somatomedin B per ml (Fig. 4). The species specificity of the assay was similar to that reported earlier by Yalow et al. (1975*a*). No detectable immunoreactive component was found in sera from rat, guinea pig, rabbit, dog, pig, cow, or horse. The mean level of somatomedin B found in normal adults aged 23 years (n = 20) was somewhat higher than that in subjects aged 40 years (n = 18), 19.8 ± 1.5 and 9.8 ± 0.7 µg/ml, respect-

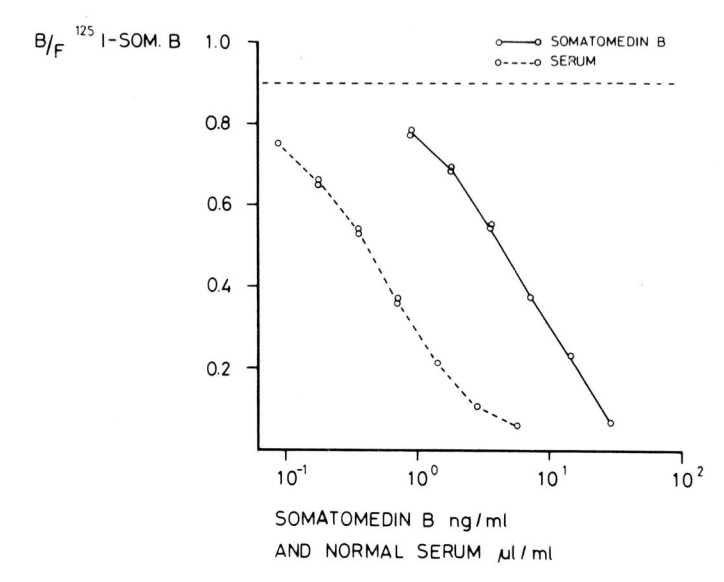

Fig. 4. Dose-response curve caused by somatomedin B and by serum. Standard diluent was 4% control guinea pig plasma in 0.02 M barbital buffer. Bound and free ^{125}I-somatomedin B were separated with charcoal.

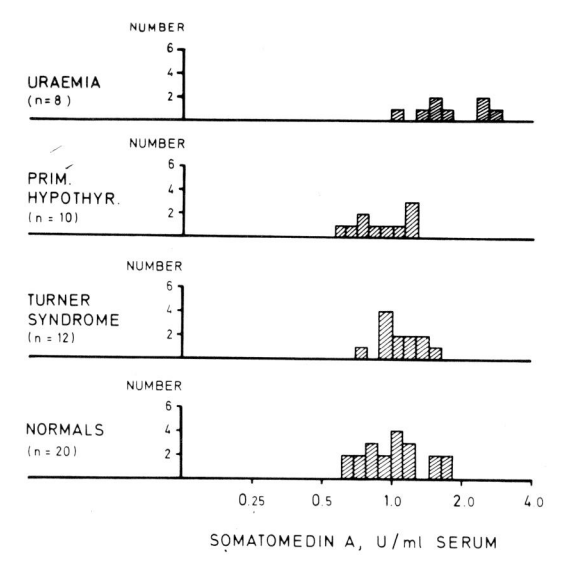

Fig. 5. Distributions of serum somatomedin A values from patients with various disorders.

Somatomedins A and B levels in serum and urine

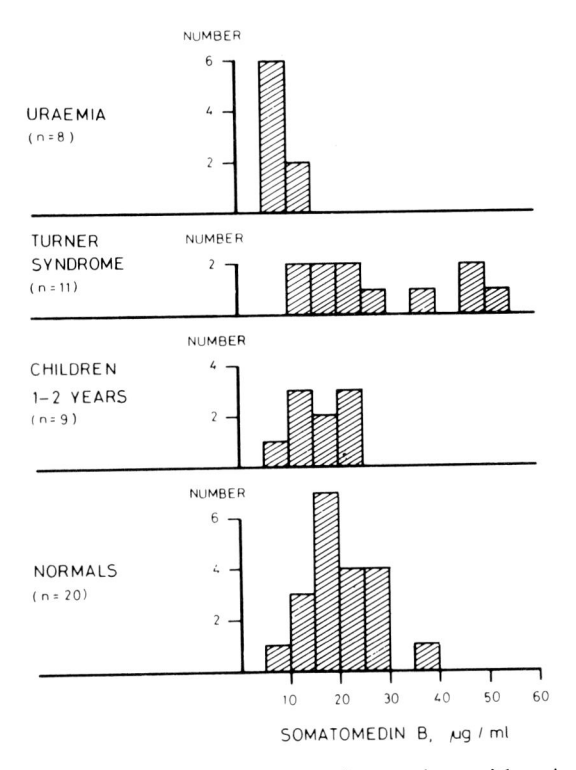

Fig. 6. Distribution of serum somatomedin B values from patients with various disorders.

ively. The somatomedin B values found in serum from patients with acromegaly (n = 20) ranged from 17.3 to 101.1 µg/ml and the mean level 36.3 ± 5.3 µg/ml differed significantly from that in normal subjects. However, no significant difference in somatomedin B levels was found between patients with pituitary dwarfism and normal subjects of the same chronological age.

In serum from patients below 2 years of age the values of somatomedin A were significantly lower than in serum from adults, whereas the somatomedin B levels were similar to those found at 20 years of age. There was no significant difference between the values found at 1 day, 1 week, and 1 year of age. Somatomedins A and B were determined in serum from patients with various disorders (Figs. 5 and 6). Normal values were found in patients with Turner's syndrome and with primary hypothyroidism (Takano et al., 1976b). Uraemic patients had significantly increased levels of somatomedin A whereas their levels of somatomedin B were decreased (Fig. 6). Earlier studies, determining somatomedin activity utilizing the sulphate uptake into cartilage, have shown decreased levels of somatomedin in serum from patients with chronic renal failure (Saenger et al., 1974). The discrepancy in values found using the radioreceptor assay for somatomedin A and the bioassay may be explained by the presence of inhibitory serum factors, which interfere in the bioassay. The presence of inhibitory factors in serum has earlier been pointed out (Salmon, 1972; Božović et al., 1974; Van den Brande and Du Caju, 1974).

THE PRESENCE OF 'BIG' SOMATOMEDIN A AND 'BIG' SOMATOMEDIN B IN SERUM

When whole plasma was separated either by gel chromatography or ultrafiltration, somatomedin A and somatomedin C as determined by bioassay were mainly recovered in a molecular weight range above 50,000 (Koumans and Daughaday, 1963; Van Wyk et al., 1969, 1971; Hall, 1972). The finding that under dissociating conditions such as acid-ethanol extraction somatomedin A appeared in a lower molecular weight range led to the assumption that somatomedin was bound to a carrier molecule (Hall and Uthne, 1971; Hall, 1972). The immunoreactive somatomedin B appeared in whole serum in the high-molecular weight range of gammaglobulins (Yalow et al., 1975a).

When acromegalic serum (containing 2.07 U of somatomedin A per ml) was applied to a column of Sephadex G-200 (5.0 × 66 cm) equilibrated with 0.05 Tris-HCl buffer, pH 7.4, all somatomedin A determined by the radioreceptor assay was recovered in a molecular weight range above 50,000 and no somatomedin A was detected in the range of 7000. Two peaks of somatomedin A were found, a major peak at Kd 0.30 and a minor peak at Kd 0.60 and the total recovery of the applied activity was 74%. The immunoreactive somatomedin B was mainly recovered immediately after the void volume and only 1-2% in the molecular weight range of 5000. Whether the biological activity of somatomedin B in vitro and in vivo parallels the total or free somatomedin B in serum and plasma has not been proved. The fractions obtained from gel chromatography at neutral pH were pooled into 3 larger fractions; I — containing mainly somatomedin B, II — containing the major peak of somatomedin A, and III — containing the smaller peak of somatomedin A (Fig. 7). These fractions were lyophilized, dissolved in 1 M acetic acid and applied separately to a column of Sephadex G-50 equilibrated with 1 M acetic acid. Under these acidic conditions immunoreactive somatomedin B was only detectable as the low-molecular weight form. Most of the somatomedin A eluted in the same position as the somatomedin A standard with a molecular size of 7000 and only 5-10% of somatomedin A remained as the high-molecular weight form (Fig. 8). The total recovery of applied somatomedin A was 200-300% indicating that 'big' somatomedin A was less potent than 'small' somatomedin A in the radioreceptor assay. The identity between this 'small' somatomedin A in serum and the purified somatomedin A has not been proved. When labelled somatomedin A was mixed with whole serum from an acromegalic patient, a normal subject, or a patient with pituitary dwarfism, the same percentage of total radioactivity (about 30%) was found in the high-molecular weight range (Fig. 9). Since the somatomedin A contents of these sera were 2.07, 1.00, and 0.50 U/ml, respectively, this finding may indicate that the amounts of binding protein were correlated to the levels of somatomedin A. Separating the protein-bound and free labelled somatomedin A with charcoal it was shown that the percentage protein-bound radioactivity increased with increasing amounts of serum. The protein-bound labelled somatomedin A was readily displaced by increasing amounts of a somatomedin A preparation (Fig. 10). Thus, somatomedin A seems to re-associate with carrier proteins in a pattern similar to that found for somatomedin C and NSILA-S (Hintz et al., 1974; Zapf et al., 1975).

As earlier shown neither labelled somatomedin B nor pure cold somatomedin B could be re-associated with serum proteins to 'big' somatomedin B (Yalow et al., 1975a). This finding does not exclude the possibility that during physiological conditions this can occur, since the purification and iodination procedures may have destroyed the binding sites of somatomedin B. It has not yet been proved whether the somatomedins or the binding proteins are regulated by growth hormone.

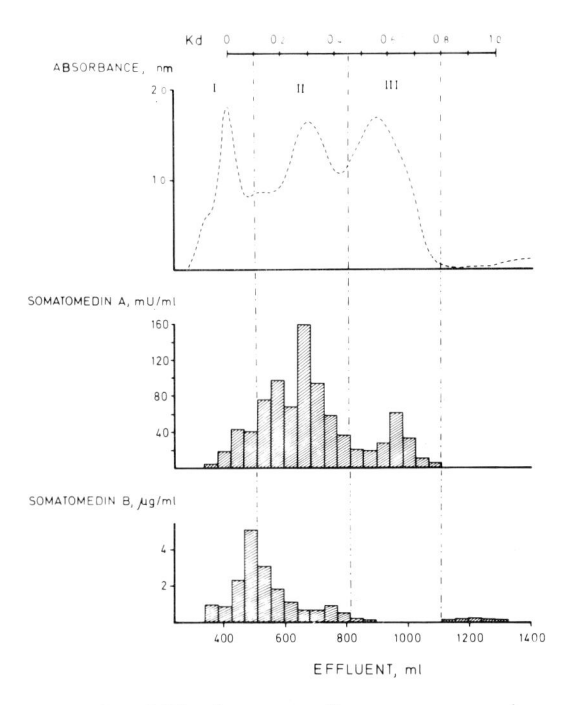

Fig. 7. Gel chromatography of 25 ml acromegalic serum on a column of Sephadex G-200 (5.0 × 66 cm) equilibrated with 0.05 M Tris-HCl buffer, pH 7.4 at +4°C. Somatomedin A and B per ml effluent were determined by radioreceptor assay and radioimmunoassay, respectively. The fractions were pooled into 3 greater fractions, peaks I, II, and III, according to the dotted lines for further purification.

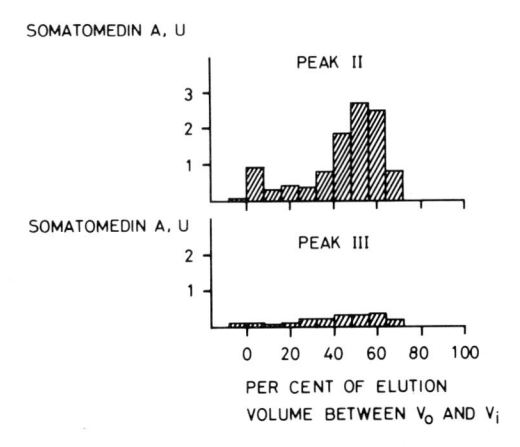

Fig. 8. The lyophilized fractions from peak II and peak III after gel chromatography on Sephadex G-200 (Fig. 7) were dissolved in 1 M acetic acid and applied separately to a column of Sephadex G-50 (1.6 × 90 cm) equilibrated with 1 M acetic acid at +4°C. Amount derived from 2.9 ml of serum was dissolved in 5 ml. After lyophilization of the Sephadex G-50 fractions the total content of somatomedin A in each fraction was determined by radioreceptor assay.

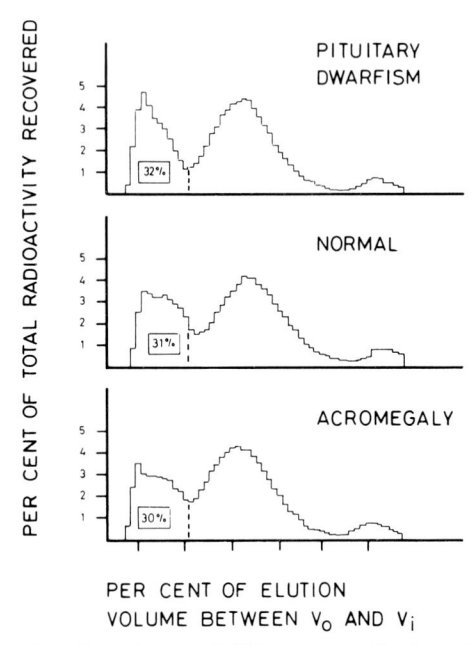

Fig. 9. Gel chromatography of a mixture of ^{125}I-somatomedin A and serum on a column of Sephadex G-50 (1.9 × 90 cm) equilibrated with 0.05 Tris-HCl buffer, pH 7.4, at +4°C. The recovery of radioactivity in various fractions is shown in per cent of total. Serum samples were taken from a patient with pituitary dwarfism (above) a normal subject (middle), and an acromegalic patient (below).

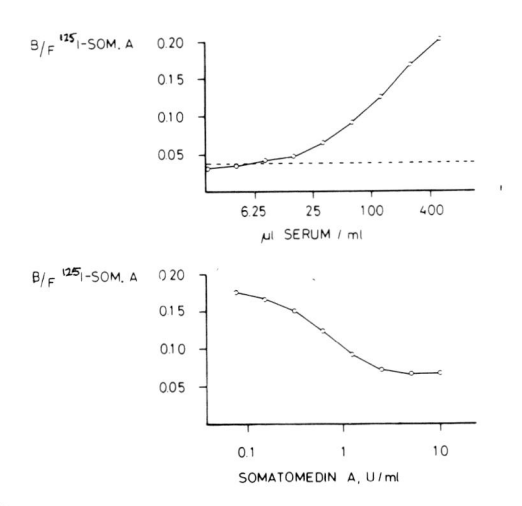

Fig. 10. Binding of ^{125}I-somatomedin A to increasing amounts of normal serum (above) and displacement of bound ^{125}I-somatomedin A by increasing amounts of somatomedin A (below) at a serum concentration of 100 µl/ml. Protein-bound radioactivity was separated from unbound with charcoal.

SOMATOMEDIN A AND B IN URINE

Somatomedin B is detectable in human urine as shown by Yalow et al. (1975b). The high concentration of salt in urine sometimes interferes with the determination of somatomedin A by the radioreceptor assay. On the average the concentrations of somatomedin A and somatomedin B in urine were about 10% of those found in serum. The 24-hour excretions of somatomedin B in urine from patients with growth-hormone deficiency and with acromegaly were 232 ± 30 µg and 791 ± 259 µg, respectively. Contrary to the findings in serum the majority of immunoreactive somatomedin B in urine appeared as the small-molecular weight form (Fig. 11). Measuring urinary somatomedin B, which may reflect the concentration of 'small' somatomedin B in serum, may prove more useful than measuring the total levels in serum. The major proportion of urinary somatomedin A seemed to be in the small, free form.

The average level of somatomedin A in amniotic fluid was high, 4.44 ± 0.60 U/ml, whereas the mean level of somatomedin B was only 0.8 ± 0.1 µg/ml.

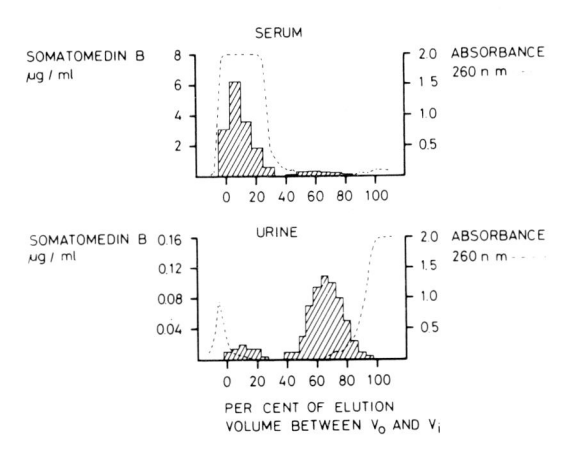

Fig. 11. Gel chromatography of 5 ml of serum and 5 ml of urine on a column of Sephadex G-50 (1.9 × 90 cm) equilibrated with 0.05 M Tris-HCl buffer, pH 7.4, at +4°C. The immunoreactive somatomedin B and the absorbance at 260 nm were determined in the fractions.

BINDING SITES OF SOMATOMEDINS A AND B IN VARIOUS TISSUES

The labelled somatomedin A bound specifically not only to chick cartilage membranes but also to membranes prepared from a variety of rat and monkey tissues, such as lung, kidney, liver, brain, thymus, spleen, pancreas, heart, and fat (Takano et al., 1976a). The percentage bound radioactivity varied between 5 and 15%. In all instances cold somatomedin A in the physiological range (0.1-5 U/ml) caused a significant displacement. Insulin in high concentrations could compete with somatomedin A for its binding sites but had no effect in physiological concentrations (10-100 µU/ml). In comparison, somatomedin A, within the physiological range could interfere with the binding of insulin. The binding of somatomedin A to membranes differed from that to particulate substance such as talc. The latter binding was non-specific since several small peptides could cause displacement and high doses of somatomedin A were necessary in order to achieve displacement. Specific binding sites of somatomedin A were also found on cells prepared from liver, kidney, and fat tissue from the rat.

Whether the specific binding for somatomedin A to a variety of tissues indicates that somatomedin A exerts biological action in these organs has to be proved. At present we only know that somatomedin A in rat adipose tissue increases the conversion of glucose to CO_2, suppresses the release of glycerol, and decreases the uptake of calcium (Hall and Uthne, 1971; Werner et al., 1974).

The labelled somatomedin B did not bind to any membranes prepared from rat and monkey tissues. Since binding studies on the presumed target cells such as glial cells and fibroblasts were unsuccessful, we assume that the biological active part of somatomedin B had been destroyed during the iodination procedure. The ability of labelled somatomedin B to stimulate DNA synthesis in glial cells and fibroblasts has not been tested.

CONCLUSION

The developments of a radioreceptor assay for somatomedin A and a radioimmunoassay for somatomedin B have simplified the work in purifying larger amounts of somatomedins for future studies of their biological effects in vitro and in vivo. At present we cannot claim that the radioreceptor assay is specific for somatomedin A, only that it measures GH-dependent factors in serum. The finding that the level of GH-immunoreactive somatomedin B in serum is 3 times higher in patients with acromegaly than in normal subjects speaks in favour of the measured substance being a GH-dependent factor. The radioreceptor assay of somatomedin A and the radioimmunoassay of somatomedin B seem to be useful and simple tools for elucidating the physiological regulation and dependence of GH and for establishing the levels in serum and urine during normal growth and development as well as in various disorders of growth.

REFERENCES

Božović, L., Božović, M. and Boström, H. (1974): In: *Abstracts: Transactions, Swedish Association of Physicians*, p. 307.
Clemmons, D.R., Hintz, R.L., Underwood, L.E. and Van Wyk, J.J. (1974): *Israel J. med. Sci.*, 10, 1254.
Fryklund, L., Uthne, K. and Sievertsson, H. (1974a): *Biochem. biophys. Res. Commun.*, 61, 957.
Fryklund, L., Uthne, K., Sievertsson, H. and Westermark, B. (1974b): *Biochem. biophys. Res. Commun.*, 61, 950.
Hall, K. (1970): *Acta endocr. (Kbh.)*, 63, 338.
Hall, K. (1972): *Acta endocr. (Kbh.), Suppl. 163*, 1.
Hall, K. and Uthne, K. (1971): *Acta med. scand.*, 190, 137.
Hall, K., Takano, K. and Fryklund, L. (1974): *J. clin. Endocr.*, 39, 973.
Haour, F. and Bertrand, J. (1974): *J. clin. Endocr.*, 38, 334.
Hintz, R.L., Orsini, E.M. and Van Camp, M.G. (1974): In: *Program, The Endocrine Society, 56th Annual Meeting*, A-71.
Koumans, J. and Daughaday, W.H. (1963): *Trans. Ass. Amer. Phycns*, 76, 152.
Lefkowitz, R.J., Roth, J. and Pastan, I. (1970): *Science*, 170, 633.
Marshall, R.N., Underwood, L.E., Voina, S.J., Foushee, D.B. and Van Wyk, J.J. (1974): *J. clin. Endocr.*, 39, 283.
Posner, B.I. (1972): *Clin. Res.*, 20, 922.
Posner, B.I. (1974): *Diabetes*, 23, 209.

Saenger, P., Wiedemann, E., Schwartz, E., Korthschnutz, S., Lewy, J.E., Riggio, R.R., Rugin, A.L., Stenzel, K.M. and New, M.I. (1974): *Pediat. Res., 8,* 163.
Salmon Jr, W.D. (1972): In: *Proceedings, II International Symposium on Growth Hormone, Milan 1971,* p. 180. Editors: A. Pecile and E.E. Müller. ICS 244, Excerpta Medica, Amsterdam.
Salmon Jr, W.D. and Daughaday, W.H. (1957): *J. Lab. clin. Med., 49,* 825.
Takano, K. (1975): *Studies on Somatomedin A.* Thesis, Karolinska Institute, Stockholm.
Takano, K., Hall, K., Fryklund, L., Holmgren, A., Sievertsson, H. and Uthne, K. (1975): *Acta endocr. (Kbh.), 80,* 14.
Takano, K., Hall, K., Fryklund, L. and Sievertsson, H. (1976a): *Hormone metab. Res., 8,* 16.
Takano, K., Hall, K., Ritzén, M., Iselius, L. and Sievertsson, H. (1976b): *Acta endocr. (Kbh.),* in press.
Thorell, J.I. and Johansson, B.G. (1971): *Biochim. biophys. Acta (Amst.), 251,* 363.
Underwood, L.E., Hintz, R.L., Voina, S.J. and Van Wyk, J.J. (1972): *J. clin. Endocr., 35,* 194.
Uthne, K. (1973): *Acta endocr. (Kbh.), Suppl. 175,* 1.
Van den Brande, J.L. and Du Caju, M.V.L. (1974): In: *Advances in Human Growth Hormone Research Symposium, Baltimore 1973,* p. 98. DHEW Publication No. (NIH) 74-612.
Van Wyk, J.J., Hall, K. and Weaver, R.P. (1969): *Biochim. biophys. Acta (Amst.), 192,* 560.
Van Wyk, J.J., Hall, K., Van den Brande, J.L. and Weaver, R.P. (1971): *J. clin. Endocr., 32,* 389.
Van Wyk, J.J., Underwood, L.E., Baseman, J.B., Hintz, R.L., Clemmons, D.R. and Marshall, R.N. (1975): In: *Advances in Metabolic Disorders, Vol. 8,* p. 127. Editors: R. Levine and R. Luft. Academic Press, New York.
Van Wyk, J.J., Underwood, L.E., Hintz, R.L., Clemmons, D.R., Voina, S.J. and Weaver, R.P. (1974): *Recent Progr. Hormone Res., 30,* 259.
Werner, S., Hall, K. and Löw, H. (1974): *Hormone metab. Res., 6,* 319.
Yalow, R.S., Hall, K. and Luft, R. (1975a): *J. clin. Invest., 55,* 127.
Yalow, R.S., Hall, K. and Luft, R. (1975b): *J. clin. Endocr., 41,* 638.
Zapf, J., Waldvogel, M. and Froesch, E.R. (1975): *Arch. Biochem., 168,* 638.
Zingg, A.E. and Froesch, E.R. (1973): *Diabetologia, 9,* 472.

SPECIFICITY, TOPOGRAPHY, AND ONTOGENY OF THE SOMATOMEDIN C RECEPTOR IN MAMMALIAN TISSUES*

A. JOSEPH D'ERCOLE, LOUIS E. UNDERWOOD, JUDSON J. VAN WYK,
CHARLES J. DECEDUE and DORETHA B. FOUSHEE

Department of Pediatrics, University of North Carolina School of Medicine, Chapel Hill, N.C., U.S.A.

Somatomedin C is a growth hormone-dependent, insulin-like, basic, serum peptide of about 7,000 daltons. Although its exact physiologic role remains to be clarified, the pure hormone fully replicates the action of serum in the stimulation of sulfate and thymidine uptake in cartilage and is a potent mitogen for fibroblasts and certain other lines of cultured cells (Van Wyk et al., 1975).

We have previously reported that the insulin-like effects of somatomedin C can be quantitatively correlated with its ability to compete with insulin for binding to cell membrane receptors for insulin, whereas the mitogenic and growth-promoting properties of somatomedin C correlate with its binding to specific somatomedin C receptors. Furthermore, evidence has been presented that the growth-promoting actions of somatomedin C are not contingent on its insulin-like properties: the concentrations of somatomedin C required to elicit 'growth' in biological assays and to inhibit ^{125}I-somatomedin C binding in radioreceptor assays are several orders of magnitude less than the concentrations of somatomedin C required to elicit insulin-like responses in adipose tissue and to inhibit ^{125}I-insulin binding to the insulin receptor (Van Wyk et al., 1975).

As a further step toward defining a possible physiologic role of somatomedin C in growth, we have now compared the distribution of somatomedin C receptors with insulin receptors in mature mammalian tissues and in tissues at different gestational ages in the fetal pig. Studies of the ontogeny of somatomedin C receptors in the fetus were undertaken as part of a continuing evaluation of hormonal mechanisms controlling fetal growth. As background for the present studies, detailed comparisons were first made between the characteristics of the somatomedin C receptor in human placental membrane preparations and the previously reported characteristics of the insulin receptor in the same tissue (Marshall et al., 1974).

---

* This work was supported by NIH Research Grants AM01022 and HD08299, NIH Training Grant AM05330, and NIH Research Fellowship HD01505 (AJD).
JJVW is an NIH Career Research Awardee Nr. 5 K06 AM14115. LEU is a Jefferson-Pilot Fellow in Academic Medicine, University of North Carolina.

MATERIALS AND METHODS

Purification and iodination of somatomedin

The 3 preparations of somatomedin C used for iodination were isolated in our laboratory by procedures described previously (Van Wyk et al., 1974, 1975). All preparations contained less than 1 part in 10,000 of radioimmunoreactive insulin. 4 µg of somatomedin C were iodinated with Na^{125}I (Amersham-Searle) using the fractional chloramine T method of Roth (1973). The iodinated hormone was purified on a 1.2 × 50 cm Sephadex G-50 column (Pharmacia). Specific activity of the iodinated preparation ranged between 170 and 220 µCi/µg. The iodinated hormone was stored at −20°C in 3-4% bovine serum albumin (Cohn fraction V; Sigma). Immediately prior to use in the competitive membrane binding assay, free iodine was removed by batch treatment with an anion exchange resin (Bio-Rad, AG 1-X8).

To conserve pure somatomedin C, 2 less pure preparations of somatomedin (E-3 and KLY) were used as standards in the competitive binding assay. E-3, prepared from human retroplacental blood by a previously described method (Marshall et al., 1974), contained 15 units of somatomedin activity/mg dry weight and less than 0.1 µU immunoreactive insulin or proinsulin/unit of somatomedin. Preparation KLY contained 6.8 units somatomedin/mg protein and less than 25 µU immunoreactive insulin or proinsulin/unit of somatomedin. One unit of somatomedin is defined as the sulfation/thymidine factor activity in 1 ml of pooled plasma from young adult males.

Preparation of cell membranes

Tissues were trimmed of extraneous connective tissue, washed in cold 0.25 M sucrose and stored in sucrose at −20°C. Upon thawing, the tissues were serially homogenized in 1-2 volumes of 0.25 M sucrose with an Omnimixer (Sorvall) and with a Polytron PT-10 (Brinkman). Membranes were prepared by differential centrifugation (Cuatrecasas, 1972). Adipocyte cell membranes were prepared by the method of Jarett (1974). Red blood cell membranes were prepared by the method of Hanahan and Ekholm (1974). The concentration of protein in each preparation was determined by the method of Lowry et al. (1951). Purification of membranes was monitored by the determination of 5′-nucleotidase activity (Gospodarowicz, 1973).

Competitive membrane binding assay

Binding of labeled hormone was assessed in 0.05 M Tris-HCl buffer at pH 7.4 in the presence of 0.25% BSA and 0.02% sodium azide. Membrane suspension, ^{125}I-somatomedin (approximately 1 fmole; 20,000 c.p.m.), and graded doses of unlabeled somatomedin or other test substances were incubated at 4° for 16-24 hours. Following incubation, samples were centrifuged at 6,000 × g for 10 minutes. The supernatant was removed by suction, and the washed, recentrifuged pellet was counted in a well type gamma spectrometer (Packard). Specific binding was determined by subtracting from each tube the counts bound in duplicate tubes containing a large excess of somatomedin. Interassay precision was approximately 15%.

RESULTS

Characteristics of the interaction of somatomedin C with human placental membrane receptor

A. Cellular localization and affinity; pH and cation dependency

Specific binding of ^{125}I-somatomedin could be demonstrated in the whole homogenate of the placenta and after each step of the membrane purification procedure. Maximum specific binding (12-20% binding/100 μg of membrane protein), however, was consistently seen in the pellet of the 40,000 × g centrifugation. This fraction also contained the highest 5′-nucleotidase specific activity. By Scatchard analysis (Scatchard, 1949; Rosenthal, 1967), the receptor appeared to have at least 2 distinct binding constants for somatomedin C (Table 1). The high affinity sites had a relatively low binding capacity, while those of lower affinity had higher capacity. Maximum binding occurred at pH 8 but changed little through a broad range between pH 7 and 9. Addition of 2 mM EDTA and various cations including 0.05 M NaCl, 2 mM calcium, magnesium, and manganese had no significant effect on binding. Unlike insulin, the binding of which improved with increasing ionic strength of buffer, binding of somatomedin declined with increasing ionic strength. At equivalent ionic strengths, specific binding of somatomedin in Tris buffer was better than binding in phosphate buffer.

TABLE 1

*Quantitation of the interaction of ^{125}I-somatomedin C with the human placental membrane receptor**

	K_{dis}	Max. somatomedin C bound (pg/μg protein)
High affinity sites	1.6×10^{-9} M	15
Low affinity sites	1.8×10^{-8} M	83

* These results were obtained using highly purified somatomedin C. The same preparation was used for iodination.

B. Rate of association and dissociation; degradation of somatomedin by membranes

The rate of binding was rapid at 37°C, but destruction or dissociation was also rapid at this temperature (Fig. 1). At lower temperatures, a steady state was reached more slowly but higher specific binding was achieved. Binding was maximal at 4°C after 24-36 hours of incubation.

To determine the rate of dissociation of somatomedin from its receptor, tubes containing ^{125}I-somatomedin C were incubated with placental membranes at 37°C for 5 minutes, at 24°C for 2 hours, and at 4°C for 20 hours. A large excess of unlabeled somatomedin was then added to the incubation tubes and at prescribed times, the membrane bound ^{125}I-somatomedin C was determined by filtration (Millipore — EHWP-02500). Dissociation of the labeled hormone from the receptor, which initially followed first order kinetics, was most rapid at 37° and progressively slower at 24° and at 4°C.

Unlike its effect on labeled insulin (Marshall et al., 1974), the placental membrane preparation exhibited little tendency to degrade somatomedin. After incubation of

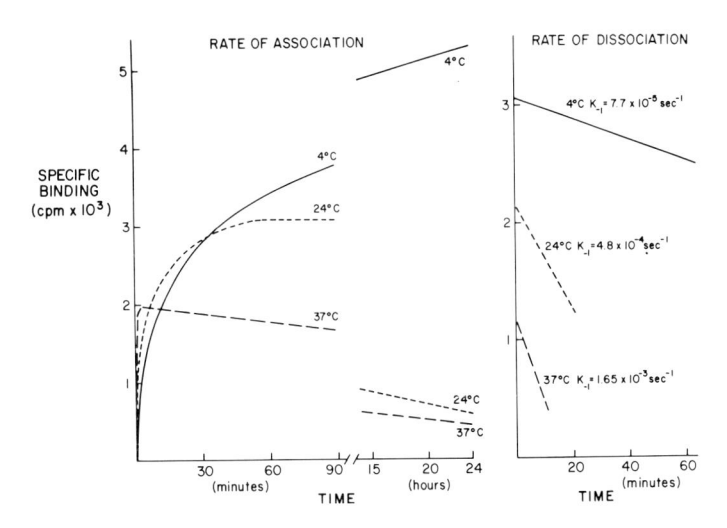

Fig. 1. Kinetics of ^{125}I-somatomedin C binding to human placental membranes. See text for experimental design.

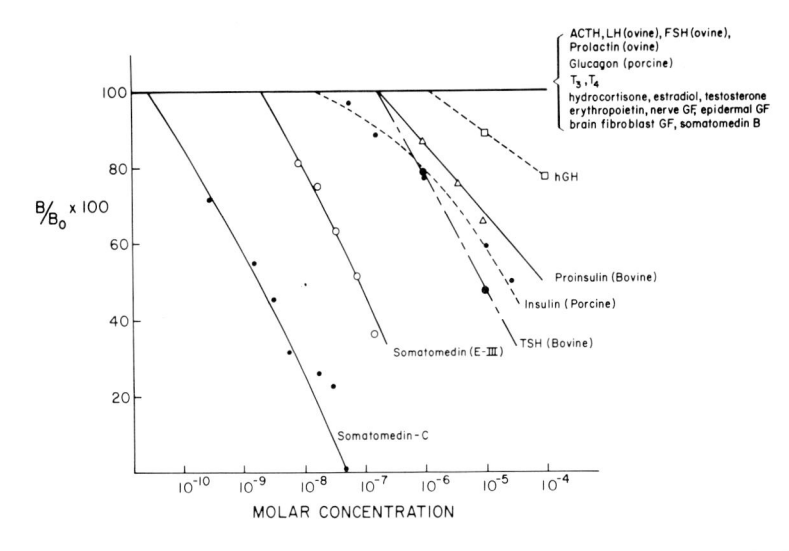

Fig. 2. Hormonal specificity of the ^{125}I-somatomedin C receptor in human placental cell membranes. The unlabeled somatomedin C (far left) was a highly purified somatomedin. The same preparation was used for iodination. Sources of other hormones are as follows: insulin (Eli Lilly, PJ 5682), proinsulin (Steiner), TSH (NIH, B6), HGH (NIH, HS 1395), ACTH (Parke-Davis, C1-305), glucagon (Eli Lilly, 258-D30-1384), prolactin (NIH, P-S-9), LH (NIH, S-18), FSH (NIH, S-10), erythropoietin (human urinary, National Lung Institute, H4H-35), nerve growth factor (Bradshaw), epidermal growth factor (S. Cohen), somatomedin B (Sievertsson), brain fibroblast growth factor (Gospodarowicz).

^{125}I-somatomedin C with 160 μg of placental membrane at 37°C for 12 hours, 90% of control (duplicate tubes without membrane) in the incubation supernatant was still precipitable with 20% TCA. This supernatant ^{125}I-somatomedin C also retained its capacity to bind to membranes. Under the same conditions, only 25% of ^{125}I-insulin remained TCA precipitable and only 10% of control could be rebound to membranes.

TABLE 2

Effect of enzymatic pretreatment of human placental cell membranes on binding of ^{125}I-somatomedin C and ^{125}I-insulin

Enzyme	Binding (% of control membranes)	
	Somatomedin	Insulin
Collagenase	82	81
Hyaluronidase	145	149
Neuraminidase	126	133
Papain	65	5
Pronase	9	1
Phospholipase C	118	110
Trypsin	84	46

Sources of enzymes and methods of membrane treatment: collagenase (crude, Worthington) at 1 mg/ml in KRB buffer, pH 6.4; hyaluronidase (type 1, Sigma), 40 μg/ml in 0.1 M phosphate, pH 5.3; neuraminidase (Worthington), 50 μg/ml in 0.1 M acetate, pH 5.1; papain (Mann), 100 μg/ml in 0.05 M acetate, pH 4.6; pronase (B grade, Calbiochem), 100 μg/ml and phospholipase C (type 1, Sigma), 60 μg/ml in 0.05 M Tris, pH 7.4; trypsin (Difco), 30 μg/ml in 0.05 M Tris-HCl, pH 8.1. All were incubated at 37°C for 30 minutes except collagenase (1 hour).
Following incubation, treated membranes were washed 3 times with Tris buffer. Specific binding was determined in duplicate tubes of membrane incubated with ^{125}I-somatomedin C at 4°C for 18 hours. Control membranes were treated in an identical manner, but no enzyme was added.

C. Hormonal specificity of the placental somatomedin C receptor
Preliminary data on the hormonal specificity of the placental somatomedin C receptor have been reported (Marshall et al., 1974). In the present study, a detailed assessment of the effect of incremental doses of a variety of hormones has been made (Fig. 2). At concentrations below 1×10^{-7} M, no inhibition of ^{125}I-somatomedin C binding was observed with ACTH, glucagon, prolactin, LH and FSH, T_3, T_4, hydrocortisone, estradiol, testosterone, erythropoietin, nerve growth factor, epidermal growth factor, somatomedin B, and brain fibroblast growth factor. A 50% reduction in binding occurred with large doses of insulin (4×10^{-5} M) and proinsulin (9×10^{-5} M). Unlabeled, pure somatomedin C reduced binding by 50% at 10^{-9} M concentration. Multiplication stimulating activity (MSA — Dr. Howard Temin) was competitive with ^{125}I-somatomedin for the somatomedin receptor but did not produce a curve which paralleled that of unlabeled somatomedin (data not shown). A preparation of somatomedin A* was approximately 6% as effective as our purest preparation of somatomedin C in inhibiting binding of ^{125}I-somatomedin C.

---

* Preparation SPE 152-1 was generously furnished by Dr. Linda Fryklund of AB Kabi, Stockholm, Sweden.

D. Preincubation of membranes with enzymes

The placental membrane preparations were preincubated with a variety of enzymes (Table 2) and subsequently tested for their capacity to bind somatomedin and insulin. In membrane pretreated with phospholipase C, neuraminidase, and hyaluronidase, the binding of both radiolabeled insulin and somatomedin C was enhanced. Binding of both hormones was markedly decreased by pronase treatment. Compared to the insulin receptor, the somatomedin receptor was relatively resistant to papain and trypsin. Binding of both hormones was reduced slightly by collagenase, but effects varied with different collagenase preparations.

E. Serum somatomedin levels in sub-human species

Using the human placental receptor competitive binding assay, somatomedin C could be detected in unextracted human serum and in a variety of sub-human species (Fig. 3). The potency of monkey, rabbit and guinea pig serum equaled that of human serum. Pig and rat serum was 1.5-2 times more active; while, compared to human serum, dog and sheep were 1/2 and 1/4 as potent, respectively. Dose response curves paralleled the partially purified somatomedin standard and the human serum. Except for the bluefish which was 2% as active as human serum, these preliminary results suggest that these sera contained comparable concentrations of substances which interact with the somatomedin receptor.

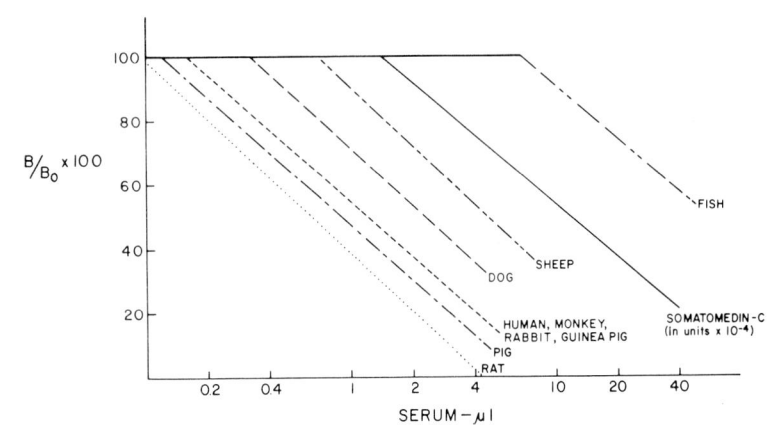

Fig. 3. Sensitivity of the ^{125}I-somatomedin C receptor of human placental cell membranes to competition by sera from various animal species. The curves for the partially purified somatomedin standard (E-3) and the human serum (pool of healthy, young adult males) were derived from several assays. The fish is represented by the Atlantic blue fish *(Pomatomus saltatrix)*.

Topography of the somatomedin C receptor in mammalian tissue

In order to determine the distribution of cells which might exhibit biological responses to somatomedin, a search for somatomedin receptors was carried out in a variety of tissues. The techniques used precluded the absolute assessment of receptors from tissue to tissue. It was believed, however, that some perspective on the distribution of somatomedin C binding sites could be gained by comparing the percentage of ^{125}I-somatomedin C specifically bound by each tissue with the percentage of ^{125}I-insulin

TABLE 3

Sensitivity of the somatomedin C and insulin receptors to unlabeled somatomedin and insulin

Tissue	Membrane protein used (μg)	^{125}I-somatomedin C receptor			^{125}I-insulin receptor		
		% specific binding	1/2 competitive concentrations		% specific binding	1/2 competitive concentrations	
			Somatomedin (U × 10^{-4}/ml)	Insulin (μU/ml)		Somatomedin (U/ml)	Insulin (μU/ml)
Human							
Kidney	200	9.3	140	—	6.4	—	44
Placenta	120	13.0	6	2.5 × 10^5	19.0	0.40	140
Red blood cells	1800	0	—	—	0	—	—
Chicken embryo							
chondrocytes*	180	2.6	—	—	4.0	—	—
Rat							
Brain	410	8.2	50	0.9 × 10^5	3.9	0.36	110
Fat	192	2.0**	—	—	15.8	—	—
Kidney	460	5.9	50	0.5 × 10^5	4.1	—	210
Liver	150	13.0	50	6.3 × 10^5	23.5	0.42	110
Lung	540	10.2	50	0.2 × 10^5	4.8	—	180
Muscle	740	4.7	100	0.2 × 10^5	4.0	—	210
Thymus	1024	6.0	200	3.1 × 10^5	3.1	—	150

*Pelvic leaflet membranes from 12-day chick embryos were provided by Dr. Raymond L. Hintz.
**Binding of ^{125}I-somatomedin C did not vary with the concentration of membrane tested.

TABLE 4

Binding of ^{125}I-somatomedin C and sensitivity to unlabeled somatomedin in the adult pig

Tissue	Membrane protein used (µg)	% specific binding	1/2 competitive concentration of somatomedin* (U x 10^{-4} ml)
Breast	225	11.3	36
Heart	200	6.0	28
Kidney	200	8.0	80
Liver	220	5.7	88
Lung	210	8.5	140
Ovary	212	12.5	20
Placenta**	214	9.0	40
Salivary gland	200	6.0	28
Testis	210	6.0	100
Thyroid	190	7.7	22

* The unlabeled somatomedin used in these studies was the partially purified preparation KLY.
** Full term, maternal portion of placenta.

specifically bound by the same membrane preparations. Such comparisons would tend to cancel out variations due to the technique of preparing membranes from dissimilar tissues. Specific binding of somatomedin was observed in a wide variety of tissues (Tables 3 and 4). The greatest specific binding was found in human placenta and rat liver membranes. In both these tissues, however, specific binding of insulin exceeded that of somatomedin. In the majority of other tissues tested, somatomedin binding exceeded insulin binding. Rat thymus and muscle membranes bound only small quantities of either labeled somatomedin or insulin unless large quantities of membrane protein were used. No binding of either ^{125}I-insulin or ^{125}I-somatomedin C to human erythrocyte membranes could be detected.

Since somatomedin exerts unequivocal insulin-like effects on fat cells (Hall and Uthne, 1971; Hintz et al., 1972; Underwood et al., 1972; Clemmons et al., 1974), it was of interest to determine whether these metabolic effects were mediated through the somatomedin or the insulin receptor. To test whether mature fat cells contained somatomedin receptors, membranes were prepared from adipocytes freed with crude collagenase from rat epididymal fat pads. These membranes failed to bind somatomedin but bound insulin in proportion to the concentration of membrane protein. Interpretation of these data initially was clouded because the collagenase preparation used destroyed the somatomedin receptor on placental membranes. A batch of collagenase was eventually found which freed adipocytes efficiently yet did not destroy the placental somatomedin receptor. Membranes from adipocytes isolated with this collagenase preparation exhibited a typical membrane dose response for insulin (Fig. 4) but only minimal somatomedin binding; furthermore, there was no increase in binding of ^{125}I-somatomedin C with increasing concentrations of membrane protein. These findings established that mature adipocytes have no specific somatomedin receptors. Therefore, the well documented insulin-like biological effects of somatomedin on adipose tissue can be entirely attributed to the capacity of somatomedin to interact with the insulin receptor.

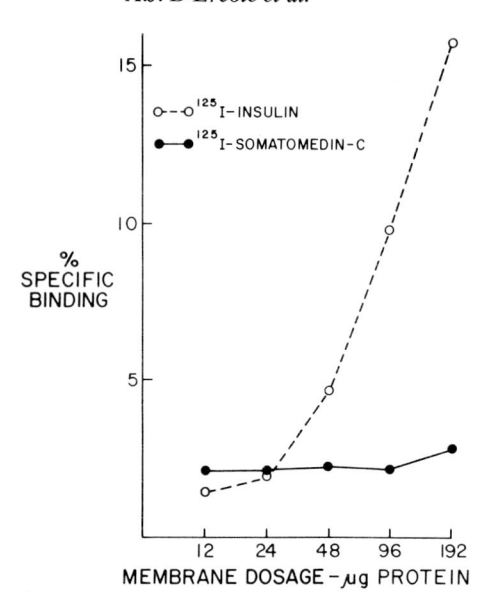

Fig. 4. Specific binding of ^{125}I-insulin and ^{125}I-somatomedin C by rat adipocyte membranes. Epididymal fat pads from 200 g rats were incubated for 1 hour at 37°C with 1 mg/ml crude collagenase (Worthington, lot Nr. 2JB) in 3% BSA and Krebs-Ringer bicarbonate buffer, pH 7.4. The 5′-nucleotidase activity in the plasma membrane fraction was 0.46 units/mg per membrane protein (1 unit = 1 μmole P_i/10 min). 5′-Nucleotidase activity and specific binding of both insulin and somatomedin were undetectable in the other cell fractions.

Ontogeny of the somatomedin C receptor in fetal tissues

In an attempt to determine the possible influence of somatomedin on the growth of the fetus, a systematic examination of fetal tissues for somatomedin C receptors was undertaken. The fetal pig was selected as a model for the study because of its ease of acquisition, its relatively long gestation (114 days), and large fetal size. In addition, the fetal portion of the placenta of the pig can be separated from the maternal portion. Litters of fetal pigs ranging in gestational age (Ullrey et al., 1965) from 27 days to 108 days were obtained. Individual organs from all the fetuses of a single sow were pooled and stored in 0.25 M sucrose at −20°C until membranes were prepared.

As in the study of receptor distribution in mature tissues, binding of ^{125}I-insulin was determined in parallel with somatomedin binding. This facilitated detection of changes in binding due to variation of membrane preparation and, at the same time, permitted assessment of binding sites for insulin which has been considered a possible stimulant of fetal growth.

Specific binding of both hormones could be demonstrated in maternal placenta, fetal placenta, liver, lung, heart, and kidney at all gestational ages tested. As with adult tissues, specific binding of ^{125}I-somatomedin C increased with increasing concentrations of membrane protein. Varying patterns of somatomedin receptor binding and sensitivity are apparent from these studies. In liver, kidney and heart, no significant differences in specific binding of ^{125}I-somatomedin C or in receptor sensitivity to unlabeled somatomedin were observed when fetal cell membranes and maternal membranes were compared. In lung, however, a striking difference was observed between

TABLE 5

Binding of ^{125}I-somatomedin and ^{125}I-insulin by pig fetal placenta

Day of gestation	^{125}I-somatomedin		^{125}I-insulin	
	% specific binding	1/2 competitive* concentration (U/ml)	% specific binding	1/2 competitive* concentration (μU/ml)
35	5.7	>0.020	6.1	10
108	13.1	0.0024	1.9	—

* One-half competitive concentration refers to the units of unlabeled somatomedin (prep KLY) or μU of unlabeled insulin necessary to decrease by 50% the ^{125}I-somatomedin C or ^{125}I-insulin specifically bound to the receptor. 200 μg of membrane protein were used in each assay tube.

fetal and maternal membranes. The fetal lung membranes exhibited greater specific binding of ^{125}I-somatomedin C and a nearly 10-fold greater sensitivity to competition by unlabeled somatomedin than did maternal lung membranes.

A third pattern of somatomedin binding has been observed in membranes from the fetal portion of the pig placenta. During the first trimester, the fetal placenta binds ^{125}I-insulin and ^{125}I-somatomedin C in approximately equal quantities. As gestation progresses, insulin binding falls while specific somatomedin binding tends to rise. Concurrent with these changes, there occurs a marked, nearly 10-fold increase in sensitivity to competition by unlabeled somatomedin for the ^{125}I-somatomedin C receptor (Table 5). In the maternal portion of the placenta, somatomedin and insulin binding paralleled each other throughout gestation. Binding of both hormones was greatest at mid-gestation.

SUMMARY AND CONCLUSIONS

These studies indicate that the somatomedin C receptor in human placenta has a high degree of hormonal specificity and is clearly different from the insulin receptor in its response to ion concentration and to treatment with certain enzymes. Somatomedin C is much more resistant than insulin to degradation by placental cell membranes. Properties similar to those of the human placental somatomedin C receptor have been reported for ^{125}I-somatomedin A in placenta (Hall et al., 1974; Takano et al., 1975) and ^{125}I-NSILA-S in rat liver (Megyesi et al., 1974). It remains to be seen whether these peptides, which exhibit biologic effects strikingly similar to somatomadin C, will compete with ^{125}I-somatomedin C for its membrane receptor.

The finding that a wide variety of tissues contain cells with specific somatomedin receptors does not prove that somatomedin has a physiologic function on those cells. Since cell membranes were usually prepared from whole organs, such preparations undoubtedly contained membranes from a variety of cell types as well as unequal numbers of cells. The absence of somatomedin receptors in mature adipocytes and red blood cells strongly suggests that, in usual concentrations, somatomedin has no physiologic effects on these cells. It appears, therefore, that the in vitro insulin-like effect of somatomedin on adipocytes observed at relatively high concentrations is mediated through the insulin receptor.

As yet, no known hormone has been shown to play a pivotal role in the stimulation of growth of the fetus. Although syndromes of fetal insulin excess are associated with fetal overgrowth, there is evidence which suggests that insulin, in physiologic concentrations, is not a primary stimulant of fetal growth (Picon, 1967; Clark et al., 1968). The possibility that somatomedin provides the impetus for fetal growth is under investigation in a number of laboratories. To date, cord blood somatomedin levels by bioassay (Kastrup and Anderson, 1975; Tato et al., 1975) and by our competitive membrane binding assay (unpublished data) have been found to be low. It is not known whether these low values are due to a deficiency of plasma proteins which bind somatomedin (Hintz et al., 1974) or to an absolute deficiency of free somatomedin.

Our findings show that somatomedin C receptors are abundant in fetal tissues and exhibit variable patterns of hormone binding from one tissue to another and from one period of gestation to another. The finding that fetal lung is more sensitive to competition by unlabeled somatomedin than maternal lung raises the possibility that somatomedin may have a special role in growth of the lung in utero. These findings are of particular interest in the light of a recent preliminary report by Sundell et al. (1975) on the possible effect of epidermal growth factor (EGF) on fetal lung growth. These investigators showed that EGF levels in serum increase as gestation progresses and that i.v. infusion of EGF, a peptide thought to be closely related to somatomedin C, hastens epithelial growth in the fetal lung and maturation of pulmonary function. The striking increase in somatomedin binding and sensitivity in fetal pig placenta which is observed as gestation advances suggests the possibility that somatomedin may exert its effect on fetal placental growth in the latter stages of gestation. Although these results obviously fall short of securing a role for somatomedin in fetal growth, they suggest that, as this hormone is further examined in this regard, consideration must be given to the possibility that it preferentially stimulates growth of tissues and acts at restricted periods of fetal life. An analogous role in fetal growth has been established for nerve growth factor (Bradshaw et al., 1974).

ACKNOWLEDGMENT

The authors are grateful to Eyvonne Bruton for technical assistance.

REFERENCES

Bradshaw, R.A., Hogue-Angeletti, R.A. and Frazier, W.A. (1974): In: *Recent Progress in Hormone Research, Vol. 30,* p. 575. Editor: R.O. Greep. Academic Press, New York.
Clark, C.M., Cahill, G.F. and Soeldner, J.S. (1968): *Diabetes, 17,* 362.
Clemmons, D.R., Hintz, R.L., Underwood, L.E. and Van Wyk, J.J. (1974): *Israel J. med. Sci., 10,* 1.
Cuatrecasas, P. (1972): *Proc. nat. Acad. Sci. (Wash.), 69,* 318.
Gospodarowicz, D. (1973): *J. biol. Chem., 248,* 5050.
Hall, K. and Uthne, K. (1971): *Acta med. scand., 190,* 137.
Hall, K., Takano, K. and Fryklund, L. (1974): *J. clin. Endocr., 39,* 973.
Hanahan, D.J. and Ekholm, J.E. (1974): In: *Methods in Enzymology, Vol. 31,* part A, p. 168. Editors: S. Fleischer and L. Packer. Academic Press, New York.
Hintz, R.L., Clemmons, D.R., Underwood, L.E. and Van Wyk, J.J. (1972): *Proc. nat. Acad. Sci. (Wash.), 69,* 2351.
Hintz, R.L., Orsini, E.M. and Van Camp, M.G. (1974): In: *Abstracts, 56th Annual Meeting of The Endocrine Society,* Abstract No. 31.
Jarett, L. (1974): In: *Methods in Enzymology, Vol. 31,* part A, p. 60. Editors: S. Fleischer and L. Packer. Academic Press, New York.
Kastrup, K.W. and Andersen, H.H. (1975): *Pediat. Res., 9,* 683 (abstract).

Lowry, O.H., Rosebrough, N.J., Farr, A.J. and Randall, R.J. (1951): *J. biol. Chem., 193,* 265.
Marshall, R.N., Underwood, L.E., Voina, S.J., Foushee, D.B. and Van Wyk, J.J. (1974): *J. clin. Endocr., 39,* 283.
Megyesi, K., Kahn, C.R., Roth, J., Froesch, E.R., Humbel, R.E., Zapf, J. and Neville Jr, D.M. (1974): *Biochem. biophys. Res. Commun., 57,* 307.
Picon, L. (1967): *Endocrinology, 81,* 1419.
Rosenthal, H.E. (1967): *Analyt. Biochem., 20,* 525.
Roth, J. (1973): *Metabolism, 22,* 1059.
Scatchard, G. (1949): *Ann. N.Y. Acad. Sci., 51,* 660.
Sundell, H., Serenius, F.S., Barthe, P., Friedman, Z., Kanarek, K.S., Escabedo, M.B., Orth, D.N. and Stahlman, M.T. (1975): *Pediat. Res., 9,* 371 (abstract).
Takano, K., Hall, K., Fryklund, L., Holmgren, A., Sievertsson, H. and Uthne, K. (1975): *Acta endocr. (Kbh.), 80,* 1.
Tato, L., Du Caju, M.V.L., Prévôt, C. and Rappaport, R. (1975): *J. clin. Endocr., 40,* 534.
Ullrey, D.E., Sprague, J.I., Becker, D.E. and Miller, E.R. (1965): *J. Animal Sci., 24,* 711.
Underwood, L.E., Hintz, R.L., Voina, S.J. and Van Wyk, J.J. (1972): *J. clin. Endocr., 35,* 194.
Van Wyk, J.J., Underwood, L.E., Hintz, R.L., Clemmons, D.R., Voina, S.J. and Weaver, R.P. (1974): In: *Recent Progress in Hormone Research, Vol. 30,* p. 259. Editor: R.O. Greep. Academic Press, New York.
Van Wyk, J.J., Underwood, L.E., Baseman, J.B., Hintz, R.L., Clemmons, D.B. and Marshall, R.N. (1975): In: *Advances in Metabolic Disorders, Vol. 8,* p. 128. Editors: R. Luft and K. Hall. Academic Press, New York.

EVIDENCE FOR A ROLE OF ADENOSINE 3':5'-MONOPHOSPHATE IN GROWTH HORMONE-DEPENDENT SERUM SULFATION FACTOR (SOMATOMEDIN) ACTION ON CARTILAGE*

HAROLD E. LEBOVITZ, MARC K. DREZNER and FRANCIS A. NEELON

Duke University Medical Center, Durham, N.C., U.S.A.

Cartilage growth is stimulated by circulating serum factors (Salmon and Daughaday, 1957). These factors are growth-hormone dependent and have insulin-like activities (Hall and Van Wyk, 1974). They are called serum sulfation factors or somatomedins, are peptides and are thought to act on chondrocytes by binding to specific plasma membrane receptors (Van Wyk et al., 1974). The mechanism by which serum sulfation factors influence chondrocyte metabolism has not been elucidated. Since the somatomedins are peptide hormones and bind to specific plasma membrane receptors (as do all protein hormones) it seems likely that their action must result from an effect at the plasma membrane. This suggests that the initial event in their actions would be either an alteration in membrane function or a change in the production of an intracellular second messenger.

Adenosine 3':5'-monophosphate (cyclic AMP) is a common second messenger for protein hormone action. In addition, cyclic AMP has been implicated in the regulation of cell growth in a number of systems (Ryan and Curtis, 1973; Pastan and Johnson, 1974). Several years ago, we became interested in the possibility that the actions of serum sulfation factors on chondrocytes might be mediated by alterations in the intracellular cyclic AMP levels and we initiated appropriate studies to investigate this thesis. This paper summarizes our data which are consistent with the hypothesis that some of the actions of serum sulfation factors are mediated by increases in chondrocyte cyclic AMP.

EXPERIMENTAL PROCEDURES

All the studies to be reported were carried out in vitro using pelvic cartilage of 10- to 12-day chicken embryos. Incubations were done in a 50 mM Tris buffer (pH 7.45) which contained essential amino acids, electrolytes, glucose (5.6 mM) and penicillin (800 IU/ml) under an atmosphere of 95% O_2-5% CO_2 (Drezner et al., 1975). Synthesis of cartilage proteoglycan, total protein and RNA was estimated by incorporation of appropriate precursors into products (Eisenbarth et al., 1973, 1974). Cartilage amino

---

* Studies supported by grants from the National Institute of Arthritis, Metabolic and Digestive Diseases of the National Institutes of Health (AM 01324, 5 T1 AM 5074, 1F 22 AM 02296-01).

TABLE 1
Metabolic effects of serum sulfation factor in vitro on embryonic chick cartilage

1. Increase amino acid transport (Adamson and Anast, 1966; Drezner et al., 1975).
2. Increase $^{35}SO_4$ incorporation into chondroitin sulfate (Adamson and Anast, 1966; Hall, 1970; Delcher et al., 1973).
3. Increase radiolabeled amino acid incorporation into proteins (Herington et al., 1972; Eisenbarth et al., 1973).
4. Increase radiolabeled proline incorporation into collagen hydroxyproline (Eisenbarth et al., 1973).
5. Increase incorporation of [^{3}H]uridine into RNA (Eisenbarth et al., 1973).
6. Increase cartilage cyclic AMP (Drezner et al., 1975).

acid transport was determined by uptake of α amino [1-^{14}C]isobutyrate (Drezner et al., 1975). Cartilage cyclic AMP was measured by a modification of the competitive binding technique (Gilman, 1970; Drezner et al., 1975). Serum pools from normal rats were the source of serum sulfation factor.

EFFECTS OF SERUM SULFATION FACTOR ON CARTILAGE METABOLISM

Serum sulfation factor influences many aspects of cartilage metabolism. Systems employing costal cartilage from hypophysectomized rats have shown that serum sulfation factor increases: (1) $^{35}SO_4$ incorporation into chondroitin sulfate (Salmon and Daughaday, 1957); (2) L-[U-^{14}C]leucine incorporation into proteoglycan (Salmon and Duvall, 1970); (3) conversion of [U-^{14}C]proline into free and collagen [^{14}C]hydroxyproline (Daughaday and Mariz, 1962); (4) [^{3}H]uridine incorporation into RNA (Salmon and Duvall, 1970); and (5) [^{3}H]thymidine into DNA (Daughaday and Reeder, 1966). Table 1 summarizes the effects of serum sulfation factor on embryonic chicken cartilage in vitro in organ culture. It should be noted that serum sulfation factor action on embryonic chicken cartilage is indistinguishable from that on rat costal cartilage.

EFFECTS OF SERUM SULFATION FACTOR ON CARTILAGE CYCLIC AMP

A series of initial studies indicated that incubation of embryonic chicken cartilage in medium containing 5% normal rat serum caused a rise in cartilage cyclic AMP within 15-30 minutes and the rise remained relatively constant for several hours. Table 2 shows the results of 10 individual experiments in which cartilages were incubated for 2 hours in medium containing 5% rat serum and the cyclic AMP content then determined. In every experiment, incubation with the serum sulfation factor causes a rise in cartilage cyclic AMP (4 of the 10 were statistically significant). When all the data are pooled it is evident that serum sulfation factor causes a mean 49.2 ± 7.6% rise in cartilage cyclic AMP.

Figure 1 shows that the rise in cartilage cyclic AMP caused by incubation in medium containing normal rat serum is growth-hormone dependent. Incubation of cartilage in medium containing 5% serum from hypophysectomized rats had no effect on cartilage cyclic AMP while incubation in medium containing 5% serum from

TABLE 2

Effect of serum on cartilage cyclic AMP content

Experiment	Cyclic AMP (pmoles/mg wet weight)		Stimulation by serum[a] (%)	T value[b]
	Medium	*Medium with 5% serum*		
1	0.48 ± 0.04	0.78 ± 0.09	62.5	3.4
2	0.40 ± 0.07	0.59 ± 0.06	47.5	2.1
3	0.40 ± 0.04	0.56 ± 0.07	40.2	2.0
4	0.47 ± 0.05	0.64 ± 0.08	36.2	1.8
5	0.47 ± 0.07	0.58 ± 0.05	23.4	1.3
6	0.11 ± 0.01	0.15 ± 0.01	36.3	2.8
7	0.10 ± 0.02	0.15 ± 0.02	50.0	1.8
8	0.34 ± 0.03	0.71 ± 0.14	108.8	2.6
9	0.58 ± 0.05	0.76 ± 0.03	31.0	3.1
10	0.39 ± 0.05	0.61 ± 0.13	56.4	1.6

[a] Mean ± S.E. of the % stimulation by serum in the combined 10 experiments is 49.2 ± 7.6 ($p < 0.001$).
[b] T value greater than 2.306 indicates significant difference of means with $p < 0.05$.

Cartilages were incubated in vitro in medium with or without 5% rat serum added. The same pool of normal rat serum was used for all experiments. After 2 hours incubation the cartilages were removed and cyclic AMP content determined. Each value is the mean ± S.E. for 5 cartilages.

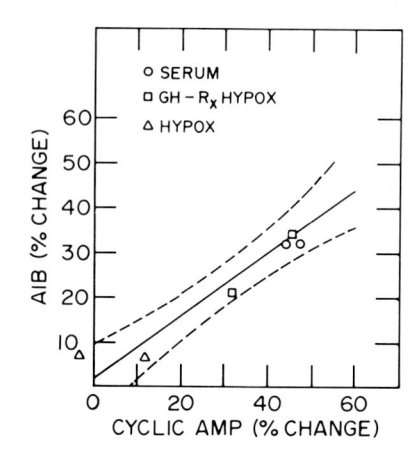

Fig. 1. Correlation between percent stimulation of cartilage α-aminoisobutyrate transport and percent increase in cartilage cyclic AMP in response to incubation in medium containing 5% serum from normal, hypophysectomized and growth hormone-treated hypophysectomized rats. In each experiment paired groups of cartilages (one for α-aminoisobutyrate uptake and one for cyclic AMP content) were incubated in medium with the appropriate serum added and compared to paired groups of control cartilages incubated in medium without any serum added. The data are expressed as the percent change from the mean values of the control cartilages. Each point is the mean of 5 observations of cyclic AMP and α-aminoisobutyrate uptake in the paired cartilage groups. The dashed lines represent the 95% confidence limits of the calculated line. α-Aminoisobutyrate uptake and cyclic AMP concentration highly correlated ($r = 0.977$; $p < 0.001$). (From Drezner et al., 1975, by courtesy of the Editors of *Biochimica et biophysica acta.*)

growth hormone-treated hypophysectomized rats caused rises comparable to those with 5% normal rat serum.

These data indicate that rat serum contains a growth-hormone dependent factor that increases cartilage cyclic AMP content.

AMINO ACID TRANSPORT

If cyclic AMP mediates an action of a hormone on a tissue it should be possible to show that: (1) the hormone changes the cyclic AMP content of the tissue, (2) there is a correlation between the change in cyclic AMP and the measured action, and (3) cyclic AMP or a derivative can mimic the action of the hormone. We carried out a series of studies to determine if we could show that cyclic AMP mediates the action of serum sulfation factor in stimulating amino acid transport. Figure 1 shows the results of a study in which paired cartilages were incubated in medium containing 5% serum from normal, hypophysectomized or growth hormone-treated hypophysectomized rats and in one of the pair, amino acid transport was determined and in the other, cyclic AMP content was measured. The percent change in each parameter compared to appropriate control tissue is plotted. It is evident that serum from hypophysectomized rats did not alter α-aminoisobutyrate uptake or cyclic AMP content. Serum from both normal rats and growth hormone-treated hypophysectomized rats increased both α-aminoisobutyrate uptake and cyclic AMP content. The correlation between the increase in cyclic AMP content and the α-aminoisobutyrate uptake was r = 0.977 (p < 0.001). Figure 2 plots similar experiments performed with the same rat serum pool at different times. As can be seen in Table 2, the cartilage cyclic AMP response to the same serum pool varies from experiment to experiment. Figure 2 shows that the effect

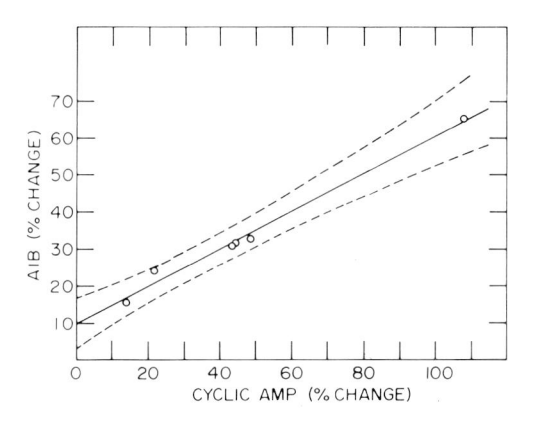

Fig. 2. Correlation between percent increase in cyclic AMP and percent stimulation of α-aminoisobutyrate transport in response to 5% serum in different experiments. The same serum pool was used in all experiments. Paired groups of cartilages were incubated in medium with 5% rat serum added and α-aminoisobutyrate uptake and cartilage cyclic AMP determined. Each experiment included paired control cartilages (incubated in medium without added serum) and the data for serum-treated cartilages are expressed as percent change from the mean value of their appropriate controls. Each point is the mean of 5 observations of cyclic AMP and α-aminoisobutyrate uptake in paired cartilages. The linear correlation of these parameters is highly significant (r = 0.993; p < 0.001). (From Drezner et al., 1975, by courtesy of the Editors of *Biochimica et biophysica acta*.)

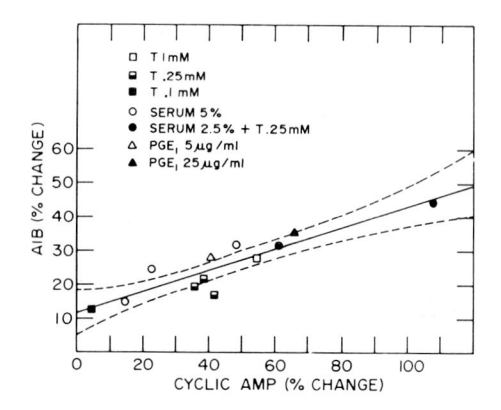

Fig. 3. Correlation between the percent increase in cartilage cyclic AMP and percent stimulation of α-aminoisobutyrate transport in response to 5% serum, 2.5% serum plus 0.25 mM theophylline (T), 0.1-1.0 mM theophylline alone, or 5 and 25 μg/ml prostaglandin E$_1$ (PGE$_1$). The experimental design is as noted in Figs. 1 and 2. The data are expressed as percent change from the mean values of the appropriate controls in each experiment. Each point is the mean of 5 observations of cyclic AMP and α-aminoisobutyrate uptake in paired cartilages. A very high correlation exists between the parameters collectively (r = 0.908; p < 0.001). (From Drezner et al., 1975, by courtesy of the Editors of *Biochimica et biophysica acta*.)

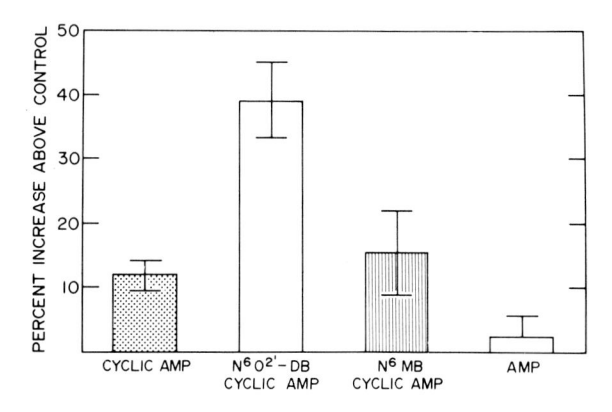

Fig. 4. Effect of cyclic AMP, N^6-monobutyryl cyclic AMP (N^6-MB cyclic AMP) and N^6,O$^{2'}$-dibutyryl cyclic AMP (N^6,O$^{2'}$-DB cyclic AMP) on in vitro cartilage α-aminoisobutyrate uptake. Cartilages were incubated in vitro in medium for 2 hours with 1 mM of the appropriate nucleotide and radiolabeled α-aminoisobutyrate. The bars represent the means and the brackets the S.E. of 5 cartilages. The data are expressed as percent stimulation above the α-aminoisobutyrate uptake of the controls. The significance of the stimulation is p <0.005 for cyclic AMP, p <0.001 for N^6,O$^{2'}$-DB cyclic AMP and p <0.05 for N^6-MB cyclic AMP.

of the serum pool on stimulating α-aminoisobutyrate uptake is very highly correlated with its effect on increasing cartilage cyclic AMP ($r = 0.993$; $p < 0.001$). Figure 3 shows that elevation of cartilage cyclic AMP by addition of agents such as prostaglandin E_1 or theophylline to the incubation medium also increases α-aminoisobutyrate uptake.

Cyclic AMP and its butyrylated derivatives, N^6-monobutyryl cyclic AMP (N^6-MB cyclic AMP) and $N^6,O^{2'}$-dibutyryl cyclic AMP ($N^6,O^{2'}$-DB cyclic AMP) can themselves stimulate α-aminoisobutyrate uptake when added to the cartilage incubation medium (Fig. 4). It is important to note that 5'-AMP has no effect on cartilage α-aminoisobutyrate uptake.

From the above data we conclude that serum sulfation factor stimulation of amino acid transport in cartilage is cyclic AMP mediated.

CARTILAGE MACROMOLECULE SYNTHESIS

We next sought to determine if the effects of serum sulfation factor on cartilage macromolecule synthesis are mediated by the alterations in cyclic AMP content. In order to determine this, we measured the effect of both exogenous and endogenous cyclic AMP on cartilage macromolecule synthesis. Figure 5 shows the effect of incubation of embryonic chicken cartilage for 12 hours in medium containing various concentrations of cyclic AMP. Radiolabeled precursors were added at the beginning of the incubation and incorporation into macromolecules determined at 12 hours. Cyclic

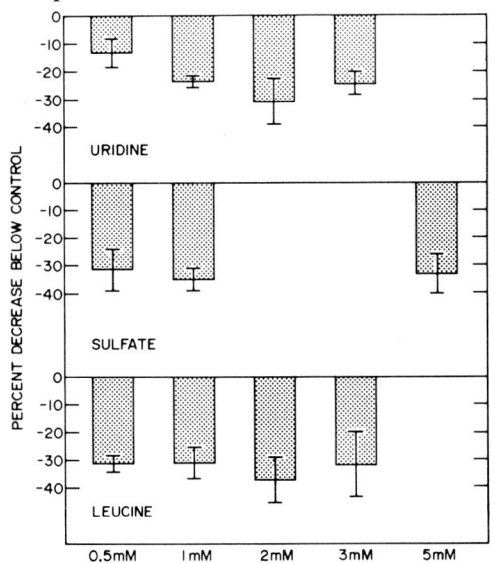

Fig. 5. Effect of cyclic AMP in the medium on cartilage macromolecule synthesis. Cartilages were incubated in vitro with medium containing the appropriate radiolabeled precursor ([^{3}H]-uridine, $^{35}SO_4$ or [^{14}C]leucine) and varying concentrations of cyclic AMP. After 12 hours the cartilages were removed and incorporation into RNA, proteoglycans and total proteins was determined. The data are expressed as percent change from control cartilages incubated in medium without added cyclic AMP. The bars represent the means and the brackets the S.E. for 5 cartilages. Significant inhibition of $^{35}SO_4$, [^{3}H]uridine and [^{14}C]leucine incorporation was observed at all cyclic AMP concentrations (0.5 to 5.0 mM) ($p < 0.01$).

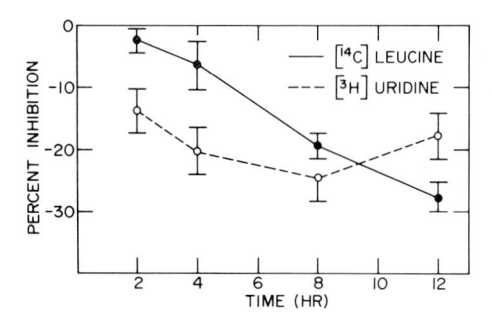

Fig. 6. Time course of cyclic AMP inhibition of [^{14}C]leucine incorporation into total proteins and [^{3}H]uridine incorporation into RNA. Cartilages were incubated in vitro in medium containing the radiolabeled precursors and 1 mM cyclic AMP. At varying time intervals the cartilages were removed and incorporation into proteins and RNA determined. The data are expressed as percent change from control cartilages (incubated without cyclic AMP). Each point is the mean and the brackets the S.E. for 5 cartilages. Cyclic AMP causes significant inhibition of [^{3}H]uridine incorporation at 4, 8 and 12 hours ($p < 0.01$) and of [^{14}C]leucine at 8 and 12 hours ($p < 0.005$).

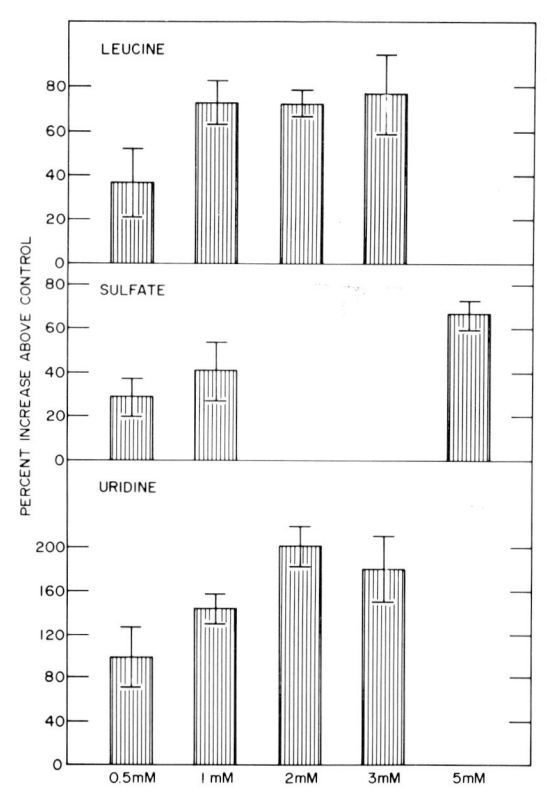

Fig. 7. Effect of N^6-MB cyclic AMP in the medium on cartilage macromolecule synthesis. The experimental design and expression of data are as described in the legend for Fig. 5. Significant stimulation of [^{14}C]leucine, ^{35}SO$_4$, and [^{3}H]uridine incorporation was observed at all N^6-MB cyclic AMP concentrations (0.5 to 5.0 mM) ($p < 0.01$).

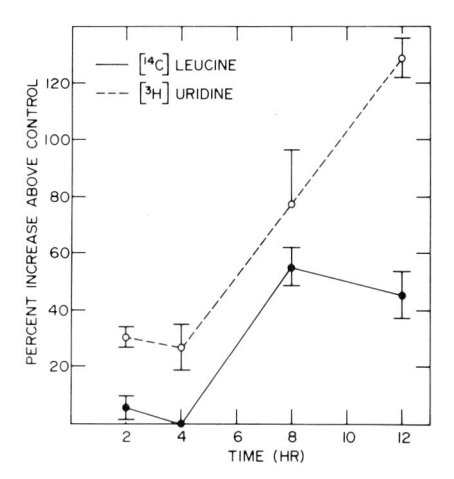

Fig. 8. Time course of N⁶-MB cyclic AMP stimulation of [¹⁴C]leucine incorporation into total proteins and [³H]uridine incorporation into RNA. Cartilages were incubated in vitro in medium containing the radiolabeled precursors and 1 mM N⁶-MB cyclic AMP. The experimental conditions and expression of data are as described in the legend for Fig. 6. N⁶-MB cyclic AMP causes significant stimulation of [³H]uridine incorporation at 2, 4, 8 and 12 hours ($p < 0.01$) and of [¹⁴C]leucine at 8 and 12 hours ($p < 0.005$).

AMP causes a significant inhibition of incorporation of radiolabeled uridine, sulfate and leucine into RNA, proteoglycans and total proteins respectively. The effect is maximal at the lower concentrations examined (0.5 mM and 1 mM). Figure 6 shows the time course for this inhibition. Radiolabeled uridine incorporation was significantly inhibited by 4 hours and no greater effect was noted on more prolonged incubation. Radiolabeled leucine incorporation was not significantly inhibited until 8 hours and was more markedly inhibited by 12 hours.

Incubation of cartilage with butyrylated cyclic AMP derivatives gave entirely opposite effects. As shown in Figure 7, N⁶-MB cyclic AMP incubated with cartilages for 12 hours caused a striking increase in the incorporation of radiolabeled precursors into RNA, total protein and proteoglycan. A significant effect was seen at 0.5 mM but greater effects occurred at higher concentrations. Similar results were observed with N⁶,O²′-DB cyclic AMP. The time course of the effect of 1 mM N⁶-MB cyclic AMP on radiolabeled leucine and uridine incorporation into cartilage proteins and RNA are presented in Figure 8. Uridine incorporation is significantly stimulated by 2 hours and after 4 hours increases in a strikingly linear fashion. Leucine incorporation is not significantly increased until 8 hours.

These paradoxical effects of cyclic AMP and its butyrylated derivatives presented a dilemma. Why should they have opposite effects and which reflects the effect of endogenous cartilage cyclic AMP?

To assess the effects of increases in endogenous cyclic AMP on cartilage macromolecule synthesis, we studied the effects of incubation of cartilage with low concentrations of theophylline, a known inhibitor of adenosine 3′:5′-monophosphate diesterase. The top panel of Figure 9 shows the increase in cyclic AMP content (relative to appropriate controls) which occurs when cartilage is incubated for 2 hours in medium containing theophylline (0.1 to 0.5 mM). This rise in cyclic AMP occurs within 30 minutes and persists throughout the incubation (data not shown). The lower

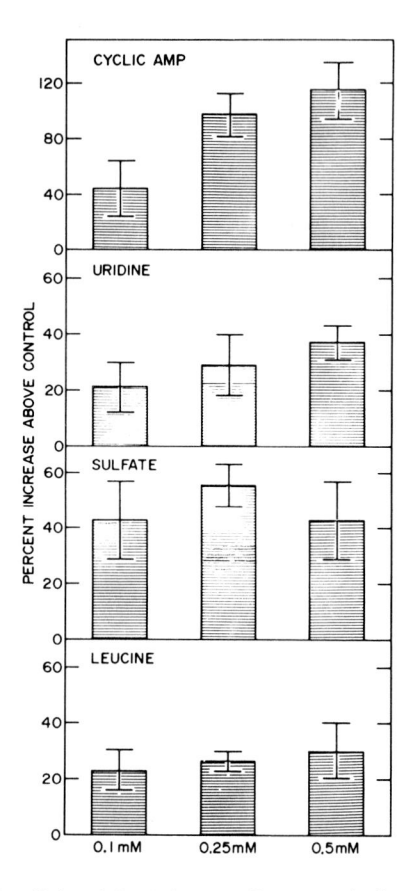

Fig. 9. Effect of theophylline (0.1 to 0.5 mM) on cartilage metabolism. The top panel represents the cartilage cyclic AMP content after a 2-hour incubation. The lower 3 panels represent [^{3}H]-uridine, ^{35}SO$_4$ and [^{14}C]leucine incorporation into RNA, proteoglycans and total proteins respectively during a 12-hour incubation. All data are expressed as percent change (mean ± S.E.) from appropriate control cartilages. Theophylline at 0.1, 0.25 and 0.5 mM significantly increase cartilage cyclic AMP content, and precursor incorporation into RNA, proteoglycans and total proteins.

3 panels present the effects of theophylline in the medium on the incorporation of radiolabeled uridine, sulfate and leucine into RNA, proteoglycans and total proteins during a 12-hour incubation. The data are expressed as the percent increase above the appropriate controls. Theophylline stimulates the incorporation of precursors into macromolecules.

Increases in endogenous cyclic AMP stimulate cartilage macromolecule synthesis and thus the butyrylated cyclic AMP derivatives, not cyclic AMP, mimic the actions of endogenous cyclic AMP.

Although the mechanism of the discordant inhibitory effects of exogenous cyclic AMP was not immediately obvious, studies in other tissues have shown that exogenous cyclic AMP may be metabolized to 5'-AMP and adenosine (Heidrick and Ryan, 1971; Froelich and Rochmüller, 1972; Kaukel et al., 1972). We explored the

TABLE 3

Degradation of exogenous cyclic AMP to extracellular adenosine during incubation with cartilage

Time of incubation (hr)	Adenosine generated (mM)	AMP generated (mM)
0	0	0
2	0.056	0
4	0.123	0
8	0.176	0
12	0.245	0
18	0.370	0

Cartilages were incubated with cyclic AMP (0.5 mM) for the designated time. Aliquots of medium were removed and chromatographed to separate the nucleosides and nucleotides. Identification was by UV absorption. Adenosine purity was proved by spectral analysis and quantitation was determined by UV absorption at 259 nm.

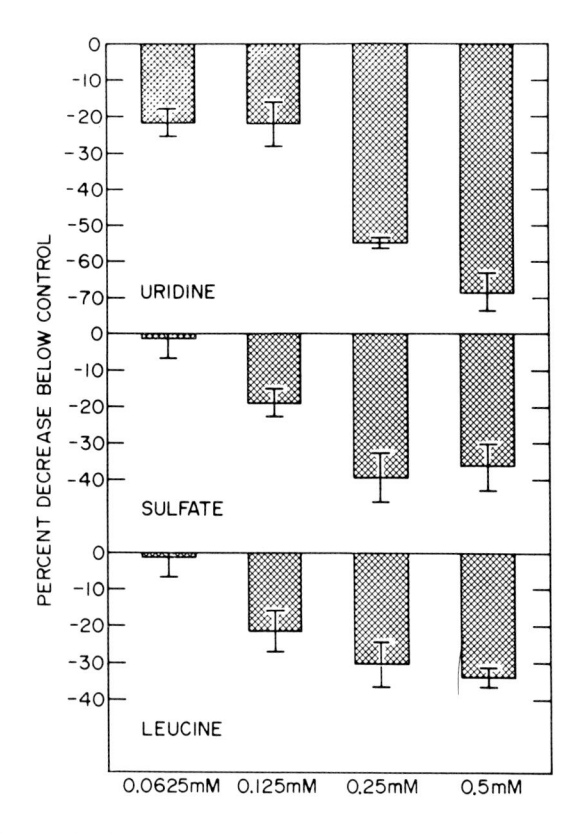

Fig. 10. Effect of adenosine in the medium on cartilage macromolecule synthesis. The experimental design and expression of the data are as described in the legend for Fig. 5. Significant inhibition of [^{3}H]uridine incorporation occurred at adenosine concentrations of 0.0625 to 0.50 mM ($p < 0.01$). Significant inhibition of $^{35}SO_4$ and [^{14}C]leucine incorporation occurred at adenosine concentrations of 0.125 to 0.50 mM ($p < 0.01$).

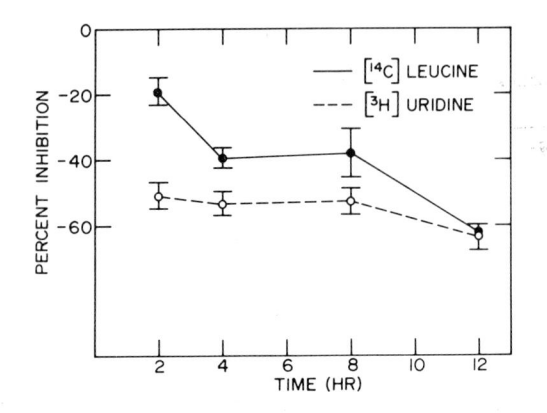

Fig. 11. Time course of adenosine inhibition of [^{14}C]leucine incorporation into total proteins and [^{3}H]uridine incorporation into RNA. Cartilages were incubated in vitro in medium containing the radiolabeled precursors and 0.5 mM adenosine. The experimental conditions and expression of data are as described in the legend for Fig. 6. Adenosine causes significant inhibition of both [^{3}H]uridine and [^{14}C]leucine at 2, 4, 8 and 12 hours ($p < 0.01$).

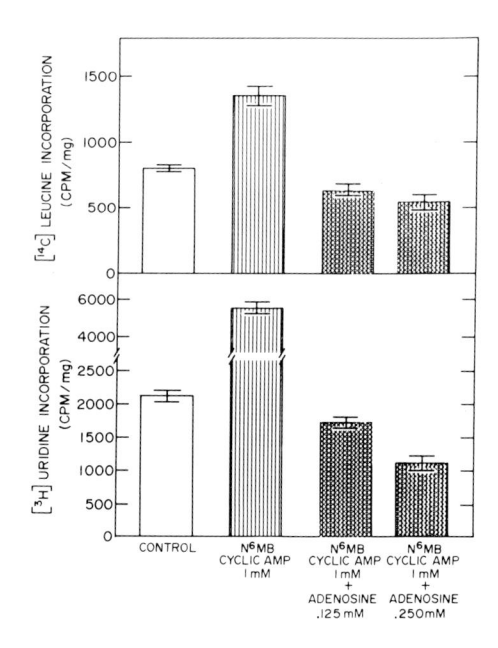

Fig. 12. Effects of adenosine on N^6-MB cyclic AMP stimulation of cartilage macromolecule synthesis. In the N^6-MB cyclic AMP and 'control' groups, cartilages were incubated for 8 hours in medium (containing [^{3}H]uridine and [^{14}C]leucine), with or without N^6-MB cyclic AMP. In the other groups, cartilages were incubated in medium (with radiolabeled precursors) containing N^6-MB cyclic AMP for 2 hours and then transferred to identical medium also containing adenosine for 6 additional hours. The bars represent the means and the brackets the S.E. for 5 cartilages. Adenosine effectively inhibits N^6-MB cyclic AMP stimulated incorporation of radiolabeled precursors into protein and RNA.

possibility that the paradoxical effects of exogenous cyclic AMP that we observed are related to its degradative metabolism.

Cartilages were incubated in medium containing cyclic AMP for 2-18 hours. Before (time 0) and after 2, 4, 8, 12 or 18 hours of incubation, aliquots of medium were removed and chromatographed on Dowex 1 × 10 (formate) columns (0.3 × 27.0 cm) developed sequentially with 0.025 N, 0.30 N and 0.50 N formic acid. Fractions were collected and nucleosides were located by UV absorbance at 259 nm. The columns were standardized with authentic samples of adenosine, 5'-AMP and cyclic AMP. The concentration of the nucleosides was determined by UV absorbance. Table 3 gives the results. There was a progressive rise in the medium concentration of adenosine and a corresponding decrease in medium concentration of cyclic AMP. Other studies (data not shown) indicate that the degradation of adenosine occurs extracellularly and with 0 order kinetics in the cyclic AMP concentration range of 0.5 to 5.0 mM. In contrast, there was no degradation of N^6-MB cyclic AMP during incubation with cartilage.

The biologic effects of adenosine on cartilage metabolism were then studied. Figure 10 demonstrates the effects of adenosine on radiolabeled uridine, sulfate and leucine incorporation into RNA, proteoglycan and total protein respectively. Uridine incorporation into RNA is inhibited by as little as 62.5 μM adenosine. Proteoglycan and total protein synthesis is inhibited by 125 μM adenosine. Adenosine inhibition of radiolabeled uridine and leucine incorporation into cartilage macromolecules occurs quite rapidly as shown by the time course of adenosine action (Fig. 11).

The data presented indicate that exogenous cyclic AMP incubated with cartilage is degraded by adenosine 3':5'-monophosphate diesterase and the phosphomonoesterases (presumably leaking out of the cartilage). The quantities of adenosine generated are sufficient to inhibit macromolecule synthesis. These concentrations of adenosine are also capable of completely abolishing the stimulatory effect of N^6-MB cyclic AMP on cartilage macromolecule synthesis (Fig. 12).

COMMENTS AND CONCLUSIONS

The data presented satisfy most of the criteria necessary to prove that cyclic AMP mediates the actions of growth hormone-dependent serum sulfation factor on cartilage. The only criterion which has not been satisfied is the demonstration that growth hormone-dependent serum sulfation factor increases adenylate cyclase activity. Studies in our own laboratory have not been carried out because of the unavailability both of a suitable purified hormone and a purified cartilage membrane adenylate cyclase preparation. Tell et al. (1973), however, reported that a partially purified somatomedin preparation, in markedly pharmacologic concentrations (10 to 20 times greater than in serum), inhibited baseline and stimulated adenylate cyclase activity in crude particulate fractions from several tissues including embryonic chicken cartilage. We have no explanation for their data and can only point out that they observed no effects with physiologic concentrations of somatomedin and that all of our data are compatible with a stimulatory effect of cyclic AMP on cartilage metabolism rather than an inhibitory effect as their data would imply.

To be objective, however, one must recognize that while our data are compatible with the thesis that serum somatomedins increase cartilage cyclic AMP which activates amino acid transport and macromolecule synthesis, they are also compatible with the thesis that all agents which stimulate cartilage growth do so by activating some other basic process and that serum somatomedins, insulin and cyclic AMP may have independent actions on this other process. Studies on the mechanism of cyclic AMP stimulation of cartilage metabolism and investigations to characterize insulin effects on this tissue should help to define whether this latter explanation is a plausible one.

Regardless of whether cyclic AMP mediates some or all somatomedin action on cartilage or whether cyclic AMP and somatomedins stimulate some other basic process, it is evident that increasing cartilage cyclic AMP content stimulates amino acid transport and macromolecule synthesis. The potent inhibitory effects of adenosine on cartilage macromolecule synthesis, however, suggest the possibility that the intracellular balance between cyclic AMP and adenosine may be the important determinant of cartilage growth rather than just the cyclic AMP concentration. Figure 13 is a theoretical model which we propose to formulate the potential interrelationship between intracellular cyclic AMP and adenosine in regulation of cartilage anabolic processes. Hormones act to stimulate the membrane-bound adenylate cyclase and promote the conversion of ATP to cyclic AMP. The cyclic AMP at the membrane stimulates amino acid transport. At an intracellular site the cyclic AMP stimulates macromolecule synthesis. Increases in intracellular cyclic AMP would lead to increases in AMP which could be dephosphorylated to adenosine which would then feed back to shut off macromolecule synthesis. Adenosine levels could be controlled independently of cyclic AMP by alterations in phosphomonoesterase, adenosine kinase or adenosine deaminase activity. Thus hormonal and regulatory effects on intracellular adenosine levels might influence macromolecule synthesis without causing changes in cyclic AMP content. In a few preliminary experiments we have shown that intracellular adenosine levels in cartilage range between 50 and 125 μM. This intriguing model can explain many unanswered questions and therefore is worthy of study to determine if it is valid.

In summary, we have presented evidence to show that cyclic AMP stimulates cartilage growth processes and that some or all of the actions of growth hormone-dependent serum sulfation factor may be mediated by the generation of cyclic AMP. We have further presented a theoretical model to suggest that the effects of cyclic AMP on cartilage macromolecule synthesis may be determined by the relative intracellular concentrations of cyclic AMP and adenosine rather than the absolute concentrations of cyclic AMP.

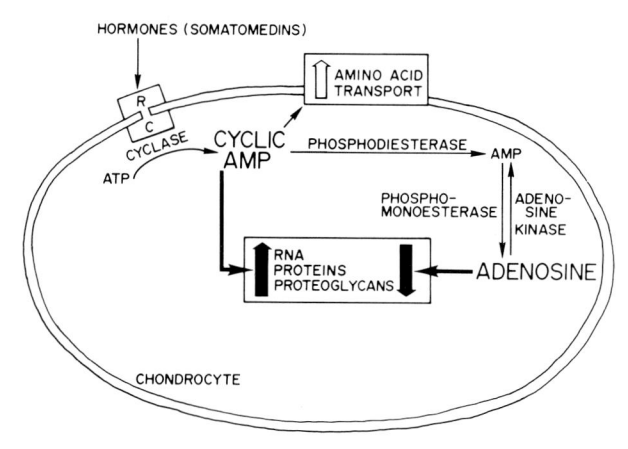

Fig. 13. Proposed model for regulation of cartilage metabolism by cyclic AMP and adenosine.

REFERENCES

Adamson, L.F. and Anast, C.S. (1966): *Biochim. biophys. Acta (Amst.), 121,* 10.
Daughaday, W.H. and Mariz, I. (1962): *J. Lab. clin. Med., 59,* 741.
Daughaday, W.H. and Reeder, C. (1966): *J. Lab. clin. Med., 68,* 357.
Delcher, H.K., Eisenbarth, G.S. and Lebovitz, H.E. (1973): *J. biol. Chem., 248,* 1901.
Drezner, M.K., Eisenbarth, G.S., Neelon, F.A. and Lebovitz, H.E. (1975): *Biochim. biophys. Acta (Amst.), 381,* 384.
Eisenbarth, G.S., Beuttel, S.C. and Lebovitz, H.E. (1973): *Biochim. biophys. Acta (Amst.), 331,* 397.
Eisenbarth, G.S., Beuttel, S.C. and Lebovitz, H.E. (1974): *J. Pharmacol. exp. Ther., 189,* 213.
Froehlich, J.E. and Rochmüller, M. (1972): *J. Cell Biol., 55,* 19.
Gilman, A.G. (1970): *Proc. nat. Acad. Sci. (Wash.), 67,* 305.
Hall, K. (1970): *Acta endocr. (Kbh.), 63,* 338.
Hall, K. and Van Wyk, J.J. (1974): *Curr. Top. exp. Endocr., 2,* 155.
Heidrick, M.L. and Ryan, W.L. (1971): *Biochim. biophys. Acta (Amst.), 237,* 301.
Herington, A., Adamson, L.F. and Bornstein, J. (1972): *Biochim. biophys. Acta (Amst.), 286,* 164.
Kaukel, E., Fuhrmann, U. and Hilz, H. (1972): *Biochem. biophys. Res. Commun., 48,* 1516.
Pastan, I. and Johnson, G.S. (1974): *Advanc. Cancer Res., 19,* 303.
Ryan, W.L. and Curtis, G.L. (1973): In: *Role of Cyclic Nucleotides in Carcinogenesis.* Editors: H. Gratzner and J. Schultz. Academic Press, New York.
Salmon, W.D. and Daughaday, W.H. (1957): *J. Lab. clin. Med., 49,* 825.
Salmon, W.D. and Duvall, M.R. (1970): *Endocrinology, 86,* 721.
Tell, G.P.E., Cuatrecasas, P., Van Wyk, J.J. and Hintz, R.L. (1973): *Science, 180,* 314.
Van Wyk, J.J., Underwood, L.E., Hintz, R.L., Clemmons, D.R., Voina, S.J. and Weaver, R.P. (1974): *Recent Progr. Hormone Res., 30,* 259.

IV. Somatostatin

SOMATOSTATIN, A HORMONE OF THE α_1-CELLS OF THE PANCREATIC ISLETS AND ITS INFLUENCE ON THE SECRETION OF INSULIN AND GLUCAGON*

S. EFENDIĆ, T. HÖKFELT and R. LUFT

Department of Endocrinology, Karolinska Hospital, and Department of Histology, Karolinska Institute, Stockholm, Sweden

Recent observations by our group have given immunohistochemical evidence that somatostatin or somatostatin-like peptides may be present not only in the central nervous system (Hökfelt et al., 1974) but also in several peripheral tissues (Luft et al., 1974; Hökfelt et al., 1975a, b). The structures containing somatostatin were of two types, and seemed to represent endocrine or endocrine-like cells and neurons.

The endocrine cells were observed in the pancreas, thyroid gland, stomach and, probably, in low numbers among the epithelial cells in the intestine. We were also able to demonstrate somatostatin-positive cell bodies in the periventricular region of the anterior hypothalamus, in all probability the origin of the extensive hypothalamic somatostatin-positive fiber plexuses. Furthermore, there was a dense somatostatin innervation of the median eminence, and dense plexuses of somatostatin-probable nerve endings in the ventromedial, arcuate and ventral premammillary nuclei. The localization of these nerve endings does not seem to indicate a release into blood vessels but more a release from nerve endings at the synapses, and could thus indicate a transmitter or modulator role of somatostatin.

In the following we shall present data concerning the localization of somatostatin or somatostatin-like immunoreactivity (SLI) in the pancreatic islets, as well as the effect of somatostatin on the function of the endocrine pancreas.

LOCALIZATION OF SOMATOSTATIN IN RAT PANCREAS

Antibodies to somatostatin were prepared by coupling somatostatin to human α-globulin as described by Arimura et al. (1975). Cryostat sections of the pancreas were incubated with somatostatin antiserum pretreated with human α-globulin, treated as usually and examined in a fluorescence microscope (Hökfelt et al., 1975b). Fluorescent structures were observed mainly in the peripheral parts of practically all islets, occasionally forming a complete ring around the islets. Most immunoreactive structures could be identified as cells with a strong cytoplasmic fluorescence and a non-fluorescent nucleus (Fig. 1). The fluorescence disappeared after pretreatment of the antibodies with somatostatin, but remained after pretreatment with glucagon. This partially confirms the specificity of the immunofluorescence.

---

* Financial support was obtained from the Swedish Medical Research Council (B75-19X-34-11), the Nordic Insulin Foundation and Knut and Alice Wallenberg's Foundation.

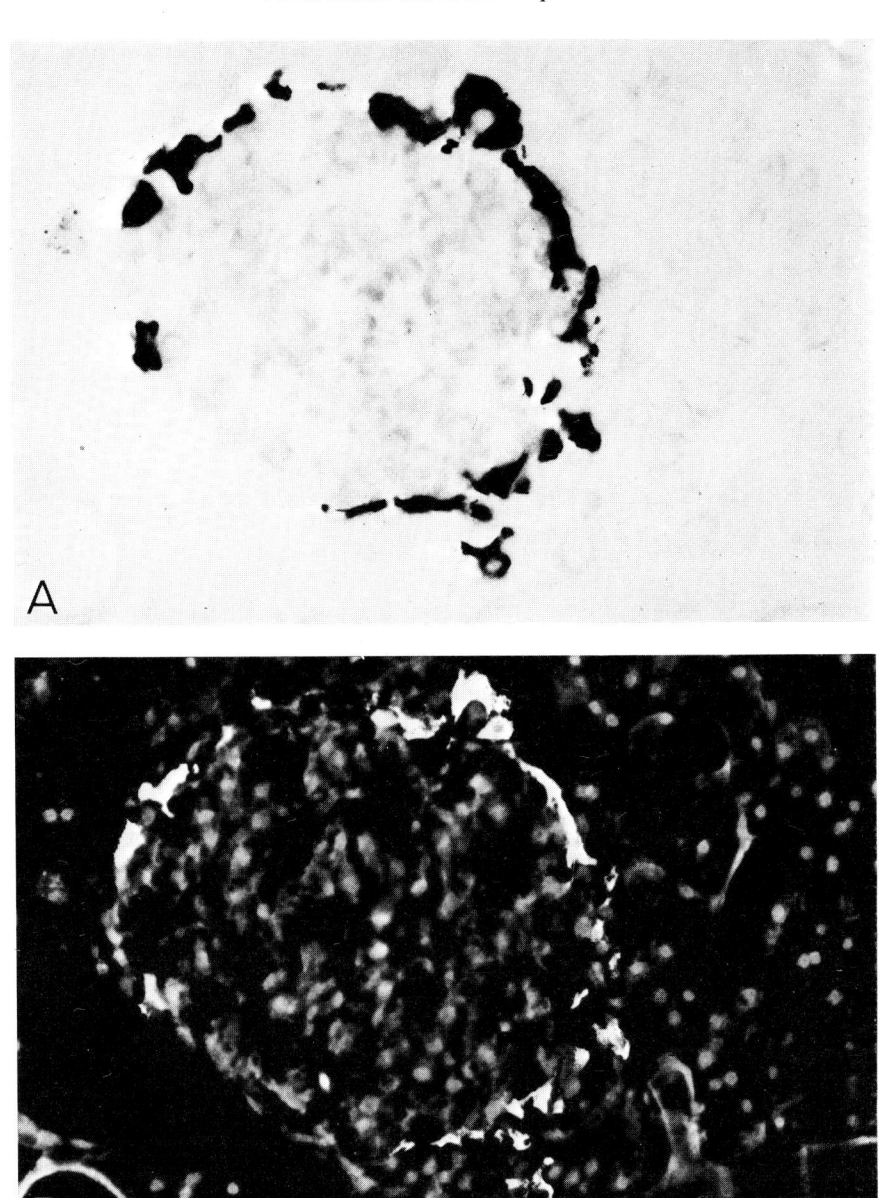

Fig. 1. Pancreatic section incubated with antibodies to somatostatin followed by fluorescein-isothiocyanate conjugated antibodies (A). After photographing the immunofluorescence picture (A) the same section was subsequently processed for silver staining according to Hellerström and Hellman (1960), and the silver-positive structures were photographed (B). Note virtually all somatostatin-positive cells (A) also are argyrophilic, and that they are the only argyrophilic structures in the section. Thus, the somatostatin-containing cells represent the so-called α_1- (or D-)cells. Magnification 400×. (From Hökfelt et al., 1975b, by courtesy of the Editors of *Acta endocrinologica.*)

The nature of the somatostatin-producing cells was further explored by silver staining of the pancreas (Hellerström and Hellman, 1960). According to these authors (Hellerström et al., 1964), this silver staining technique allows a classification of the pancreatic α-cells into 2 distinct groups depending on the presence (= α_1-cells) or absence (=α_2-cells) of cytoplasmic argyrophilia. It has conclusively been shown that the α_2-cell is the source of glucagon (Lundquist et al., 1970), whereas great controversy exists as to the biological significance of the α_1-cell.

By consecutive treatment of the same sections with immunofluorescence technique and silver staining it could be shown that virtually all somatostatin-positive cells were also silver-positive (Fig. 1), whereas the glucagon cells were not. Thus, somatostatin in the rat islets is localized in the α_1-cells.

Effect of somatostatin on glucose-induced insulin release in the isolated perfused rat pancreas

Sprague-Dawley rats, weighing 200-250 g and fasted for 24 hours were used for the preparation of the perfused isolated pancreas. The pancreas was isolated by a slight modification of the technique of Loubatières et al. (1969).

Figure 2 demonstrates that linear somatostatin in a concentration as low as 1 ng/ml, when infused for 10 minutes prior to and then during the infusion with 3.0 mg/ml of glucose, induced a significant suppression of glucose-stimulated insulin release. By increasing the somatostatin concentration to 100 ng/ml, an almost complete inhibition of basal as well as of glucose-induced insulin release was obtained.

Effect of somatostatin on arginine-induced insulin and glucagon release from the isolated perfused rat pancreas

Insulin release induced by arginine (5 mg/ml/min) showed a biphasic pattern, the primary peak being very pronounced (Fig. 3). Somatostatin (10 ng/ml/min) markedly inhibited this insulin release. The pattern of arginine-induced glucagon release resembled that of insulin release (Fig. 4). Somatostatin in the above dose significantly inhibited basal as well as stimulated glucagon release.

DISCUSSION

The isolation and synthesis of the growth hormone release-inhibiting hormone (GH-RIH or somatostatin) was soon followed by the demonstration of its inhibitory effect on insulin and glucagon release in man (Alberti et al., 1973; Gerich et al., 1974; Mortimer et al., 1974; Efendić and Luft, 1975b), in the baboon (Koerker et al., 1974), as well as in the isolated pancreas of the dog (Alberti et al., 1973; Iversen 1974) and rat (Efendić et al., 1974; Efendić and Luft, 1975a). As demonstrated in the present report, somatostatin-inhibited glucose induced insulin release from the perfused rat pancreas in a dose as low as 1 ng/ml. Furthermore, it appeared from these studies — where arginine-stimulated secretions of insulin and glucagon were measured simultaneously — that approximately the same dose of somatostatin was required for the suppression of the functions of the β-cells as well as α_2-cells of the islets.

Even if 1 ng/ml may be considered a small dose, this is still far beyond the amount that could possibly reach the pancreas from the hypothalamus by the systemic circulation. In this connection it is of special interest that we have been able to demonstrate that somatostatin is also produced in the pancreatic islets, more precisely in their α_1-

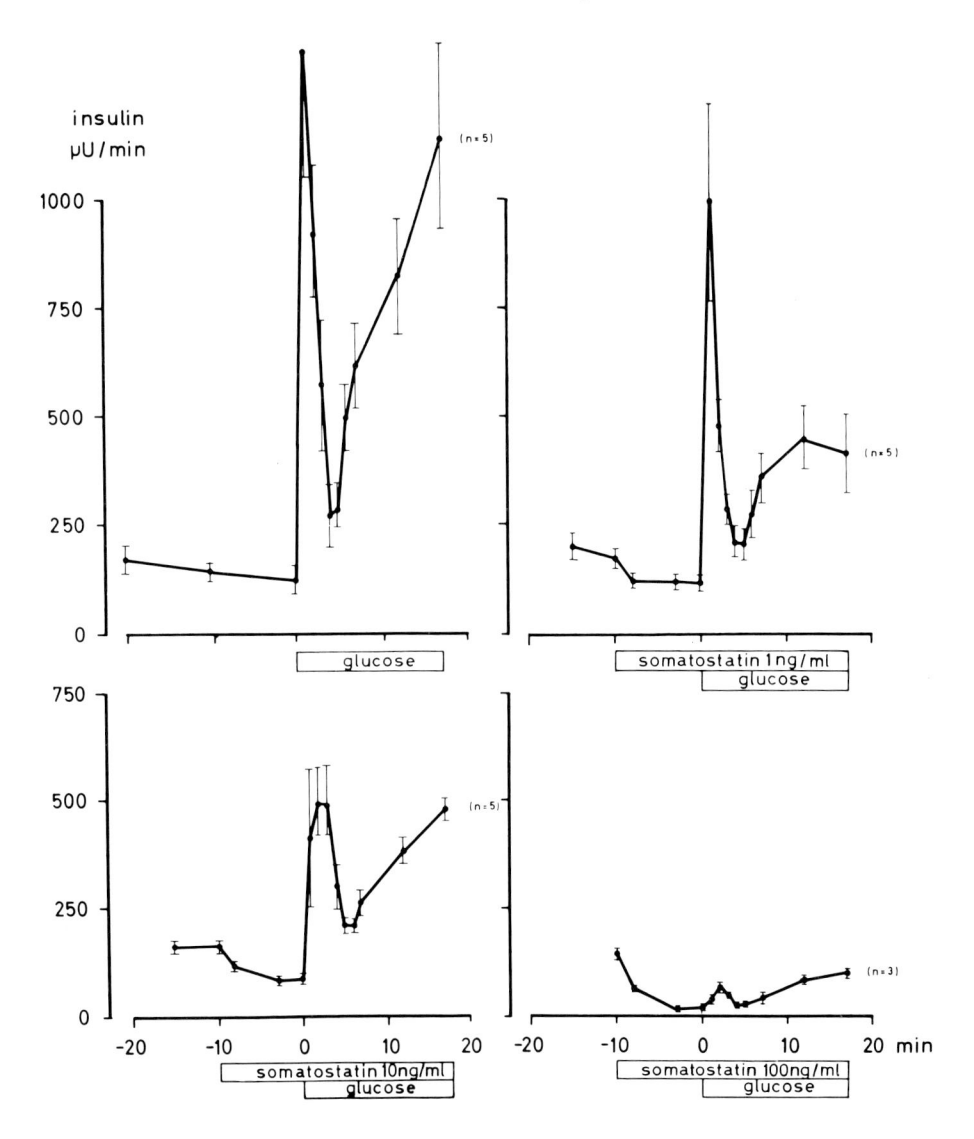

Fig. 2. Effect of linear somatostatin (1-10-100 ng/ml of perfusate) on glucose (3.0/mg/ml)-induced insulin release from the isolated perfused rat pancreas. Following isolation, the pancreas was equilibrated for 30 minutes with 1.5 mg/ml of glucose in the perfusate. The glucose stimulus was applied between 0 and 17 minutes, somatostatin 10 minutes prior to and during the glucose stimulation. Results are expressed as the mean ± SEM of 5 experiments. (From Efendić et al., 1974, by courtesy of the Editors of the *Federation of European Biochemical Societies Letters*.)

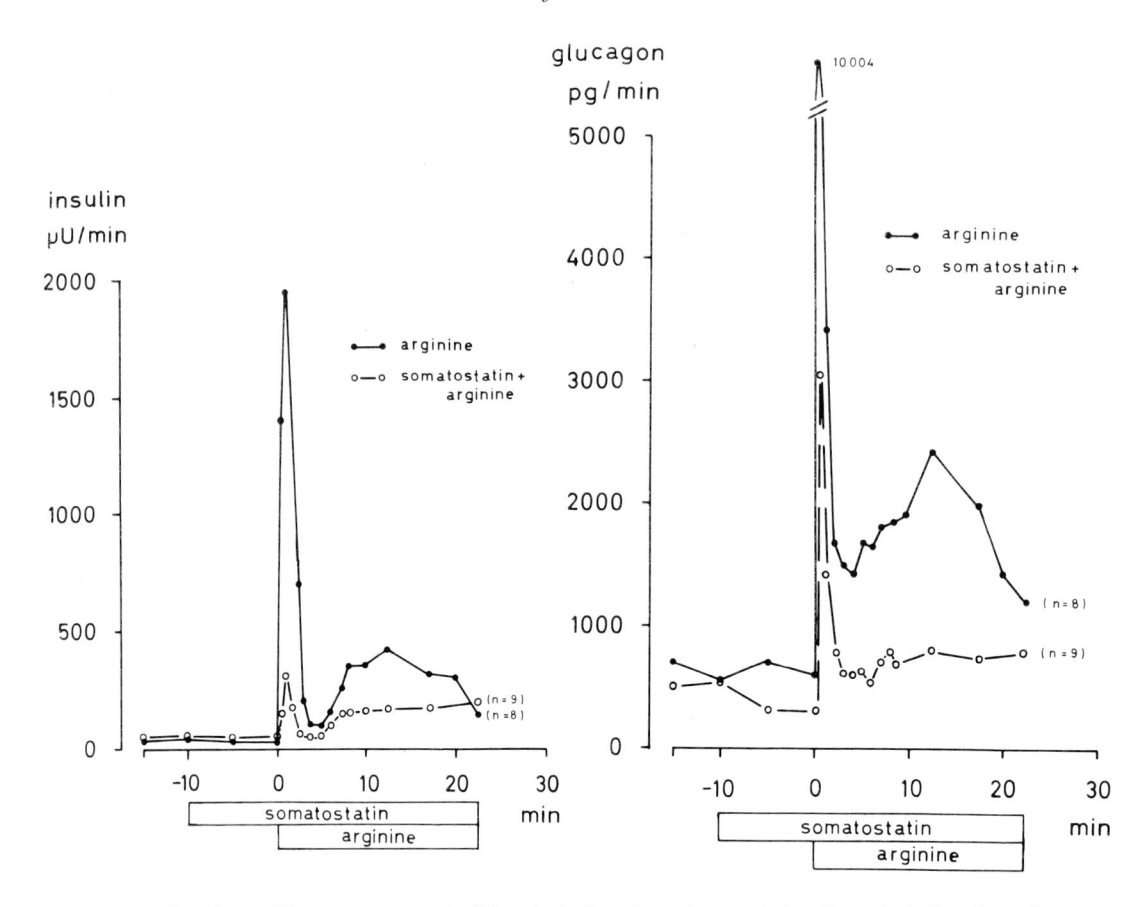

Fig. 3. Effect of linear somatostatin (10 ng/ml of perfusate) on arginine (5 mg/ml of perfusate)-induced insulin release from the isolated perfused rat pancreas. Following isolation the pancreas was equilibrated for 30 minutes with 0.8 mg/ml of glucose in the perfusate. The arginine stimulus was applied between 0 and 22 minutes, somatostatin 10 minutes prior to and during the arginine stimulation. The results are expressed as the mean ± SEM.

Fig. 4. Effect of linear somatostatin (10 ng/ml of perfusate) on arginine-induced glucagon release from the isolated perfused rat pancreas. For further details, see legend Figure 3.

cells (Luft et al., 1974; Hökfelt et al., 1975b). We may therefore consider somatostatin — or a similar peptide — as a pancreatic hormone, and postulate that its inhibitory action on insulin and glucagon release are physiological events.

REFERENCES

Alberti, K.G.M., Christensen, N.J., Christensen, A., Prange-Hansen, A., Iversen, J., Lundbaek, K., Seyer-Hansen, K. and Ørskov, H. (1973): *Lancet, 2,* 1299.

Arimura, A., Sato, H., Coy, D.H. and Schally, A.V. (1975): *Proc. Soc. exp. Biol. (N.Y.), 148,* 784.

Efendić, S. and Luft, R. (1975a): *Acta endocr. (Kbh.), 78,* 510.

Efendić, S. and Luft, R. (1975b): *Acta endocr. (Kbh.), 78,* 516.
Efendić, S., Luft, R. and Grill, V. (1974): *FEBS Letters, 42,* 169.
Gerich, J.E., Lorenzi, M., Schneider, V., Kwan, C.W., Karam, J.H., Guillemin, R. and Forsham, P.H. (1974): *Diabetes, 23,* 876.
Hellerström, C. and Hellman, B. (1960): *Acta endocr. (Kbh.), 35,* 518.
Hellerström, C., Hellman, B., Petersson, B. and Alm, G. (1964): In: *The Structure and Metabolism of the Pancreatic Islets, Vol. I,* p. 117. Editors: S. Brolin, B. Hellman and H. Knutsson. Pergamon Press, London.
Hökfelt, T., Efendić, S., Hellerström, C., Johansson, O., Luft, R. and Arimura, A. (1975b): *Acta endocr., (Kbh.), 80, Suppl. 200,* 1-41.
Hökfelt, T., Efendić, S., Johansson, O., Luft, R. and Arimura, A. (1974): *Brain Res., 80,* 165.
Hökfelt, T., Johansson, O., Efendić, S., Luft, R. and Arimura, A. (1975a): *Experientia (Basel), 31,* 852.
Iversen, J. (1974): *Scand. J. clin. Lab. Invest., 33,* 125.
Koerker, D.J., Ruch, W., Chideckel, E., Palmer, J., Goodner, C.J., Ensinck, J. and Gale, C.C. (1974): *Science, 184,* 482.
Loubatières, A., Mariani, M.M., Ribes, G., de Malbosc, H. and Chapal, J. (1969): *Diabetologia, 5,* 1.
Luft, R., Efendić, S., Hökfelt, T., Johansson, O. and Arimura, A. (1974): *Med. Biol., 52,* 428.
Lundquist, G., Brolin, S.E., Unger, R.H. and Eisentraut, A.M. (1970): In: *The Structure and Metabolism of the Pancreatic Islets, Vol. II,* p. 115. Editors: S. Falkmer, B. Hellman and I.B. Täljedal. Pergamon Press, Oxford.
Mortimer, C.H., Carr, D., Lind, T., Bloom, S.R., Mallinson, C.N., Schally, A.V., Tunbridge, W.M.G., Yeomanas, L., Coy, D.H., Kastin, A., Besser, G.M. and Hall, R. (1974): *Lancet, 1,* 697.

V. GH. Clinical investigations

SEROTONINERGIC CONTROL OF HUMAN GROWTH HORMONE SECRETION: THE ACTIONS OF L-DOPA AND 2-BROMO-α-ERGOCRYPTINE*

GEORGE A. SMYTHE, PAUL J. COMPTON and LESLIE LAZARUS

The Garvan Institute of Medical Research, St. Vincent's Hospital, Sydney, Australia.

That neural mechanisms of growth hormone (GH) release are controlled by the brain catecholamines dopamine or norepinephrine has been a widely held assumption since the finding that α-adrenergic blockade suppresses the human growth hormone (HGH) responses to insulin-induced hypoglycaemia (Blackard and Heidingsfelder, 1968). The observation by Boyd et al. (1970) that the administration of L-dopa, the amino acid precursor of dopamine and norepinephrine, to human subjects caused a release of HGH seemed to confirm the catecholamine hypothesis of HGH secretion involving α-adrenergic stimulation. The finding of Sherman et al. (1971) showing that the phenothiazine, chlorpromazine could inhibit HGH release was taken as further support for the catecholamine hypothesis. More recently, experiments using the ergot alkaloid 2-bromo-α-ergocryptine (CB-154, Sandoz) have shown that as well as having inhibitory effects on prolactin (PL) release, it exerts profound inhibitory effects on GH secretion in acromegalic subjects (Liuzzi et al., 1974; Thorner et al., 1975) but has stimulatory effects in normal subjects (Camanni et al., 1975). Since 2-bromo-α-ergocryptine has been shown to interact with and stimulate dopamine receptors (Corrodi et al., 1973), this ergot derivative has been assumed also to act on GH secretory mechanisms via dopaminergic pathways (Camanni et al., 1975).

It is clear, however, that certain important evidence is not consistent with the original hypothesis that GH release is stimulated via α-adrenergic mechanisms. The dopamine analogue apomorphine exerts similar actions to those of L-dopa in stimulating GH release in normal subjects (Lal et al., 1972) but blocks GH secretion in subjects with acromegaly (Chiodini et al., 1974). Similarly, patients with acromegaly were found by Liuzzi et al. (1972) to respond to L-dopa administration with a suppression of GH secretion. However, in apparent contradiction, Cryer and Daughaday (1974) demonstrated that serum GH levels in acromegalic subjects are suppressed by α-adrenergic blockade with phentolamine. Relevant to the effects of L-dopa in acromegaly is the observation that intrahypothalamic infusion of dopamine in the baboon resulted in suppression of the relatively high pre-infusion levels of serum GH (Toivola and Gale, 1970). A similar effect occurs in the rat (Collu et al., 1972), but under the same conditions these workers showed that infusion of serotonin increases GH secretion. It is thus apparent that stimulatory dopaminergic pathways for GH secretion cannot satisfactorily explain all the experimental results obtained in normal subjects, subjects with acromegaly, or in animals. It does not help our understanding of GH

---

* This work was supported in part by grants from the National Health and Medical Research Council of Australia and Sandoz Research Fund.

TABLE 1
Examples of effects of adrenergic-active compounds on activity of serotonin (5-HT)

Compound	Effect	Reference
Apomorphine	Reduces rat brain 5-HT	Paasonen and Giarman (1958)
Apomorphine	Alters 5-HT activity, hyperthermia and hypothermia	Grabowska et al. (1973) Quock and Horita (1974)
Chlorpromazine	Alters 5-HT activity in vivo and in vitro (various systems)	Bradley (1963) (review)
	Blocks 5-HT on uterine contractions	Costa (1956)
Dopamine, L-dopa	Displaces 5-HT from receptors	Ng et al. (1970)
	Depletes brain 5-HT (rat)	Algeri and Cerletti (1974)
	Decreases brain 5-HT turnover (man)	Goodwin et al. (1971)
	Decreases 5-HT synthesis	Karobath et al. (1972)
LSD	Agonist and antagonist of 5-HT at 5-HT receptors	Gaddum and Hameed (1954) Costa (1956) Berridge and Prince (1974)
Phentolamine	Reverses 5-HT pressor activity	Page and McCubbin (1953)
(Regitin)	Antagonizes 5-HT on umbilical vessels	Aström and Samelius (1957)

secretory pathways to simply say that they differ in these three situations. Perhaps they do, but a preferable hypothesis would be a neural mechanism for GH release that was *common* to these different cases and which could accommodate all the experimental findings.

Recently, strong evidence favouring a major role for the indoleamine, serotonin, in GH stimulatory mechanisms has been presented (Bivens et al., 1973; Smythe and Lazarus, 1973b, 1974a; Smythe et al., 1975). We have emphasised the close structural analogy between dopamine, serotonin and the simple O-methylated derivatives of these amines (Smythe and Lazarus, 1973a,b, 1974a; Smythe et al., 1975) and have shown that O-methylation of dopamine (Smythe and Lazarus, 1973a) or serotonin (Smythe et al., 1975) provides derivatives which still interact with the dopamine or serotonin receptor in competition with their parent amines but which have opposite actions. It is because of the close structural relationship between dopamine and serotonin and consequent requirements for interaction at their respective receptors, that compounds such as L-dopa, apomorphine, D-lysergic acid diethylamide (LSD), the α-adrenergic blocking agents (e.g. phentolamine), the phenothiazines (e.g. chlorpromazine), apart from their adrenergic effects, profoundly alter serotonin activity and brain serotonin levels (see Table 1). On this basis the original evidence for the catecholamine hypothesis of GH release must be reassessed, as the data can be applied equally well to serotoninergic pathways. Similarly data obtained with 2-bromo-α-ergocryptine may be applied to serotoninergic as well as dopaminergic pathways. 2-Bromo-α-ergocryptine is a derivative of lysergic acid and, structurally, is closely analogous to the LSD derivative 2-bromo-D-lysergic acid diethylamide (BOL). Like 2-bromo-α-ergocryptine, LSD and BOL have been shown to interact with dopamine receptors (Pieri et al., 1974; Von Hungen et al., 1974) and LSD has been shown to inhibit PL release in the rat (Quadri and Meites, 1971).

In relation to GH secretory mechanisms, it is well known that LSD, BOL and the simpler analogue, methysergide, are able to act as serotonin agonists and antagonists. In fact, Bivens et al. (1973) have shown methysergide to inhibit GH stimulation in normal subjects. LSD and its derivatives possess the indole ring as part of their structure and the close structural analogy between serotonin and the LSD molecule has been commented on by Chothia and Pauling (1969), Kang and Green (1970) and Baker et al. (1973). On structural grounds alone, 2-bromo-α-ergocryptine must be considered able to interact with serotonin receptors. We have suggested that the action of melatonin on the serotonin receptor is analogous to that of LSD (Smythe and Lazarus, 1974b) because of the apparent similarity of action of the two compounds with respect to both agonist and antagonist effects on serotoninergic pathways.

Our findings with respect to the effects of the pineal hormone, melatonin, on GH secretion (Smythe and Lazarus, 1974a,b) indicate that melatonin and L-dopa may have virtually identical effects on GH secretory pathways. Melatonin stimulates HGH release in normal, or 'basal' subjects (Smythe and Lazarus, 1974b) but blocks the stimulation of HGH following hypoglycaemic doses of insulin (Smythe and Lazarus, 1974a). It thus seemed likely that dopamine derived from L-dopa might induce HGH release in basal subjects but inhibit it in 'stimulated' subjects by interactions with hypothalamic serotonin receptors in the way we proposed for melatonin rather than via adrenergic pathways. In order to test this hypothesis the HGH response to L-dopa, hypoglycaemic doses of insulin, a combination of L-dopa and insulin and a combination of L-dopa and serotoninergic blockade using cyproheptadine were examined in normal subjects. The HGH and human prolactin (HPL) responses to various protocols (including 2-bromo-α-ergocryptine administration) were also examined in normal and acromegalic subjects with a view to establishing a unifying concept of GH release in terms of serotoninergic stimulation or blockade.

SUBJECTS AND PATIENTS STUDIED

These studies were performed on 2 groups of volunteers from whom informed consent had been obtained. The first group comprised 15 normal male and 1 normal female subject aged from 21-28 years. The second group comprised 6 male and 7 female patients with acromegaly aged from 25-62 years. Three of the patients had not received any prior treatment for their acromegaly at the time of study. The other 10 patients had been treated with various techniques including transfrontal hypophysectomy, transsphenoidol cryogenic pituitary destruction (Bleasel and Lazarus, 1974), external irradiation, or oral chlorpromazine. In one of the patients (M.M.) the acromegaly was part of a pluriglandular syndrome together with pheochromocytoma and hyperparathyroidism.

MATERIALS AND METHODS

The tests began at 8.30 a.m. after an overnight fast. Blood was sampled from an indwelling catheter in an antecubital vein. The patency of the catheter was maintained by a slow isotonic saline infusion. Baseline blood samples were collected before stimulation tests and at regular intervals thereafter. Each subject rested on a bed for the duration of the test. Tests on the same subject were carried out at least 1 week apart.

Study 1

Insulin-induced hypoglycaemia control study
Six normal male and 1 normal female subject participated in this study. After collection of basal blood samples, insulin (Novo Industri A/S Denmark, neutral porcine crystalline; 0.1 U/kg) was given intravenously at time '0' minutes. Blood samples were collected during 135 minutes following the insulin administration.

Study 2

Insulin-induced hypoglycaemia after L-dopa administration
At 30 minutes prior to insulin administration each subject from Study 1 was administered 1 500-mg tablet of L-dopa (Roche Products). Insulin 0.1 U/kg was administered at time 0 minutes using the same protocol as described in the control study.

Study 3

L-dopa stimulation test
Thirteen normal male subjects and 1 normal female subject participated in this study. After collection of basal blood samples each subject was administered 1 500-mg tablet of L-dopa at 0 minutes and blood was collected during the following 135 minutes.

Study 4

L-dopa stimulation test after cyproheptadine administration
Seven of the male subjects from Study 3 (above) took part in this study. Five days before the commencement of an L-dopa stimulation test, each subject began a regimen of oral cyproheptadine (Periactin; Merck, Sharp and Dohme). On the 1st and 2nd days 4 mg was taken twice daily and on the 3rd, 4th and 5th days 4 mg was taken 3 times daily. On the day of testing (6th day) a final 4 mg was taken 1-2 hours before the commencement of the test. An L-dopa stimulation test was then carried out as described in Study 3 (above).

Study 5

Melatonin stimulation test
Oral melatonin (1 g) was administered to the 9 normal male subjects who participated in this study. The protocol previously described (Smythe and Lazarus, 1974b) was used.

Study 6

2-Bromo-α-ergocryptine suppression test
The dose used was 2.5 mg orally. Initially this was given to fasting patients but due to the high incidence of nausea and vomiting in the initial studies the majority of the studies were performed after a light breakfast. This avoided the unpleasant side effects.

Studies carried out on individual subjects rather than groups are described in the following section. Plasma glucose was estimated on a technicon Auto-analyser using the ferricyanide method for reducing sugars. Serum GH was measured by the radioimmunoassay technique of Molinatti et al. (1969) and the results are expressed in microunits of the WHO International Reference Preparation for HGH radioimmunoassay (2 μU = 1 ng HGH). Serum prolactin (PL) was measured by radioimmunoassay using materials supplied by the NIAMDD. Human PL (HPL-V-L-S # 1) was used as assay standard and, radioiodinated, as the radioactive tracer in the assay. The antiserum was rabbit antiserum to V-L-S # 1. Results are expressed as ng/ml of HPL V-L-S # 1. Statistical analysis was carried out using either Wilcoxon's test for pair differences or Student's t-test.

RESULTS

The serum GH response to insulin-induced hypoglycaemia and the effect of the prior administration of L-dopa on this response is illustrated in Figure 1. Following L-dopa administration, the mean serum HGH levels were significantly reduced at all times after 45 minutes. L-dopa highly significantly reduced the mean peak level of HGH from 109.2 ± 8.1 to 70.1 ± 6.6 μU/ml ($p < 0.0025$). The total amount of HGH secreted by each subject during the test was estimated by integrating the area under the individual response curves with and without L-dopa. L-dopa was found also to cause a highly significant reduction in this parameter ($p < 0.0025$). The effect of L-dopa on the mean peak and integrated HGH responses to insulin-induced hypoglycaemia is shown in Figure 2. The stippled area in the histogram of the mean integrated response for insulin + L-dopa in this Figure represents the contribution to the HGH rise shown by the same subjects when L-dopa alone was administered. One of the subjects who failed to exhibit an HGH response to L-dopa alone also failed to show any suppression of his HGH response to insulin following L-dopa. He was eliminated from this study on the assumption that he did not satisfactorily absorb the dose of L-dopa. L-dopa did not have any significant effect on the degree of hypoglycaemia induced by the insulin in these studies.

The prior administration of the serotonin antagonist cyproheptadine resulted in suppression of the HGH response to L-dopa in each of the 6 subjects who responded to

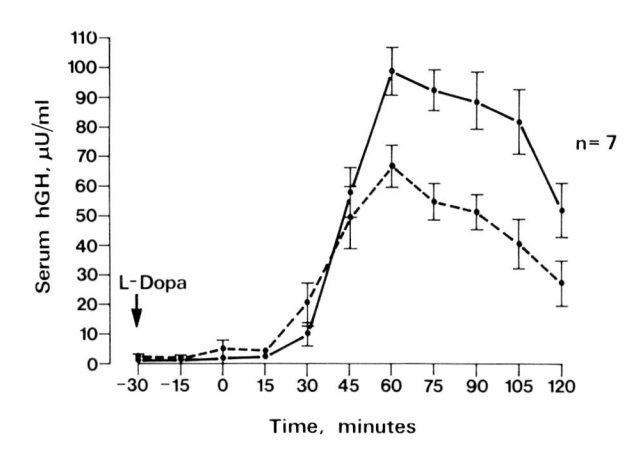

Fig. 1. The effect of L-dopa on the serum HGH response to insulin-induced hypoglycaemia in 7 normal subjects. The broken line shows their response after L-dopa and the continuous line shows their control response. Means ± SEM are shown.

L-dopa alone, with the mean peak HGH response following L-dopa being reduced from 31.0 ± 8.3 to 15.5 ± 7.5 μU/ml ($p < 0.05$) by the cyproheptadine regimen. Of these subjects only 4 achieved peak serum GH responses greater than 20 μU/ml after L-dopa alone and statistical analyses were carried out on these subjects as a group to avoid the effect of high standard deviation. The integrated area under the GH response curve for these subjects was significantly reduced by the cyproheptadine as illustrated in Figure 3. One of the subjects in this study showed a total inhibition of his normally good GH response to L-dopa by the cyproheptadine regime but his PL response remained intact (Fig. 4) indicating that the GH response to L-dopa is activated by a different mechanism (serotoninergic) to the PL inhibition (dopaminergic) which is not affected by the serotonin antagonist cyproheptadine. In our hands many of the subjects who show good GH response to L-dopa also respond with serum GH rises following oral administration of the serotonin precursors L-tryptophan (50 mg/kg) and 5-hydroxy-L-tryptophan (5-HTP, 150 mg/kg) but L-dopa is the only one of these amino acids which alters PL secretion. The subject whose various GH responses to these amino acids are shown in Figure 5, illustrates this point. The failure of indoleamine precursors and derivatives to alter PL secretion when they do alter GH secretion is further shown in the study with melatonin (Fig. 6). Following the administration of melatonin (1 g, oral) the subjects respond with a typical rise in serum GH (Smythe and Lazarus 1974b) but there was no effect on serum PL levels.

In confirmation of reports from other groups (Liuzzi et al., 1974; Thorner et al., 1975) 2 bromo-α-ergocryptine (CB-154) was found to have a profound effect on HGH secretion in the patients with acromegaly. Ten of 13 subjects showed a marked and prolonged suppression of CB-154. The individual responses to CB-154 are shown in Figure 7. The lack of response to CB-154 in 2 of the 3 patients is explainable in terms of their other drug therapy at the time of testing. One subject was being chronically treated with a phenothiazine, thioridazine, which can act as a serotonin antagonist (Bradley, 1963). The other subject was being chronically treated with the serotonin antagonist cyproheptadine but this subject did respond to the CB-154 with a significant suppression of his serum PL levels indicating dopamine receptors were not blocked by the cyproheptadine. A further patient failed to respond to CB-154 with a suppression of serum HGH when being treated with oral chlorpromazine but did respond to CB-154 with a typical fall in GH secretion after chlorpromazine therapy was discontinued.

In a very limited trial mixed responses were found following the acute administration of melatonin (1 g, oral) to 3 acromegalic patients (Fig. 8). One of these 3 subjects demonstrated a significant rise in serum HGH levels by 120 minutes, similar to that seen in normal subjects. However, the other 2 subjects responded with significant falls to about 30% of their basal levels by 120 minutes. The effect observed following melatonin in 1 of these subjects was comparable to her response following CB-154 administration but blood collection was not continued for a sufficiently long time to determine whether the inhibitory effect was transitory, as it was in the other subject.

DISCUSSION

The results of this study show that the effect of L-dopa in inducing GH release is inhibited by serotonin receptor antagonism and also that in normal subjects the administration of L-dopa blocks the GH secretory mechanisms normally strongly stimulated by hypoglycaemic doses of insulin.

That L-dopa inhibits PL release via a different pathway to that by which it stimu-

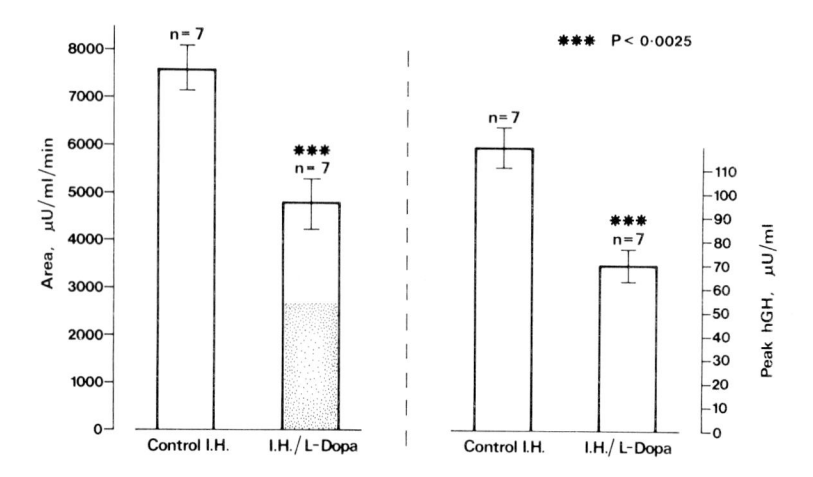

Fig. 2. The effect of L-dopa on the mean integrated and peak HGH responses to insulin-induced hypoglycaemia (IH) in normal subjects. The stippled area shown in the histogram for the mean area of IH/L-dopa represents the contribution to the HGH response when L-dopa, alone, was administered. Means ± SEM are shown.

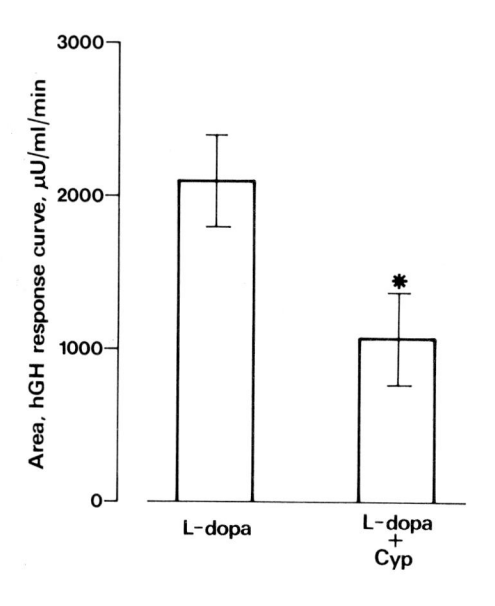

Fig. 3. The effect of cyproheptadine (Cyp) on the mean integrated HGH response to L-dopa in normal subjects. Means ± SEM are shown (*); p <0.05.

Serotonergic control of GH secretion

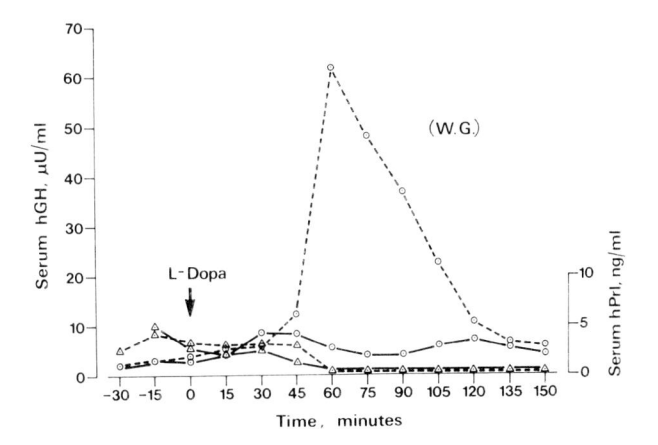

Fig. 4. The effect of cyproheptadine on the HGH and HPL response to L-dopa in subject W.G. The broken line shows the control responses and the continuous line the responses after cyproheptadine. The curves connected by circles indicate the HGH response and those connected by triangles indicate the HPL response.

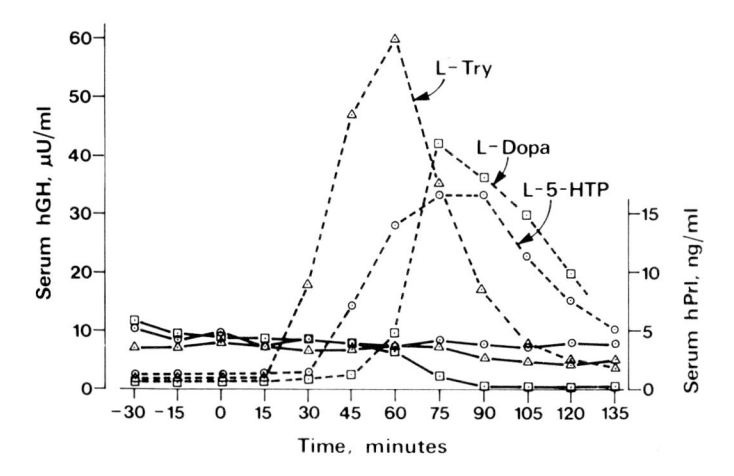

Fig. 5. The effect of the amino acids L-dopa, L-tryptophan (L-Try) and 5-hydroxy-L-tryptophan (5-HTP) on HGH and HPL secretion in subject R.S. The broken lines connect the HGH responses and the continuous lines connect the HPL responses.

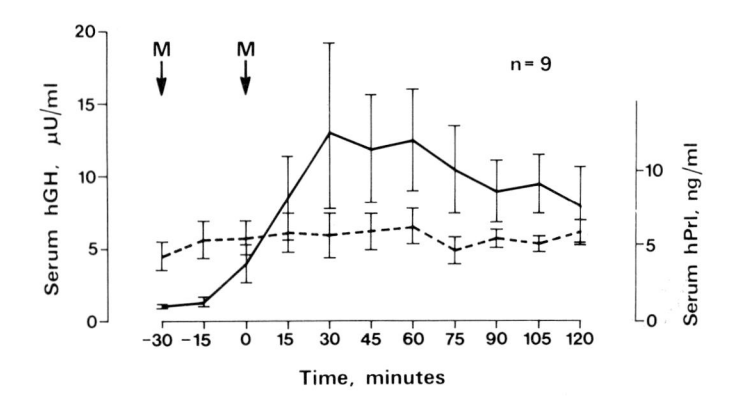

Fig. 6. The mean response of serum HGH (continuous line) and HPL (broken line) to oral melatonin (2 × 500 mg at points indicated by M) in normal subjects. Means ± SEM are shown.

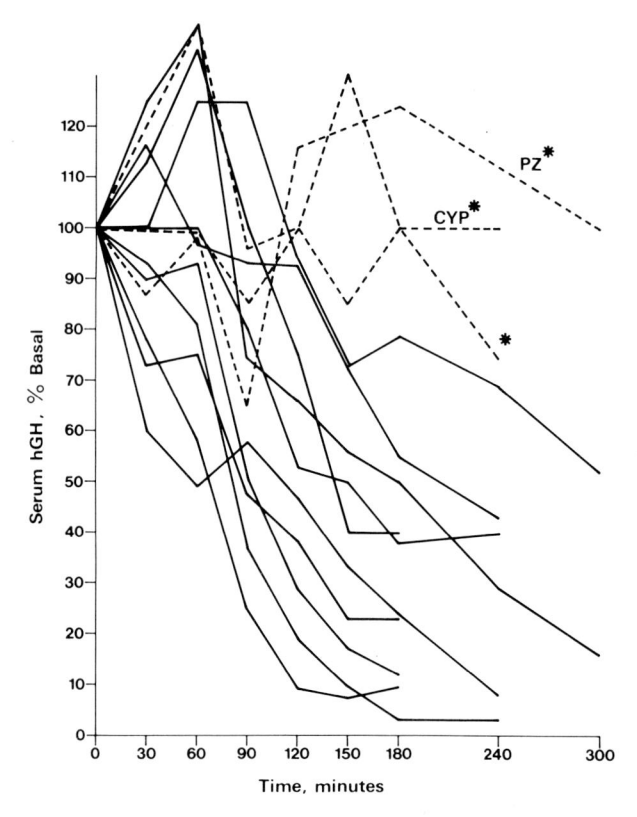

Fig. 7. The response of serum HGH to oral 2-bromo-α-ergocryptine (CB-154, 2.5 mg) administered at time '0' in 13 patients with acromegaly. The asterisks indicate no significant response (3 subjects). The curves labelled PZ and CYP represent the responses of subjects receiving chronic doses of a phenothiazine (thioridazine) and cyproheptadine, respectively, at the time of the test.

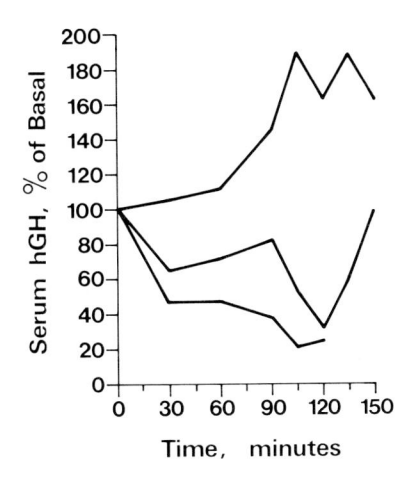

Fig. 8. The response of serum HGH to oral melatonin (1 g) administration at time '0' in 3 patients with acromegaly.

lates GH release in both normal subjects and in subjects with acromegaly is indicated by the experiments with cyproheptadine. This serotonin antagonist blocks the GH response to L-dopa but not the dopaminergic inhibition of PL in normal subjects. This is in accord with the fact that L-dopa administration to mice results in a displacement of brain serotonin which is proportional to the increasing levels of dopamine, indicating competitive interactions of dopamine with serotonin receptors (Algeri and Cerletti, 1974). The data similarly show that in the acromegalic subject being treated chronically with cyproheptadine the HGH response to 2-bromo-α-ergocryptine was prevented but PL was inhibited in the usual way.

The actions of L-dopa reported here are identical to those reported for melatonin (Smythe and Lazarus, 1974a,b) and are analogous to those reported for apomorphine (Lal et al., 1972; Chiodini et al., 1974). These results suggest a similar mechanism of action of these diverse compounds on GH secretory pathways. The effects of these and other compounds on GH secretory mechanisms can be explained in terms of a common pathway whether we consider results obtained in normal subjects or patients with acromegaly or animal experiments. We propose the common pathway to be serotoninergic, involving serotonin receptors in the hypothalamus, and perhaps other areas of the central nervous system. Such a hypothesis can accommodate all the experimental data and is perhaps best explained in terms of competitive interactions of antagonists, or partial agonists, with serotonin receptors which mediate in the neural release of GH. Thus, a subject in a basal state whose serum GH levels are low may be considered to have these serotonin receptors in a state of low stimulation with 'spare' or unoccupied receptors. When the competitive antagonists such as dopamine or melatonin enter this system they may interact with, or bind to, unoccupied receptors causing a stimulatory response and, ultimately, an increase in GH secretion from the pituitary gland. On the other hand, in a subject whose GH secretory mechanisms are maximally stimulated, such as in acromegaly or following insulin-induced hypoglycaemia, the serotonin receptors may be considered as being fully occupied at any instant and also in a state of flux, with serotonin continually binding with and leaving its receptors. In this case, when an antagonist (e.g. dopamine or melatonin) binds at the receptor and is not readily displaced by serotonin there is a net suppression of stimula-

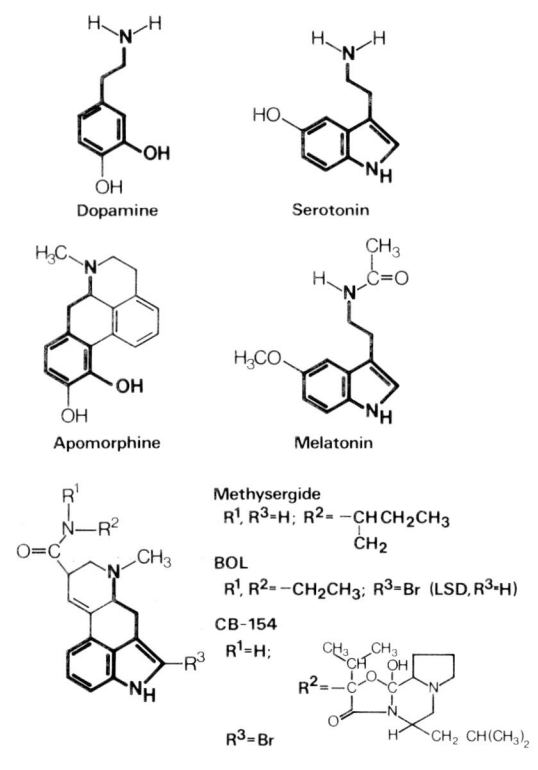

Fig. 9. Structural formulae of compounds which interact with serotonin receptors. The heavy lines indicate the regions of the molecules proposed to be involved in receptor interactions.

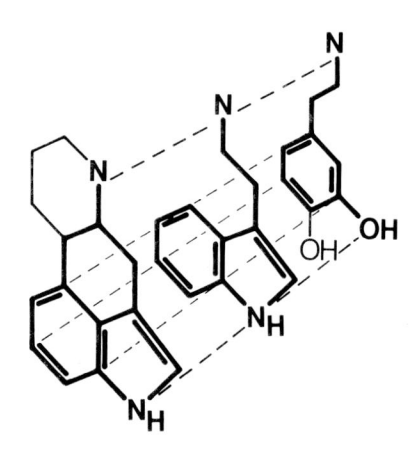

Fig. 10. Superimposition of the dopamine, serotonin and ergoline ring systems.

tion and a reduced release of GH. In systems where a close study of receptor binding by agonists and antagonists is possible, such as that of transmitter function at the neuromuscular junction, the effect of antagonists acting initially as agonists is well known (Paton, 1970). In order to explain the fact that some compounds first excite, then block; that antagonists are always relatively slow to wash out from a tissue while their relative agonists can be rapidly removed, Paton (1970) has suggested that the concept of 'spare' or unoccupied receptors, which he terms 'effigacy', be used.

The same concept can be used to explain the actions of 2-bromo-α-ergocryptine. Figure 9 shows the structural formulae for some of the compounds used to obtain data relevant to the neural regulation of GH secretion. The regions of each compound, which we propose interact with serotonin receptors, are shown with bold lines and the way the common structural features of the dopamine, indoleamine, and ergoline skeletons may be superimposed is shown in more detail in Figure 10. The nature of the binding between LSD and a salivary gland serotonin receptor has recently been studied by Berridge and Prince (1974). These workers showed that LSD competes with serotonin and tryptamine for the same receptor, it is only slowly displaced by other agonists and antagonists, and its action is inhibited by pretreatment with serotonin. Berridge and Prince (1974) conclude that the interaction of LSD with the serotonin receptor is very stable and reverses much more slowly than does the serotonin complex with the receptor. These workers further suggest that the reason LSD remains attached to the serotonin receptor a far longer time than other agonists and antagonists is due to interactions of the 'non-indoleamine' portion of the molecule with regions peripheral to the actual receptor, which act as semi-permanent attachment points, enabling the alkaloid molecule to flap up and down on the serotonin receptor. This is an important concept and may explain the difference in magnitude and duration of action on GH secretion of the various compounds under discussion (see Fig. 9). Dopamine has the simplest structure and has no groups available to interact with 'regions peripheral' to the serotonin receptor and thus would be expected to have the shortest duration of action. Indeed, the short time for which L-dopa induces suppression of GH in subjects with acromegaly has been commented on (Chiodini et al., 1974). Also, Lal et al. (1975) have shown L-dopa to be less effective in altering GH secretion than the structurally more complicated analogue, apomorphine. At the other extreme, 2-bromo-α-ergocryptine has the greatest 'non-indoleamine' component of any of the other compounds and therefore could most easily bind to 'regions peripheral' to the serotonin receptor.

The evidence to date certainly indicates that 2-bromo-α-ergocryptine is the most potent of these compounds and has the longest duration of action. Of the compounds under consideration, melatonin is closest structure-wise to serotonin, of which it is a simple derivative, and might perhaps be expected to be relatively potent, if short term, in its action. The potential of melatonin as an investigative or therapeutic tool remains to be fully explored. It has the advantage of being devoid of toxic side effects and in our hands is a more potent and consistent acute stimulus for GH release in normal subjects than is L-dopa. It also appears that in certain patients with acromegaly melatonin is as potent as 2-bromo-α-ergocryptine but its effect seems short-lived.

In conclusion, the results of this study are in total accord with the hypothesis that serotonin is the major neural aminergic stimulus for growth hormone secretion and provide an explanation for what were previously considered anomolous or paradoxical results in patients with acromegaly.

SUMMARY

New data presented here, together with a reassessment of some of the original data relevant to the mechanism controlling release of GH, strongly support the hypothesis that neural regulation of GH secretion is controlled via brain serotonin receptors. It is shown that L-dopa, like the pineal hormone melatonin, can inhibit as well as stimulate the release of HGH in normal subjects. The actions of L-dopa on GH release, but not on PL release, are inhibited by serotoninergic blockade. The effects of compounds such as L-dopa and 2-bromo-α-ergocryptine on GH release are proposed to result from interaction at serotonin receptors, with the nature of the response being dependent on the relative numbers of these receptors which are occupied at any one time. On the basis of common serotoninergic pathways, the 'paradoxical' effects of α-adrenergic blocking drugs and dopaminergic drugs in normal subjects versus patients with acromegaly are, in fact, predictable ones.

ACKNOWLEDGEMENTS

We wish to thank Sandoz (Australia) Pty. Ltd. for the generous gift of 2-bromo-α-ergocryptine (CB-154, Sandoz) used in these studies.

REFERENCES

Algeri, S. and Cerletti, C. (1974): *Europ. J. Pharmacol., 27,* 191.
Aström, A. and Samelius, U. (1957): *Brit. J. Pharmacol., 12,* 410.
Baker, R.W., Chothia, C., Pauling, P. and Weber, H.P. (1973): *Molec. Pharmacol., 9,* 23.
Berridge, M.J. and Prince, W.T. (1974): *Brit. J. Pharmacol., 51,* 269.
Bivens, C.H., Lebovitz, H.E. and Feldman, J.M. (1973): *New Engl. J. Med., 289,* 236.
Blackard, W.G. and Heidingsfelder, S.A. (1968): *J. clin. Invest., 47,* 1407.
Bleasel, K.F. and Lazarus, L. (1974): *Aust. N.Z.J. Surg., 44,* 250.
Boyd III, A.E., Lebovitz, H.E. and Pfeiffer, J.B. (1970): *New Engl. J. Med., 283,* 1425.
Bradley, P.B. (1963): In: *Physiological Pharmacology, Vol. I,* Chapter E, p. 417. Editors: W.S. Root and F.G. Hofmann. Academic Press, New York.
Camanni, F., Massara, F., Belforte, L. and Molinatti, G.M. (1975): *J. clin. Endocr., 40,* 363.
Chiodini, P.G., Liuzzi, A., Botalla, L., Cremascoli, G. and Silvestrini, F. (1974): *J. clin. Endocr., 38,* 200.
Chothia, C. and Pauling, P. (1969): *Proc. nat. Acad. Sci. (Wash.), 63,* 1063.
Collu, R.F., Fraschini, F., Visconti, P. and Martini, L. (1972): *Endocrinology, 90,* 1231.
Corrodi, H., Fuxe, K., Hökfelt, T., Lidbrink, P. and Ungerstedt, U. (1973): *J. Pharm. Pharmacol., 25,* 409.
Costa, E. (1956): *Proc. Soc. exp. Biol. (N.Y.), 91,* 39.
Cryer, P.E. and Daughaday, W.H. (1974): *J. clin. Endocr., 39,* 658.
Gaddum, J.H. and Hameed, K.A. (1954): *Brit. J. Pharmacol., 9,* 240.
Goodwin, F.K., Dunner, D.L. and Gershon, E.S. (1971): *Life Sci., 10(I),* 751.
Grabowska, M., Michaluk, J. and Antkiewicz, L. (1973): *Europ. J. Pharmacol., 20,* 133.
Kang, S. and Green, J.P., (1970): *Proc. nat. Acad. Sci. (Wash.), 67,* 62.
Karobath, M., Diaz, J.L. and Huttunen, M. (1972): *Biochem. Pharmacol., 21,* 1245.
Lal, S., De la Vega, C.E., Sourkes, T.L. and Friesen, H.G. (1972): *Lancet, 2,* 661.
Lal, S., Martin, J.B., De la Vega, C.E. and Friesen, H.G. (1975): *Clin. Endocr., 4,* 277.
Liuzzi, A., Chiodini, P.G., Botalla, L., Cremascoli, G. and Silverstrini, F. (1972): *J. clin. Endocr., 35,* 941.
Liuzzi, A., Chiodini, P.G., Botalla, L., Cremascoli, G., Müller, E.E. and Silverstrini, F. (1974): *J. clin. Endocr., 38,* 910.

Molinatti, G.M., Massara, F., Pennisi, F., Scassellati, G.A., Strumia, E. and Vancheri, L. (1969): *J. nucl. Biol. Med., 13,* 26.
Ng, K.Y., Chase, T.N., Colburn, R.W. and Kopin, I.J. (1970): *Science, 170,* 76.
Paasonen, M.K. and Giarman, N.J. (1958): *Arch. int. Pharmacodyn., 114,* 189.
Page, I.H. and McCubbin, J.W. (1953): *Amer. J. Physiol., 174,* 436.
Paton, W.D.M. (1970): In: *Molecular Properties of Drug Receptors,* p. 3. Editors: R. Porter and M. O'Conner. J. and A. Churchill, London.
Pieri, L., Pieri, M. and Haefely, W. (1974): *Nature (Lond.), 252,* 586.
Quadri, S.K. and Meites, J. (1971): *Proc. Soc. exp. Biol. (N.Y.), 137,* 1242.
Quock, R.M. and Horita, A. (1974): *Science, 183,* 539.
Sherman, L., Kim, S., Benjamin, F. and Kolodny, H.D. (1971): *New Engl. J. Med., 284.* 72.
Smythe, G.A. and Lazarus, L. (1973a): *Endocrinology, 93,* 147.
Smythe, G.A. and Lazarus, L. (1973b): *Nature (Lond.), 244,* 230.
Smythe, G.A. and Lazarus, L. (1974a): *J. clin. Invest., 54,* 116.
Smythe, G.A. and Lazarus, L. (1974b): *Science, 184,* 1373.
Smythe, G.A., Brandstater, J.F. and Lazarus, L. (1975): *Neuroendocrinology, 17,* 245.
Toivola, P. and Gale, C.C. (1970): *Neuroendocrinology, 6,* 210.
Thorner, M.O., Chait, A., Aitken, M., Benker, G., Bloom, S.R., Mortimer, C.H., Sanders, P., Stuart Mason, A. and Besser, G.M. (1975): *Brit. med. J., 1,* 299.
Von Hungen, K., Roberts, S. and Hill, D.F. (1974): *Nature (Lond.), 252,* 588.

NEUROENDOCRINE CONTROL OF GROWTH HORMONE SECRETION: EXPERIMENTAL AND CLINICAL STUDIES*

A. LIUZZI,[2] A.E. PANERAI,[1] P.G. CHIODINI,[2] C. SECCHI,[3] D. COCCHI,[1] L. BOTALLA,[2] F. SILVESTRINI[2] and E.E. MÜLLER[1]

[1] 2nd Department of Pharmacology, [2] Center of Endocrinology, Ospedale Maggiore, [3] Department of Biochemistry, Institute of Animal Physiology and [1] Department of Experimental Endocrinology, University of Milan, Milan, Italy

Growth hormone (GH) secretion is regulated by central nervous (CNS) mechanisms including a GH-releasing factor (GRF), which is incompletely identified (see Müller, 1973), and a GH-inhibiting factor (GIF or somatostatin), which has been isolated, identified and synthesized (Blackwell and Guillemin, 1973; Brazeau et al., 1973). Parallel advances have also been made in establishing and understanding main pathways of brain monoaminergic neurotransmitters (Falck et al., 1962; Fuxe, 1965) and their role in the control of hypothalamic neurohormones (McCann et al., 1974). Many studies have unequivocally demonstrated that in some mammalian species, α- and β-adrenergic receptors respectively stimulate or inhibit GH secretion (Martin, 1973; Müller, 1973).

Results pertaining to the role of specific monoamines, dopamine (DA), norepinephrine (NE) and serotonin (5-HT) in the GH control system (see Müller, 1973) are contradictory, however, and recent pathologic findings (e.g. acromegaly) have cast doubt as to the site(s) at which monoamines act to regulate GH secretion (Liuzzi et al., 1974a).

The present studies concern experimental and clinical findings related to these two particular aspects of GH regulation. Details on material and methods have been incorporated into the Results section (see also Liuzzi et al., 1974b; Müller, 1976; Müller et al., 1976). Double antibody radioimmunoassays for rat GH (RGH), canine GH (CGH), human GH (HGH) and prolactin (HPL) were used to measure plasma GH in rats (Schalch and Reichlin, 1966), dogs (Cocola et al., 1976) and GH (Molinatti et al., 1969) and plasma PL in humans (Kit Biodata, Rome).

RESULTS AND DISCUSSION

Dopamine and serotonin and GH release

Activation of DA receptors and GH release
In man, direct or inferential evidence that supports the existence of a stimulatory role of DA receptor activation in the release of GH includes: the rise of HGH induced by

---

* These studies were supported by C.N.R. grants No. CT 74.00232.04 and CT 74.00140.04.

apomorphine (Lal et al., 1972), a direct stimulant of DA receptors (Andén et al., 1967); the increased HGH levels following administration of DA-β-hydroxylase inhibitors (Aidaka et al., 1973); the effectiveness of amphetamine derivatives in stimulating HGH secretion (Rees et al., 1970), in view of the fact that some of the central effects of these drugs result from an increase of DA receptor activity (Carlsson, 1970).

2-Br-α-ergocryptine (CB-154, Sandoz), an ergot derivative inducing a long-lasting activation of DA receptor sites in the rat striatum (Corrodi et al., 1973) has been used. The drug was administered orally (2.5 mg) between 8.00 and 9.00 a.m. to 13 healthy male and female subjects and its effect on HGH evaluated and compared with that induced in the same individuals by oral administration of L-dopa (Larodopa, 500 mg). CB-154 increased HGH levels after 210 minutes, an effect still present 300 minutes after drug ingestion; this pattern contrasted with that of L-dopa, which induced a more prompt but short-lived increase of HGH levels (Fig. 1).

The observation that L-dopa increased GH is not proof of a stimulatory action of DA, since the drug might act to increase NE levels in the hypothalamus or limbic system and thence mediate GH release.

To more fully understand the mechanism involved in the GH-releasing effect of L-dopa, this drug was given acutely (500 mg p.o.) to 6 normal individuals pretreated for 1 week with pimozide (4 mg/day), a selective blocker of DA receptors (Andén et al., 1970) and the GH-releasing effect of L-dopa alone or after pimozide pretreatment compared. The HGH elevation induced by L-dopa was almost completely abolished by pimozide (data not presented).

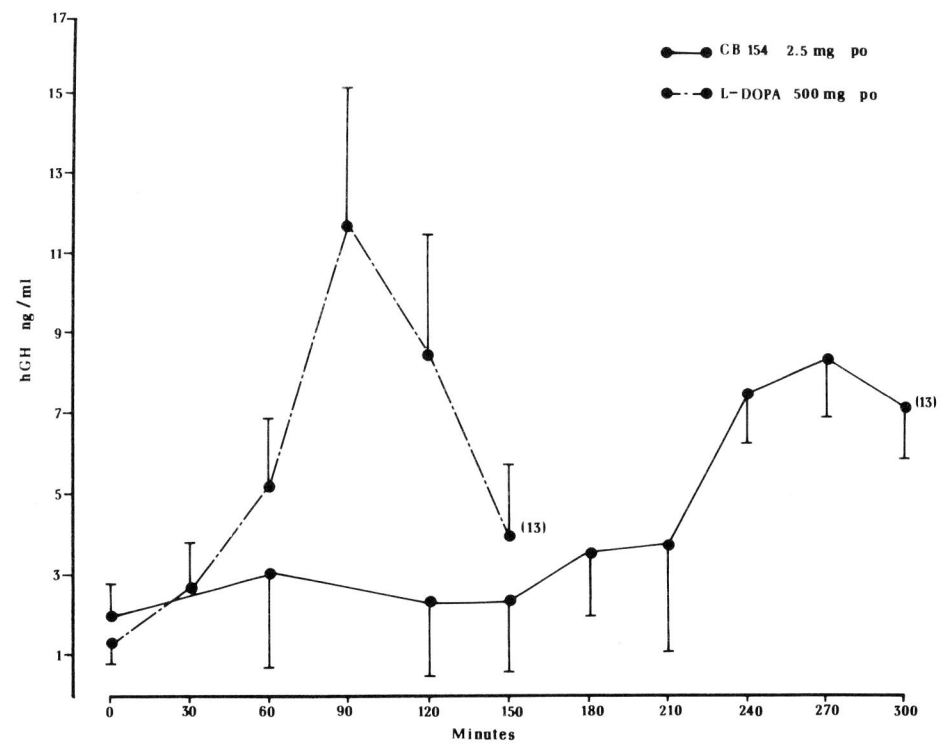

Fig. 1. Plasma HGH levels (mean $\pm$ SE) in 13 male and female healthy subjects after administration of CB-154 (2.5 mg p.o.) or L-dopa (500 mg p.o.).

TABLE 1

Effect of CB-154 alone or in combination with pimozide on plasma GH levels of infant rats

Treatments		Plasma GH levels (ng/ml)	
−2 hr	0	2 hr	4 hr
Diluent	Diluent	17.0 ± 1.1	18.0 ± 1.0
Diluent	CB-154 2.5 mg/kg b.w.s.c.	32.5 ± 3.0**	24.6 ± 1.2*
Pimozide 500 µg/kg b.w.s.c.	Diluent	12.0 ± 1.6*	22.0 ± 4.5
Pimozide 500 µg/kg b.w.s.c.	CB-154 2.5 mg/kg b.w.s.c.	16.5 ± 2.5***	17.7 ± 2.6

Each point is the mean ± SE of 5 determinations. 15-day-old male and female rats were used.
*P < 0.05 vs respective diluent. **P < 0.01 vs respective diluent. ***P < 0.01 vs CB-154.

Similar studies were also performed in the unanesthetized infant rat. Administration of CB-154 to 15-day-old male and female rats induced a rise in RGH levels at both 2 and 4 hours; pretreatment with pimozide reduced the increased GH levels present 2 and 4 hours after CB-154, although at 4 hours the effect was not statistically significant. Pimozide itself significantly decreased base line RGH levels 4 hours later (Table 1).

5-HT system and GH release
While the role of brain catecholamines (CA) in the neuroendocrine mechanism(s) controlling GH secretion has been extensively studied (Müller, 1973; Martin, 1973), less attention has been paid to the involvement of the indoleaminergic system and results to date are conflicting (Table 2). Recently, the availability of a rapid and sensitive homologous RIA for canine plasma GH (Cocola et al., 1976) has enabled the effects of functional suppression or activation of the 5-HT system on the GH response evoked by insulin hypoglycemia to be studied. Hypothalamic 5-HT content increases in the rat during insulin-induced hypoglycemia (Gordon and Meldrum, 1970) and HGH responses to insulin-induced hypoglycemia or physical exercise are inhibited by cyproheptadine and methysergide, 2 alleged 5-HT antagonists (Bivens et al., 1973; Smythe and Lazarus, 1974). It has therefore been suggested that 5-HT is a stimulant in hypoglycemia-induced HGH release (Bivens et al., 1973).

In unanesthetized beagles, the insulin-induced hypoglycemia CGH release was not reduced — rather it was magnified — after pharmacologic blockade of 5-HT by DL-parachlorphenylalanine (PCPA) (Müller et al., 1976). On the other hand, in dogs previously fed a tryptophan (TP)-deficient diet for 40 days, infusion of TP induced striking reductions in CGH after insulin hypoglycemia (Müller et al., 1976).

To elucidate further the role of 5-HT in hypoglycemia-stimulated release of CGH, the hypoglycemic stimulus was given following pretreatment with 8-β-carbobenzyl-ossiaminomethyl 1-6-dimethyl-10-α-ergoline (Methergoline or Liserdol, Farmitalia), a drug thought to specifically block central 5-HT receptors (Beretta et al., 1965; Ferrini and Glässer, 1965). Figure 2 shows that like PCPA, Methergoline potentiated the CGH rise evoked by insulin hypoglycemia, without modifying the fall in blood glucose induced by the insulin.

TABLE 2

Serotoninergic system and GH release

Drug	Animal species	Dosage and route of administration	Effect	References
5-HT	Rat	1 μg i.v.	↑	Collu et al., 1972
	Rat	1 μg i.v.	→	Müller et al., 1973
	Monkey	10 μg microinjected into ventromedial nucleus	→	Toivola and Gale, 1972
	Human*		↑	Feldman and Lebovitz, 1972
TP	Human	70 mg/kg p.o.	Slight increase	Müller et al., 1974
	Human	10 g i.v.	→	McIndoe and Turkington, 1973
5-HTP	Rat	10-50 mg/kg i.p.	↑	Smythe and Lazarus, 1973a
	Human	150 mg p.o.	↑	Imura et al., 1973
	Human	150 mg p.o.	→	Müller et al., 1974
	Human	150 mg p.o.	→	Benkert et al., 1973
	Human	200 mg p.o.	→	Handwerger et al., 1975
Cyproheptadine	Human	32 mg p.o.	Blunts insulin hypoglycemia	Bivens et al., 1973
	Human	5 mg i.v.	Blocks 5-HTP rise	Nakai et al., 1974
	Human	60 mg p.o.	Blunts exercise-induced	Smythe and Lazarus, 1974
Melatonin	Human	500 mg p.o.	Blunts insulin hypoglycemia	Smythe and Lazarus, 1974
Methysergide	Human	16 mg p.o.	Blunts insulin hypoglycemia	Bivens et al., 1973

* Patients with carcinoid syndrome. → No effects. ↑ Stimulation.

Also in man, evidence suggests an inhibitory role for 5-HT in hypoglycemia-induced GH release. Combined treatment with L-TP (70 mg/kg p.o.) and insulin blunted the HGH response after administration of insulin alone in 10 male and female subjects (Fig. 3).

Recently, on the basis of a hypothetical hypothalamic receptor-site model (Smythe and Lazarus, 1973b), it has been proposed that L-dopa induces GH release by an interaction with 5-HT receptors (Smythe and Lazarus, *This Volume*, p. 222). To test this hypothesis, Methergoline (8 mg/day for 4 days) was administered to 6 male and female subjects and the GH-releasing effect of L-dopa was determined before and after Methergoline administration. Pretreatment with Methergoline did not modify the L-dopa-induced HGH rise (Fig. 4). Collectively, the data presented indicated that in the species examined and under the experimental conditions reported activation of DA receptor sites stimulates GH release, while, conversely, activation of the 5-HT system inhibits the hypoglycemia-induced GH release.

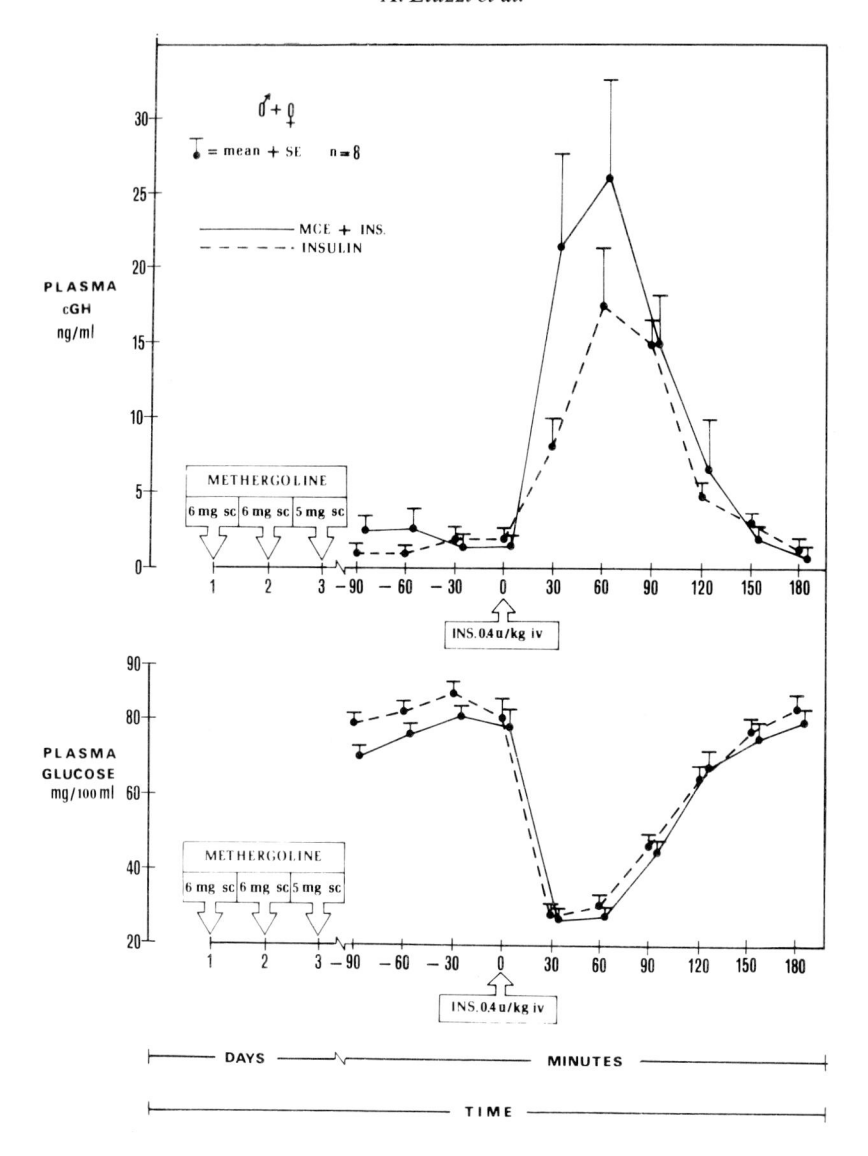

Fig. 2. Effect of insulin-induced hypoglycemia on plasma CGH and blood glucose levels of unanesthetized male and female beagles pretreated or not with Methergoline. Means ± SE. A factorial analysis of variance of the 2 responses showed that the increase in plasma CGH by the combined treatment was significantly different from that evoked by insulin alone (F = 3.95, p < 0.05).

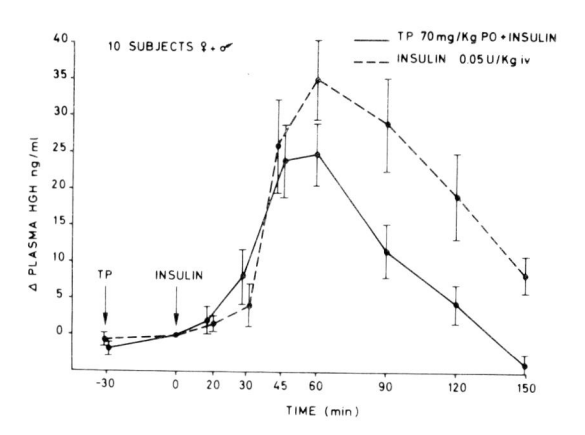

Fig. 3. Effect of oral tryptophan (TP) preloading on insulin-induced HGH release in normal subjects (From Müller et al., *Journal of Clinical Endocrinology and Metabolism* (1974), by courtesy of Charles C Thomas, Publisher.)

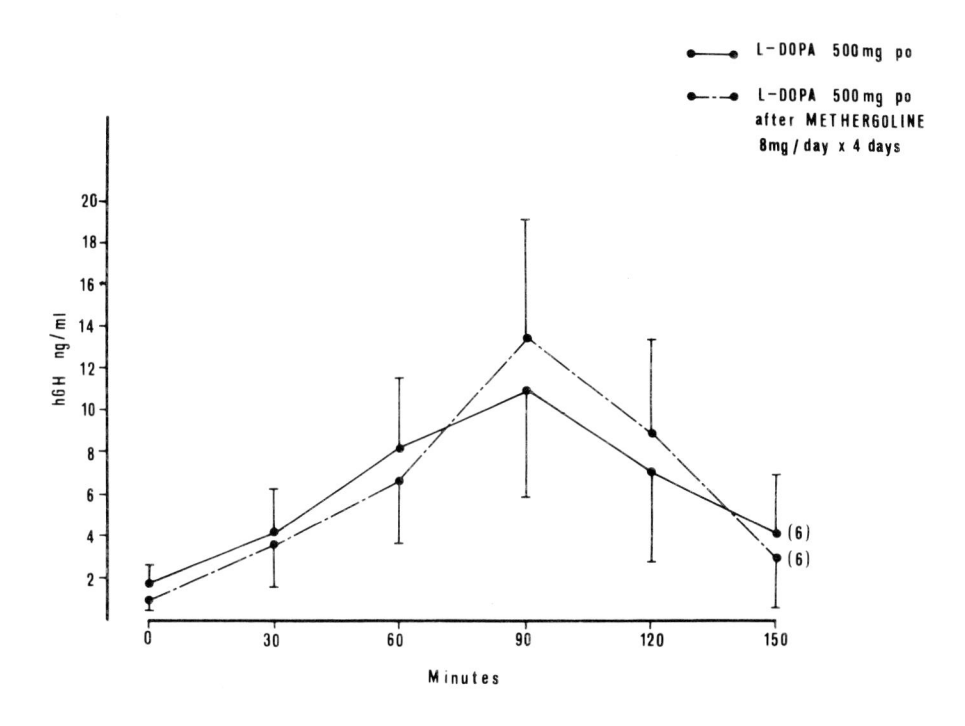

Fig. 4. GH-releasing effect of L-dopa (500 mg p.o.) in 6 male and female healthy subjects pretreated or not with Methergoline (mean ± SE).

Like apomorphine (Lal et al., 1972), CB-154, a long-lasting DA-stimulant drug apparently devoid of any action on brain 5-HT turnover (Corrodi et al., 1973, 1975) induced a rise in HGH levels. This time course of drug action on HGH correlated well with the reported time course for activation of DA receptors (Corrodi et al., 1973). A delayed rise in HGH levels following acute CB-154 administration had also been reported by Camanni et al. (1975). Suppression of L-dopa induced HGH elevation by pimozide, a specific DA receptor blocker, further supports the existence of a dopaminergic mechanism in GH control and indicates that L-dopa in man acts mainly through an activation of DA receptor sites. Such a conclusion agrees with the finding that, in man, only 5% of infused L-dopa is converted to NE or metabolic products of NE (Goodal and Alton, 1972).

The rise in plasma GH levels induced by CB-154 in the infant rat and its suppression by pimozide contrasts with the reported inhibitory role for DA in the adult rat (Collu et al., 1972; Kato et al., 1973); the effect was, however, obtained without the use of urethane anesthesia, in an animal preparation which is characterized by spontaneously low and uniform plasma GH levels. The stimulatory role of the dopaminergic system in the infant rat is supported by the finding that, like CB-154, systemic administration of L-dopa also increases plasma GH levels (Irit Gil-Ad, unpublished results). Clarification of the role of DA in the adult rat requires the use of specific agonists or antagonists of the dopaminergic system and the avoidance of the disturbing effect of anesthesia (Martin et al., 1975; Müller, 1975).

Recently, it has been proposed by Smythe and Lazarus (*This Volume*, p. 222) that serotonin should be considered the principal monoamine in relation to GH-releasing mechanisms in both man and the rat, and that L-dopa induces GH release by 5-HT receptor interaction and not via adrenergic pathways. Overall, the present data do not support these conclusions, rather they suggest that 5-HT plays an inhibitory role in hypoglycemia-induced GH release.

Most of the data in the literature favoring a stimulatory role for 5-HT neurons, are based on the use of the 5-HT precursor, 5-hydroxytryptophan (5-HTP), and of cyproheptadine, methysergide or melatonin as 5-HT antagonists (Smythe and Lazarus, 1973a, 1974; Smythe et al., 1975; Bivens et al., 1973). 5-HTP administered systemically may lead to substantial 5-HT accumulation in cells that do not ordinarily contain the indoleamine and may also interfere with CA and their precursors for transport, storage and metabolism within the CNS (Johnson et al., 1968; Ng et al., 1972; Okada et al., 1972). The specificity and site of action of cyproheptadine and methysergide is also questionable. Cyproheptadine was originally developed as an antihistaminic agent and subsequently shown to possess both antihistaminic, anticholinergic and anti-5-HT activity to a relatively high degree (Stone et al., 1961). Recently, it has been reported to block apomorphine-induced HGH release (Winkelmann et al., 1975). Methysergide, being strongly hydrophylic is poorly, if at all, transported across the blood-brain barrier (Dewhurst, 1970) and its actions appear to be confined to peripheral receptor-site blockade of 5-HT (Cerletti et al., 1960). Melatonin has been proposed as an antagonist of 5-HT at a receptor-site level on the assumption that all O-methylated derivatives of monoamines give derivatives that inhibit monoamine action at hypothalamic receptor sites (Smythe and Lazarus, 1973b; Smythe et al., 1975). The possibility exists, however, that the action of melatonin on GH secretion far from being due to 5-HT receptor blockade, may result, instead, from 5-HT receptor stimulation. In this context, it is noteworthy that both 5-HT and melatonin block gonadotropin (Kamberi et al., 1970) and stimulate prolactin secretion (Lu and Meites, 1971).

The possible interaction of L-dopa with serotonin receptor site(s) to induce HGH release is not supported by the inability of Methergoline to alter the HGH rise induced by L-dopa, in contrast to pimozide. Until more convincing experimental evidence is produced, it would appear that distinct DA and 5-HT receptor sites are present, which exert opposite actions on the GH-releasing mechanism(s).

GH-lowering effect of dopaminergic compounds in acromegaly

In recent years, neurotransmitter dysfunction has been suggested in the etiology of specific neuroendocrine disorders (Sherman and Kolodny, 1971); as a corollary neuropharmacologic approaches were introduced to treat disorders of neuroendocrine function. In the case of GH in acromegaly phenothiazines (Sherman et al., 1971) and, more recently, DA-stimulant drugs (see below) or α-adrenergic blockers alone or in combination with β-stimulants have been used (Cryer and Daughaday, 1974). In 1972, we showed, rather unexpectedly, that acute administration of L-dopa (500 mg p.o.) markedly lowered HGH levels in plasma of some acromegalic subjects (Liuzzi et al., 1972). The time course of the L-dopa effect involved a short-lived, about 2 hours, depressant effect on GH secretion and a rebound increase in plasma GH ensued from 3-5 hours after administration (Chiodini et al., 1974a). Like L-dopa, apomorphine (0.75 mg s.c.), a more selective stimulant of DA receptors (Andén et al., 1967), reduced HGH levels in L-dopa-responsive acromegalic subjects, although, the HGH-lowering effect of apomorphine was shorter (about 1 hour) than that of L-dopa (Chiodini et al., 1974a).

A more striking and long-lasting effect was obtained with CB-154 (2.5 mg p.o.) which more directly activates DA receptors for longer (Corrodi et al., 1973). The inhibition with CB-154 occurred rather slowly (about 2 hours), but the effect was sustained for up to 5 hours. Figure 5 compares the HGH-lowering effect of L-dopa and CB-154

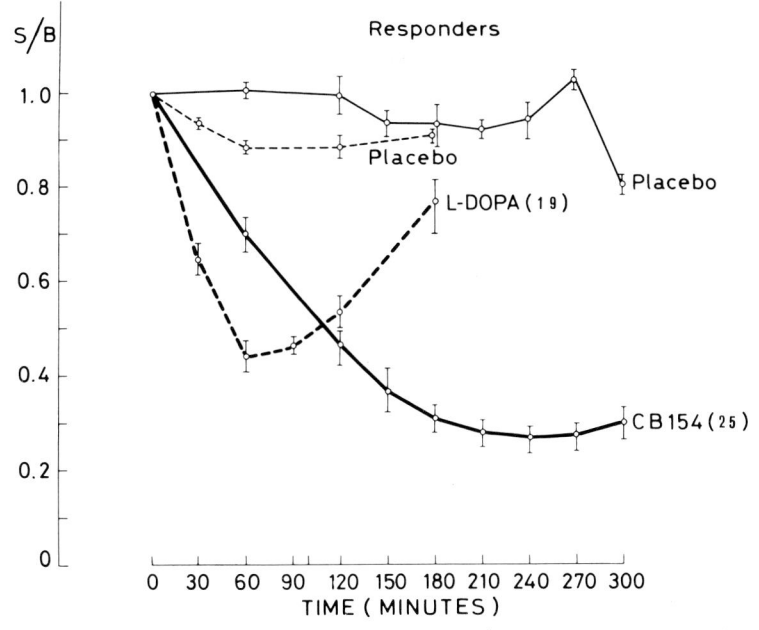

Fig. 5. Plasma HGH values after administration of L-dopa (500 mg p.o.) or CB-154 (2.5 mg p.o.) in 'responder' acromegalic patients. Values are expressed as ratio of suppressed (S) to base line (B) ± SE.

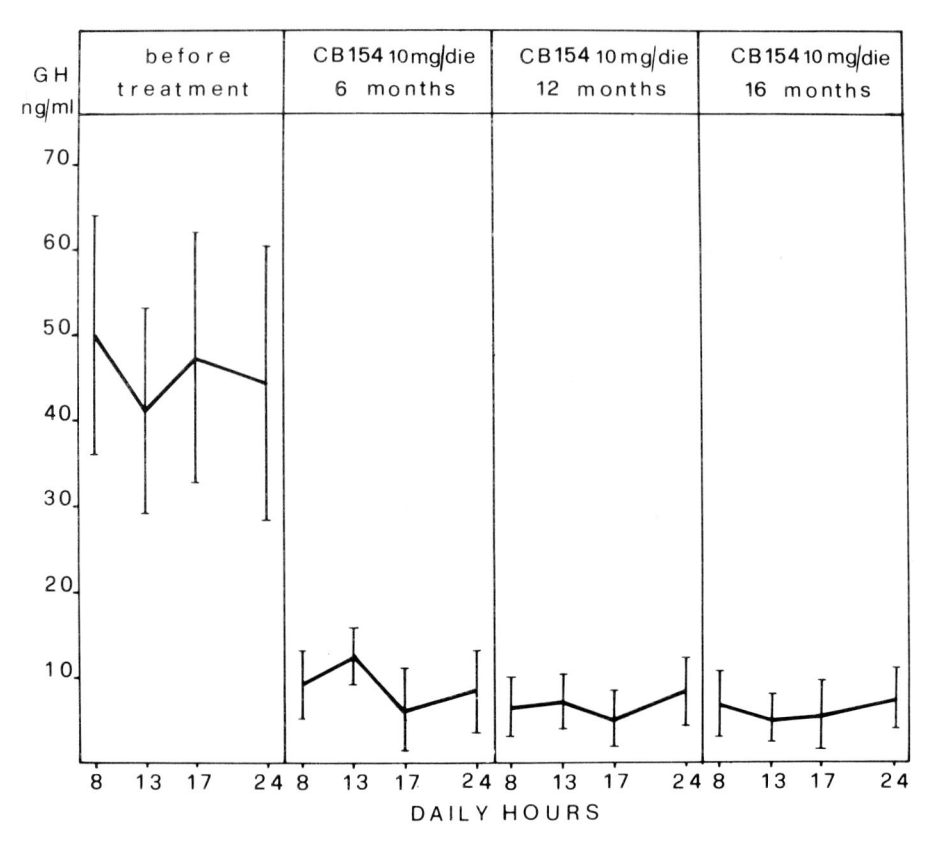

Fig. 6. Plasma HGH levels (mean ± SE) in 7 acromegalic patients, responsive to acute CB-154 (2.5 mg) administration, during 16 months of uninterrupted treatment with CB-154 (10 mg/day).

in a population of 'acromegalic responders'. The inhibition of HGH levels present in many acromegalic patients following administration of dopaminergic compounds has been confirmed by others (Althoff et al., 1975; Camanni et al., 1975; Thorner et al., 1975).

The long duration of CB-154 effect on GH levels and its relative safety (side effects: nasal stuffiness, nausea, orthostatic hypotension) suggested its chronic treatment (Chiodini et al., 1974*b*, 1975). In 7 acromegalic subjects, in whom plasma HGH concentration had dropped acutely after a single dose of CB-154 (2.5 mg p.o.), chronic CB-154 treatment (10 mg p.o. for 30 days) was accompanied by a significant and stable reduction in HGH levels. Withdrawal of the drug was followed by a rapid return of HGH to pre-treatment values. Re-instatement of oral CB-154 treatment again suppressed HGH levels as measured 30 and 60 days later. Figure 6 shows the depressant effect on HGH levels over 16 months of uninterrupted therapy in the same subjects. Patients unresponsive to acute administration of the drug (2.5 mg p.o.) showed no appreciable variation in HGH levels after 30 days of treatment. Moreover, in acute experiments, doubling of the dose (5.0 mg) did not produce better results in the unresponsive patients, suggesting different, individual sensitivities to dopaminergic stimulation.

Long-term administration of CB-154 not only suppresses circulating GH levels, but produces marked clinical and metabolic improvements. In an overtly diabetic acromegalic, during the initial 15 days of therapy, chronic CB-154 suppressed HGH levels, reduced blood glucose and the glycosuria disappeared; oral anti-diabetic therapy was thus stopped (data not presented). That activation of dopaminergic mechanisms underlies the HGH-lowering effect of CB-154 is suggested by the fact that pimozide (4.0 mg p.o. for 7 days) given to 7 patients before CB-154 administration reduced the HGH-lowering effect (Chiodini et al., 1974b). Thus, the consistently suppressive effect of dopaminergic stimulation on GH levels in certain acromegalic patients suggests a new rational medication for this disease. Its chronic clinical and metabolic efficacy remains to be established.

Site(s) of action of dopaminergic compounds: CNS or pituitary level?

The physiopathology of acromegaly is incompletely understood (Müller et al., 1975) both with regard to some anomalous responses to dopaminergic stimulation and the unresponsiveness in other patients. It is generally accepted that dopaminergic compounds affect the GH control system at loci within the CNS (Müller, 1973). If these compounds act exclusively at suprapituitary sites, the paradoxical fall observed in some patients could result from a functional imbalance between excitatory (GRF) and inhibitory (GIF) inputs which regulate the pituitary GH release.

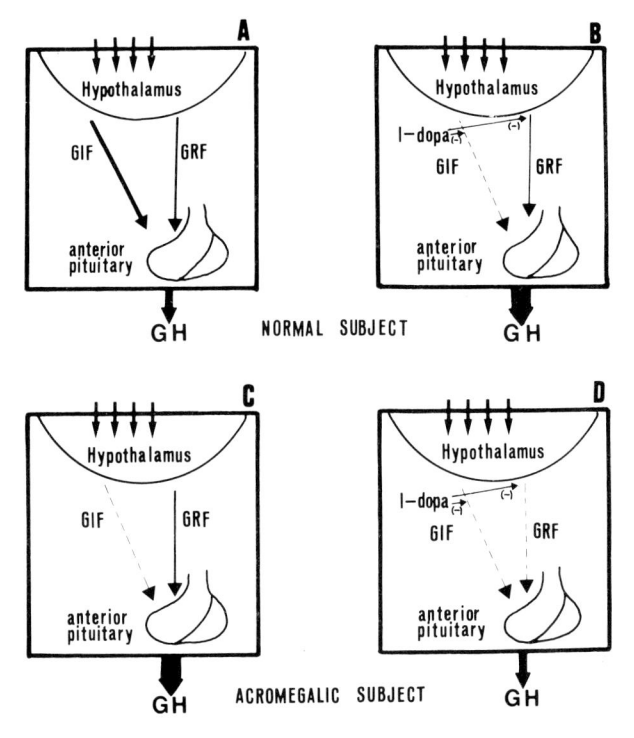

Fig. 7. Hypothetical mechanism of action of L-dopa in increasing or decreasing respectively, GH levels in normal or acromegalic subjects. A. normal subject; B. normal subject given L-dopa; C. acromegalic subject; D. acromegalic subject given L-dopa (see text).

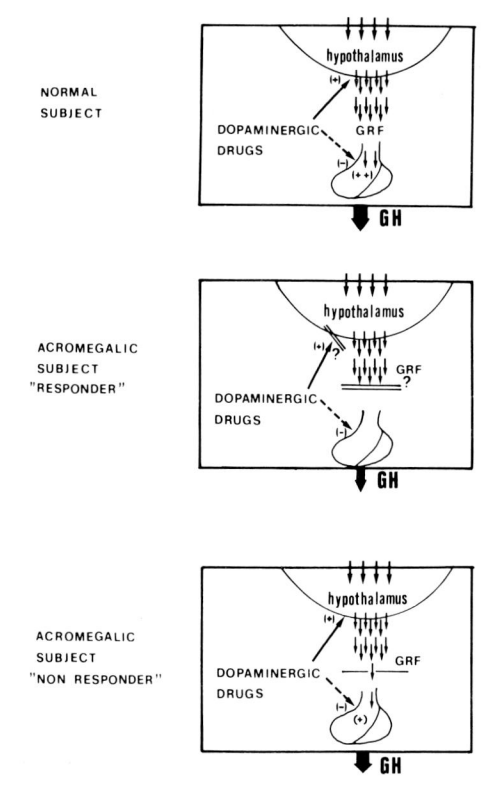

Fig. 8. Hypothetical model to explain the double site of action of dopaminergic drugs at suprapituitary and pituitary levels in 'responder' or 'non-responder' acromegalics (see text).

Electrophysiological evidence also points towards mainly depressive DA effects within the CNS (McLennan and York, 1967), so that L-dopa and DA-stimulant drugs might also inhibit both GRF- and GIF-secreting structures. These drugs might stimulate GH release in the healthy human by depressing the GIF center(s), whose tone normally overrides the tone of the antagonistic GRF center(s). If in some acromegalics, the high circulating levels of GH result from a primary defect in the GIF-secreting structures, L-dopa would act principally on the remaining GRF center(s) to reduce GH levels (Fig. 7).

Although DA-like compounds are generally considered to act within the CNS, direct effects of CA and CB-154 on pituitary prolactin secretion have been reported (Birge et al., 1970; MacLeod et al., 1970; Del Pozo and Fluckiger, 1974). By analogy with prolactin secretion, it could be postulated that DA-like compounds have two sites of action: at suprapituitary and at pituitary levels. In healthy subjects, the stimulatory action of these compounds on the CNS overrides direct effects on the pituitary. In acromegaly where a functional disconnection exists between the CNS and the anterior pituitary, the stimulatory CNS action on GH release would be completely (responder acromegalics) or partially (non-responder acromegalics) lacking; the inhibitory effect on the pituitary would thus predominate (Fig. 8). No effect or HGH increases after dopaminergic compounds in some acromegalics would reflect a more efficient functional connection between the CNS and the pituitary gland.

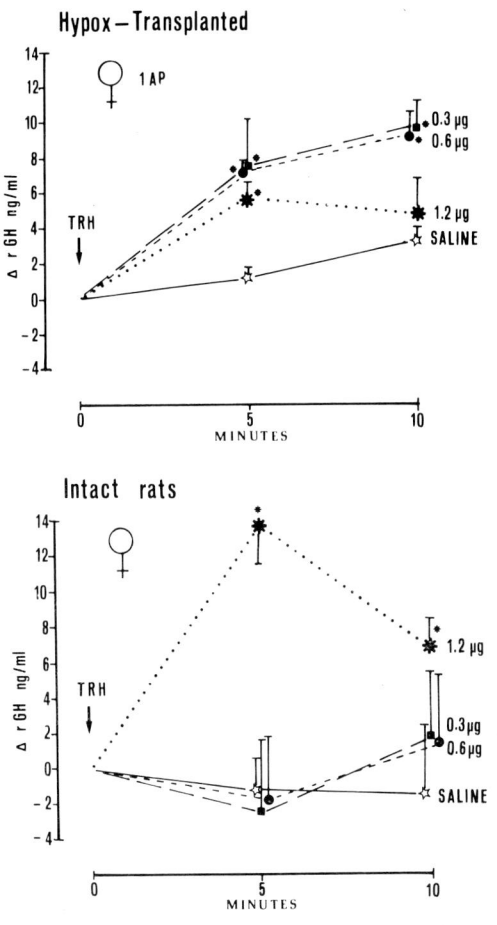

Fig. 9. Effect of TRH on plasma GH levels in hypophysectomized (hypox) rats bearing 1 AP under the kidney capsule or in intact weight-matched controls. Data are expressed as Δ values from base line. Means ± SE of 5 to 11 determinations for each point are shown. Asterisks indicate statistically significant differences from base line values. All TRH doses were effective in the hypox rats, while in the intact controls only the hightest dose (1.2 μg) was effective. (From Udeschini et al., *Endocrinology* (1976), by courtesy of Charles C Thomas, Publisher.)

In many acromegalics a functional disconnection between the CNS and the anterior pituitary (AP) explains the nonspecific GH-releasing effect of TRH (Irie and Tsushima, 1972; Schalch et al., 1972; Faglia et al., 1973). Under physiologic conditions, the specific neuroendocrine influence for GH secretion (GRF) would impair TRH-induced GH release; suppression or diminution of the specific neuroendocrine influence would produce preferential release of GH after TRH. The peculiar sensitivity to the GH-releasing effect of TRH in the hypophysectomized rat bearing an ectopic pituitary, in which the functional and anatomical links between the CNS and the pituitary have been interrupted, supports this hypothesis (Fig. 9). The analogy between this animal model and the responder acromegalic, however, cannot be extended, since in the former CB-154 does not lower plasma rat GH (unpublished results). Thus, functional

TABLE 3

Homogeneity in the GH responses to TRH and CB-154 in acromegaly (46 subjects)

Number of subjects	TRH	CB-154
21	+	+
21	−	−
2	+	−
2	−	+

TRH +: increment in plasma HGH $\geqslant$ 100%. CB-154 +: decrement in plasma HGH $\geqslant$ 50%.

CNS-AP disconnections may explain the anomalous GH response to TRH in acromegaly, but it does not account for the HGH-lowering actions of dopaminergic drugs. These drugs are inactive on otherwise normal pituitaries devoid of CNS influences; thus, CB-154 fails to alter GH secretion of the rat AP in vitro (unpublished results). On the other hand, ergot derivatives suppress directly GH secretion from pituitary tumors in the rat (Quadri and Meites, 1973).

A pituitary tumor in acromegaly would thus be a prerequisite for the GH-depressing effect of dopaminergic compounds. Recently, histological evidence has shown a large percentage of chromophobe adenomas in cases of HGH overproduction (see Daughaday, 1974) and chromophobes may indeed represent degranulated somatotrophs and lactotroph cells (Guyda et al., 1973). Thus, chromophobe cells without the histology of prolactin or GH cells may occur in the pituitary tumors of 'responder' acromegalics and respond as PRL and GH cells (mammosomatotroph cell). Such bivalent cells in the AP would account for the nonspecific stimulation of GH release by TRH and the paradoxical HGH-lowering effect of DA-like compounds, stimuli which affect PL secretion in the same way (Frantz et al., 1974). The homogeneity of the 2 anomalous GH responses in a population of acromegalic subjects supports this hypothesis: in 42 out of 46 patients, responses (21 subjects) or otherwise (21 subjects) were seen to either TRH or CB-154 respectively (Table 3).

Such mammosomatotrophic cells are consistent with the observation that GIF (250 or 500 μg i.v.) neither blocks the TRH-induced PL release nor significantly lowers the TRH-induced GH release (data not reported). A note of caution with regard to this hypothesis is the failure of CB-154 to block TRH-induced GH release, although the same drug inhibits TRH-induced PL release (data not reported). The more than qualitative difference may only reflect differently sized secretory pools for the 2 hormones within mammosomatotroph cells — a smaller easily suppressible pool of PL, and a major GH pool, which is not so.

The paradoxical responses of GH secretion to dopaminergic stimulation have been used to formulate hypotheses explaining the patho-physiology of acromegalic diseases. Each has attractive aspects although none fully explain the variety of responses seen in individual cases following traditional stimuli, DA-like drugs and TRH (Liuzzi et al., 1974a). Acromegaly does not seem to be a CNS disease, a pituitary disease or a combination, rather it is a syndrome possibly involving the CNS and dedifferentiated pituitary tumors.

CONCLUSIONS

The specific role of dopamine and serotonin in the neuroendocrine control of growth hormone release has been discussed and possible reason(s) for discrepant results considered. Evidence for activation of DA receptor sites and thence stimulation of GH release in both man and rodents (infant rat) is available; conversely, functional activation of 5-HT system inhibits hypoglycemia-induced GH release in the dog and man. Finally, on considering the suppressive effect of dopaminergic drugs on GH levels in acromegalic subjects, some hypotheses as to the site(s) of action of these compounds are presented.

ACKNOWLEDGEMENTS

The participation to these studies of Dr. I. Gil-Ad, Dr. F. Cocola and Dr. G. Cremascoli is gratefully acknowledged. Thanks are given to the NIAMDD Rat Pituitary Hormone Program for supplying the NIAMDD-Rat GH kit, to Dr. A.E. Wilhelmi for pure dog GH, to Sandoz A.G., Basel and Farmitalia, Milan for CB-154 and Methergoline, respectively.

REFERENCES

Aidaka, H., Nagasaka, A. and Takeda, A. (1973): *J. clin. Endocr., 37,* 145.
Althoff, P.H., Neubauer, M., Handzel, R., Bechstein, V. and Schöffling, K. (1975): In: *Abstracts, International Symposium on Growth Hormone and Related Peptides, Milan, Suppl. 1,* p. 102.
Andén, N.E., Rubenson, A., Fuxe, K. and Hökfelt, T. (1967): *J. Pharm. Pharmacol., 19,* 627.
Andén, N.E., Butcher, S.G., Corrodi, H., Fuxe, K. and Ungerstedt, U. (1970): *Europ. J. Pharmacol., 11,* 303.
Benkert, O., Laackmann, G., Souvatzoglou, A. and Von Verder, K. (1973): *J. Neural Transm., 34,* 291.
Beretta, C., Glässer, A.H., Nobili, M.B. and Silvestri, R. (1965): *J. Pharm. Pharmacol., 17,* 423.
Birge, C.A., Jacobs, L.S., Hammer, C.T. and Daughaday, W.H. (1970): *Endocrinology, 86,* 120.
Bivens, C.H., Lebovitz, H.E. and Feldman, J.M. (1973): *New Engl.J.Med., 289,* 236.
Blackwell, R.E. and Guillemin, R. (1973): *Ann. Rev. Physiol., 35,* 357.
Brazeau, P., Vale, W., Burgus, R., Ling, N., Butcher, M., Rivier, J. and Guillemin, R. (1973): *Science, 179,* 77.
Camanni, F., Massara, F., Belforte, L. and Molinatti, G.M. (1975): *J. clin. Endocr., 40,* 363.
Carlsson, A. (1970): In: *Amphetamine and Related Compounds,* p. 289. Editors: E. Costa and S. Garattini. Raven Press, New York.
Cerletti, A., Berde, B., Doepfner, W., Emmeneger, H., Konzett, H., Schalch, W.R., Taeschler, M. and Weidman, H. (1960): Scientific Exhibit, 6th International Congress on Internal Medicine, Basel.
Chiodini, P.G., Liuzzi, A., Botalla, L., Cremascoli, G. and Silvestrini, F. (1974a): *J. clin. Endocr., 38,* 200.
Chiodini, P.G., Liuzzi, A., Cremascoli, G., Botalla, L., Silvestrini, F. and Müller, E.E. (1974b): In: *Abstracts, 56th Meeting, Endocrinological Society, Atlanta, Ga.,* Abstr. Nr. 206.
Chiodini, P.G., Liuzzi, A., Botalla, L., Oppizzi, G., Müller, E.E. and Silvestrini, F. (1975): *J. clin. Endocr., 40,* 705.
Cocola, F., Udeschini, G., Secchi, C., Neri, P. and Müller, E.E. (1976): *Proc. Soc. exp. Biol. (N.Y.), 151,* 140.
Collu, R., Fraschini, F., Visconti, P. and Martini, L. (1972): *Endocrinology, 90,* 1231.
Corrodi, H., Farnebo, L.-O., Fuxe, K. and Hamberger, B. (1975): *Europ. J. Pharmacol., 30,* 172.

Corrodi, H., Fuxe, K., Hökfelt, T., Lidbrink, P. and Ungerstedt, U. (1973): *J. Pharm. Pharmacol., 25,* 409.
Cryer, P.E. and Daughaday, W.H. (1974): *J. clin. Endocr., 39,* 658.
Daughaday, W.H. (1974): In: *Textbook of Endocrinology,* p. 31. Editor: R.H. Williams. W.B. Saunders Co., Philadelphia.
Del Pozo, E. and Fluckiger, E. (1974): In: *Human Prolactin,* p. 291. Editors: J.L. Pasteels and C. Robyn. ICS 308, Excerpta Medica, Amsterdam.
Dewhurst, W.G. (1970): In: *Principles of Psychopharmacology,* p. 109. Editor: W.G. Clark. Academic Press, New York.
Faglia, G., Beck-Peccoz, C., Ferrari, C., Travaglini, P., Ambrosi, B. and Spada, A. (1973): *J. clin. Endocr., 36,* 1259.
Falck, B., Hillarp, N.A., Thieme, G. and Torp, A. (1962): *J. Histochem. Cytochem., 10,* 348.
Feldman, J.M. and Lebovitz, H.E. (1972): In: *Abstracts, 4th International Congress of Endocrinology, Washington,* p. 35. ICS 256, Excerpta Medica, Amsterdam.
Ferrini, R. and Glässer, A. (1965): *Psychopharmacologia (Berl.), 8,* 271.
Frantz, A.G., Habif, D.V., Hyman, G.A., Suh, H.K., Sassin, J.F., Zimmerman, E.A., Noel, G.L. and Kleinberg, D.L. (1974): In: *Human Prolactin,* p. 273. Editors: J.L. Pasteels and C. Robyn. ICS 308, Excerpta Medica, Amsterdam.
Fuxe, K. (1965): *Acta physiol. scand., 64, Suppl. 247,* 39.
Goodal, M.C. and Alton, H. (1972): *Biochem. Pharmacol., 21,* 2401.
Gordon, A.E. and Meldrum, B.S. (1970): *Biochem. Pharmacol., 19,* 3042.
Guyda, H., Robert, F., Colle, E. and Hardy, J. (1973): *J. clin. Endocr., 36,* 531.
Handwerger, S., Plonk, J.W., Lebovitz, H.E., Bivens, C.H. and Feldman, J.M. (1975): *Hormone metab. Res., 7,* 214.
Imura, H., Nakai, I. and Yoshimi, T. (1973): *J. clin. Endocr., 36,* 204.
Irie, M. and Tsushima, T. (1972): *J. clin. Endocr., 35,* 97.
Johnson, G.A., Kim, E.G. and Boumka, S.J. (1968): *J. clin. Endocr., 36,* 204.
Kamberi, I.A., Mical, R.S. and Porter, J.C. (1970): *Endocrinology, 87,* 1.
Kato, Y., Dupré, J. and Beck, J.C. (1973): *Endocrinology, 93,* 135.
Lal, S., De La Vega, C.E., Sourkes, T.L. and Friesen, H.G. (1972): *Lancet, 2,* 661.
Liuzzi, A., Chiodini, P.G., Botalla, L., Cremascoli, G. and Silvestrini, F. (1972): *J. clin. Endocr., 35,* 941.
Liuzzi, A., Chiodini, P.G., Botalla, L., Silvestrini F. and Müller, E.E. (1974a): *J. clin. Endocr., 39,* 871.
Liuzzi, A., Chiodini, P.G., Botalla, L. Cremascoli, G., Müller, E.E. and Silvestrini, F. (1974b): *J. clin. Endocr., 38,* 910.
Lu, K.-H. and Meites, J. (1971): *Endocrinology, 91,* 1314.
McCann, S.M., Fawcett, C.P. and Krulich, I., (1974): In: *Endocrine Physiology,* p. 31. Editor: S.M. McCann. Butterworths and University Park Press, London and Baltimore.
McIndoe, J.H. and Turkington, R.W. (1973): *J. clin. Invest., 52,* 1972.
McLennan, H. and York, D.H. (1967): *J. Physiol. (Lond.), 189,* 393.
MacLeod, R.M., Fontham, E.H. and Lehmeyer, J.E. (1970): *Neuroendocrinology, 6,* 283.
Martin, J.B. (1973): *New Engl. J. Med., 288,* 1384.
Martin, J.B., Audet, J. and Daunders, A. (1975): *Endocrinology, 96,* 839.
Molinatti, G.M., Massara, F., Strumia, E., Pennisi, F., Scassellati, G.A. and Vancheri, L. (1969): *J. nucl. biol. Med., 13,* 26.
Müller, E.E. (1973): *Neuroendocrinology, 11,* 338.
Müller, E.E. (1976): In: *Proceedings, VIth International Congress of Pharmacology, Helsinki, Vol. 3.* Editors: J. Tuomisto and M. Paasonen. Pergamon Press, Oxford. In press.
Müller, E.E., Cocchi, D., Jalanbo, H. and Udeschini, G. (1973): In: *Abstracts, 55th Meeting, Endocrinological Society, Chicago, Ill.,* Abstr. Nr. 399.
Müller, E.E., Brambilla, F., Cavagnini, F., Peracchi, M. and Panerai, A. (1974): *J. clin. Endocr., 39,* 1.
Müller, E.E., Udeschini, G., Secchi, C., Zambotti, F., Panerai, A.E., Cocola, F. and Mantegazza, P. (1976): *Acta endocr. (Kbh.),* in press.
Müller, E.E., Liuzzi, A., Chiodini, P.G. and Silvestrini, F. (1975): *Neuroendocrinology,* submitted for publication.

Nakai, Y., Imura, H., Sakurai, H., Kurahachi, H. and Yoshimi, T. (1974): *J. clin. Endocr., 38,* 446.
Ng, L.K.Y., Chase, T.N., Colburn, R.W. and Kopin, I.J. (1972): *Brain Res., 45,* 499.
Okada, F., Saito, Y., Fujeda, T. and Yamashita, I. (1972): *Nature (Lond.), 238,* 355.
Quadri, S.K. and Meites, J. (1973): *Proc. Soc. exp. Biol. (N.Y.), 142,* 837.
Rees, L., Butler, P.W.P., Gosling, C. and Besser, G.M. (1970): *Nature (Lond.), 228,* 565.
Schalch, D.S., Gonzales-Barcena, D., Kastin, A.J., Schally, A.V. and Lee, L.A. (1972): *J. clin. Endocr., 35,* 609.
Schalch, D.S. and Reichlin, S. (1966): *Endocrinology, 79,* 275.
Sherman, L. and Kolodny, H.D. (1971): *Lancet, 1,* 682.
Sherman, L., Kim, S., Benjamin, F. and Kolodny, H.D. (1971): *New Engl. J. Med., 284,* 72.
Smythe, G.A. and Lazarus, L. (1973a): *Nature (Lond.), 244,* 1973.
Smythe, G.A. and Lazarus, L. (1973b): *Endocrinology, 93,* 147.
Smythe, G.A. and Lazarus, L. (1974): *J. clin. Invest., 54,* 116.
Smythe, G.A., Brandstater, J.F., and Lazarus, L. (1975): *Neuroendocrinology, 17,* 245.
Stone, C.A., Wengler, H.C., Lidden, C.T., Stavorski, J.M. and Ross, C.A. (1961): *J. Pharm. exp. Ther., 131,* 73.
Thorner, M.O., Chait, A., Aitken, M., Benker, G., Bloom, S.M., Mortimer, C.H., Sanders, P., Stuart-Mason, A. and Besser, G.M. (1975): *Brit. med. J., 1,* 299.
Toivola, P.T.K. and Gale, C.C. (1972): *Endocrinology, 90,* 895.
Udeschini, G., Cocchi, D., Pancrai, A.E., Gil-Ad, I., Rossi, G.L., Chiodini, P.G., Liuzzi, A. and Müller, E.E. (1976): *Endocrinology, 98,* 727.
Winkelmann, W., Schorm, H., Hadam, W.R., Heesen, D. and Mies, R. (1975): In: *Abstracts, International Symposium on Growth Hormone and Related Peptides, Milan, Suppl. 1,* p. 99.

GROWTH HORMONE RELEASE IN HUNTINGTON'S DISEASE*

STEPHEN PODOLSKY and NORMAN A. LEOPOLD

Medical Service, Boston VA Outpatient Clinic, Department of Medicine, Boston University School of Medicine, Boston, Mass., and Department of Neurology, Hahnemann Medical College, Philadelphia, Pa., U.S.A.

Huntington's disease (hereditary chorea) is a fatal disease of the central nervous system. The major clinical features of progressive dementia and choreiform movements were described over a century ago by Huntington (1872). The primary pathology includes neuronal atrophy and excessive accumulation of lipofuscin deposits in the caudate nucleus, cerebral cortex and hypothalamus (Bruyn, 1968, 1973; Klintworth, 1973).

Although some symptoms of Huntington's disease such as cachexia, hyperphagia and hyperhidrosis have been attributed to involvement of various portions of the hypothalamus (Facon et al., 1957; Bruyn, 1968), neuroendocrine studies of patients with this disease have been quite few. We recently reported that many patients with Huntington's disease have abnormalities of glucose, insulin and growth hormone dynamics, perhaps related to hypothalamic pathology (Podolsky et al., 1972*b*; Podolsky and Leopold, 1973, 1974; Leopold and Podolsky, 1975). In this report we will review our studies of growth hormone secretion in patients with Huntington's disease.

MATERIALS AND METHODS

Studies were carried out on 17 randomly selected male patients with Huntington's disease. All had a progressive dementia of varying degree, choreiform movements and a positive family history of the disease. Only one had a positive family history of diabetes mellitus. The age range of the patients was 27-79 years with a mean age of 43.5 years. The average length of documented clinical signs was 6.9 years (range 4-17 years). All patients were nonobese, in good nutritional status, and off all medication for at least 1 week prior to the studies. Patients who had been receiving haloperidol were taken off this drug several weeks before the studies.

All 17 patients underwent a standard 5-hour oral glucose tolerance test (GTT), performed after an overnight fast. A second GTT was done in each patient after 3 days of priming with 0.5 g oral 1-dihydroxyphenylalanine (L-dopa) administered 3 times a day plus 0.5 g L-dopa administered 30 minutes prior to the repeat test. Ten of the patients also received L-dopa by the same routine over 3 days, with hourly blood samples being taken over a 5-hour period after the final morning dose of 0.5 g L-dopa. In addition, 10 of the patients were given a 2.5-hour intravenous arginine tolerance test after an over-

---

* This work was supported by grants from the Veterans Administration, including a Clinical Investigatorship (S.P.), and from the Foundation for Research in Hereditary Disease and the Huntington's Chorea Foundation.

night fast (Martin et al., 1968; Podolsky and Sivaprasad, 1972). The arginine was infused over 30 minutes in a dosage of 0.5 g/kg as 10% arginine monohydrochloride in sterile water (Cutter Laboratories, Berkeley, Calif.).

During all these studies blood samples were withdrawn through an indwelling needle. Patients were supine, comfortable and free of stress during the tests. Plasma glucose was measured by the glucose oxidase method (Saifer and Gerstenfeld, 1958). Plasma growth hormone was measured by radioimmunoassay (Lau et al., 1966). Plasma prolactin was also measured by radioimmunoassay (Hwang et al., 1971).

RESULTS

During the oral glucose tolerance test mean fasting plasma glucose level was 81.7 ± 2.9 mg/100 ml (mean $\pm$ SEM) and at 2 hours was 123.0 ± 7.6 mg/100 ml, in the 17 patients with Huntington's disease (Fig. 1). In 30 control subjects mean fasting plasma glucose was 74.3 ± 2.5 mg/100 ml and at 2 hours was 100.3 ± 3.0 mg/100 ml. Nine of the neurological patients had impaired carbohydrate tolerance and 8 had completely normal carbohydrate tolerance.

Mean fasting plasma growth hormone level was slightly but not significantly higher in the patients with Huntington's disease than in normal controls. During the baseline glucose tolerance test growth hormone levels rose from 3.6 ± 0.6 ng/ml to 11.1 ± 4.5 ng/ml at 1 hour in these patients, while the controls showed a suppression of growth hormone from 2.5 ± 0.4 ng/ml to 1.6 ± 0.5 ng/ml at 1 hour (Fig. 2). The growth hormone level at 1 hour was significantly different from controls ($p = <0.04$). There was also an exaggerated growth hormone rise late in the course of the GTT, with a level of 18.6 ± 5.6 ng/ml in the patients compared to 9.1 ± 2.0 ng/ml in the controls. Figure 3 shows a replotting of this data as percent rise above fasting growth hormone level.

No alteration of the involuntary movements occurred in any of our patients with Huntington's disease after L-dopa priming. After L-dopa administration, mean fasting growth hormone levels rose considerably in these patients (from 3.6 ± 0.6 ng/ml to 17.9 ± 2.3 ng/ml), but showed a lesser rise in control subjects (from 2.5 ± 0.4 ng/ml to 6.4 ± 1.3 ng/ml). When the GTT was repeated, control subjects no longer showed a glucose-induced growth hormone suppression but instead had a sharp rise from 6.4 ± 1.3 ng/ml to 16.6 ± 3.4 ng/ml at 0.5 hour (Fig. 4). In the patients with Huntington's disease the opposite effect occurred, with a complete suppression of the elevated fasting growth hormone levels. After L-dopa there were significant differences between control and Huntington's disease growth hormone values at 0 time ($p = <0.01$) and at 1 hour ($p = <0.01$). Figure 5 shows a replotting of this data as percent rise above fasting growth hormone level.

Arginine infusion in 10 of the patients with Huntington's disease resulted in an increase in plasma glucose levels from 77.6 ± 3.0 mg/100 ml to 97.3 ± 5.3 mg/100 ml at 30 minutes, compared to 20 controls where the rise was from 73.1 ± 1.9 mg/100 ml to 86.3 ± 3.7 mg/100 ml at 30 minutes (Fig. 6). The magnitude of glucose elevation after arginine infusion was unrelated to the state of carbohydrate tolerance in the neurological patients.

In patients with Huntington's disease arginine infusion resulted in a significantly greater increase in growth hormone levels from 2.6 ± 0.5 ng/ml to 28.3 ± 3.7 ng/ml at 60 minutes, compared to the peak response in controls from 3.2 ± 0.6 ng/ml to 17.6 ± 2.7 ng/ml at 60 minutes (Fig. 7). Arginine-stimulated growth hormone response was significantly higher than normal in Huntington's disease at 30 minutes ($p = <0.01$) and at 60 minutes ($p = <0.05$).

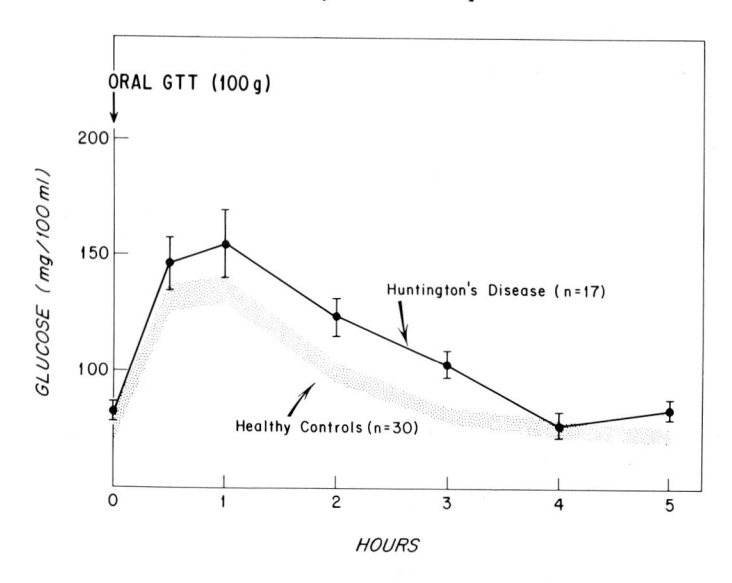

Fig. 1. Glucose response to oral glucose load in 17 patients with Huntington's disease and in 30 normal control subjects (mean ± SEM).

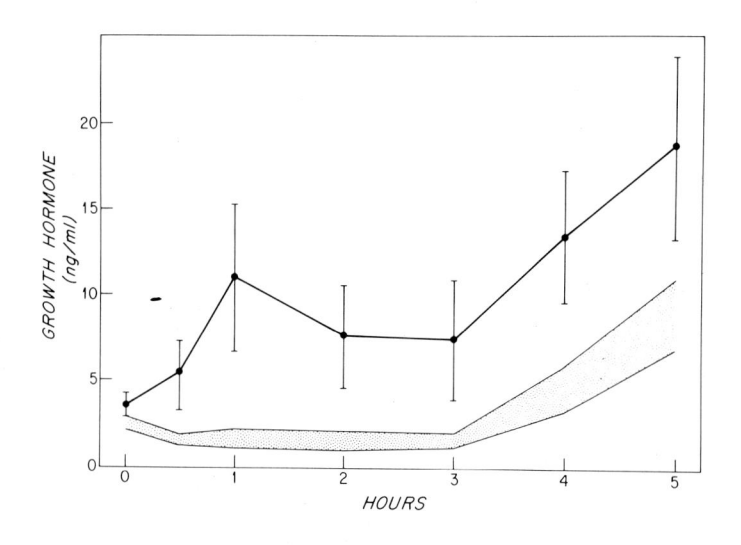

Fig. 2. Growth hormone response to oral glucose load in 14 patients with Huntington's disease and in 15 control subjects (mean ± SEM). Results in controls are indicated by shaded area.

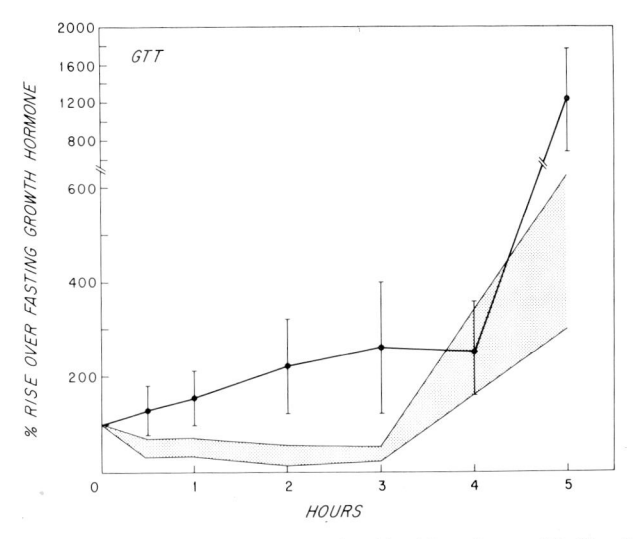

Fig. 3. Growth hormone response to oral glucose load in 14 patients with Huntington's disease and in 15 control subjects (plotted as percent rise above fasting growth hormone level).

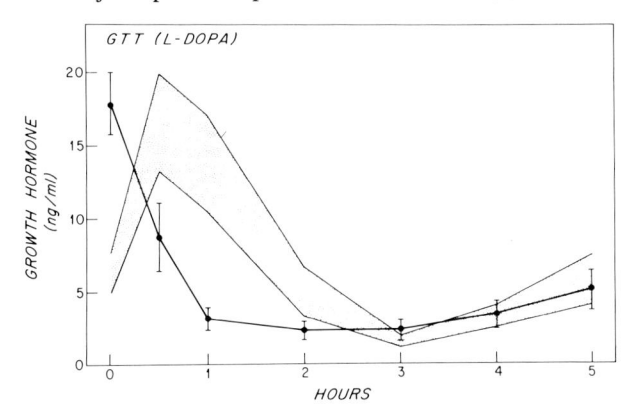

Fig. 4. Effect of L-dopa administration on growth hormone response to oral glucose load in the same 14 patients with Huntington's disease and in the same 15 control subjects (mean ± SEM).

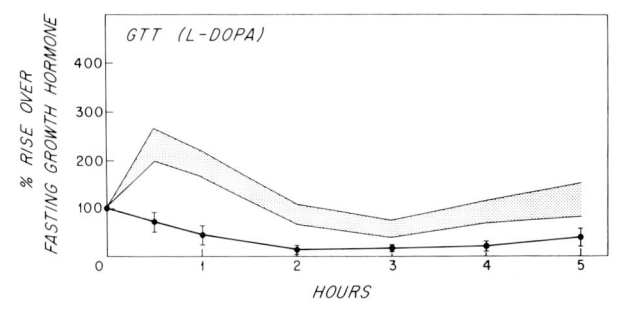

Fig. 5. Effect of L-dopa on growth hormone response to oral glucose load in 14 patients with Huntington's disease and in 15 control subjects (plotted as percent rise above fasting growth hormone level).

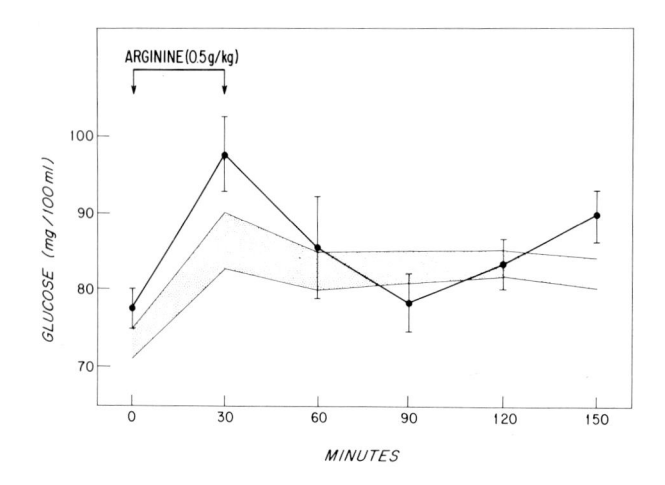

Fig. 6. Glucose response to intravenous arginine infusion in 10 patients with Huntington's disease and in 20 control subjects (mean ± SEM).

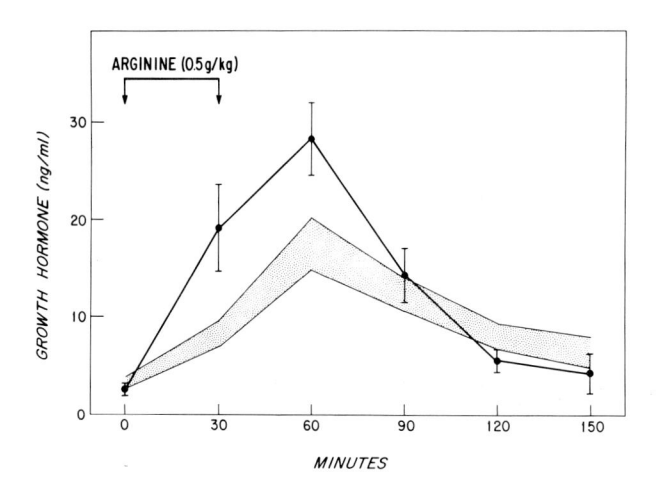

Fig. 7. Growth hormone response to intravenous arginine infusion in 10 patients with Huntington's disease and in 20 control subjects (mean ± SEM).

Administration of L-dopa alone to 10 of the patients with Huntington's disease resulted in a rise of growth hormone levels from a mean fasting value of 1.4 ± 0.5 ng/ml to a peak level of 15.5 ± 6.1 ng/ml at 1 hour (Fig. 8). However, plasma prolactin levels fell in response to L-dopa administration, dropping from 4.4 ± 0.6 ng/ml to 2.0 ± 0.6 ng/ml at 3 hours (Fig. 9). L-dopa alone was administered in a single dose of 0.5 g prior to blood sampling, after 3 days of L-dopa 0.5 g 3 times daily, as in the previous studies.

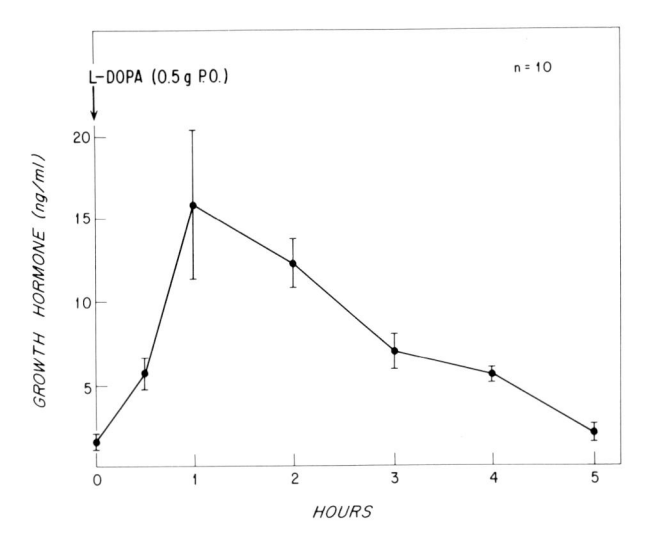

Fig. 8. Growth hormone response to L-dopa administration alone in 10 patients with Huntington's disease (mean ± SEM).

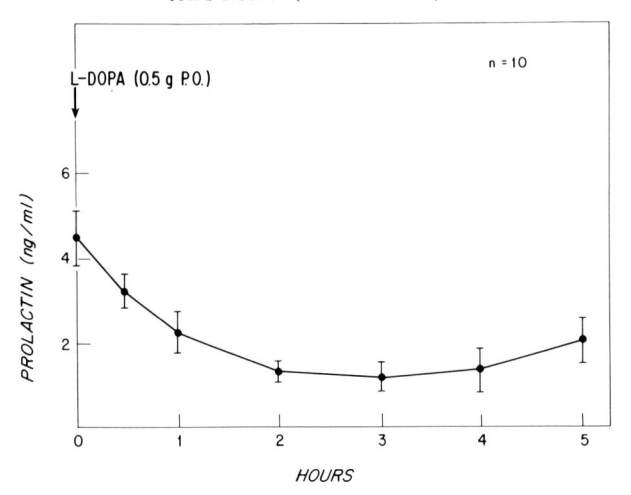

Fig. 9. Prolactin response to L-dopa administration alone in 10 patients with Huntington's disease (mean ± SEM).

DISCUSSION

As in Huntington's disease, impaired carbohydrate tolerance or frank diabetes mellitus frequently occur in other hereditary neurological diseases, including Friedreich's ataxia (Podolsky et al., 1964; Podolsky and Sheremata, 1970), myotonic dystrophy (Huff et al., 1967) and ataxia telangiectasia (Schalch et al., 1970). None of our patients with Huntington's disease had vascular complications of diabetes mellitus such as retinopathy (Krall and Podolsky, 1971), or had any evidence of ketosis, but their hyperglycemia was not severe. We were struck by the similarity of the appearance of patients with Lawrence's syndrome (lipoatrophic diabetes), where the neurological manifestations are associated with a very unusual form of insulin-dependent but ketoacidosis-resistant diabetes with hyperlipemia and apparent absence of all subcutaneous adipose tissue (Podolsky, 1971).

During the GTT, growth hormone did not suppress normally and also rose to abnormally high levels at the end of the test. Growth hormone response to an oral glucose load is one of the standard methods of evaluating the integrity of the hypothalamic-pituitary axis (Glick et al., 1965). Lack of suppression of growth hormone during a GTT, although certainly not specific for hypothalamic involvement, is seen in conditions associated with high circulating levels of growth hormone such as acromegaly (Hartog et al., 1964), malnutrition (Alvarez et al., 1972), cirrhosis (Hernandez et al., 1969; Podolsky et al., 1973) and also in precocious puberty associated with polyostotic fibrous dysplasia (Podolsky and Bryan, 1973). Although liver function tests were normal in our patients, Bolt and Lewis (1973) recently reported abnormal liver biopsies in patients with Huntington's disease.

Administration of the biogenic amine precursor L-dopa causes an acute increase of growth hormone levels in normal and parkinsonian patients (Boyd et al., 1970; Eddy et al., 1971). This rise cannot be suppressed by glucose administration. The mechanism of action has been postulated to be due in part to elevated levels of dopamine in the tuberoinfundibular system of the hypothalamus which then stimulates the release of growth hormone-releasing factor (GRF) by the median eminence. This hypothesis is supported by the presence of high concentrations of dopamine within the tuberoinfundibular system (Fuxe, 1963; Fuxe and Hokfelt, 1966). Adrenergic mechanisms are involved in growth hormone release produced by hypoglycemia (Blackard and Heidingsfelder, 1968) and perfusion of catecholamines through the central nervous system leads to release of growth hormone in animals (Müller et al., 1970). Phenothiazines block the excitatory actions of catecholamines and impair the release of growth hormone (Sherman et al., 1971), as does phentolamine, an α-adrenergic blocking agent. Propranolol, a β-adrenergic blocking agent, is associated with a transient rise of growth hormone levels as well as abnormalities of carbohydrate metabolism including both hypoglycemia (Kotler et al., 1966) as well as hyperosmolar nonketotic diabetic coma (Podolsky and Pattavina, 1973).

Repeat GTTs after L-dopa administration (including a dose 30 minutes prior to the repeat test) demonstrated suppression of elevated growth hormone levels in our patients with Huntington's disease, but not in the controls. There were significant differences between the 2 groups at 0 time, 0.5 hour and 1 hour. These data could be a reflection of direct hypothalamic involvement in Huntington's disease. Alternately, since L-dopa has also been reported to cause suppression of elevated growth hormone levels in acromegaly (Sherman et al., 1973), this finding may be a general feature of states of accelerated growth hormone secretion. Perhaps excessive stimulation or hyperresponsiveness of the tuberoinfundibular system leads to a suppression of GFR, or to the release of growth hormone-inhibiting factor (somatostatin) (Martin, 1973). However, growth hormone release appeared to be normal after L-dopa alone, without a GTT. Prolactin levels also fell as expected (Frantz, 1973) after L-dopa alone.

Arginine infusion in patients with Huntington's disease stimulates excessive growth hormone secretion, compared to healthy control subjects, with significantly higher growth hormone levels at 30 minutes and at 60 minutes. All but 1 patient in our series responded to this provocative test, although some reports indicate inconsistent responses of growth hormone in normals (Best et al., 1968). Growth hormone hypersecretion following arginine infusion also occurs in severe protein-calorie malnutrition (Smith et al., 1974) and in nonpotassium-depleted patients with hepatic cirrhosis

(Podolsky et al., 1972a, 1973), two conditions with elevated fasting growth hormone levels. Cirrhosis may be associated with gliosis of the caudate nucleus (Victor et al., 1965), the primary area of basal ganglia involvement in Huntington's disease (Bruyn, 1968). In addition, chorea and abnormal catecholamine metabolism may occur during episodes of hepatic coma (Fisher, 1974; Knell et al., 1974). Whether there exists a common disturbance of CNS catecholamines that might account for similar abnormalities of growth hormone secretion in Huntington's disease and cirrhosis is not yet known.

Arginine infusion stimulates both excessive growth hormone and insulin secretion in Huntington's disease (Podolsky et al., 1972b). Although the insulin hypersecretion precedes the growth hormone hypersecretion in time, an underlying disturbance of growth hormone regulation may itself be responsible for the insulin hypersecretion occurring in patients with Huntington's disease following arginine infusion. Arginine also stimulates hypersecretion of insulin in acromegaly (Beck et al., 1965), cirrhosis (Podolsky et al., 1973) and in growth hormone pretreated dogs (Pierlussi and Campbell, 1973), suggesting that hyperinsulinemia is in part dependent upon elevated basal growth hormone levels. However, basal growth hormone levels were normal in this series of patients with Huntington's disease.

Increased sensitivity to intracerebral dopamine has been proposed to explain the exacerbation of adventitious movements in these patients after several weeks of L-dopa administration (Klawans and Rubovits, 1972). It is not clear whether there is a relationship between the increased choreiform movements observed by others and the lowering of growth hormone levels following acute administration of L-dopa plus glucose to patients with Huntington's disease. We did not observe any worsening of the involuntary movements at the low doses of L-dopa employed in our studies.

Neuronal loss in several hypothalamic regions including the paraventricular and ventromedial nuclei has been described in Huntington's disease (Bruyn, 1968, 1973), but the consistency of this pathology is unknown. Whether this pathology alone can account for the elevated growth hormone responses to arginine and an apparent abnormal growth hormone-glucose interrelationship awaits further studies. The growth hormone secretory disturbances may be a nonspecific finding, or they may be due to an imbalance among the biogenic amines involved in hypothalamic secretion of growth hormone-inhibiting (somatostatin) and growth hormone-releasing factors (Martin, 1973; Podolsky and Leopold, 1973).

ACKNOWLEDGMENTS

Dr. Alfred E. Wilhelmi, Atlanta, Ga., supplied the highly purified human growth hormone (HS 1147) used for radioimmunoassay standards, through the National Pituitary Agency and the National Institute of Arthritis and Metabolic Diseases. Dr. Andrew Frantz, New York, N.Y., kindly performed the prolactin measurements by radioimmunoassay. Ms. Kay Pattavina provided excellent technical assistance.

REFERENCES

Alvarez, L.C., Dimas, C.O., Castro, A., Rossman, D.G., Vanderlaan, E.F. and Vanderlaan, W.P. (1972): *J. clin. Endocr., 34,* 400.
Beck, P., Schalch, D.S., Walker, J.L., Kipnis, D.M. and Daughaday, W.H. (1965): *J. Lab. clin. Med., 66,* 366.
Best, J., Catt, K.J. and Burger, H.G. (1968): *Lancet, 2,* 124.
Blackard, W.G. and Heidingsfelder, S.A. (1968): *J. clin. Invest., 47,* 1407.
Bolt, J.M.W. and Lewis, G.P. (1973): *Quart. J. Med., 42,* 151.
Boyd, A.E., Lebovitz, H.E. and Pfeiffer, J.B. (1970): *New Engl. J. Med., 283,* 1425.

Bruyn, G.W. (1968): In: *Handbook of Clinical Neurology, Vol. 6*, p. 268. Editors: P.J. Vinken and G.W. Bruyn. North-Holland Publishing Co., Amsterdam.
Bruyn, G.W. (1973): In: *Advances in Neurology. Vol. 1. Huntington's Chorea*, p. 399. Editors: A. Barbeau, T.N. Chase and G.W. Paulson. Raven Press, New York.
Eddy, R.L., Jones, A.L., Chakmakjian, Z.H. and Silverthorne, M.C. (1971): *J. clin. Endocr., 33*, 709.
Facon, E., Steriade, M., Cortez, P. and Voinesco, S. (1957): *Acta neurol. belg., 57*, 898.
Fisher, J.E. (1974): *Arch. Surg., 108*, 325.
Frantz, A.G. (1973): *Progr. Brain Res., 39*, 311.
Fuxe, K. (1963): *Acta physiol. scand., 58*, 383.
Fuxe, K. and Hokfelt, T. (1966): *Acta physiol. scand., 66*, 245.
Glick, S.M., Roth, J., Yalow, R.S. and Berson, S.A. (1965): *Recent Progr. Hormone Res., 21*, 241.
Hartog, M., Gaafar, M.A., Meisser, B. and Fraser, R. (1964): *Brit. med. J., 1*, 1229.
Hernandez, A., Zorilla, E. and Gershberg, H. (1969): *J. Lab. clin. Med., 73*, 25.
Huff, T.A., Horton, E.S. and Lebovitz, H.E. (1967): *New Engl. J. Med., 277*, 837.
Huntington, G. (1872): *Med. Surg. Reporter, 26*, 317.
Hwang, P., Guyda, H. and Friesen, H. (1971): *Proc. nat. Acad. Sci. (Wash.), 68*, 1902.
Klawans, H.L. and Rubovits, R. (1972): *Neurology (Minneap.), 22*, 107.
Klintworth, G.K. (1973): In: *Advances in Neurology. Vol. I. Huntington's Chorea*, p. 353. Editors: A. Barbeau, T.N. Chase and G.W. Paulson. Raven Press, New York.
Knell, A.J., Davidson, A.R., Williams, R., Kantamaneni, B.D. and Curzon, G. (1974): *Brit. med. J., 1*, 549.
Kotler, M.N., Berman, L. and Rubenstein, A.H. (1966): *Lancet, 2*, 1389.
Krall, L.P. and Podolsky, S. (1971): In: *Diabetes*, p. 268. Editors: R.R. Rodriguez and J. Vallance-Owen. ICS 231, Excerpta Medica, Amsterdam.
Lau, K.S., Gottlieb, C.W. and Herbert, V. (1966): *Proc. Soc. exp. Biol. (N.Y.), 123*, 126.
Leopold, N.A. and Podolsky, S. (1975): *J. clin. Endocr., 41*, 160.
Martin, J.B. (1973): *New Engl. J. Med., 288*, 1384.
Martin, M.M., Gaboardi, F., Podolsky, S., Raiti, S. and Calcagno, P.L. (1968): *New Engl. J. Med., 279*, 273.
Müller, E.E., Pecile, A., Felici, M. and Cocchi, O. (1970): *Endocrinology, 86*, 1376.
Pierlusi, J. and Campbell, J. (1973): *Fed. Proc., 32*, 265.
Podolsky, S. (1971): In: *Joslin's Diabetes Mellitus, 11th Ed.*, p. 722. Editors: A. Marble, P. White, R.F. Bradley and L.P. Krall. Lea and Febiger, Philadelphia.
Podolsky, S. and Bryan, R.S. (1973): In: *Clinical Aspects of Metabolic Bone Disease*, p. 484. Editors: B. Frame, A.M. Parfitt and H. Duncan. ICS 270, Excerpta Medica, Amsterdam.
Podolsky, S., Burrows, B.A., Zimmerman, H.J. and Pattavina, C.G. (1972a): In: *Growth and Growth Hormone*, p. 402. Editors: A. Pecile and E.E. Müller. ICS 244, Excerpta Medica, Amsterdam.
Podolsky, S. and Leopold, N.A. (1973): *Progr. Brain Res., 39*, 225.
Podolsky, S. and Leopold, N.A. (1974): *J. clin. Endocr., 39*, 36.
Podolsky, S., Leopold, N.A. and Sax, D.S. (1972b): *Lancet, 1*, 1356.
Podolsky, S. and Pattavina, C.G. (1973): *Metabolism, 22*, 685.
Podolsky, S., Pothier, A. and Krall, L.P. (1964): *Arch. intern. Med., 114*, 533.
Podolsky, S. and Sheremata, W.A. (1970): *Metabolism, 19*, 555.
Podolsky, S. and Sivaprasad, R. (1972): *J. clin. Endocr., 35*, 580.
Podolsky, S., Zimmerman, H.J., Burrows, B.A., Cardarelli, J.A. and Pattavina, C.G. (1973): *New Engl. J. Med., 288*, 644.
Saifer, S. and Gerstenfeld, S. (1958): *J. Lab. clin. Med., 51*, 448.
Schalch, D.S., McFarlin, D.E. and Barlow, M.H. (1970): *New Engl. J. Med., 282*, 1396.
Sherman, L., Kim, S., Benjamin, F. and Kolodny, H.D. (1971): *New Engl. J. Med., 284*, 72.
Sherman, L., Kolodny, H.D., Singh, A., Deutsch, S. and Benjamin, F. (1973): *Clin. Res., 20*, 867.
Smith, S.R., Edgar, P.J., Pozefsky, T., Chhetri, M.K. and Prout, T.E. (1974): *J. clin. Endocr., 39*, 53.
Victor, M., Adams, R.D. and Cole, M. (1965): *Metabolism, 44*, 345.

VI. GH. Clinical investigations. Selected topics

HUMAN GROWTH HORMONE CHANGES WITH AGE*

T.L. BAZZARRE, A.J. JOHANSON, C.A. HUSEMAN, M.M. VARMA and R.M. BLIZZARD

Department of Pediatrics, University of Virginia Hospital, Charlottesville, Va., U.S.A.

Senescent rat pituitaries respond minimally to extracts of hypothalami (presumably containing growth hormone-releasing factor (GHRF)) which induce much greater GH release in young rats; and growth hormone-releasing activity detectable in hypothalami of young rats is decreased in senescent rats (Pecile et al., 1965). Lifespan, as well as growth, of hypophysectomized animals is much shortened, implicating the importance of the pituitary and its secretions in longevity (Verzar and Spichtin, 1966).

Although there is a decrease of about 20% in human pituitary size with advanced age (Verzar, 1966) pituitary content of HGH is not different in children and adults (Gershberg, 1957) and does not change with increasing age (Daughaday, 1974).

Human growth hormone secretion and metabolism at different ages has been assessed by various methods. Metabolic clearance rates of HGH in adults are unchanged with increasing age (Taylor et al., 1969; McGillivray et al., 1970). There also is no difference among normal, acromegalic and hypopituitary individuals in metabolic clearance of HGH. Finkelstein et al. (1972) found greatly diminished secretion of growth hormone in older individuals, 40-62 years, compared with younger adults, determined by serum levels obtained at 20-minute intervals for 24 hours. Thompson et al. (1972) found lower integrated concentrations of growth hormone in adults 30-50 years old than in children and adolescents 8-16 years old. Absent or diminished number and level of sleep-related HGH surges have been observed in adults over 50 years of age (Carlson et al., 1972; Blichert-Toft, 1975) (Table 1).

Responses of HGH in older individuals to provocative stimuli have been found diminished in some studies or unchanged in others, compared with those in younger adults. Late elevations in HGH following glucose administration to older subjects have been determined to be normal in several studies (Root and Oski, 1969; Dudl et al., 1973) or diminished in others (Danowski et al., 1969; Sandberg et al., 1973; Vidalon et al., 1973). Responses to insulin-induced hypoglycemia have been reported to be normal in some studies (Root et al., 1969; Cartlidge et al., 1970; Kalk et al., 1973) while diminished in another (Laron et al., 1966). Responses of HGH to arginine have been normal in 2 of 3 older individuals tested (Root and Oski, 1969) and not statistically different from those of younger adults (Dudl, 1973; Blichert-Toft, 1975). Buckler (1969) found the HGH response to Bovril, a protein drink, greatly diminished in older males, but not females.

Root and Oski (1969) gave HGH to 3 elderly males for a 10-day period demonstrating nitrogen retention, but no change in hydroxyproline excretion. Increased calcium excretion occurred in 1 patient only. The responses were interpreted as deficient

* Supported by USPHS RR 847 and HG 08926.

TABLE 1
Growth hormone responses in older people

Reference	Status	Sex	BW	Basal	p-Glucose	p-Insulin	p-Arginine	Other
Root and Oski (1969)	Ill	M	↓		N2/3	N	N2/3	
Danowski et al. (1969)	?	F	N	↓	↓			
Buckler (1969)	N N	M F						↓ p-Bovril N-Bovril
Cartlidge et al. (1970)	Hosp.	M&F	↓			N		
Laron et al. (1970)	N	M&F					↓	
Carlson et al. (1972)	N	M&F	±10%					↓ 4/6 sleep
Vidalon et al. (1973)	?	F	N	↓	↓			
Sandberg et al. (1973)	Amb			↑	↓			
Dudl et al. (1973)	N	M	?N		N(IV)		↓ 50% (but not statistically)	
Kalk et al. (1973)	N	M&F	1/2↑	N		N		
Blichert-Toft (1975)	N	M&F		↑			N	↓ sleep ↓ c̄ anesthesia
Thompson et al. (1972)	N	M						↓ 24-hr ICGH
Finkelstein et al. (1972)	N	M&F						↓ 24-hr secretion
Present study	N	M(1F)	N			N		↓ exercise, ↓ ICGH ↓ sleep, ↓ L-dopa

N = normal. ICGH = integrated concentrations of human growth hormone.

relative to those in younger normal subjects. Rudman et al. (1971) administered HGH for 7 days to 4 adult males (52-79 yr) and found no difference in metabolic response from that of 4 normal children (9-13 yr), who were less responsive than hypopituitary patients.

We wished to extend these studies including larger numbers of subjects in several age categories to determine if there were significant differences in responsiveness to stimuli for HGH release and if there were changes in growth hormone responsiveness with increased age. Changes in androgen concentration and binding to plasma protein, and in adrenocortical response to insulin-induced hypoglycemia among the different age groups were also evaluated. This is a report of our preliminary observations.

MATERIALS AND METHODS

Subjects

Twenty-two subjects were studied (Table 2).
Group I: 6 hypopituitary subjects 9-29 years old. The criterion for definition of hypopituitarism was the failure to respond with HGH levels of >5 ng/ml to any of the provocative tests for growth hormone release used in our study (Table 3).
Group II: 4 normal males 23-29 years old.
Group III: 6 normal males 35-51 years old.
Group IV: 6 normal subjects 61-69 years old.

All normal subjects were within ± 10% of their ideal body weight for height, age and sex except WH in group II and HW in group IV who were between 10 and 20% greater than their ideal weight (New York Metropolitan Life Insurance Tables, 1959).

Evaluation of HGH secretion

Exercise (Keenan et al., 1972), L-dopa (Eddy et al., 1971), and insulin-induced hypoglycemia (Penny et al., 1969) were used as provocative tests for HGH release. For the exercise stimulation test blood samples were drawn at time 0 after overnight fasting, after 20 minutes of exercise (consisting of running or walking on a graded treadmill), and after 20 minutes rest following the exercise period. The L-dopa and insulin stimulation tests were conducted in tandem. Blood was drawn at time 0 after an overnight fast, 250-500 mg L-dopa (depending on body weight) was given by mouth, and blood samples were collected at 30, 60 and 90 minutes post L-dopa ingestion. Immediately after drawing the 90-minute post L-dopa blood sample, 0.1 U insulin/kg body weight was given intravenously, which in all subjects caused a 50% or greater decrease of blood sugar, except in HW in group IV. Blood samples were drawn at 15, 30, 45 and 60 minutes post insulin injection.

Twenty-four-hour integrated concentrations of HGH (ICGH) (Kowarski et al., 1971) were determined on day 5 of the control period and on day 5 of the HGH administration period during the metabolic balance experiments. Blood was withdrawn at a constant rate and samples collected for half-hour periods during the 24 hours employing a constant withdrawal pump. Occasionally the continuous withdrawal technique was discontinued and blood samples were drawn every 30 minutes to complete the 24-hour period. Peaks in HGH associated with sleep and awake periods were examined from these data. Twenty-four-hour ICGH for each period was determined separately from pooled 100-μl aliquots from each half-hour sample, and as the mean of the half-hour samples, assayed individually. In all but one pair of data, the difference between HGH levels by the 2 methods was less than 0.3 ng/ml.

Assay methods

Levels of HGH were determined on all samples in triplicate by double antibody radioimmunoassay (Schalch and Parker, 1964). The lowest point on the standard curve was 0.2 ng/ml and the highest 4.0 ng/ml.

Plasma corticoids were determined by competitive binding (Murphy, 1967) on samples drawn at the same times as those for HGH immediately before and following intravenous insulin. From the pooled half-hour collections, integrated concentrations for testosterone (Varma et al., 1975), testosterone binding (Forest et al., 1968), and prolactin (Thorell and Johansson, 1971), were also determined.

TABLE 2

Clinical data of subjects

Group	Subject	Sex	Age	Height (cm)	Weight (kg)	HGH (dose/day)
I. Hypopituitary	RP	M	15	123	24.1	1.76
	AJ	F	12	115	19.6	1.57
	EF	M	29	167	52.0	3.25
	MF	M	18	137	37.2	2.38
	WC	M	11	137	47.5	3.03
	JD	F	9	97	13.3	1.18
II. 23- to 29-year-old males	TB	M	25	170	66.2	3.90
	DS	M	29	169	56.8	3.44
	JB	M	23	186	71.1	4.10
	WH	M	25	181	80.0	4.41
III. 35- to 51-year-old males	FM	M	48	171	69.1	
	RB*	M	51	180	70.0	
	MA*	M	39	170	70.9	
	RD*	M	35	174	69.8	
	JH*	M	50	173	71.8	
	MV*	M	36	174	70.3	
IV. 61- to 69-year-old subjects	HW	F	61	158	64.1	3.80
	JW	M	66	181	73.7	4.16
	JJ	M	69	185	71.7	4.16
	GA*	M	67	171	72.2	4.16
	FB	M	66	157	58.4	3.43
	WB	M	61	176	74.8	4.28

* No metabolic balance data.

TABLE 3

Clinical summary of hypopituitary subjects

Subject	Sex	Etiology	CA*	BA[†]	TSH deficiency	ACTH deficiency**
RP	M	Idiopathic	15	7	−	−
AJ	F	Idiopathic	12.5	6	−	+
JD	F	Idiopathic	9	8.5	−	−
EF	M	Idiopathic traumatic delivery	29	18	+	+
MF	M	Idiopathic traumatic delivery	18	11	+	+
WC	M	Postoperative craniopharyngioma	11	10	+	+

*CA = chronological age. [†] BA = bone age. ** as detected by metapyrone testing.

Short-term metabolic balance studies

Fourteen subjects participated in the metabolic balance studies which consisted of a 3-day dietary adjustment period, a 5-day control period and a 5-day growth hormone treatment period. The growth hormone dose administered was 0.168 U/kg body weight$^{3/4}$/day i.m. (Rudman et al., 1971). This amount of HGH, but not a lower dose, was observed to exhibit slight but definite responses of nitrogen retention in normal adults (Rudman et al., 1971).

Two daily meal plans alternated throughout the study. The metabolic diets were constant, low hydroxyproline diets, prepared from the same lots of food. The energy and protein requirements were determined according to recommended dietary allowances (NAS-NRC, 1974). Dietary protein was 0.8 g/kg for adults and 1.2-1.5 g/kg for hypopituitary subjects 9-18 years old. Energy intake was altered during the adjustment period and first day of the control period when necessary to maintain a constant weight during the control period for each individual. Diets for the hypopituitary subjects were based on the requirements of children of similar weight and height rather than chronologic age. Subjects were weighed daily. Continuous 24-hour urines were collected for determination of creatinine (Technicon, 1970), nitrogen (Technicon, 1970), hydroxyproline (Hosley et al., 1970), and calcium (by spectrophotometer). Student's t test was used for statistical comparisons.

TABLE 4

Mean maximum HGH response in ng/ml

Group		Exercise	L-dopa	Insulin	Sleep peak	Control period (24-hour ICGH)	HGH period (24-hour ICGH)
I. Hypopituitary	n	5	5	4	–	5	4
	Mean	<2^a	<2^a	<2^A		<2^a	5.2**
	SE	–	–	–		–	0.8
II. 23- to 29-year-old males	n	4	4	4	4	4	4
	Mean	22^b	23^b	30^B	21^a	3.5^b	3.7
	SE	11	6.5	9	6.7	0.7	1.0
III. 35- to 51-year-old males	n	6	6	6	6	6	1
	Mean	4.0^{bc}	4.9^c	29^B	4.8	<2^a	4.3
	SE	0.6	1.8	12	2.8	–	–
IV. 61- to 69-year-old subjects	n	6	6	6	6	6	4
	Mean	3.0^{ac}	11^d	15^B	3.5^b	<2^a	3.7**
	SE	0.9	1.5	5	1.0	–	0.4

a, b, c, d = means in each column with different small letter superscripts are statistically different (P < 0.05).

A, B = means in each column with different capital letter superscripts are statistically different (P < 0.01).

** different from control period (P < 0.01).

RESULTS

The results of growth hormone determinations are presented in Table 4. Values <2 ng/ml were computed as 2 ng/ml and >40 ng/ml as 40 ng/ml. None of the hypopituitary subjects responded to any provocative stimulation with levels greater than 2 ng HGH/ml. The mean peak HGH concentrations following exercise, L-dopa, and insulin of the 23- to 29-year-old group were higher than for any of the other groups. Peak HGH levels during sleep and ICGH prior to HGH administration were also higher in the 23- to 29-year-old group.

In response to exercise, mean HGH levels of the older subjects, group IV, were similar to group I, the hypopituitary subjects. Group II had statistically higher levels than group I and IV ($p < 0.05$), but not group III, and group III had levels higher than group I, but not group IV ($p < 0.05$). Only 1 of the 6 group IV subjects had a growth hormone level >5 ng/ml in response to exercise.

The responses to L-dopa were greatest in group II ($p < 0.05$) compared with the other 3 groups. Group IV, however, had greater responses than group I ($p < 0.05$) and group III ($p < 0.05$). In response to insulin groups II, III and IV had statistically similar responses which were greater than those of the hypopituitary subjects ($p < 0.01$). Although not statistically lower, the mean peak response in group IV was about half that of group II or III.

Mean peak sleep responses were higher in group II than group IV ($p < 0.05$). Individual 30-minute levels were not determined in the hypopituitary subjects. Only 1 of the 3 subjects in group III and 1 of 6 subjects in group IV had a nocturnal HGH peak greater than 5 ng HGH/ml whereas 3 of 4 men in group II had peaks >10 ng/ml.

The mean 24-hour ICGH of group II, 3.5 ng/ml, was significantly greater ($p < 0.05$) than that of any of the other groups. The range of the 24-hour integrated concentrations of the young men was from 2 to 5.2 ng/ml. The 24-hour ICGH was less than 2 ng/ml for all subjects in groups I, III and IV. The 24-hour ICGH were not statistically changed with HGH administration in group II. However, groups I and IV had significantly increased mean ICGH with HGH administration ($p < 0.01$).

The results of determinations of 24-hour integrated concentrations of other hormones and corticoid responses to insulin-induced hypoglycemia are presented in Table 5. Mean 24-hour integrated prolactin concentration of the hypopituitary subjects was higher ($p < 0.05$) than in groups II or III. WC, who had had removal of a craniopharyngioma had a level of 27 ng/ml. The mean 24-hour integrated concentrations for prolactin of groups II, III and IV were not statistically different.

Mean 24-hour integrated testosterone concentrations were higher ($p < 0.05$) in young men, group II, than in the older men, group IV. Group III was not statistically different from either group II or IV. Mean testosterone binding to plasma protein was greater in group IV than in groups II and III ($p < 0.025$).

The mean change in plasma corticoids in response to insulin-induced hypoglycemia was significantly less in group IV than all other groups ($p < 0.025$).

Metabolic response to HGH

Metabolic study data was available for only 1 individual in group III and therefore comparisons were made among groups I, II and IV only. The metabolic responses of the subjects to the administration of growth hormone are presented in Table 6.

During growth hormone administration a mean weight loss of 0.28 kg/day was observed in group II which was statistically different from groups I and IV ($p < 0.05$) who gained means of 0.17 and 0.33 kg/day respectively. Three of the 4 young men in

TABLE 5
Other hormones

Group			24-hour integr. conc. prolactin (ng/ml)	24-hour integr. conc. testosterone (ng/dl)	24-hour integr. conc. testosterone binding (%)	Maximum Δ corticoids post insulin (μg/dl)
I. Hypopituitary		n	5			5
	X̄		16.7[a]			12.2[b]
	SE		0.8			4.9
II. 23- to 29-year-old males		n	4	4	4	4
	X̄		13.4[b]	962[a]	89.8[a]	16.3[b]
	SE		1.0	77	0.74	3.8
III. 35- to 51-year-old males		n	6	6	5	5
	X̄		12.7[b]	811	88.7[a]	14.6[b]
	SE		1.1	53	0.79	5.1
IV. 61- to 69-year-old males		n	6	5	5	5
	X̄		14.5	675[b]	91.8[b]	6.8[a]
	SE		1.5	72	0.86	2.0

a, b = means in each column with different superscripts are statistically different ($P < 0.05$).

TABLE 6
Metabolic response to 0.168 U HGH/kg body weight $(BW)^{3/4}$/day

Group			Mean Δ BW (kg)	Mean Δ creatinine (mg/kg BW/day)	Mean Δ calcium (mg/kg BW/day)	Mean Δ hydroxy-proline (μg/kg BW/day)	Mean Δ nitrogen (mg/kg BW/day)
I. Hypopituitary		n	6	5	5	5	5
	Mean		+0.17[a]	+1.56	+2.94[A]*	+495[A]	−38.6[a]*
	SE		0.17	0.98	0.18	77	7.3
II. 23- to 29-year-old men		n	4	4	4	4	4
	Mean		−0.28[b]	−0.38	+0.78[B]	+83[B]	−20.8[a]
	SE		0.14	1.21	0.28	19	7.2
IV. 61- to 69-year-old subjects		n	6	5	5	5	5
	Mean		+0.33[a]	+0.49	+1.05[B]*	+89[B]	−24.5[b]*
	SE		0.28	0.49	0.22	18	1.8

A, B = means in each column with different capital letter superscripts are statistically different ($P < 0.01$). a, b = means in each column with different small letter superscripts are statistically different ($P < 0.05$). * = significant differences between control and HGH periods ($p < 0.05$).

group II lost weight during the HGH administration period, while 1, WH, who was 15% above ideal weight maintained his body weight. Four of the 6 subjects in group IV gained 0.7 kg/day while given HGH.

Mean urinary creatinine excretion changes with HGH treatment paralleled mean body weight changes in each group. Mean urinary creatinine excretion changes were not statistically different among the 3 groups, nor were the changes within each group from the control period to HGH treatment period significantly different.

The increase in calcium excretion with HGH treatment was greater ($p < 0.01$) in group I, the hypopituitary subjects, than in group II or group IV. The mean calcium excretion increased significantly with HGH administration in group I ($p < 0.05$) and IV ($p < 0.01$) but not in group II.

The mean increase of hydroxyproline excretion was greater ($p < 0.01$) in the hypopituitary subjects than in group II or group IV. The mean hydroxyproline excretion of the HGH period was not statistically different from the control period within any of the groups.

A decrease in urinary nitrogen excretion (increased nitrogen retention) was observed in all subjects in all groups during HGH treatment. The greatest mean decrease in nitrogen excretion was observed in the hypopituitary subjects which was statistically different ($p < 0.05$) from the older subjects, group IV but not group II, probably because of the great variability in the latter group. With HGH administration nitrogen excretion was increased significantly ($p < 0.05$) in the hypopituitary group and group IV ($p < 0.01$) compared with the control period. In group II there was no significant change in nitrogen excretion with HGH administration.

DISCUSSION

Even with the small numbers in each of the subject groups, significantly decreased mean responses to known stimuli for HGH release — exercise, L-dopa and sleep — were observed when 61- to 69-year-old subjects were compared with 23- to 29-year-old men. The probable effect of increased weight on diminishing HGH responses is exemplified by those responses of WH, who was 15% greater than ideal weight and had the least response in group II to exercise, L-dopa, and sleep. Also, significantly less growth hormone was secreted by the older subjects as determined by the ICGH method (Kowarski et al., 1971). Mean response to insulin hypoglycemia, although half that of the younger men, was not significantly lower. By exercise testing and ICGH, the older subjects were similar in their responses to hypopituitary patients tested.

Although numerous publications have considered the question of alteration with increased age of growth hormone release in response to provocative stimuli, previously cited conflicting studies had left the question unresolved. In several of the studies, hospitalized, or sick subjects were studied (Root and Oski, 1969; Cartlidge et al., 1970; Sandberg et al., 1973). Normalcy of the body weight for height rarely was considered, although obesity is known to be a factor in diminished responses of HGH (Dudl et al., 1973). The subjects in our study were carefully selected for their normal weight and good health.

We conclude, therefore, that our preliminary studies are most suggestive of a waning responsiveness of HGH release from pituitaries of aging individuals to known provocative stimuli.

If, as we have shown, there may be a degree of partial growth hormone deficiency in older individuals, it might be expected that their responses to exogenous HGH administration would mimic those observed in hypopituitarism. Indeed, many of the observations of this study are consistent with this thesis. HGH administration resulted in comparable mean weight gain in both the hypopituitary patients and in the older age group while there was a mean weight loss in the younger age group. Creatinine excretion changed in parallel with weight, indicating accretion of lean body mass in both groups that gained weight when given HGH. Significantly increased calcium excretion and decreased nitrogen excretion during HGH administration was observed for both the hypopituitary and older subjects but not younger males. There were no significant changes in any groups of urinary hydroxyproline. The mean magnitude of change of both urinary calcium and nitrogen was significantly greater in the hypopituitary than in all normal subjects as was the increase of hydroxyproline. The responses in the older group, although not statistically different from those of the younger men, were intermediate between theirs and those of the hypopituitary subjects.

It seems, therefore, that the metabolic response of older individuals is greater than that of younger males, but less than that in hypopituitarism, and is consistent with a degree of HGH deficiency. This is in contrast to the observations of Rudman et al. (1971) who studied 4 normal males (52-79 yr) and 4 normal children (9-13 yr) and found 'that no definite relationships between age and responsiveness were apparent'. Root and Oski (1969) studied the effects of a larger dose of HGH administration to 3 old men, but had no control group for comparison. They found increased nitrogen retention which was similar to that of younger subjects in other studies but no changes in urinary calcium and hydroxyproline. This failure of response was stated to be different from that in younger subjects and was interpreted by implication as evidence for tissue unresponsiveness to HGH in the older individuals.

The increase in 24-hour ICGH in the hypopituitary and older subjects in our study with HGH administration may suggest tissue unresponsiveness, reflected by a lower metabolic clearance rate. However, it is more probably an achievement of the 'normal' integrated concentrations observed in the young men both pre and post HGH treatment. The ICGH values observed by us in the young men are lower than those found by Thompson et al. (1972) in boys 8-16 yr and higher than the normal of 1.8 ng/ml for 30- to 50-year-old men in their study. The lack of difference among the groups in mean ICGH during HGH administration suggests no differences in metabolic degradation.

The decreased plasma testosterone and increased testosterone binding to plasma proteins in older men that we have observed are similar to observations of Vermeulen et al. (1972) and Rubens et al. (1974), which are made from determination in single plasma samples. In contrast the values we obtained were from 24-hour integrated blood samples. These lower levels could be a reflection of declining gonadotropin function or of lack of responsiveness to gonadotropins by the gonads.

The observation of significantly decreased corticoid response to insulin-induced hypoglycemia in the older subjects as compared to all other groups is most provocative since it implies either a lesser ACTH release or an impairment of adrenal responsiveness to ACTH with age. Cartlidge et al. (1970) have reported plasma cortisol responses in hospitalized 80- to 95-year-old subjects to be 4.5 to 19.4 μg/dl in response to insulin-induced hypoglycemia; 50% of the responses were abnormally low.

We conclude that our studies suggest that, with increasing age, there may be declining function of the pituitary in its secretory responses to stimuli for HGH and ACTH and possibly gonadotropins. Responsiveness to exogenous HGH in older subjects is intermediate between that of hypopituitary and normal young men. Whether this declining function is primary or secondary to the aging process is an unanswered question.

We recognize that our conclusions are based on data from a small number of individuals. However, the measurement of hormones and other substances in samples of blood withdrawn continuously for periods of 24 hours from these individuals should provide more reliable conclusions than those based on single blood samples from large numbers of individuals. By adopting similar techniques in larger numbers of individuals and by monitoring various other parameters, we wish to extend our observations to permit more definite conclusions.

REFERENCES

Blichert-Toft, M. (1975): *Acta endocr. (Kbh.), Suppl., 195,* 78.
Buckler, J. (1969): *J. clin. Sci., 37,* 765.
Carlson, H., Gillin, J., Gorden, P. and Snyder, F. (1972): *J. clin. Endocr., 34,* 1102.
Cartlidge, N., Black, M., Hall, M. and Hall, R. (1970): *Geront. clin. (Basel), 12,* 65.
Danowski, T., Tsai, T., Morgan, C., Sieracki, J., Alley, R., Robbins, T., Sabeh, G. and Sunder, J. (1969): *Metabolism, 18,* 811.
Daughaday, W. (1974): In: *Textbook of Endocrinology.* Editor: R.H. Williams. W.B. Saunders Company, Philadelphia.
Dudl, R., Ensinck, J., Palmer, H. and Williams, R. (1973): *J. clin. Endocr., 37,* 11.
Eddy, R.L., Jones, A.L., Chakmakjian, Z.H. and Silverthorne, M.C. (1971): *J. clin. Endocr., 33,* 709.
Finkelstein, J., Roffwarg, H., Boyar, R., Kream, J. and Hellman, C. (1972): *J. clin. Endocr., 35,* 665.
Forest, M.G., Rivarola, M.A. and Migeon, C.J. (1968): *Steroids, 12,* 323.
Gershberg, H. (1957): *Endocrinology, 61,* 160.
Hosley, H.F., Olson, K.B., Horton, J., Michelsen, P. and Atkins, R. (1970): *Advanc. automated Anal., 105.*
Kalk, W., Vinick, A., Pimstone, B. and Jackson, W. (1973): *J. Geront., 28,* 431.
Keenan, B.S., Killmer Jr, L.B. and Sode, J. (1972): *Pediatrics, 50,* 760.
Kowarski, A., Thompson, R.G., Migeon, C.J. and Blizzard, R.M. (1971): *J. clin. Endocr., 32,* 356.
Laron, A., Doron, M. and Arnikan, B. (1966): *Med. and Sport, 40,* 162.
McGillivray, M., Frohman, L. and Doe, J. (1970): *J. clin. Endocr., 30,* 632.
Murphy, B.E.P. (1967): *J. clin. Endocr., 27,* 973.
National Academy of Sciences — National Research Council (1974): *Recommended Dietary Allowances.* Washington, D.C.
Pecile, A., Müller, E., Falconi, G. and Martini, L. (1965): *Endocrinology, 77,* 241.
Penny, R., Blizzard, R.M. and Davis, W.T. (1969): *J. clin. Endocr., 29,* 1499.
Root, A. and Oski, F. (1969): *J. Geront., 24,* 97.
Rubens, R., Dhont, M. and Vermeulen, A. (1974): *J. clin. Endocr., 40,* 40.
Rudman, D., Chyatte, S., Patterson, J., Gerron, G., O'Beirne, I., Barlow, J., Ahmann, P., Jordan, A. and Mosteller, R. (1971): *J. clin. Invest., 51,* 1941.
Sandberg, H., Hashimine, N., Maeda, S., Symons, D. and Zarodnick, J. (1973): *J. Amer. Geriat. Soc., 21,* 433.
Schalch, D. and Parker, M. (1964): *Nature (Lond.), 203,* 1141.
Taylor, A., Zinster, J. and Mintz, D. (1969): *J. clin. Invest., 48,* 2349.
Technicon® Auto Analyzer® (1970): *Instruction Manual.* Technicon Instruments Corp., Tarrytown, N.Y.
Thorell, J.I. and Johansson, G.B. (1971): *Biochim. biophys. Acta (Amst.), 251,* 363.
Thompson, R., Rodriquez, A., Kowarski, A., Migeon, C.J. and Blizzard, R.M. (1972): *J. clin. Endocr., 35,* 334.
Varma, M., Varma, R.R., Johanson, A.J., Kowarski, A. and Migeon, C.J. (1975): *J. clin. Endocr., 40,* 868.
Vermeulen, A., Rubens, R. and Verdonck, L. (1972): *J. clin. Endocr., 34,* 730.
Verzar, F. (1966): In: *The Pituitary Gland.* Editors: G.W. Harris and B.T. Donovan. University of California Press, Los Angeles.
Verzar, F. and Spichtin, H. (1966): *Gerontologia, 12,* 48.
Vidalon, C., Khurana, R., Chae, S., Gegick, C., Stephans, T., Nolan, S. and Danowski, T. (1973): *J. Amer. Geriat. Soc., 21,* 253.

PLASMA SOMATOMEDIN. CLINICAL OBSERVATIONS*

J.L. VAN DEN BRANDE

Department of Pediatrics, Erasmus University, Rotterdam, The Netherlands

The success of the treatment of pituitary dwarfs with human growth hormone is another example of the classical lag between the empirical observation and the understanding of the mechanism involved. Many years of research on the mode of action of growth hormone has not entirely clarified the problem. The hypothesis of Salmon and Daughaday (1957) that growth hormone does not exert its growth-promoting action directly but rather through an intermediate agent, which was called first the sulfation factor and more recently somatomedin (SM) (Daughaday et al., 1972), remains as yet unproved.

This review summarizes the data on somatomedin activity in plasma in clinical conditions. Indirectly, the evidence supports the view that SM has a role in normal growth regulation as well as in certain growth disturbances and in the adaptive mechanisms which operate during challenges inflicted upon the organism by the environment.

METHODS

Cartilage bioassays

In their original studies, Salmon and Daughaday (1957) utilized costal cartilage from the hypophysectomized rat for measurement of the incorporation of ^{35}S-sulfate in vitro. Dissatisfaction with the precision of this technique prompted Almqvist (1961) to change the design to eliminate variation between animals. The observation of Daughaday and Reeder (1966) that normal plasma also stimulates DNA replication in rat costal cartilage enabled us to develop a technique for measurement of thymidine and sulfate incorporation in a single assay (Van den Brande et al., 1971). Cartilage from fasted normal rats, which is somewhat less sensitive than that from hypophysectomized rats, has been used (Almqvist, 1961; Yde, 1968; Alford et al., 1972). These changes of design did not much improve the precision of the method. Nevertheless, the assays using rat costal cartilage remain excellent in conditions where sensitivity is of primary importance.

The observations of Adamson and Anast (1966) that serum could stimulate the incorporation of sulfate by pelvic rudiments of chick embryos led to the development of a bioassay using this tissue (Hall, 1970). This assay has a somewhat higher precision than that based on the hypophysectomized rat, but is less sensitive. We have developed a method for measuring SM activity based on the incorporation of ^{35}S-sulfate

---

* Studies by the author mentioned in this review were supported in part by the Dutch Foundation for Medical Research (FUNGO), grant nr. 13-24-12.

and ³H-methyl-thymidine into identical cylindrical tissue fragments of costal cartilage from young pigs (Van den Brande and Du Caju, 1974a). Though less sensitive than the hypophysectomized rat assay, the method has the advantage of being more precise and much less time-consuming for large numbers of samples. Phillips et al., (1974a) have confirmed its utility and have suggested alternative incubation schemes when only sulfate incorporation is measured.

Cartilage from other species can be used for the bioassay of SM. We have used monkey rib cartilage, both to measure semipurified SM preparations (Van Wyk et al., 1971) and SM in whole plasma (Van den Brande et al., 1974). Bala et al. (1975) have described a sensitive and accurate method using rib cartilage from young rabbits.

Finally, the elegant technique of Herbai (1970) in which the sulfation activity of mouse costal cartilage is measured in vivo, should be mentioned.

Bioassays using cultured cells

The stimulation of DNA replication or cell division by SM-rich preparations or pure SMs has been demonstrated for many cell types (Salmon and Hosse, 1971; Westermark, 1971, and others). However, few of these systems have been developed for SM assay. The bioassay based on DNA replication in human glia-like cells (Westermark, 1971) has been used by Uthne (1973) for measuring somatomedin B (SM B). Garland et al. (1972) have used isolated chondrocytes from chicken embryo cartilage. Serum stimulates DNA replication in these cells but the applicability of this system for routine SM bioassay is still uncertain (Daughaday and Garland, 1972).

Recently, Wästeson et al. (1973) have described a method for measuring SM activity using the sulfation of polysaccharides in human fetal lung fibroblasts. This system appears to be highly sensitive to the stimulating effect of normal serum as well as of SM A.

Radioreceptor assays

Based upon the finding that human placental membranes possess highly specific receptors for SM C, Van Wyk and his co-workers have developed a competitive binding assay (Marshall et al., 1974; Van Wyk et al., 1974). Hall et al. (1974) have developed a similar assay for SM A. The application of these relatively simple yet highly specific and sensitive techniques will undoubtedly give a great impetus to SM research.

Radioimmunoassay

Recently, Yalow et al. (1975) described a sensitive and precise radioimmunoassay for SM B.

STUDIES ON PLASMA SOMATOMEDIN LEVELS IN NORMAL INDIVIDUALS

The newborn

All the available evidence indicates low SM activity in cord blood in comparison with that of plasma from normal adults (1 U/ml by definition). There is, however, a large discrepancy between the results of different investigators.

Chesley (1962), using the hypophysectomized rat assay found only a slight lowering.

Andersen et al. (1974) obtained values within the limits of the reference preparation in the chick embryo assay. Using the pig cartilage assay Tato et al. (1975) reported a wide variation between 0.1 and 1.3 U/ml and Hintz et al. (1974) obtained similar data (range 0.22 to 0.94 U/ml). This variation is greater than might be expected from assay variability and other factors must contribute to it.

The maternal values at delivery have been found normal by Chesley (1962), Giordano et al. (1975) (hypophysectomized rat assay) and Andersen et al. (1974) (chick embryo assay). In contrast, both Tato et al. (1975) and Hintz et al. (1974) found low maternal values using the pig cartilage assay. The disparity in the results may reflect differences due to the bioassay as well as the clinical circumstances. It has not been established whether the low SM values in cord blood truly reflect a low SM or are an effect of inhibition. An excess of inhibiting substance is apparently not present, since mixing cord plasma with normal plasma did not lower the total SM activity (Tato et al., 1975). The same authors reported uniformly low values 2-10 hours after birth, and a remarkable return to a normal average of 0.98 (range 0.43-1.55) 4-5 days after birth, before stabilizing at the low value (av. 0.42, range 0.10-0.89) characteristic of infants (Almqvist and Rune, 1961; Van den Brande and Du Caju, 1974b). The significance of these findings is difficult to evaluate since many hormonal changes, for example, in estrogens, cortisone and the thyroid hormones, occur during the immediate postnatal period, and some of these may well influence the results.

Postnatal life

The early studies of Almqvist and Rune (1961) and Kogut et al. (1963) suggested that SM progressively increases during early childhood. We have confirmed this and determined the normal limits for different age groups (Van den Brande and Du Caju, 1974b) (Table 1). The low values in young children have recently been confirmed by Takano and Hall (1975), using their radioreceptor assay for SM A. Studies by Daughaday et al. (1959) have not disclosed any variations with age during adult life. Sex differences have never been recorded.

The remarkable fact that children have the lowest SM levels at the time of their fastest growth is quite intriguing. A possible explanation is that cartilage sensitivity decreases with age, as in rats (Heinz et al., 1970) and rabbits (Beaton and Singh, 1975). For practical purposes it is important to apply a correction factor when comparing patients of different ages.

TABLE 1

SM activity (U/ml) (mean values and SD) in different age groups in normal subjects

	\multicolumn{6}{c}{Age group (years)}
	0-2	2-4	4-6	6-10	10-14	Adults
Mean						
SM	0.54	0.73	0.86	0.99	1.01	1.09
SD	0.10	0.09	0.13	0.12	0.16	0.13
N	7	5	10	14	9	6

TABLE 2

Morning–afternoon difference of plasma SM activity (U/ml) in a random population of schoolchildren

	Total (n = 67)	Morning (n = 46)	Afternoon (n = 21)	P
Uncorrected	1.27 ± 0.43*	1.20 ± 0.41	1.45 ± 0.44	< 0.05
Corrected for plasma cortisol**	1.00 ± 0.31	0.95 ± 0.28	1.09 ± 0.34	n.s.

*Standard deviation. **Individual value divided by the mean of the appropriate class.

Diurnal variation

No systematic variation was noted in 3 normal adults studied by Daughaday et al. (1959). In a random population of 67 schoolchildren aged from 9 to 12 years, we obtained blood either during the morning or the afternoon and found a significantly higher mean value in the afternoon samples (Table 2). There was a negative correlation between plasma SM activity and cortisol levels. An average SM of 1 U/ml was obtained in samples with 12-20 μg% cortisol (Fig. 1). This is in accordance with the normal average for this age group obtained from morning samples. After correction for the difference due to cortisol the morning-afternoon difference vanished (Van den Brande et al., 1975). This observation again stresses the need for standardization of the conditions under which blood samples are obtained. In all our clinical studies, unless otherwise stated, we obtained blood between 8 and 10 a.m. after an overnight fast.

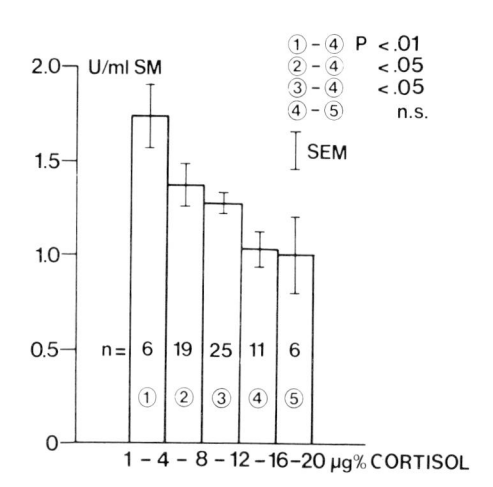

Fig. 1. Relationship between plasma SM activity (U/ml) and cortisol levels in 67 schoolchildren. The data are grouped in classes of equal width. SM by porcine cartilage assay, cortisol by competitive protein binding. (From Van den Brande et al., 1975, by courtesy of the Editor of *Advances in Metabolic Disorders*.)

Plasma somatomedin

Nutritional status

Our early measurements of SM under different conditions gave low values in severely malnourished Dutch children with normal plasma proteins (Du Caju and Van den Brande, 1973). Grant et al. (1973) reported similar findings in patients with kwashiorkor and also correlated plasma protein levels with SM during recovery. In a number of plasma samples from children with marasmic kwashiorkor a heat-labile inhibitor was demonstrated (Van den Brande et al., 1975) which was similar to the inhibitor described in fasting rats by Salmon (1972). After heating, a potency ratio of 0.60 and 0.71 in relation to the standard control was found (Van den Brande et al., 1975). Since the inhibiting material may not all have been destroyed by the heating procedure, as suggested by the shallowness of the slope in the lower region of the curve, these patients appear to have had at least normal SM levels. Teleologically such an inhibitor would appear to provide an ideal protection for the organism in times of prolonged food deprivation, by preventing wastage of energy in the process of growth.

In obesity we have found normal SM levels (Van den Brande and Du Caju, 1974b). More recently Chaussain et al. (1975) have confirmed this. In a few very obese children, however, they found reduced SM activity.

CONDITIONS CHARACTERIZED BY ABNORMAL GH SECRETION

Hypopituitarism

The low SM levels in this condition have been demonstrated by all SM assay techniques. Dose- and time-related changes in plasma SM during acute administration of

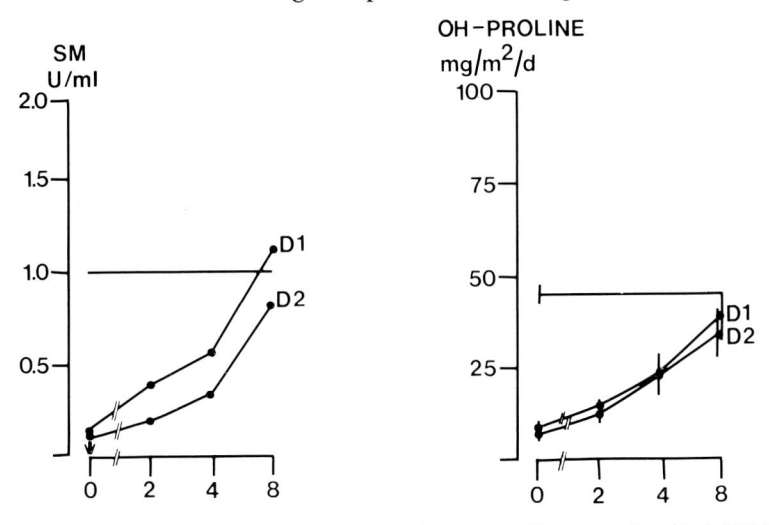

Fig. 2. Plasma SM levels and average total urinary hydroxyproline excretion ($\pm$ 1 SEM) during consecutive 7-day periods of HGH administration at increasing dosages to 2 patients with isolated GH deficiency. SM was measured on days 4 or 5 of each period and is expressed as the potency ratio of the plasma of the patient to the average obtained in plasma of children of the same age (horizontal line). Hydroxyproline values are compared with the average $\pm$ 1 SEM obtained from 9 children of the same age range. HGH was administered at 8 p.m.; the dose is expressed as mg/m²/day in abscissa.

human growth hormone (HGH) have been documented in limited numbers of patients (Daughaday et al., 1959; Almqvist, 1960; Hall, 1971; Van den Brande, 1973).

Hall and Olin (1972) correlated growth velocity and plasma SM during chronic HGH treatment as follows: K = 0.61 × Units SM + 0.008, in which K is an age-independent index of growth velocity. They also found that the relationschip between SM, dose of HGH in mg/kg/week (D) and duration of treatment in years (Y) could be expressed as follows: Units SM = 4.35 × D (1-0.617 ^{10}log Y) + 0.22. This SM level is much higher than those obtained after 4-5 days of daily injection of HGH at different dosages (Van den Brande, 1973).

Taking into account the differences in potency of the preparations used, the doses needed to obtain the same SM levels are 4-5 times higher during the initial days of therapy than after long-term treatment. Although quantitative comparison with the data of Almqvist (1960) is not possible, a similar discrepancy appears to exist. Not only does SM remain unexpectedly low in relation to the dose during short-term GH administration, hydroxyproline excretion, considered a good index of collagen turn-over, follows the SM activity, reaching a normal value only after a very high dose (Fig. 2). The data suggest that in hypopituitarism the target for HGH initially needs larger amounts of this hormone to produce SM. It is also conceivable that the turnover of SM is initially faster, resulting in low SM levels in spite of a normal or high production rate. The low hydroxyproline excretion is surprising. From the known high sensitivity of cartilage in hypophysectomized animals rapid growth and hence high hydroxyproline excretion would be expected. Quantitative studies linking SM, metabolic parameters and growth velocity are needed to explain these apparent discrepancies.

Craniopharyngioma

The enigma that some postoperative craniopharyngioma patients exhibit 'catch-up' growth while having no demonstrable GH in plasma has not been solved. It has been well documented that such children have normal SM activity (Weldon et al., 1972; Finkelstein et al., 1972; Kenny et al., 1973; Van den Brande and Du Caju, 1974*b*).

Since prolactin levels are normal (Kenny et al., 1973) it is unlikely that this hormone is responsible for this phenomenon. It seems more probable (Kenny et al., 1973), that the concomitant obesity of these children plays an important role. Irrespective of its etiology, severe obesity is characterized by low GH levels, hyperinsulinism and a normal to slightly increased growth rate. Our finding of normal SM values in 'simple' obesity completes the parallelism. It is quite possible that the high insulin levels play a role in this condition. This opinion is supported by the recent observation of Phillips et al. (1975) that high doses of insulin stimulate SM release from the perfused liver.

Emotional deprivation short stature

It has been well documented that a number of children with emotional deprivation short stature present with low GH levels which return to normal following admission to hospital. It is under debate whether or not GH deficiency is the cause of their growth failure (Powell et al., 1967; Tanner et al., 1971, and others). In 5 children, with the diagnosis of emotional deprivation short stature on the basis of their history, short stature and spontaneous catch-up growth during hospitalization, we found uniformly low SM levels (Van den Brande and Du Caju, 1974*b*). All other parameters studied varied widely (Table 3). This included the degree of concomitant underweight, as well as the plasma GH and insulin levels.

Plasma somatomedin

TABLE 3

Summary of clinical and laboratory findings in children with emotional deprivation short stature

No.	Sex	Age	Skeletal age	Height (SDS)	Weight (% of median for height)	Growth velocity following admission (cm/year)*	Max. insulin (μU/ml)**	Max. GH (μU/ml)**	Plasma SM (corr. for age) (U/ml)
1	M	1^1	0^6	-2.1	84	12.0	53	8.5	0.50
2	M	1^4	0^6	-3.7	103	11.8	4	40.0	0.30
3	M	1^9	0^9	-3.9	75	11.5	68	10.0	0.09
4	M	8^9	6^3	-4.7	82	13.2	22	5.0	0.31
5	F	8^9	5^0	-4.2	109	8.3	1	14.0	0.45

*Growth velocity is based on measurements during the first month of observation. **Maximum plasma level recorded during an arginine infusion test (0.5 g/kg over 30 minutes i.v.).

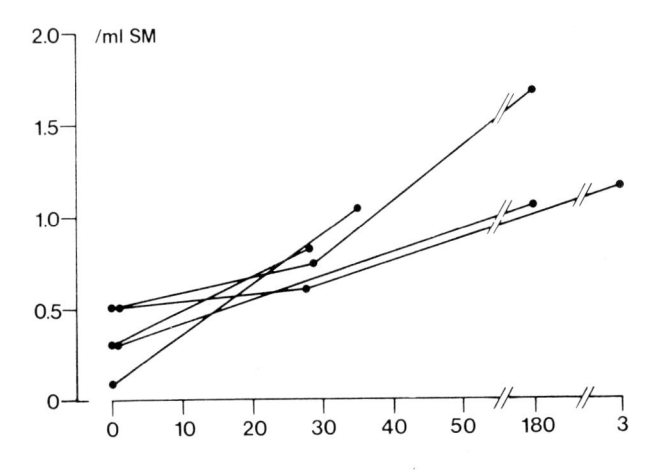

Fig. 3. Plasma SM activity (U/ml) during treatment in hospital and subsequently in a foster home in children with emotional deprivation short stature expressed in abscissa as days after admission. (From Van den Brande et al., 1975, by courtesy of the Editor of *Advances in Metabolic Disorders.*)

The data suggest that a decrease in SM activity rather than in GH production is involved in the failure of growth of these children. Two remarkable phenomena were observed during catch-up growth. One was the slow return to normal of SM in spite of normal GH levels and fast growth. The other was a uniformly low insulin response to arginine over the same period (Table 3, Figs. 3 and 4). The former supports the notion that the SM-deprived tissues are extremely sensitive; the latter, inexplicable in itself, is interesting because it is an exception to the general concordance between insulin levels and growth (Van den Brande et al., 1975).

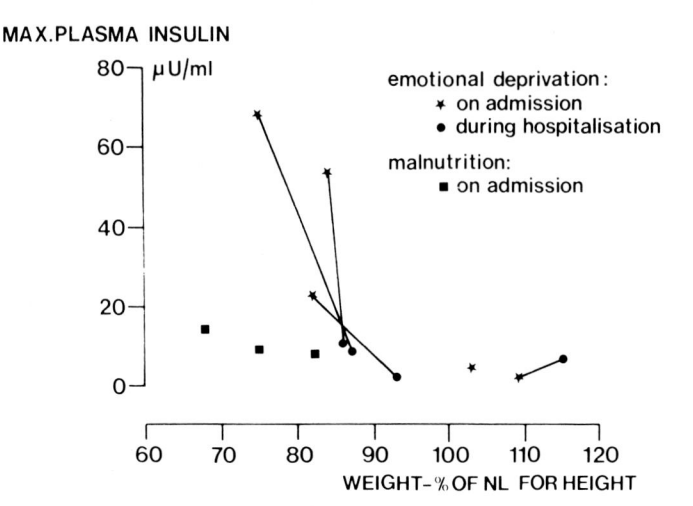

Fig. 4. Relationship between maximal plasma insulin levels during arginine infusion and weight for height in children with emotional deprivation and malnutrition. None of these children had low plasma protein or edema. (From Van den Brande et al., 1975, by courtesy of the Editor of *Advances in Metabolic Disorders.*)

Acromegaly

The mean SM level in untreated acromegalics has invariably been found to be high. Whereas there is no demonstrable relationship with plasma GH levels, a better though not perfect correlation is found with clinical parameters of activity (Daughaday et al., 1959; Takano et al., 1975). SM levels by bioassays are lower than by receptor assay (Takano et al., 1975). This discrepancy remains yet unexplained.

Decreased GH responsiveness

Patients with short stature presenting with the full clinical picture of hyposomatotropism, but with high immunoreactive GH, were first described by Laron et al. (1966) and subsequently by others. They have insulinopenia and their SM levels are invariably low. Acute GH administration fails to induce any important changes in the metabolic indices of GH effect. Chronic treatment does not significantly affect the insulinopenia, the SM levels or the endogenous GH secretion, and growth velocity is barely, if at all, increased (Merimee et al., 1968; Elders et al., 1971; Laron et al., 1971; Najjar et al., 1971; Tanner et al., 1971; New et al., 1972; Van den Brande et al., 1974; Kastrup et al., 1975; Takano et al., 1975). Neither immunologically nor physicochemically does the GH molecule of these patients appear to differ from normal (Elders et al., 1973; Van den Brande et al., 1974). A peripheral specific diminished responsiveness to GH seems the most likely cause of the disorder.

OTHER ENDOCRINE DISTURBANCES

Hypothyroidism

It has been known for a long time that thyroxine, when injected into hypophysectomized rats, stimulates sulfation in costal cartilage, whereas its direct addition to the incubation medium does not (Murphy et al., 1956; Salmon and Daughaday, 1957). Audhya and Gibson (1975) demonstrated that L-triiodothyronine and to a lesser degree L-thyroxine stimulate sulfation in chick embryo cartilage in vitro only in the presence of a non-pituitary dependent factor in serum. We have reported a rather wide range of SM activity in the plasma of hypothyroid children, averaging 0.82 U/ml (n = 7) which increased on treatment to 1.30 (n = 6) (Van den Brande and Du Caju 1974b). Recently Takano et al. (1975) reported normal values in untreated hypothyroid patients using the SM A radioreceptor technique. It appears, therefore, that SM production is unimpaired in most hypothyroid patients and that the increased values after treatment are due to the direct effect of thyroid hormone in the bioassay.

Hypoinsulinism

Yde (1964) was the first to report slightly decreased SM activity in diabetics. Lecornu (1973) reported an average of 0.43 U/ml with some increase during treatment (average 0.56 U/ml). In 5 untreated juvenile diabetics we found values in the low-normal range (Van den Brande and Du Caju, 1974b). The data need to be extended, but they support the view that insulin is not an absolute prerequisite for SM production.

Hypercorticism

The suppressive effect of glucocorticoids on sulfation in cartilage was first demonstrated by Layton (1951). Recently Tessler and Salmon (1975) showed that the inhibition of sulfate incorporation by cartilage from normal rats, both in vitro and in vivo, is related to the glucocorticoid activity of the substance used. Phillips et al. (1974b) concluded from their experiments with hypophysectomized rats that glucocorticoids in physiological concentrations may not interfere with SM action but rather depress SM generation.

Elders et al. (1973), studying a patient on alternate day treatment with prednisone, came to a similar conclusion. We recorded low SM activity in a patient with Cushing's syndrome (0.41 U/ml) normalizing during catch-up growth (0.95 U/ml) (Van den Brande and Du Caju, 1974b), a finding similar to those of Lecornu (1973). The negative correlation between SM and plasma cortisol in a random population of children mentioned above further suggests that cortisol may have an important modulating effect on plasma SM.

Influence of sex hormones

Phillips et al. (1974b) have demonstrated that testosterone affects neither in vitro incorporation of sulfate in pig and chick cartilage, nor SM generation in hypophysectomized rats. This is in agreement with our observation in 3 boys: neither testosterone treatment nor spontaneous puberty had any influence on their SM levels (Van den Brande and Du Caju, 1974b).

Estradiol appears to have no direct effect on sulfation of cartilage from hypophysectomized rats, chick embryos or young pigs (Wiedemann and Schwartz, 1972; Phillips et al., 1974b). In contrast, in the rat as well as in man SM activity is depressed by treatment with estradiol. Wiedemann and Schwartz (1972) showed such depression in estrogen-treated acromegalics and also demonstrated that the GH-induced rise of SM in hypopituitary adults was prevented by concomitant treatment with ethinylestradiol. They made the interesting observation that in GH-treated hypopituitary patients only the increases in SM levels, calciuria and, perhaps, hydroxyprolinuria were prevented by estradiol while the effect on N retention was unimpaired. They concluded from this that the metabolic effects of circulating SM may be limited to the supporting tissues.

NONENDOCRINE DISORDERS

Constitutionally short and tall stature

In view of the absence of hormonal deviations, morphological abnormalities or any other obvious disorders in such children, they are mostly regarded as at the extremes of the normal variation.

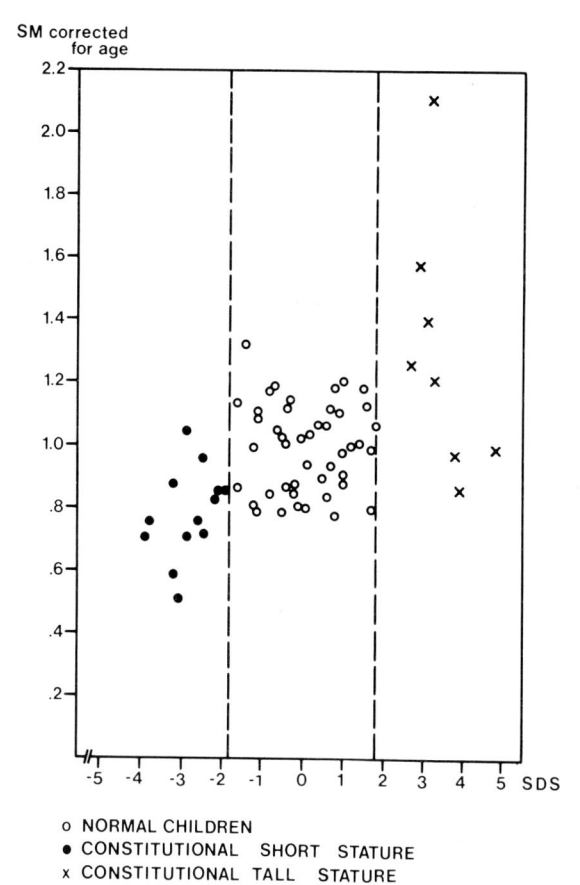

Fig. 5. Relationship between plasma SM (U/ml) and standard deviation score (SDS) for height in normals and constitutionally short and tall children.

We have shown that SM values in short children are slightly but significantly less than normal (0.80 ± 0.15 U/ml), whereas the opposite is true of tall children (1.34 ± 0.41 U/ml) (Van den Brande and Du Caju, 1974b). This is even clearer when the data are plotted as a function of SDS for height (SDS = standard deviation score = (observed height-mean height)/SD)) (Fig. 5). Similar observations in respect of short children were made by Lecornu (1973). An indication that SM activity is correlated with growth velocity in normal children had already been obtained by Hall (1972). More recently Hall and Filipson (1975) established a nice overall correlation with growth velocity at identical dental age in children with various growth disorders, including the constitutionally short and tall. It thus appears that in these children growth velocity is determined by the circulating SM activity rather than by the amount of GH available to the SM-producing tissues.

Short dysmorphic children (so-called 'primordial dwarfism')

This is a heterogeneous group of children with minor or major congenital deformities and short stature, in which, in contrast to the former group, there is practically no positive family history for short stature. They are usually small at birth; frequently they have only slightly retarded skeletal age and height prognosis is poor. We found their SM values to be low with a rather wide variation. We also observed that in the afternoon such children have normal SM values (Van den Brande and Du Caju, 1974b). However, since normal children also significantly increase their levels in the afternoon, the dysmorphic children appear simply to exhibit a parallel change, remaining far below the normals. When their SM activities are plotted against SDS for height or against growth velocity, or just corrected for body surface area, they always fall below the range obtained for constitutionally short children (Du Caju and Van den Brande, in preparation). It is puzzling that such children appear to attain the same height and exhibit the same growth velocity as constitutionally short children with less SM activity in their plasma.

By definition, patients with Turner's syndrome belong to this group. Over the years many authors have reported on SM activity in such patients. Their figures have varied from low (Wright et al., 1965; Lecornu, 1973) to very high (Daughaday et al., 1969). It is difficult to interpret these data since age differences and time of sampling may have influenced the results. Recently Takano et al. (1975) have reported normal SM A levels by radioreceptor assay in 12 patients between the age of 10 and 30 years, but time of sampling was not mentioned. It is therefore not possible at present to conclude whether or not Turner patients differ from the other dysmorphic short children.

Achondroplasia

We have reported high SM levels in afternoon samples from 2 patients with achondroplasia (Van den Brande and Du Caju, 1974b). However, in fasting morning samples from 7 children values were within the normal range (Du Caju and Van den Brande, in preparation). Relative to the slow growth of these patients the values are high in comparison with those in constitutionally short children, suggesting a normal SM production and a diminished target organ responsiveness.

Cerebral gigantism

This syndrome is characterized by gigantism, a peculiar pear-shaped head and nonprogressive neurological disturbances. Only 2 reports have been forthcoming on SM in this rare condition. In both low SM values were recorded (Du Caju and Van den Brande, 1973; Lecornu, 1973). It is very tempting to interpret this remarkable dissociation between SM and growth rate as suggesting high sensitivity to SM and either abnormally fast turnover or low production of SM. However, the participation of an inhibitor in the total SM activity has not yet been excluded.

Renal insufficiency

The high levels of SM A measured by radioreceptor assay by Takano et al. (1975) strongly contrast with the low SM activity demonstrated by Saenger et al. (1974), Stuart et al. (1974) and Bala et al. (1975) by bioassay. This discrepancy suggests the presence of an inhibiting factor. Bożovíc et al. (1975) have recently reported the existence of such an inhibitor in serum from uremic patients. Bala et al. (1975) found no change in SM activity after hemodialysis in contrast to Stuart et al. (1974) who reported a rise. The view that the presence of growth-inhibiting material, rather than a fall in SM proper, is responsible for the slow growth of children with renal failure is also in keeping with the observation that nephrectomy does not reduce SM activity in man (Stuart et al., 1974).

Liver cirrhosis

Low SM levels in patients with cirrhosis of the liver were found by Wu et al. (1974), Stuart and Lazarus (1975) and Bala et al. (1975). This is in agreement with McConaghey's finding, amply confirmed, that the liver produces an SM-like material (McConaghey and Sledge, 1970; Hall and Uthne, 1971; Uthne and Uthne, 1972; Hintz et al., 1972, and others). However, whether or not low SM activity in plasma represents low SM proper or is due to the presence of an inhibiting substance remains to be established.

DISCUSSION

In spite of the wide variety of bioassays used, the data have been surprisingly consistent. This is particularly true in those conditions in which large changes of SM activity are found, such as GH deficiency, acromegaly or malnutrition. Smaller deviations from normal may have been obscured in the past by the imprecision of the techniques used.

Two further sources of variability, age-dependent changes and the cortisol-related diurnal fluctuations in children when using the pig assay should be considered when interpreting data. It is not at present known whether such diurnal changes will also become apparent with other bioassays.

While the bioassay is valuable for measuring total SM-like activity, it does not distinguish between the different components. More specific assay systems should greatly facilitate this much needed discrimination. In spite of these limitations some tentative conclusions may be formulated.

The clinical data have not revealed any situations which are incompatible with the hypothesis of Salmon and Daughaday (1957). Growth hormone appears to be the main regulator of SM levels, which in most instances are nicely correlated with growth velocity. There are, however, some exceptions to this general rule. In some situations the relationship between GH and SM is disturbed. In the newborn, GH levels are high while SM is low, and in the first 6 years of life GH remains constant while SM levels increase. While the former appears to favor the existence of a feedback influence of SM on GH secretion as suggested by Hintz et al. (1974), the latter appears to argue against it. So does the variation of SM with growth velocity in normal children and without concomitant variation in maximal GH levels on stimulation (Hall, 1972; Du Caju and Van den Brande, in preparation). However, the chronicity of the situation may blunt all visible feedback signals. The association of high GH levels in kwashiorkor with low SM activity similarly represent no argument for a feedback mechanism — which then should be effectuated through an inhibitor — since marasmic children with normal or low GH have equally low SM (Du Caju and Van den Brande, 1973). The situation in chronic renal failure and cirrhosis of the liver, both of which are characterized by high GH and low SM activity should be considered with caution. In renal failure SM A has been shown to be normal and hence the feedback if present should be accomplished by an inhibitor rather than by SM proper. Secondly, so many profound metabolic disturbances accompany both conditions that GH could be high through a quite different pathway. In conclusion, there is at present no strong argument in favor of a direct influence of SM on GH secretion.

Another intriguing question is whether the mass of liver tissue involved in protein-synthetic and other 'anabolic' functions is related to the amount of SM found in plasma, as suggested by the liver regeneration experiments of Uthne and Uthne (1972). Such a relationship is also compatible with the variations of SM with age and with body size, and the findings in cirrhosis of the liver. It is conceivable that the production of protein carriers, in parallel with the general protein-synthetic activity of the liver, specifically constitutes the limiting factor, by protecting only the protein-bound SM against early destruction. It is therefore possible that any substance which can increase protein synthesis, such as insulin or prolactin, or decrease it, such as high doses of estrogen or cortisone, may similarly influence plasma SM.

SM levels and growth, be it metabolic indices or actual growth in length, follow, each other closely in most conditions. In a few situations the correlation fails. One of them is emotional deprivation, in which fast growth is accompanied by low SM activity in plasma. A similar situation may exist in the initial phase of GH administration to hypopituitary patients, but there the relationship between metabolic parameters and growth velocity remains to be established. A third such situation is found in cerebral gigantism. The normal SM values do not correlate with the growth failure in achondroplasia which represents an opposite situation. Differences in sensitivity to SM might explain the apparent discrepancies. The question should be raised whether a varying rate of SM disappearance from plasma due to differences in binding at the peripheral sites contributes to the actual SM levels found, or whether other mechanisms are involved.

From this review it has also become apparent that both stimulatory (thyroid hormones) and inhibitory substances (in malnutrition, probably in renal failure, and perhaps cortisol) change the total SM activity in plasma and probably influence its effect on the target. How much of the overall regulation of growth is due to the influence of such peripheral modulators rather than to SM proper remains to be established.

ACKNOWLEDGMENTS

Thanks are due to C. Hoogerbrugge, A. van Male and A. van Rooyen for their technical assistance and to Mrs. M. de Bruijne, Mrs. M. Vooys and the audiovisual center of the Sophia Children's Hospital, Rotterdam, for help in the preparation of this manuscript.

REFERENCES

Adamson, L.F. and Anast, C.S. (1966): *Biochim. biophys. Acta (Amst.), 121,* 10.
Alford, F.P., Bellair, J.T., Burger, H.G. and Lovett, N. (1972): *J. Endocr., 54,* 365.
Almqvist, S. (1960): *Acta endocr. (Kbh.), 35,* 381.
Almqvist, S. (1961): *Acta endocr. (Kbh.), 36,* 31.
Almqvist, S. and Rune, I. (1961): *Acta endocr. (Kbh.), 36,* 566.
Andersen, H.J., Kastrup, K.W. and Lebech, P.E. (1974): *Acta paediat. scand., 63,* 328.
Audhya, T.K. and Gibson, K.D. (1975): *Proc. nat. Acad. Sci. (Wash.), 72,* 604.
Bala, R.M., Hankins, C. and Smith, G.R. (1975): *Canad. J. Physiol. Pharmacol., 53,* 403.
Beaton, G.R. and Singh, V. (1975): *Pediat. Res., 9,* 683.
Božović, Lj.,Boström, H. and Bozóvić, M. (1975): *Acta endocr. (Kbh.), Suppl. 199,* 177.
Chaussain, J.L., Binet, E., Schlumberger, A. and Job, J.C. (1975): *Pediat. Res., 9,* 667.
Chesley, L.C. (1962): *Amer. J. Obstet. Gynec., 84,* 1075.
Daughaday, W.H. and Garland, J.T. (1972): In: *Growth and Growth Hormone,* p. 168. Editors: A. Pecile and E.E. Müller. ICS 244, Excerpta Medica, Amsterdam.
Daughaday, W.H., Hall, K., Raben, M.S., Salmon, W.D., Van den Brande, J.L. and Van Wyk, J.J. (1972): *Nature (Lond.), 235,* 107.
Daughaday, W.H., Laron, Z., Pertzelan, A. and Heins, J.N. (1969): *Trans. Ass. Amer. Phycns, 82,* 129.
Daughaday, W.H. and Reeder, C. (1966):*J. Lab. clin. Med., 68,* 357.
Daughaday, W.H., Salmon, W.D. and Alexander, F. (1959):*J. clin. Endocr., 19,* 743.
Du Caju, M.V.L. and Van den Brande, J.L. (1973): *Acta paediat. scand., 62,* 96.
Elders, M.J., Garland, J.I., Daughaday, W.H., Fisher, D.A. and Hughes, E.R. (1971): *Pediat. Res., 5,* 398.
Elders, M.J., Wingfield, B.S., McNatt, L.M. and Hughes, E.R. (1973): *Pediat. Res., 7,* 324.
Finkelstein, J.W., Kream, J., Ludan, A. and Hellman, L. (1972):*J. clin. Endocr., 35,* 13.
Francis, M.J.O. and Hill, D.J. (1975): *Nature (Lond.), 255,* 167.
Garland, J.T., Lottes, M.E., Kozak, S. and Daughaday, W.H. (1972): *Endocrinology, 90,* 1086.
Giordano, G., Foppiani, E., Minuto, F. and Perroni, D. (1975): *Acta endocr. (Kbh.), Suppl. 199,* 118.
Grant, D.B., Hambley, J., Becker, D. and Pimstone, B.L. (1973): *Arch. Dis. Childh., 48,* 596.
Hall, K. (1970): *Acta endocr. (Kbh.), 63,* 338.
Hall, K. (1971): *Acta endocr. (Kbh.), 66,* 491.
Hall, K. and Filipson, R. (1975): *Acta endocr. (Kbh.), 78,* 234.
Hall, K. and Olin, P. (1972): *Acta endocr. (Kbh.), 69,* 417.
Hall, K., Takano, K. and Fryklund, L. (1974):*J. clin. Endocr., 39,* 973.
Hall, K. and Uthne, K. (1971): *Acta med. scand., 190,* 137.
Heinz, J.N., Barland, J.T. and Daughaday, W.H. (1970): *Endocrinology, 87,* 688.
Herbai, G. (1970): *Acta physiol. scand., 80,* 470.
Hintz, R.L., Clemmons, D.R. and Van Wyk, J.J. (1972): *Pediat. Res., 6,* 353.
Hintz, R.L., Seeds, J.M. and Johnsonbaugh, R.E. (1974): *Pediat. Res., 8,* 369.
Kastrup, K.W., Andersen, H. and Hanssen, K.F. (1975): *Acta paediat. scand., 64,* 613.
Kenney, F.M., Guyda, H.J., Wright, J.C. and Friesen, H.G. (1973):*J. clin. Endocr., 36,* 378.
Kogut, M.D., Kaplan, S.A. and Shimizu, C.S.N. (1963): *Pediatrics, 31,* 538.
Laron, Z., Pertzelan, A., Karp, M., Kowadlo-Silbergeld, A. and Daughaday, W.H. (1971): *J. clin. Endocr., 33,* 332.
Laron, Z., Pertzelan, A. and Mannheimer, S. (1966): *Israel J. med. Sci., 2,* 152.
Layton, L.L. (1951): *Proc. Soc. exp. Biol. (N.Y.), 76,* 596.
Lecornu, M. (1973): *Arch. franç. Pédiat., 30,* 595.
Marshall, R.N., Underwood, L.E., Voina, S.J., Foushee, D.B. and Van Wyk, J.J. (1974): *J. clin. Endocr., 39,* 283.
McConaghey, P. and Sledge, C.B. (1970): *Nature (Lond.), 225,* 1249.
Merimee, T.J., Hall, J., Rabinowitz, D., McKusick, V.A. and Rimoin, D.L. (1968): *Lancet, 2,* 191.
Murphy, W.R., Daughaday, W.H. and Hartnett, C. (1956):*J. Lab. clin. Med., 47,* 715.

Najjar, S.S., Khachadurian, A.K., Ilbawi, M.N. and Blizzard, R.M. (1971): *New Engl. J. Med., 284,* 809.
New, M.I., Schwartz, E., Parks, G.A., Landey, S. and Wiedemann, E. (1972): *J. Pediat., 20,* 620.
Phillips, L.S., Herington, A.C. and Daughaday, W.H. (1974a): *Endocrinology, 94,* 856.
Phillips, L.S., Herington, A.C. and Daughaday, W.H. (1974b): In: *Advances in Human Growth Hormone Research Symposium, Baltimore 1973,* p. 50. Editor: S. Raiti. DHEW Publication No. (NIH)74-612. Government Printing Office, Washington, D.C.
Phillips, L.S., Herington, A.C., Mueller, M.C. and Daughaday, W.H. (1975): In *Program, 51st Annual Meeting of the Endocrine Society,* p. 116.
Powell, G.F., Brasel, J.A., Raiti, S. and Blizzard, R. (1967): *New Engl. J. Med., 276,* 1279.
Saenger, P., Wiedemann, E., Schwartz, E., Korth-Schutz, S., Lewy, J.E., Riggio, R.R., Rubin, A.L., Stenzel, K.H. and New, M.I. (1974): *Pediat. Res., 8,* 163.
Salmon, W.D. (1972): In: *Growth and Growth Hormone,* p. 180. Editors: A. Pecile and E.E. Müller. ICS 244, Excerpta Medica, Amsterdam.
Salmon, W.D. and Daughaday, W.H. (1957): *J. Lab. clin. Med., 49,* 825.
Salmon, W.D. and Hosse, B.R. (1971): *Proc. Soc. exp. Biol. (N.Y.), 136,* 805.
Stuart, M.C. and Lazarus, L. (1975): *Med. J. Aust., 1,* 216.
Stuart, M.C., Lazarus, L. and Hayes, J. (1974): *ICRS med. Sci., 2,* 1102.
Takano, K., Hall, K., Ritzén, M., Iselius, L. and Sievertsson, H. (1975): In: *Studies on Somatomedin A with Special Reference to its Measurement by a Radioreceptor Assay.* Thesis, Kabi, Stockholm.
Tanner, J.M., Whitehouse, R.H., Hughes, P.C.R. and Vince, F.P. (1971): *Arch. Dis. Childh., 46,* 745.
Tato, L., Du Caju, H.V.L., Prévôt, C. and Rappaport, R. (1975): *J. clin. Endocr., 40,* 534.
Tessler, R.H. and Salmon, W.D. (1975): *Endocrinology, 96,* 898.
Uthne, K. (1973): *Acta endocr. (Kbh.), Suppl. 175.*
Uthne, K. and Uthne, T. (1972): *Acta endocr. (Kbh.), 71,* 255.
Van den Brande, J.L. (1973): *Plasma Somatomedin. Studies on some of its Characteristics and on its Relationship with Growth Hormone.* Thesis. Gemeentedrukkerij, Rotterdam.
Van den Brande, J.L. and Du Caju, M.V.L. (1974a): *Acta endocr. (Kbh.), 75,* 233.
Van den Brande, J.L. and Du Caju, M.V.L. (1974b): In: *Advances in Human Growth Hormone Research Symposium, Baltimore 1973,* p. 98. Editor: S. Raiti. DHEW Publication No. (NIH)74-612. Government Printing Office, Washington, D.C.
Van den Brande, J.L., Du Caju, M.V.L., Visser, H.K.A., Schopman, W., Hackeng, W.H.L. and Degenhart, H.J. (1974): *Arch. Dis. Childh., 49,* 297.
Van den Brande, J.L., Van Buul, S., Heinrich, U., Van Roon, F., Zurcher, T. and Van Steistegem, A.C. (1975): *Advanc. metab. Disorders, 8,* 171.
Van den Brande, J.L., Van Wyk, J.J., Weaver, R.P. and Mayberry, H.E. (1971): *Acta endocr. (Kbh.), 66,* 65.
Van Wyk, J.E., Underwood, L.E., Hintz, R.L., Voina, S.J. and Weaver, R.P. (1974): *Recent Progr. Hormone Res., 30,* 259.
Van Wyk. J.J.. Van den Brande, J.L. and Weaver, R.P. (1971): *J. clin. Endocr., 32,* 389.
Wästeson, Å., Uthne, K. and Westermark, B. (1973): *Biochem. J., 136,* 1069.
Weldon, V.V., Jacobs, L.S., Pagliara, A.S. and Daughaday, W.H. (1972): In: *Abstracts, IV International Congress of Endocrinology, Washington 1972,* Abstr. No. 40. ICS 256, Excerpta Medica, Amsterdam.
Westermark, B. (1971): *Exp. Cell Res., 69,* 259.
Wiedemann, E. and Schwartz, E. (1972): *J. clin. Endocr., 34,* 51.
Wright, J.C., Brasel, J.A., Aceto, I., Finkelstein, J.W., Kenney, F.M., Spaulding, J.S. and Blizzard, R.M. (1965): *Amer. J. Med., 38,* 499.
Wu, A., Grant, D.B., Hambley, J. and Levi, A. (1974): *Clin. Sci. metab. Med., 47,* 359.
Yalow, R.S., Hall, K. and Luft, R. (1975): *J. clin. Invest., 55,* 127.
Yde, H. (1964): *Lancet, 2,* 624.
Yde, H. (1968): *Acta endocr. (Kbh.), 57,* 557.

INTERRELATIONS OF THE EFFECTS OF GROWTH HORMONE AND TESTOSTERONE IN HYPOPITUITARISM

M. ZACHMANN, A. AYNSLEY-GREEN and A. PRADER

Department of Pediatrics, University of Zurich, Zurich, Switzerland

The multiple interactions between growth hormone and androgens are of physiological importance in normal subjects, but also have clinical and therapeutic implications in patients with growth hormone (GH) deficiency. In animal experiments, it was shown many years ago that the effectiveness of exogenous testosterone depends on an intact pituitary function (Simpson et al., 1944) and that in hypophysectomized rats, testosterone does not increase the growth of any tissue except the accessory sexual muscles (Scow and Hagan, 1965). In normal human males, several authors have observed an enhancing effect of testosterone on the maximum GH levels after various stimuli (Deller et al., 1966; Martin et al., 1968; Illig and Prader, 1970).

Recently, we have shown that in boys with pituitary insufficiency testosterone and anabolic steroids are less active than in normal boys, not only in accelerating growth but also in stimulating the development of the secondary sex characteristics.

We would now like to discuss three aspects of the testosterone effect and its modification by GH: the growth-accelerating effect, the androgenic effect, and the metabolic effect.

GROWTH PROMOTION AS MODIFIED BY HGH

The growth response to androgens in hypopituitary patients varies individually. Some GH-deficient patients respond quite well to androgens alone, but others not at all. Figure 1 is the growth velocity chart of a girl with isolated partial GH deficiency who showed a marked response. The first peak of 9.5 cm per year was due to the anabolic steroid. The second peak, induced by HGH, was only slightly greater than the first one. The difference, of course, which is not evident from the Figure, was due to the effect on bone maturation, which was faster than normal during the first phase of treatment, but normal with HGH. We made this observation many years ago; at present we would not consider prepubertal treatment with anabolic steroids in hypopituitarism. Such a good response to an anabolic steroid is an exception. In the majority of GH-deficient patients the response of growth and bone maturation is absent or small. In a patient with operated craniopharyngioma, for example (Fig. 2) anabolic steroid was started at an age of about 11 years and had no effect on growth velocity, while the addition of HGH resulted in a marked response with a peak of 9 cm per year.

It is our impression that GH-deficient patients, both boys and girls, in general show a much smaller acceleration in growth and bone age in response to androgens than do patients with a normal endogenous GH production. The differences between individual patients are probably due to different degrees of GH deficiency. To investigate

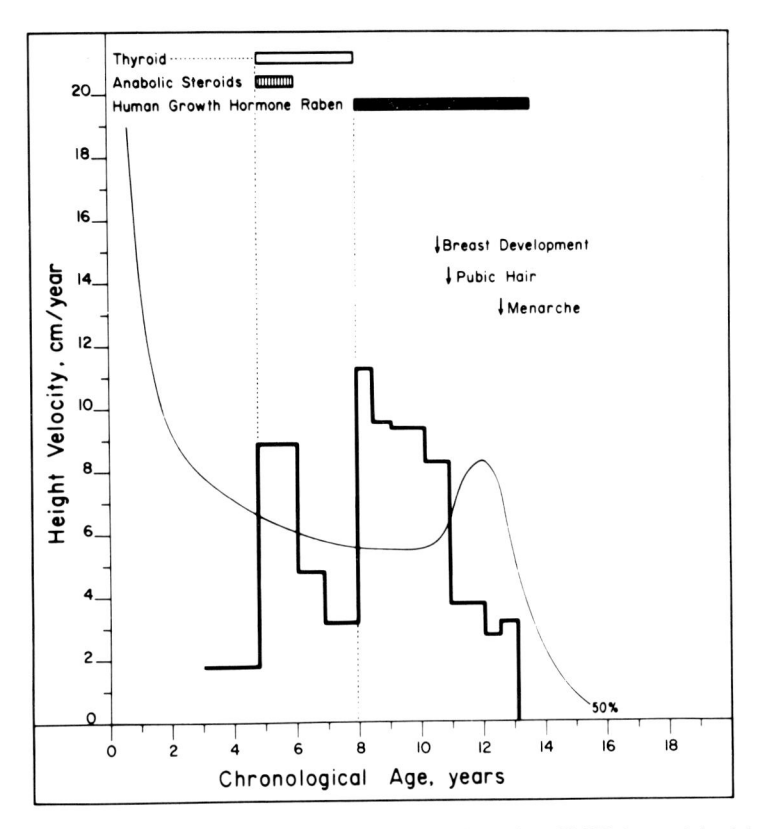

Fig. 1. Growth velocity in response to anabolic steroids and to HGH in a girl with isolated partial GH deficiency.

this further, we have studies the growth pattern in a relatively large group of male patients with pubertal bone age treated with a long-acting preparation, a mixture of testosterone propionate, valerate and undecylenate, administered monthly by intramuscular injection in a replacement or somewhat higher dose. Our aim was to compare the growth response to testosterone in boys having a normal GH production with that in boys suffering from GH deficiency on and off HGH treatment. We could not make a similar analysis for estrogen-treated girls, because of the lower frequency of GH deficiency.

The *number of patients* treated with testosterone was 41. Fifteen had normal GH secretion, 6 of them suffering from isolated gonadotropin deficiency and 9 from congenital anorchia. Twenty-six had a GH and a gonadotropin deficiency, of which 14 were on testosterone alone and 12 on simultaneous HGH treatment. The bone age at the start of treatment was similar in all groups and ranged from 12.5 to 13.5 years as measured by the method of Greulich and Pyle (1959). This is the stage when the pubertal growth spurt starts in normal boys. The control group consisted of 15 normal untreated boys with comparable bone age from the Zurich longitudinal growth study.

For purposes of comparison, we have calculated the *ratio between observed and expected growth velocity for bone age*. For expected growth velocity, we took the standards of Tanner et al. (1966). It was necessary to calculate these ratios for bone age

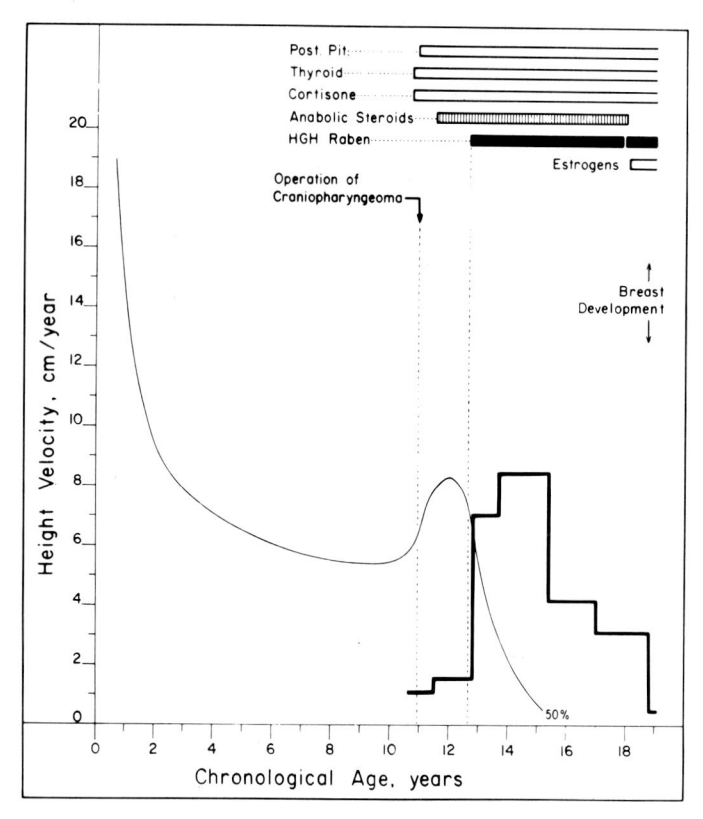

Fig. 2. Growth velocity in response to anabolic steroids and to anabolic steroids + HGH in a girl with operated craniopharyngioma.

rather than chronological age, because many patients with gonadotropin and GH deficiency showed a large difference between bone and chronological age.

This ratio should be 1 or near 1 in normal subjects throughout puberty as well as in patients who lack spontaneous puberty, but have normal GH secretion and adequate testosterone replacement therapy. If our hypothesis that there is a diminished response to testosterone in GH deficiency without HGH treatment is correct, the ratio should be below 1 in patients treated with testosterone alone and 1 or more in patients treated with both testosterone and HGH.

The *untreated normal boys* had a ratio very close to 1 throughout the whole observation period (Fig. 3). This shows that Tanner's standards are valid also for Swiss boys. The 0 point was centered to the bone age of 13.2 years, which was the mean bone age of our patients when testosterone treatment was started.

In *isolated gonadotropin deficiency* (Fig. 4) the ratio before testosterone treatment was only 0.5. This was because the mean bone age at the start of treatment was 13.5 years and there was no spontaneous pubertal growth spurt, which normally starts at a bone age of about 12 years. During the first 6 months of treatment the ratio was 1.37 or above normal, probably because the mean testosterone dose (218 mg/m^2/month) was more than that for physiological replacement therapy in early puberty. After that the ratio became close to 1.

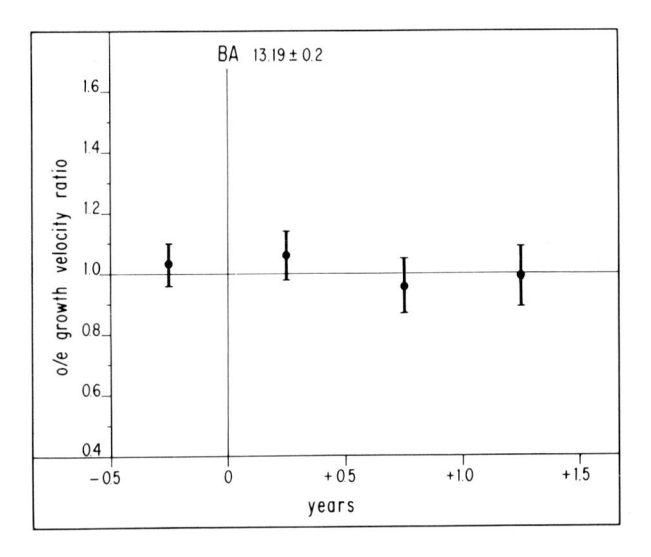

Fig. 3. Observed to expected (o/e) growth velocity ratios in 15 normal untreated boys (for explanations see text).

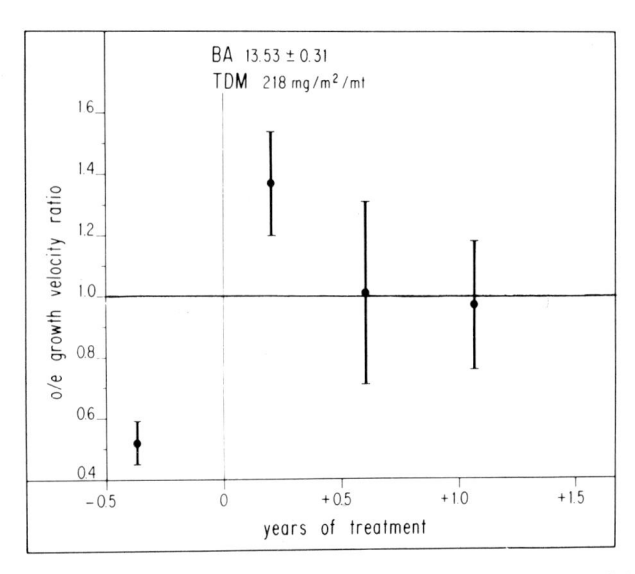

Fig. 4. o/e growth velocity ratios in 6 patients with isolated gonadotropin deficiency before and after testosterone treatment (TDM = mean monthly testosterone dose, BA = bone age at start of treatment).

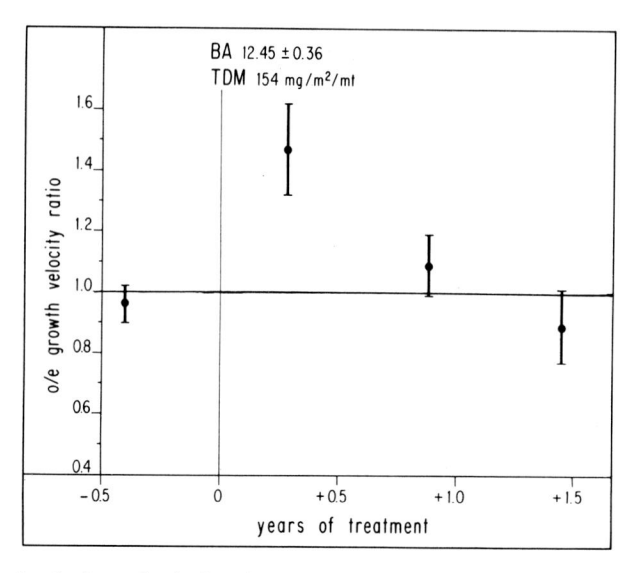

Fig. 5. o/e growth velocity ratios in 9 patients with congenital anorchia.

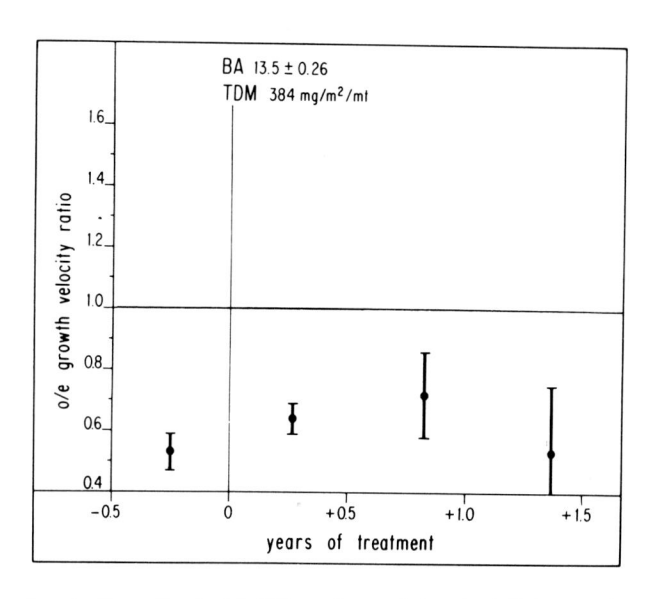

Fig. 6. o/e growth velocity ratios in 14 GH- and gonadotropin-deficient patients not receiving HGH.

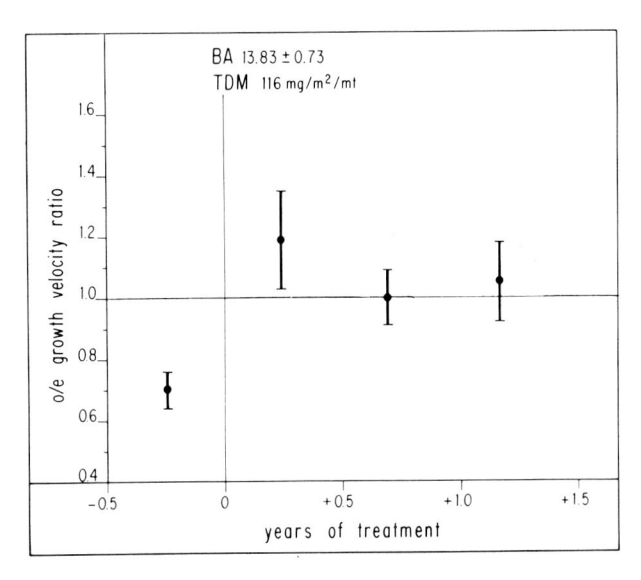

Fig. 7. o/e growth velocity ratios in 12 GH- and gonadotropin-deficient patients on HGH.

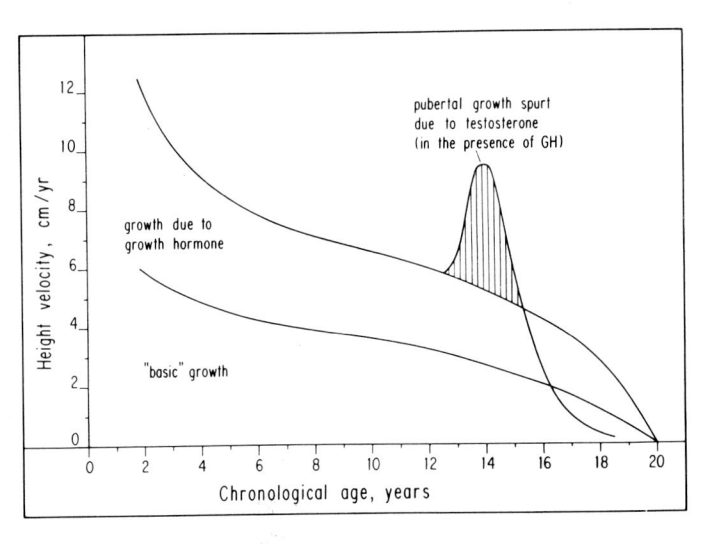

Fig. 8. Schematic presentation of the contribution of 'basic' growth, GH and testosterone to the male pubertal growth spurt (see text).

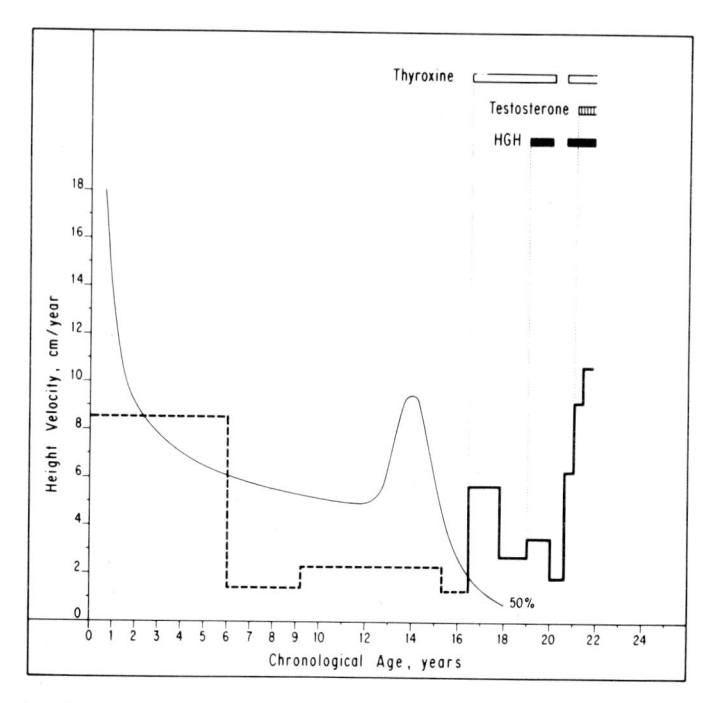

Fig. 9. Growth velocity in response to thyroxin, thyroxin + HGH, thyroxin + HGH + testosterone in a patient with isolated GH, TSH and gonadotropin deficiency.

In *congenital anorchia* (Fig. 5) the results were similar. However, the pretreatment ratio was normal, because in these patients the condition had already been known at a prepubertal bone age and treatment was started earlier at a mean bone age of 12.4 years. During the first 6 months the ratio was again above normal with a value of 1.47, while with subsequent treatment it levelled off close to 1.

In the patients with *combined GH and gonadotropin deficiency receiving testosterone treatment alone* (Fig. 6) the results were quite different. As expected, the ratio was very low before treatment. But even with testosterone at a dose much higher than in the previous 2 groups (384 mg/m²/month) it increased only very little or not at all, indicating that testosterone has a very poor growth-stimulating effect in the absence of GH.

By contrast, a very good response was observed in the *GH- and gonadotropin-deficient patients on HGH treatment* (Fig. 7). Before testosterone, but on HGH alone, the ratio was somewhat higher than in the patients without HGH, but still much below normal. On HGH and testosterone it increased to 1.19 during the first 6 months and then remained at 1. Under HGH treatment, these patients thus responded in a similar way to those with normal GH secretion. This excellent result was achieved even though the dose of testosterone (116 mg/m²/month) was less than half of that given to the patients not receiving HGH.

In summary, these results show that in GH deficiency testosterone stimulates growth much less than in subjects with normal GH secretion. Evidently, the normal pubertal growth spurt in boys is due not only to the enhanced production of testosterone but also to a synergistic action between testosterone *and* GH. Schematically this could be presented in the following way (Fig. 8). Before puberty, the normal growth rate is due partly to a 'basic' growth tendency, as seen in patients with isolated complete GH deficiency, and partly to GH, provided that thyroid hormone secretion is normal. For a normal pubertal growth spurt, GH or testosterone alone is not sufficient, both hormones are required. In other words, the pubertal acceleration in boys is due to testosterone in the presence of GH as a permissive factor. A practical conclusion from this is that in boys with GH deficiency whose bone age corresponds to the beginning of puberty, HGH treatment should not be discontinued and replaced by testosterone, but both hormones should be given simultaneously until final height has been reached.

This is illustrated by a patient with combined idiopathic GH, TSH and gonadotropin deficiency, whose treatment was started late (Fig. 9). Thyroid replacement induced a small transitory growth stimulation and the subsequent addition of HGH at a bone age of 12 years had no decisive influence on the growth rate. However, when testosterone was added to these 2 hormones a dramatic growth spurt with a peak of 11 cm/year was observed. McGillivray et al. (1974) have also recently obtained a better growth rate in GH-deficient patients treated with HGH and fluoxymesterone than in patients treated with HGH alone. Most of their patients had a prepubertal bone age and they were very cautious in their conclusions, stating that their 18 months' trial was insufficient to evaluate the effect on bone maturation. We would not advocate treatment with anabolic steroids in addition to HGH in hypopituitary patients, either in general or before puberty. We rather recommend therapy first with HGH alone, and then with the addition of testosterone, only if there is also a deficiency of gonadotropin and when bone age approaches puberty.

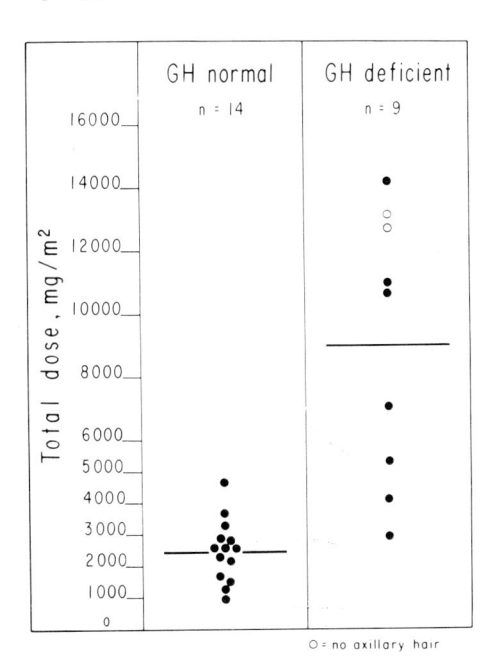

Fig. 10. Total testosterone dose necessary to induce axillary hair in 14 patients without and in 9 patients with GH deficiency (From Zachmann and Prader, 1970, by courtesy of the Editors of the *Journal of Clinical Endocrinology and Metabolism.*)

ANDROGENIC EFFECT OF TESTOSTERONE AS MODIFIED BY HGH

Our next point of discussion is the modifying influence of GH on the testosterone-induced development of secondary sex characteristics. Again from clinical experience, we and others believed the effectiveness of testosterone in the induction of secondary sex characteristics to be lower in boys with untreated GH deficiency. Because accurate quantitative criteria were not available in all cases, the assessment of the androgenic effect was more difficult than the analysis of growth. In some patients, however, we were able to compare the testosterone dosages required to induce axillary hair (Zachmann and Prader, 1970). In Figure 10, we have computed the total testosterone dosage in mg/m^2, which was required to produce axillary hair in patients with normal GH secretion (on the left) and in patients with untreated GH deficiency (on the right). In the former there was little variation from a mean of about 2500 mg/m^2; in the GH-deficient patients not receiving HGH, on the other hand, the mean dose was over 9000 mg/m^2, and there was a much wider scatter, presumably due to different degrees of GH deficiency.

The time to produce axillary hair is shown in Figure 11. In patients with normal GH production (left), there was an inverse relationship between the monthly testosterone dose and the time needed for axillary hair to appear. There was no such relationship in the untreated GH-deficient subjects (right) in whom axillary hair developed much later or not at all.

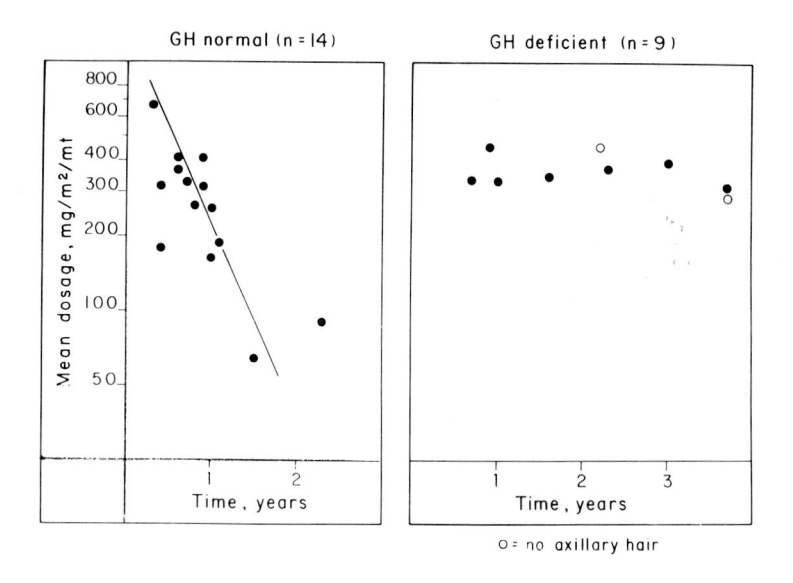

Fig. 11. Relationship between mean dosage of testosterone and time of appearance of axillary hair (same patients as in Fig. 10). (From Zachmann and Prader, 1970, by courtesy of the Editors of the *Journal of Clinical Endocrinology and Metabolism.*)

METABOLIC EFFECT OF TESTOSTERONE AS MODIFIED BY HGH

Recently, we have been able to confirm biochemically that the anabolic effect of testosterone, which is, of course the basis for its growth-promoting effect, is reduced in GH-deficient patients not receiving HGH. We were interested to see whether testosterone-induced nitrogen retention would be reduced in GH-deficient patients without HGH and whether it could be enhanced by HGH. Instead of the classic nitrogen balance technique which requires about 3 weeks in hospital, we used a short test without diet based on the stable isotope ^{15}N (Zachmann et al., 1973). The patients were given 0.2 g/kg of ^{15}N-labeled ammonium chloride, divided into 3 doses, before and after 6 daily doses of 15 mg/m^2 of intramuscular testosterone propionate. Twenty-four hour urines were collected on the test days before and while on testosterone. The total nitrogen was determined by the Kjeldahl method and the percentage of ^{15}N was determined either by mass spectrometry or on a ^{15}N analyzer. From this, the excreted amount of ^{15}N was calculated and, knowing the ingested quantity, a true urinary ^{15}N balance could be determined. We have so far studied only a few patients with this technique, because the labeled compounds are expensive and the analysis is time-consuming.

Figure 12 shows the metabolic response to testosterone without and with HGH treatment in the patient with GH, TSH and gonadotropin deficiency, whose growth velocity chart was presented in Figure 9. In 2 normal prepubertal boys (left column), the positive change of the nitrogen balance induced by testosterone ranged from 40-60% of the pretreatment balance. In the GH-deficient patient not receiving HGH, there was a definite but weak response of +33%. The response was markedly stronger (+143%) when the test was repeated using the same testosterone dose, but at the end of 3 months of treatment with HGH. These results should be regarded as preliminary, since the number of cases is very small. However, they provide biochemical support for our clinical analysis of growth.

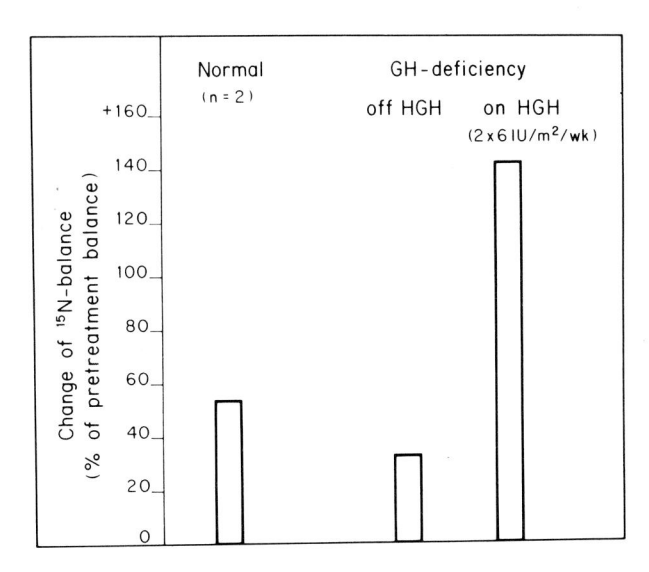

Fig. 12. Effect of testosterone (6 × 15 mg/m^2) on ^{15}N balance in 2 normal prepubertal boys and in a patient with GH deficiency off and on HGH.

CONCLUSIONS

In conclusion, our results suggest that GH is necessary as a permissive factor to allow testosterone to be fully effective with respect to protein anabolism, growth promotion and androgenicity. Hopefully, the study of such hormone interrelations will not only prove useful for a more adequate treatment of hypopituitary patients but also for the better understanding of the physiological mechanisms of puberty.

REFERENCES

Deller, J.J., Plunket, D.C. and Forsham, P.H. (1966): *Calif. Med., 104,* 359.
Greulich, W.W. and Pyle, S.I. (1969): *Radiographic Atlas of Skeletal Development of the Hand and Wrist, 2nd Ed.* Stanford University Press, Stanford, Calif.
Illig, R. and Prader, A. (1970): *J. clin. Endocr., 30,* 615.
MacGillivray, M.H., Kolotkin, M. and Munschauer, R.W. (1974): *Pediat. Res., 8,* 103.
Martin, L.G., Clark, J.W. and Connor, T.B. (1968): *J. clin. Endocr., 28,* 425.
Scow, R.O. and Hagan, S.N. (1965): *Endocrinology, 77,* 582.
Simpson, M.E., Marx, W., Becks, H. and Evans, H.M. (1944): *Endocrinology, 35,* 309.
Tanner, J.M., Whitehouse, R.H. and Takaishi, M. (1966): *Arch. Dis. Childh., 41,* 454.
Zachmann, M. and Prader, A. (1970): *J. clin. Endocr., 30,* 85.
Zachmann, M., Völlmin, J.A. and Prader, A. (1973): *Acta paediat. scand., 62,* 97.

INTERMITTENT VERSUS CONTINUOUS HGH TREATMENT OF HYPOPITUITARY DWARFISM

ZVI LARON* and ATHALIA PERTZELAN

Institute of Paediatric and Adolescent Endocrinology, Beilinson Medical Center, Petah Tikva, and Sackler School of Medicine, Tel-Aviv University, Tel-Aviv, Israel

Treatment with HGH was first reported by Raben who treated successfully a 17-year-old hypopituitary male in 1958. Since then many centres have accumulated clinical experience with HGH administration over longer periods of time (Prader et al., 1964, 1967, 1972; Wright et al., 1965; Rosenbloom, 1966; Seip and Trygstad, 1966; Tanner and Whitehouse, 1967; Goodman et al., 1968; Henneman, 1968; Parker and Daughaday (a review of mainly other authors), 1968; Westphal, 1968; Ferrandez et al., 1970; Job and Canlorbe, 1970; Läsker, 1970; Shizume et al., 1970, 1974; Soyka et al., 1970; Tanner et al., 1971; Aceto et al., 1972; Hall and Olin, 1972; Root, 1972; MacGillivray et al., 1974; Braunstein et al., 1975; Joss, 1975). The therapeutic schemes differ from centre to centre, varying in the selection of the patient material according to diagnosis and age at initiation of therapy, number of patients, the extraction procedure of HGH, dose and mostly duration of therapy. The above groups employed continuous

TABLE 1

Effect of human growth hormone on linear growth in patients with pituitary insufficiency

Authors	Year	Number of patients	Authors	Year	Number of patients
Raben	1958	1	Root	1972	19
Wright et al.	1965	12	Aceto et al.	1972	71
Seip and Trygstad	1966	12	Hall and Olin	1972	20
Rosenbloom	1966	9	Prader et al.	1972	44
Goodman et al.	1968	35	Raiti et al.	1973	6
Henneman	1968	50	Kirkland et al.	1973	18
Parker and Daughaday	1968	10	MacGillivray et al.	1974	12
Westphal	1968	9	Shizume et al.	1974	11
Job and Canlorbe	1970	12	Braunstein et al.	1975	5
Läsker	1970	10	Guyda et al.	1975	151
Shizume et al.	1970	14	Joss	1975	21
Soyka et al.	1970	15	Laron et al.	1976	51
Tanner et al.	1971	57			

Attempts were made to avoid citing the same group twice, assuming that repeated reports deal with the same patient material.

* Established Investigator of The Chief Scientist's Bureau Ministry of Health.

therapy of 2 to 3 weekly injections, and reported good effects, but many of these patients were treated actually for short periods. Although it seems advisable to administer HGH therapy on a continuous basis for the whole period of growth, intermittent therapeutic schemes have been tried due to the shortage of HGH (Kirkland et al., 1973; Laron and Pertzelan, 1974; Guyda et al., 1975; Pertzelan et al., 1976).

Table 1 summarizes the most known reports which deal with the therapeutic effect of human growth hormone. When considering the question as to whether long-term intermittent treatment gives results which are equal to or better than those obtained with continuous therapy, we concluded that such a critical appraisal cannot be made indiscriminately on the basis of all published cases, simply lumped together. Many of these were treated for relatively short periods of time. Furthermore, most authors failed to distinguish between the various diagnostic entities. Other important factors which must be taken into consideration include the age at initiation of HGH therapy and any previous treatment administered. It was therefore decided to analyze only a group selected on the basis of clearly defined criteria, with the aim of obtaining reliable guide-lines for future use.

SUBJECTS AND METHODS

We screened the 51 patients treated with HGH by our group since 1963 for those who fulfilled the following criteria: (1) a duration of at least 3 years of intermittent HGH therapy; (2) regular monthly follow-up; (3) absence of overt intracranial lesion; (4) no signs of puberty.

Twelve patients of 2 diagnostic entities were found to meet the above criteria: 7 patients with isolated HGH deficiency (IGHD) (Table 2), and 5 patients with multiple pituitary hormone deficiencies (MPHD) (Table 3). It is of note that the patients in the second group were older than those in the first. Patient No. 10 was born prematurely, with a birth weight of 1000 g and at age of 3 years suffered from tuberculous meningitis (in Egypt). The patients with IGHD received only HGH, whereas those with MPHD received also 1-thyroxine and 1 (Pt. No. 10) also hydrocortisone replacement therapy in a dose of 15 mg/day. None received sex hormones or their derivatives.

The laboratory diagnosis of HGH deficiency was established when the level of plasma HGH as measured by radioimmunoassay (using the charcoal separation of Herbert et al. (1960) with a method described from our laboratory (Laron and Mannheimer, 1966)), did not exceed 2 ng/ml during stimulation by insulin hypoglycaemia, L-arginine infusion and L-dopa ingestion (Laron et al., 1973). All patients were subjected to a complete investigation for their endocrine profile adjusted to age (Laron, 1969). In addition, all patients underwent a skull X-ray, eye fundi and visual fields examination and an electroencephalogram.

The patients underwent monthly clinical examinations and body measurements with a Harpenden stadiometer, both during HGH treatment and intervals. The measurements were always performed by the same person. Linear growth was plotted on Tanner and Whitehouse's growth charts (Tanner et al., 1966). Plasma was tested for HGH antibodies in the standard radioimmunoassay system, and for PBI and/or T_4 every 1-6 months.

The growth hormone for clinical use (HGH) was extracted in our laboratory by Mrs. Sara Assa from pituitaries collected at autopsies, according to Raben's method (Raben, 1957). The hormone was administered intermittently in courses of various lengths of time interspaced by intervals of various duration, dictated by the availability of the hormone. The HGH was injected intramuscularly 3 times per week. The

TABLE 2
Pertinent data of patients with isolated HGH deficiency (IGHD)

No.	Patient Name	Sex	Remarks	At referral CA (yr:mths)	BA	Ht. (cm)	At beginning of HGH CA (yr:mths)	BA	Ht. (cm)	Dose of HGH (mg/injection*)	At last examination CA (yr:mths)	BA	Ht. (cm)
1	Z.D.	M	Hereditary	1:0	0:3	59.8	2:5	0:9	66.0	2(24)	5:5	3:0	92.2
2	B.B.-S.	F	Sporadic	7:6	4:0	98.0	9:1	4:6	103.4	2(7)3(25)	13:7	11:0	130.5
3	H.A.	F	Hereditary	6:4	3:0	87.2	6:8	3:0	88.5	2(28)	11:10	9:0	120.0
4	L.I.	F	Sporadic	0:8	0:3	53.7	3:2	0:9	64.2	2(38)	8:11	6:6	110.0
5	J.V.	M	Sporadic	2:9	0:9	80.0	4:0	1:6	86.1	2(13)3(34)	10:10	6:6	121.3
6	N.E.	M	Sporadic	5:4	2:3	88.5	6:8	2:9	93.6	2(13)3(33)	13:9	8:0	127.7
7	E.I.	M	Hereditary	2:5	0:6	68.7	4:5	1:6	74.6	2(61)	11:11	10:0	129.8

CA = chronological age; BA = bone age; Ht. = height.
* = HGH was administered by 3 intramuscular injections per week; in brackets, the number of months of net treatment.

TABLE 3
Pertinent data of patients with multiple pituitary hormone deficiencies (MPHD)

No.	Patient Name	Sex	Hormone deficiency in addition to HGH	At referral CA (yr:mths)	BA	Ht. (cm)	At beginning of thyroxine treatment CA (yr:mths)	BA	Ht. (cm)	Additional replacement therapy (mg/day)	At beginning of HGH treatment CA (yr:mths)	BA	Ht. (cm)	Dose of HGH (mg/injection*)	At last examination CA (yr:mths)	BA	Ht. (cm)
8	E.S.	F	TSH, Gn	2:3	0:9	73.0	4:0	—	83.0	l-thyroxine 0.05	8:8	7:6	109.8	2(21)	12:7	10:0	130.6
9	O.L.	F	ACTH, TSH, Gn	7:0	4:0	106.0	10:0	5:6	115.0	l-thyroxine 0.10	13:3	8:0	131.7	2(30)	17:2	13:6	154.6
10	S.D.	M	ACTH, TSH, Gn	13:10	6:0	120.0	14:5	6:0	120.5	l-thyroxine 0.10 hydrocort. acetate 15	18:8	8:0	124.2	2(26)	22:9	12:6	139.0
11	A.A.	M	ACTH, TSH, Gn	11:10	5:0	99.1	13:6	5:9	106.6	l-thyroxine 0.10	15:1	7:3	113.7	3(34)	19:10	11:0	137.5
12	D.H.	M	TSH, Gn	9:7	4:9	111.6	10:3	5:0	113.9	l-thyroxine 0.10	12:0	6:9	118.9	2(32)	17:4	10:0	143.5

CA = chronological age; BA = bone age; Ht. = height.
* = HGH was administered by 3 intramuscular injections per week; in brackets, the number of months of net treatment.

initial dose was 2 mg per injection; this dose was augmented to 3 mg whenever the monthly height increment declined to half that of the previous month. The biological activity in rats of several of our batches was found to be 1-1.5 IU/mg.

RESULTS

The effectiveness of the intermittent therapeutic schedule is presented separately for each of the 2 diagnostic groups, and the patients in each group are arranged in serial order in accordance with the length of the whole intermittent HGH treatment period. Comparison was made of the data obtained during the 'basal period', i.e. the pre-HGH treatment period (in the IGHD group no treatment was given during this period, whereas in the MPHD thyroxine replacement therapy was instituted) and 'the intermittent HGH treatment period', i.e. the periods of actual HGH administration (net HGH) as well as the interspaced intervals during which no HGH treatment was given. Furthermore, in an attempt to determine the effectiveness of HGH during repeated courses of treatment and the growth during the intervals between courses, the growth velocity cm/year was calculated for each course and for each interval separately and the results compared.

ISOLATED HGH DEFICIENCY (IGHD)

From Table 4 is it evident that the highest growth velocity was achieved in the first course. In 3 patients a trial dose of 3 mg per injection was subsequently administered; in patient No. 2 there was a consequent rise to the velocity seen during the first course, while in patients No. 5 and 6 no such rise was observed. In patients No. 3 and 6, the growth velocity exceeded that of the first course when treatment was reinstituted after an intermission longer than 1 year (course V and VII, respectively). There was no apparent decline in the response to HGH during subsequent courses. The growth rate in between courses of HGH treatment were practically always less than during the basal period.

Table 5 shows that the actual duration of HGH administration was approximately 1/2 to 2/3 of the whole HGH treatment period. The growth velocity during the period of HGH administration ranged from 1.6 to 3.4 times that of the basal period, except in patient No. 5, in whom it was less (1.5). The growth stimulation by HGH however was offset by the reduced growth rate in between courses of HGH, so that the rate of growth over the whole HGH treatment period varied from 104-243% of that registered in the basal period.

Figures 1 and 2 illustrate the effect of intermittent HGH treatment on the growth curve of patients No. 4 and 7. Figure 3 shows the external change in the latter patient.

MULTIPLE PITUITARY HORMONE DEFICIENCIES (MPHD)

In this group (Table 4) there were no marked differences in the growth velocity between the first and subsequent courses. Here again, as in the IGHD group there was no progressive decline in the effectiveness of the HGH treatment. During the intervals the growth velocity was again less than during the basal period.

In Table 5 it is seen that the actual duration of HGH administration (21-34 months), was also 1/2 to 2/3 of the whole HGH treatment period (47-64 months). The

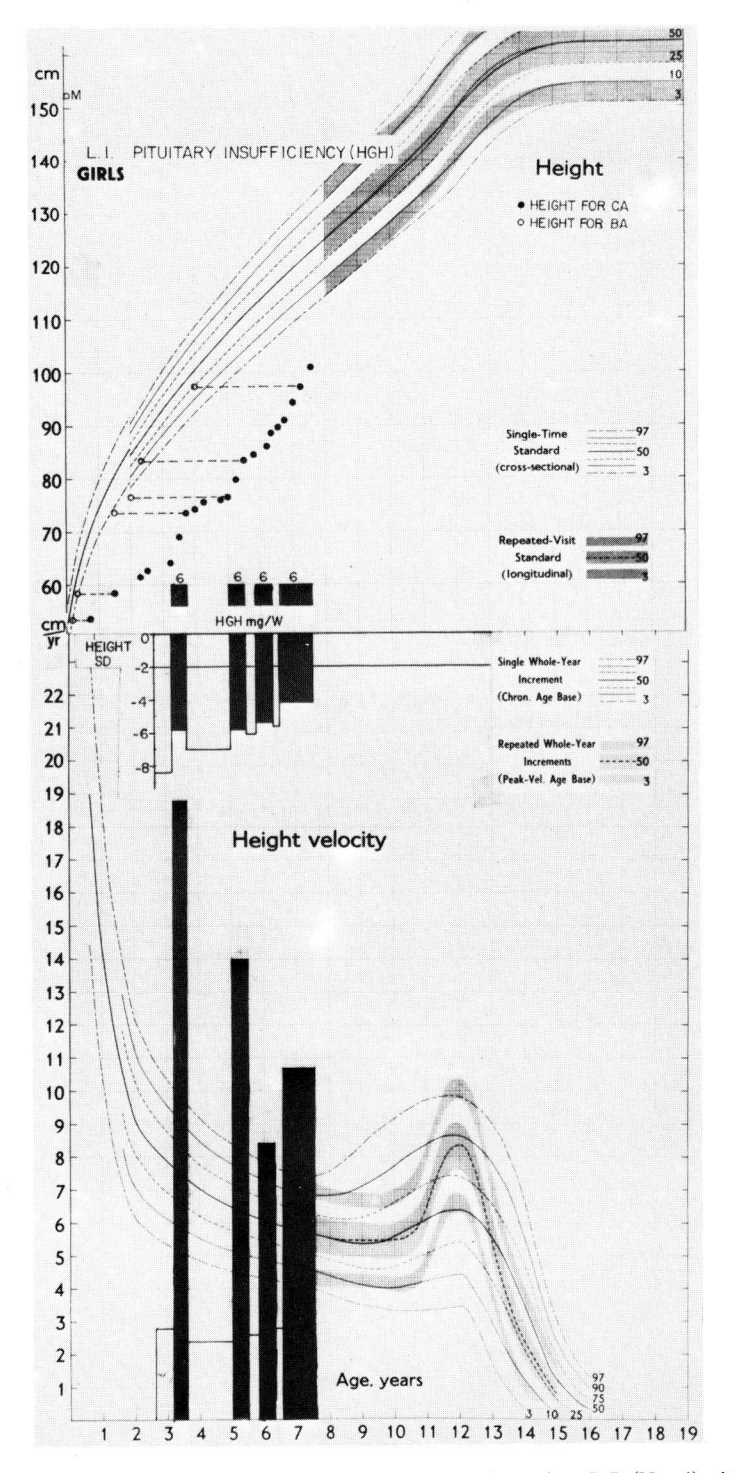

Fig. 1. The long-term effect of intermittent HGH treatment in patient L.I. (No. 4) with isolated HGH deficiency. Within 5 9/12 years she grew 46 cm. The upper part of the figure shows the height plotted on the Tanner Growth Chart. The open circles reflect height for bone age (BA). The lower part illustrates the growth velocity during and between courses of HGH treatment. The middle part depicts the changes in height SD induced by the treatment.

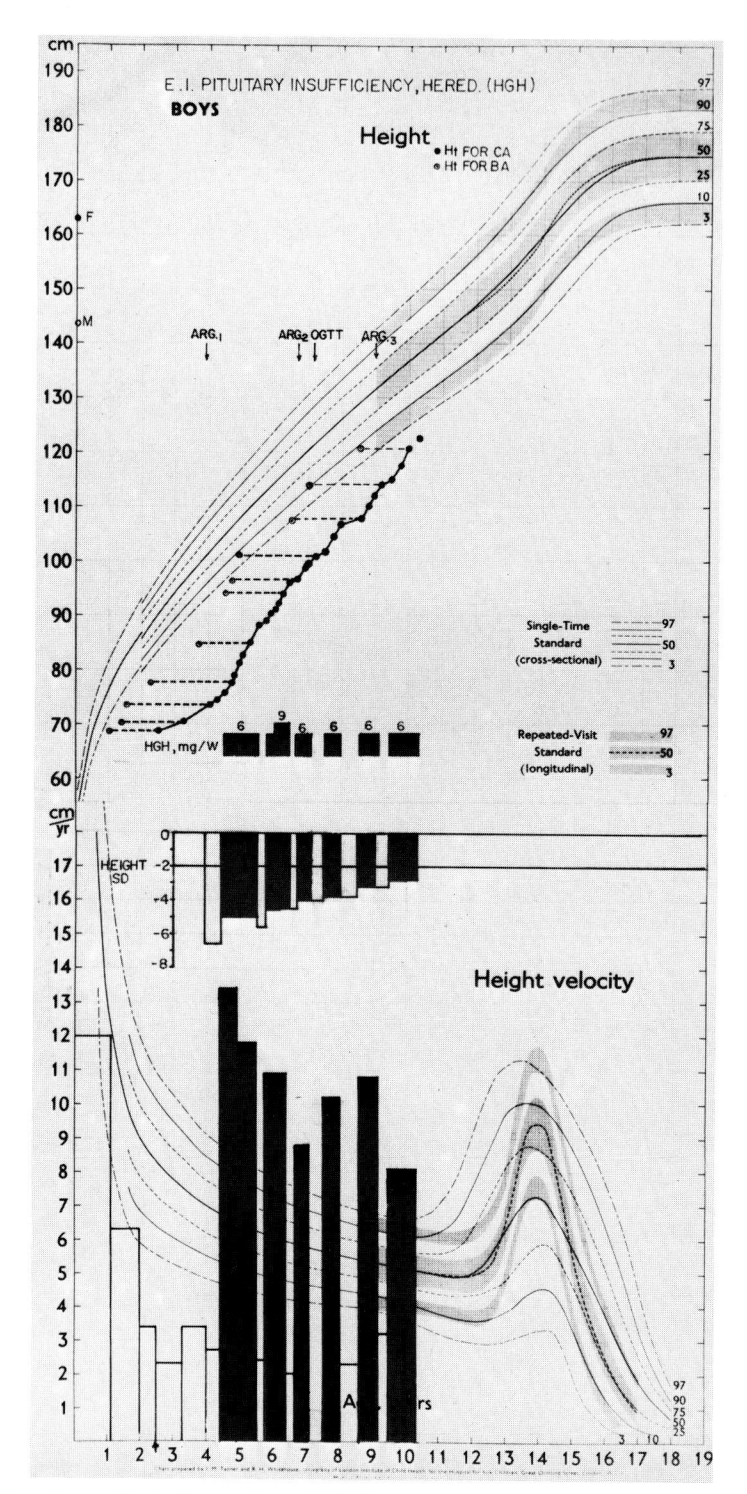

Fig. 2. The long-term effect of intermittend HGH treatment in patient E.I. (No. 7) with isolated HGH deficiency. Within 7 6/12 years he grew 55.2 cm. The upper part of the figure shows the height plotted on the Tanner Growth Chart. The open circles reflect height for bone age (BA). The lower part illustrates the growth velocity during and between courses of HGH treatment. The middle part depicts the changes in height SD induced by the treatment.

TABLE 4
Details of linear growth, calculated as growth velocity (cm/year) during and between repeated courses of HGH therapy*

Patient No.	Basal period	Course I On	Course I Off	Course II On	Course II Off	Course III On	Course III Off	Course IV On	Course IV Off	Course V On	Course V Off	Course VI On	Course VI Off	Course VII On	Course VII Off
Isolated HGH deficiency
1	4.3	15.2(6)	2.8(3)	12.4(6)	4.1(5)	9.1(12)	2.7(4)								
2	3.4	9.6(6)	1.8(2)	5.8(5)	3.2(8)	10.2(6)	0(10)	10.5(5)	0(2)	8.7(10)					
3	3.9	13.6(6)	1.5(4)	11.0(6)	2.3(10)	8.6(6)	0(3)	11.0(6)	2.3(17)	14.0(4)					
4	3.9	18.8(6)	2.4(5)	14.0(6)	2.6(4)	8.4(6)	2.8(3)	10.7(12)	2.9(9)	9.6(8)					
5	4.8	10.1(16)	3.1(11)	5.0(5)	1.8(4)	8.0(6)	1.9(7)	7.0(6)	0.9(4)	5.6(7)	0(9)	7.9(7)			
6	3.8	9.0(11)	5.1(4)	7.6(6)	3.2(3)	5.0(6)	1.4(8)	5.2(6)	0(7)	8.0(6)	0(2)	5.4(6)	1.1(15)	9.2(5)	
7	3.0	12.6(13)	2.4(3)	10.9(8)	2.0(3)	10.2(6)	0(4)	10.2(6)	2.4(6)	10.8(7)	3.2(4)	8.1(11)	0(9)	8.8(10)	
Multiple pituitary hormone deficiencies
8	5.7	7.4(6)	3.6(12)	8.3(8)	2.4(14)	7.9(7)									
9	5.1	10.4(6)	0(6)	10.2(6)	1.2(5)	8.5(8)	2.6(6)	6.1(10)							
10	0.9	5.2(6)	0.4(7)	7.6(6)	0.8(3)	6.0(10)	1.0(10)	6.5(4)							
11	4.4	8.0(12)	2.7(4)	7.6(6)	0(5)	8.4(3)	0.8(10)	7.4(6)	1.8(4)	6.9(7)	2.9(5)				
12	4.2	10.2(6)	2.8(7)	6.4(6)	2.0(13)	9.2(3)	0.6(2)	6.9(12)	1.2(10)	5.7(5)					

On = HGH therapy; off = no HGH therapy; in brackets, duration of course or interval in months.
* = Usual dose was 2 mg 3 times weekly; when in italics dose was 3 mg 3 times weekly.

TABLE 5
Effect of intermittent HGH administration on linear growth

Patient No.	Follow-up (months)				Growth						
		Intermittent treatment				Intermittent treatment					
	Basal*	HGH	Nil	Total	Basal* (cm/yr)	HGH	Nil (cm/yr)	Total	HGH	Nil (% basal)	Total

Isolated HGH deficiency

1	17	24	12	36	4.3	11.3	3.5	8.7	262	81	200
2	19	32	22	54	3.4	9.0	1.6	6.0	264	47	176
3	4	28	34	62	3.9	11.3	1.7	6.0	289	43	154
4	31	38	30	68	3.9	11.9	3.1	8.0	305	79	205
5	15	47	35	82	4.8	7.8	1.6	5.1	162	33	104
6	16	46	39	85	3.8	6.4	2.5	4.7	213	83	156
7	24	61	29	90	3.0	10.2	1.2	7.3	340	40	243

Multiple pituitary hormone deficiencies

8	56	21	26	47	5.7	8.3	2.6	5.3	146	46	93
9	39	30	17	47	5.1	8.4	1.2	5.8	165	25	114
10	51	26	23	49	0.9	6.3	0.7	3.6	700	77	400
11	19	34	28	62	4.4	7.6	1.2	4.9	172	27	111
12	21	32	32	64	4.2	7.4	1.7	4.6	176	40	109

*Basal period = pre HGH treatment (patients with IGHD on no treatment, patients with MPHD on T_4 ± hydrocortisone.

growth velocity during HGH treatment varied from 1.5-7.0 times that of the basal period, except for patient No. 10. Calculation showed that in 4 out of 5 patients there was no significant change in growth velocity during the whole intermittent HGH treatment period when compared to that in the basal period. The effect of HGH is visible in the period of actual HGH treatment.

Patient No. 10 grew 0.9 cm/year during the basal period of 51 months. Although the measurable response of HGH therapy was smaller than that in the other patients, his overall gain in height was remarkable. Figures 4 and 5 illustrate the linear growth induced by HGH in patients No. 11 and 12.

EFFECT ON BONE MATURATION

Isolated HGH deficiency (IGHD)

At start of therapy the bone age (BA) ranged from 9/12 to 4 6/12 year (Table 2). In no patient did the bone age exceed the chronological age (CA) during HGH therapy and at the end of the study the lag remained from 2 to 6 years.

Multiple pituitary hormone deficiencies (MPHD)

At the start of HGH therapy the bone age ranged from 6 9/12 to 8 years, the retardation being more marked than in the IGHD group. In no patient did the bone age exceed the chronological age during treatment and the lag remained from 2.5 to 10 years.

Fig. 3. The same patient as in Figure 2. (E.I.) illustrating the actual change in appearance, especially the change in the facial features induced by the HGH therapy. Right: age 3 years; left: age 11 years.

ANTIBODIES TO HGH

The plasma of all the patients was examined once every 1-6 months for the presence of anti-HGH antibodies. Only patient No. 4 revealed a very low titre after 6 months of treatment. This titre remained largely unchanged during the ensuing years and did not affect the response to HGH.

DISCUSSION

In the present paper which describes the results of intermittent HGH treatment in 12 patients deficient of HGH, there were 2 major findings:
1. The treatment was effective in accelerating growth velocity and there was no apparent progressive decline in the response to it over long periods of treatment.
2. There was a difference in response between the patients with isolated HGH deficiency (IGHD) and those with idiopathic multiple pituitary hormone deficiencies (MPHD) (i.e. not of tumoral origin).

To the best of our knowledge, only Kirkland et al. (1973) and recently Guyda et al. (1975) have reported on the use of an intermittent therapy with HGH. Kirkland and co-workers employed 4 courses a year, 1 course every 3 months, injecting a dose of 2-5 IU of HGH every day for 15-25 days, Guyda and co-workers administered HGH for 6 months (2-5 IU 3 times a week), followed by 6 months without HGH therapy. Part of

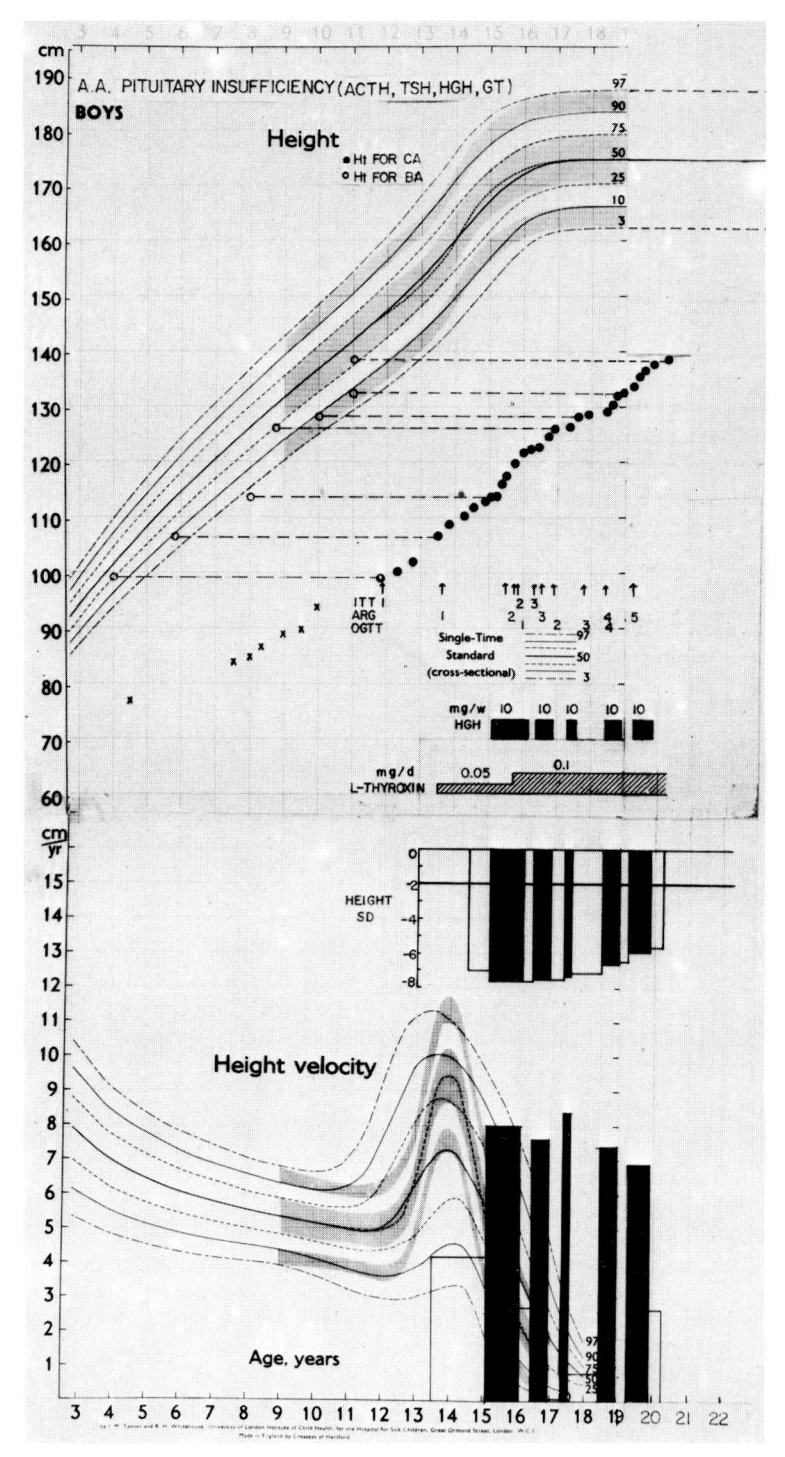

Fig. 4. The long-term effect of intermittent HGH treatment in patient A.A. (No. 11) with multiple pituitary hormone deficiencies. Within 4 9/12 years he grew 24 cm. The upper part of the figure shows the height plotted on the Tanner Growth Chart. The open circles reflect height for bone age (BA). The lower part illustrates the growth velocity during and between courses of HGH treatment. The middle part depicts the changes in height SD induced by the treatment.

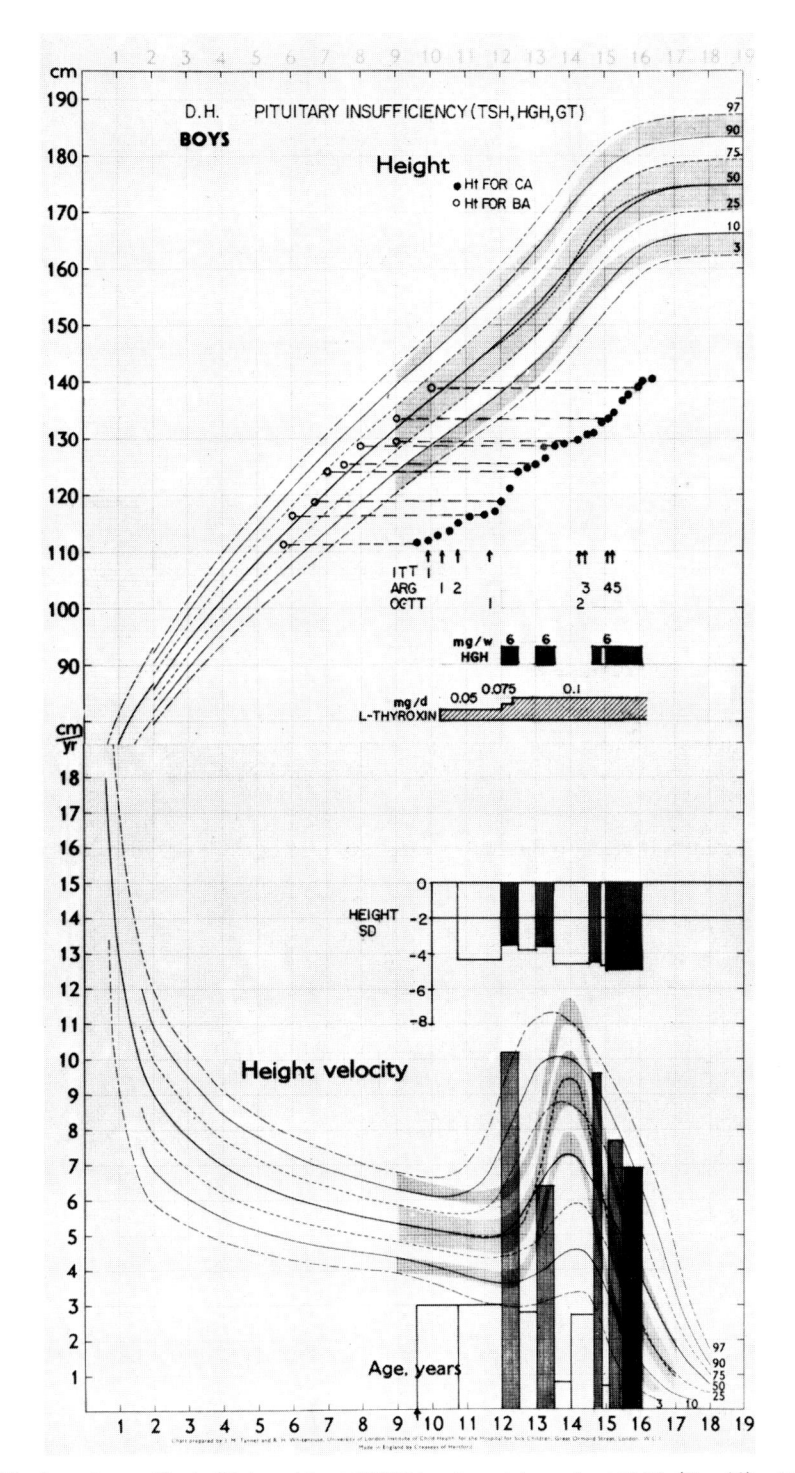

Fig. 5. The long-term effect of intermittent HGH treatment in patient D.H. (No. 12) with multiple pituitary hormone deficiencies, within 5 4/12 years he grew 24.6 cm. The upper part of the figure shows the height plotted on the Tanner Growth Chart. The open circles reflect height for bone age (BA). The lower part illustrates the growth velocity during and between courses of HGH treatment. The middle part depicts the changes in height SD induced by the treatment.

the patients were followed-up to 5 years. In our investigation, as well as in Kirkland's report, the response to HGH was satisfactory over a long period of time (despite variability in the response of individual patients), in contradistinction to previous papers reporting continuous HGH therapy (Henneman, 1968; Soyka et al., 1970; Ferrandez et al., 1970; Job and Canlorbe, 1970; Läsker, 1970; Tanner et al., 1971; Hall and Olin, 1972; Prader et al., 1972), which describe a progressive decrease in response over several years of treatment. The long-term effect of HGH in the patients reported by Guyda et al. (1975) is also satisfactory, but due to the large intervals without therapy the overall effect is relatively diminished. Like other authors (Henneman, 1968; Tanner et al., 1971; Hall and Olin, 1972; Guyda et al., 1975) we found that during the intervals between courses of treatment, the growth velocity decreased markedly to levels even lower than those recorded before initiation of therapy; this was not due to seasonal variation, duration of treatment, length of intervals between courses, nor to anti-HGH antibody formation.

There were 2 main differences in response to HGH between the IGHD and MPHD groups. In the patients with IGHD the acceleration of growth during the first course of treatment was more pronounced than that during subsequent courses, while in those with MPHD the growth velocity showed no clear diminution in later courses. This marked initial response to HGH has been recorded by other authors (Prader et al., 1964, 1967, 1972; Seip and Trygstad, 1966; Ferrandez et al., 1970; Shizume et al., 1970; Soyka et al., 1970; Tanner et al., 1971; Hall and Olin, 1972), none of whom, however, differentiated between these 2 groups of patients. Secondly the overall growth response was better in the patients with IGHD than in the patients with MPHD (Table 5). In considering this difference between the 2 groups of patients it may be of importance that our patients with IGHD were, on the whole, younger than those with MPHD and their bone age was lower. In addition to this age difference, it seems possible that HGH induces less of a response in patients with MPHD because their replacement therapy with other hormones is not optimal.

It is of note that in patients No. 3 and 6 (IGHD) the response to HGH when it was reinstituted after an interval longer than 1 year was as good as that during the first course. This observation stresses once more the efficacy of the intermittent regime. Also of interest is patient No. 10 (MPHD), who showed a different pattern of growth (Table 5). His rate of growth before therapy was much less than in the others in his group, but his response to HGH was relatively much higher. This pattern is similar to that seen in patients with craniopharyngioma (Karp and Laron, 1967, Prader et al., 1972; Laron et al., 1976).

In order to further assess the effectiveness of our treatment, we have attempted to compare our results to those reported by other groups. We found only 2 reports which fulfilled our criteria of at least 3 years follow-up and which could be analyzed to distinguish between IGHD and MPHD patients; Kirkland et al. (1973), who used a different intermittent schedule, and Prader et al. (1972), who used prolonged continuous treatment. A summary of this comparison is shown in Table 6. We included in this Table also the findings of Braunstein et al. (1975) who treated for 2 years only 5 patients with Hand-Schüller-Christian disease who suffered also from diabetes insipidus. It would be of great interest to similarly analyze the data of the Canadian group (Guyda et al., 1975), as they used an intermittent regime for 5 years, but these authors do not distinguish between diagnostic entities. In the patients with IGHD we found that during a mean overall treatment period of 5 3/4 years, there was a mean growth of 36.4 cm. Kirkland et al. (1973) reported a mean growth of 28.8 cm for a mean treatment period of 5.5 years whilst Prader et al. (1972) obtained a result of 22.9 cm for a mean of 3 years of continuous therapy.

TABLE 6

Comparative effectiveness of 3 therapeutic schedules on linear growth in patients with isolated HGH deficiency and multiple pituitary hormone deficiencies (mean and range)

Authors and mode of HGH administration	Number of patients	Age years:months (range)	Treatment period	Growth during treatment period cm (range)	Effect of HGH (calculated) years:months/ 10 cm growth
Isolated HGH deficiency					
Laron et al. (1976) Intermittent	7	5:2 (2:5-9:1)	5:9 (3-7:6)	36.4 (26.2-55.2)	1:7
Kirkland et al. (1973)* Intermittent	4	7:3 (3:2-10:6)	5:3 (4:0-9:0)	28.8 (22.8-43.5)	1:9
Prader et al. (1972)** Continuous	4		3:0	22.9	1:4
Multiple pituitary hormone deficiencies					
Laron et al. (1976) Intermittent	5	13:6 (8:8-18:8)	4:6 (3:11-5:4)	21.4 (14.8-24.6)	2:1
Kirkland et al. (1973)* Intermittent	3	5:7 (2:4-8:3)	6:0 (4:0-8:0)	28.6 (21.6-40.0)	2.0
Prader et al. (1972)** Continuous	7		3:0	18.3	1:7
Braunstein et al. (1975)*** Continuous	5	12:3 (9:4-16.6)	2:0	12.6	1:8
Mixed group (?) (proportions unknown)					
Guyda et al. (1975)† Intermittent	7	8 (5-10)	5	30	1:8

The material from Kirkland and Prader was classified and recalculated by us.
* HGH was injected in a dose of 2-5 IU for 15-25 days, every 3 months.
** HGH was injected in a dose of 5 mg/sq.m twice weekly continuously.
*** HGH was injected in a dose of 2 IU 3 times a week continuously.
† The number of patients treated for 3 years was 27, for 4 years 13, and 5 years 7. HGH was administered for 6 months per year in a dose of 2-5 IU 3 times a week.

In the MPHD group the mean duration of intermittent HGH treatment in our patients was 4.5 years with a mean height gain of 21.4 cm; in Kirkland's patients over a mean period of 6 years the gain was 28.6 cm, and with Prader's regime over a mean period of 3 years the gain was 18.3 cm. A rough calculation of the overall gain of the mixed group of patients below age 12 years of Guyda et al. (1975) from their Figure 2A, reveals a growth of 30 cm in 5 years of intermittent treatment.

Another method with which to compare these studies was to calculate the time needed for the patient to grow 10 cm. Under our regime, the IGHD patients needed a period of 1 7/12 years. The patients of Kirkland et al. (1973) grew 10 cm in 1 9/12 years. If we assume that 1 IU is approximately 1.0-1.3 mg depending upon the method of extraction and batch, our therapeutic regime used less hormone and less time to achieve the same growth as Kirkland's patients. The patients of Prader et al. (1972) grew 10 cm in only 3 months less than ours, but with a far larger amount of hormone used. The same is true for the patients of Braunstein et al. (1975). It is apparent that in the patients with MPHD all 3 therapeutic approaches produced less of a response than in IGHD patients. Judging from the results it is probable that a great part of the Canadian patients with long-term therapy belonged to the IGHD group.

The advantage of intermittent HGH therapy lies in the fact that a smaller quantity of HGH leads to the same gain in linear height as that resulting from the continuous administration of HGH. As the availability of this hormone is limited and the commercial preparation very expensive, this would seem to be of great practical importance.

In conclusion we recommend the following regime of HGH therapy. Replacement therapy should be started as early as possible. The first course should be 1 year in the patients with IGHD and 2 years in those with MPHD, followed by an intermittent regime of 8-9 months effective treatment interspaced with 3-4 months of no treatment. This scheme seems to prevent the progressive decline in growth velocity, and enables long-term therapy. It has been suggested (Rudman et al., 1973) that the HGH injection be given in the evening as the anabolic response to exogenous HGH is inversely related to the plasma cortisol concentration at the time of its injection. Further observation on this subject should be of practical interest.

ACKNOWLEDGEMENT

The authors thank Mrs. Ruth Keret, M.Sc., from our laboratory, for the diagnostic HGH determinations and the assays of HGH antibodies in the plasma.

REFERENCES

Aceto Jr, T., Douglas Frasier, S., Hayles, A.B., Meyer-Bahlburg, H.F.L., Parker, M.L., Munschauer, R. and DiChiro, G. (1972): *J. clin. Endocr., 35,* 483.
Braunstein, G.D., Raiti, S., Hansen, J.W. and Kohler, P.O. (1975): *New Engl. J. Med., 292,* 332.
Ferrandez, A., Zachmann, M., Prader, A. and Illig, R. (1970): *Helv. paediat. Acta, 25,* 566.
Goodman, H.G., Grumbach, M.M. and Kaplan, S.L. (1968): *New Engl. J. Med., 278,* 57.
Greulich, W.W. and Pyle, S.I. (1959): *Radiographic Atlas of Skeletal Development of the Hand and Wrist, 2nd Ed.* Stanford University Press.
Guyda, H., Friesen, H., Bailey, J.D., Leboeuf, G. and Beck, J.C. (1975): *Canad. med. Ass. J., 112,* 1301.
Hall, K. and Olin, P. (1972): *Acta endocr. (Kbh.), 69,* 417.
Henneman, Ph.H. (1968): *J. Amer. med. Ass., 205,* 828.
Herbert, V., Lau, K.S., Gottlieb, C.W. and Bleicher, S.I. (1960): *J. clin. Invest., 39,* 1157.
Job, J.C. and Canlorbe, P. (1970): *Ann. Endocr. (Paris), 31,* 89.
Joss, E.E. (1975): *Monogr. Paediat., No. 5,* 27.
Karp, M. and Laron, Z. (1967): *Harefuah, 73,* 41.
Kirkland, R.I., Kirkland, J.L., Librik, L. and Clayton, G.W. (1973): *J. clin. Endocr., 37,* 204.
Laron, Z. (1969): In: *The Hypothalamus and the Pituitary Gland (Hypophysis). Paediatric Endocrinology,* pp. 35-111. Editor: D.W. Hubble. Blackwell, Oxford.
Laron, Z., Josefsberg, Z. and Doron, M. (1973): *Clin. Endocr., 2,* 1.
Laron, Z. and Mannheimer, S. (1966): *Israel J. med. Sci., 2,* 115.
Laron, Z. and Pertzelan, A. (1974): In: *Advances in Human Growth Hormone Research Symposium, Baltimore 1973,* pp. 75-178. Editor: S. Raiti. DHEW Publication No. (NIH) 74-612, Washington.
Laron, Z., Pertzelan, A., Kiwity, Sh., Livneh-Zirinsky, M. and Keret, R. (1976): *Growth without Growth Hormone: Myth or Fact?* Serono Symposia, Academic Press, New York. In press.

Läsker, G. (1970): *Dtsch. Gesundh.-Wes., 49,* 2328.
MacGillivray, M.H., Kolotkin, M. and Munschauer, R.W. (1974): *Pediat. Res., 8,* 103.
Parker, M.L. and Daughaday, W.H. (1968): In: *Growth Hormone,* p. 398. Editors: A. Pecile and E.E. Müller. ICS 158, Excerpta Medica, Amsterdam.
Pertzelan, A., Kauli, R., Assa, S., Greenberg, D. and Laron, Z. (1976): *Clin. Endocr.,* in press.
Prader, A., Illig, R., Széky, J. and Wagner, H. (1964): *Arch. Dis. Childh., 39,* 535.
Prader, A., Zachmann, M., Poley, J.R., Illig, R. and Széky, J. (1967): *Helv. paediat. Acta, 22,* 440.
Prader, A., Ferrandez, A., Zachmann, M. and Illig, R. (1972): In: *Growth and Growth Hormone,* p. 452. Editors: A. Pecile and E.E. Müller. ICS 244, Excerpta Medica, Amsterdam.
Raben, M.S. (1957): *Science, 125,* 883.
Raben, M.S. (1958): *J. clin. Endocr., 18,* 901.
Raiti, S., Trias, E., Levitsky, L. and Grossman, M. (1973): *Amer. J. Dis. Child., 126,* 597.
Root, A.W. (1972): In: *Human Pituitary Growth Hormone,* p. 124. Charles C Thomas, Springfield, Ill.
Rosenbloom, A.L. (1966): *J. Amer. med. Ass., 198,* 130.
Rudman, D., Freides, D., Patterson, J.H. and Gibbas, D.L. (1973): *J. clin. Invest., 52,* 912.
Seip, M. and Trygstad, O. (1966): *Acta paediat. scand., 55,* 287.
Shizume, K., Matsuzaki, F., Irie, M. and Osawa, N. (1970): *Endocr. jap., 17,* 297.
Shizume, K., Demura, R., Ichikawa, K., Takano, K., Odagini, E., Maeda, T., Suda, T. and Demura, H. (1974): *Endocr. jap., 21,* 485.
Soyka, L.F., Bode, H.H., Crawford, J.D. and Flynn Jr, F.J. (1970): *J. clin. Endocr., 30,* 1.
Tanner, J.M., Whitehouse, R.H. and Takaishi, M. (1966): *Arch. Dis. Childh., 41,* 454, 613.
Tanner, J.M. and Whitehouse, R.H. (1967): *Brit. med. J., 2,* 69.
Tanner, J.M., Whitehouse, R.H., Hughes, P.C.R. and Vince, F.P. (1971): *Arch. Dis. Childh., 46,* 745.
Westphal, O. (1968): *Acta paediat. scand., Suppl. 182.*
Wright, J.C., Brasel, J.A., Aceto Jr, T., Finkelstein, J.W., Kenny, F.M., Spaulding, J.S. and Blizzard, R.M. (1965): *Amer. J. Med., 38,* 499.

METABOLIC AND THERAPEUTIC STUDIES ON 246 PATIENTS WITH ACROMEGALY TREATED WITH HEAVY PARTICLES

JOHN H. LAWRENCE

Donner Laboratory and Lawrence Berkeley Laboratory, University of California, Berkeley, Calif., U.S.A.

Acromegaly is a disorder usually caused by an eosinophilic tumor of the pituitary gland, with the resulting hypersecretion of growth hormone leading to the development of its many signs and symptoms. The life expectancy of these patients is reduced; one study reports that in 100 cases of acromegaly, 50% had died before the age of 50 years, and 79% by the age of 60 years (Evans et al., 1966); another study reports that the number of deaths in 194 patients was almost twice that expected from a matched general population (Wright et al., 1970). Surgical hypophysectomy was not possible until relatively recent years because of the need for cortisone; however, with the availability of replacement therapy, the neurosurgeons are now treating acromegaly by various techniques of hypophysectomy, usually followed by replacement therapy (Rand, 1966; Ray, 1971; Wilson et al., 1972; Hardy, 1973). The work reported here is based on the use of high energy heavy particles, with their special radiobiological properties making it possible to overcome the relative radioresistance of the pituitary gland and to treat this pituitary disorder successfully by a non-invasive method (Lawrence et al., 1970; Linfoot et al., 1975).

Since 1958 we have treated 246 patients with acromegaly, and have observed the subsequent return of growth hormone levels to normal with the relief of signs and symptoms in most patients. Significant clinical and metabolic improvement is generally seen when the growth hormone level has dropped to 10 ng/ml or less. This level was reached in 68% of the patients within 2 years, and in 90% within 5 years. Headache, the most frequent and troublesome symptom encountered, has either markedly improved or disappeared in most patients within 1 year. Lethargy and weakness improve following treatment. The acral enlargement has not progressed, but has decreased in one-third of the patients within 4 years after completing therapy. The typical coarse heavy facial appearance of these patients had undergone satisfying changes in one-third of the group followed for at least 4 years. Paresthesias, which were present in about 50% of the patients, improved in about one-half within 1 year, and in nearly all within 3 years. In addition, abnormalities in carbohydrate metabolism, including insulin resistance, diabetic-type glucose tolerance curves, and the presence of diabetes mellitus, disappear following treatment. The remaining normal pituitary gland continues to function and, thus far, only 38% of the patients have developed some degree of hypopituitarism as a result of achieving adequate control of their disease.

Thirty of the 246 patients (12%) have had previous pituitary surgery with recurrence of signs and symptoms, and 12 (5%) had previous pituitary surgery and were sent to us for prophylactic postoperative irradiation. It is also of interest to note that many of the patients referred to us for treatment have been judged by those referring them to us as being poor surgical risks. We have turned away no patients, or pre-selected patients on the basis of their general condition or the duration of their disease, provided that pneumoencephalographic and arteriographic studies have shown no supra- or extra-sellar extension.

There are 28 patients in our group (11%) whose growth hormone level did not fall below 10 ng/ml within 3 years and who either have subsequently undergone a second pituitary procedure (12 surgery and 3 pituitary irradiation) or still have elevated HGH levels. Of the latter 13 patients, 1 died with HGH level still 40 ng/ml (4 years after treatment); 7 have HGH levels still in the 23-50 ng/ml range (4-11 years after treatment); and 5 have HGH levels in the 11-17 ng/ml range (3-8 years after treatment). In an additional group of 11 patients whose HGH levels were between 13 and 26 ng/ml 3 years after treatment, we observed the subsequent reduction to less than 10 ng/ml; we anticipate a similar result in most of these patients with fasting HGH levels in the 11-17 ng/ml range (some already have HGH values suppressed to 10 ng/ml or less during the glucose tolerance test).

Our total experience of treating 246 patients over a period of 17 years demonstrates that good control of acromegaly can be achieved by this safe method, with a low incidence of side effects. Only 21 patients have died, most of the deaths being the result of cardiovascular and cerebrovascular complications in patients with cardiomegaly and long-standing hypertensive cardiovascular disease at the time of treatment. Also, as noted above, many of the patients were referred because they were considered to be poor surgical risks. Long-term studies would be remarkably better, we think, if the patients were treated early in the onset of their disease (average duration of acromegaly is 9 years before we treated them), before complications such as cardiovascular disease became advanced, and if our series of patients did not include a large number of 'failed' cases after other methods of treatment.

With passage of time of follow-up, the precent survival after treatment in terms of median survival times, and survival as compared to a sex-and-age matched cross section of the population is increasing. It is important that when large series of patients are treated by other methods, clinical and metabolic follow-up is carried out at regular intervals, and in 100% of the patients; thus, ultimately, chronological extension of comfortable life can be compared. Finally, if patients are treated earlier in their disease, survival should approach that of a sex-and-age matched group from the general population.

In addition to our work in Berkeley, this modality is currently employed for treating pituitary disorders at Harvard University (Kjellberg, 1975) and at centers in Moscow and Dubna in Russia (Abazov et al., 1971). Other centers which could treat pituitary disorders are the Brookhaven National Laboratory in Upton, Long Island, the Fermi National Accelerator Laboratory in Batavia, Illinois, the NASA-Langley Research Center at Langley Field in Virginia, the University of Uppsala in Sweden, the University of Zurich in Switzerland, Columbia University in New York (under construction), and Indiana University in Bloomington (under construction). The group at the Atomic Energy Research Establishment (AERE) in Harwell, England is planning to start their program soon (Hockaday et al., 1975).

REFERENCES

Abazov, V.I., Astrakhan, B.V., Blokhin, N.N., Blokhin, S.I., Bugarchov, B.B., Dzhelepov, V.P., Goldn, L.L., Kiseleva, V.N., Komarov, V.I., Kleinbock, Y.L., Khoroshkov, V.S., Lamonov, M.F., Minakova, E.I., Molokanov, A.G., Onosovsky, K.K., Pavlonsky, L.M.,

Ruderman, A.I., Reshetnikov, G.P., Salamoy, R.F., Shmakova, N.L., Savchenko, O.V., Stekolnikov, V.P., Shimchuck, G.G., Vajnberg, M.S., Vajnson, A.A. and Yarmonenko, S.P. (1971): *Use of Proton Beams in the USSR for Medical and Biological Purposes*. Joint Institute for Nuclear Research, E-5854, Dubna.

Evans, H.M., Briggs, J.H. and Dixon, J.S. (1966): In: *The Pituitary Gland, Vol. I*, pp. 439-491. Editors: G.S. Harris and B.T. Donnovan. University of California Press, Berkeley.

Hockaday, T.D.R., Laing, A.H., Welbourn, R.B. and Hartog, M. (1975): *Brit. med. J., 1*, 457.

Hardy, J. (1973): In: *Diagnosis and Treatment of Pituitary Tumors*, pp. 179-194. Editors: P.O. Kohler and G.T. Ross. ICS 303, Excerpta Medica, Amsterdam.

Kjellberg, R.N. (1975): In: *Tumors of the Nervous System*, pp. 145-174. Editor: H.G. Sedel. John Wiley and Sons, New York.

Lawrence, J.H., Tobias, C.A., Linfoot, J.A., Born, J.L., Lyman, J.T., Chong, C.Y., Manougian, E. and Wei, W.C. (1970): *J. clin. Endocr., 31*, 180.

Linfoot, J.A., Chong, C.Y., Lawrence, J.H., Born, J.L., Tobias, C.A. and Lyman, J.T. (1975): In: *Hormonal Proteins and Peptides, Vol. III*, pp. 191-246. Editor: C.H. Li. Academic Press, New York.

Rand, R.W., Solomon, D.H., Dashe, A.M. and Heuser, G. (1966): *Trans. Amer. neurol. Ass., 91*, 324.

Ray, B.S. (1971): *Surgical Treatment of Acromegaly*. The Leo Davidson Lecture, Albert Einstein College of Medicine, New York.

Wilson, C.B., Rand, R.W., Heuser, G., Levin, S., Goldfield, E., Schneider, V., Linfoot, J. and Hosobuchi, Y. (1972): *Calif. Med., 117/5*, 1.

Wright, A.D., Hill, D.M., Lowy, C. and Fraser, R. (1970): *Quart. J. Med., 39*, 1.

VII. Human chorionic somatomammotropin

ISOLATION AND CHARACTERIZATION OF BOVINE AND OVINE PLACENTAL LACTOGEN*

ROBERT E. FELLOWS, FRANKLYN F. BOLANDER, THOMAS W. HURLEY and STUART HANDWERGER

Departments of Physiology and Pharmacology, Medicine, and Pediatrics, Duke University Medical Center, Durham, N.C., U.S.A.

Although the discovery in 1962 by Josimovich and MacLaren of a hormone immunologically related to growth hormone in the human placenta was instrumental in bringing the subject of primate placental lactogens (chorionic somatomammotropins) into sharp focus, pioneering studies on lactogens of placental origin had been carried out even earlier in subprimate species, including the rat (Lyons, 1944), and mouse (Cerruti and Lyons, 1960). These set the stage for more recent investigations which have clearly demonstrated the presence of placental lactogenic hormones in a number of subprimate species including the rat (Matthies, 1967; Gusdon et al., 1970; Shiu et al., 1973; Kelley et al., 1975), mouse (Kohomoto and Bern, 1970), guinea pig (Kelley et al., 1973), hamster (Kelley et al., 1973), goat (Buttle et al., 1972), sheep (Forsyth, 1974; Kelley et al., 1973; Handwerger et al., 1974), cow (Forsyth and Buttle, 1972) and fallow deer (Forsyth, 1974). Although Gusdon et al. (1970) have reported immunological evidence for a placental protein similar to human placental lactogen (HPL) in the rabbit, pig, dog, cat and horse, data from other laboratories are contradictory, and the question of placental lactogens in these species remains unresolved.

We have undertaken the isolation and purification of placental lactogen from the sheep and cow for the purposes of chemical characterization (Hurley et al., 1975; Bolander and Fellows, 1976a) including sequence analysis, investigation of biological activities both in vivo and in vitro (Handwerger et al., 1974, 1976), and delineation of hormone receptor interaction (Bolander and Fellows, 1976b). Particular emphasis has been placed on development of a model system in the sheep for study of the physiological role of placental lactogen in pregnancy (Handwerger et al., 1975).

PURIFICATION OF OVINE PLACENTAL LACTOGEN (OPL)

Ovine placentas were obtained surgically from late-pregnant animals and stored at −20° until used. After partial thawing and careful rinsing in distilled water, cotyledons were homogenized in absolute ethanol and centrifuged at 9000 × g for 1 hour. All preparative steps were carried out at 4°. The dried precipitate was extracted with 0.1 M ammonium bicarbonate, pH 9.5, containing 1 mM phenylmethylsulfonyl fluoride. The soluble extract was precipitated with 35% ammonium sulfate, pH 6.5 and

* Supported in part by grants AM-12861 and HD-08722 from the National Institutes of Health, USPHS, and a grant from the National Foundation.

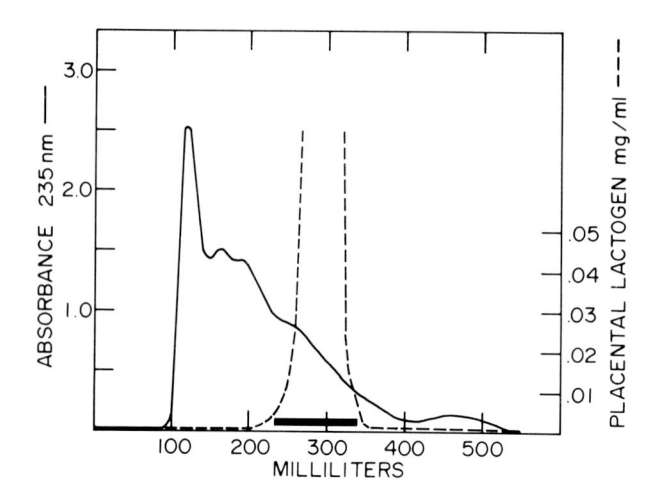

Fig. 1. Chromatography of ovine placental lactogen on Sephadex G-150. The 35-65% ammonium sulfate precipitate was dissolved in and eluted with 0.1 M ammonium bicarbonate, pH 9.5, at 4°. Material active in the mammary radioreceptor assay was pooled as indicated by the bar.

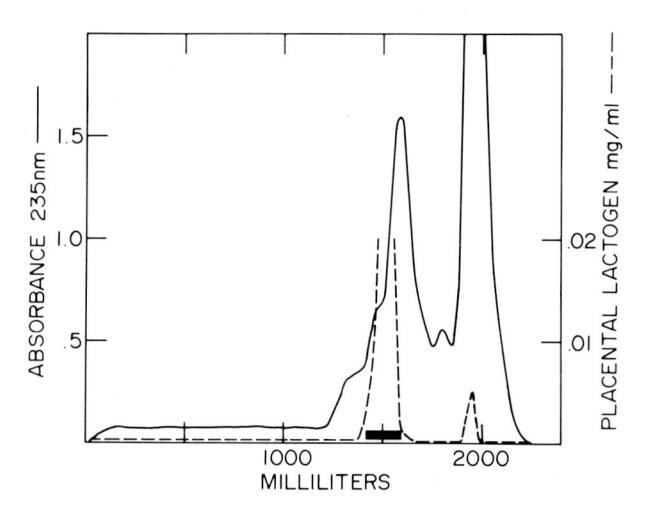

Fig. 2. Chromatography of ovine placental lactogen on DEAE-cellulose. The active fraction from Figure 1 was applied to a column, 2.5 × 35 cm, in 0.01 M Tris-HCl, pH 9.0, and eluted with a sodium chloride gradient.

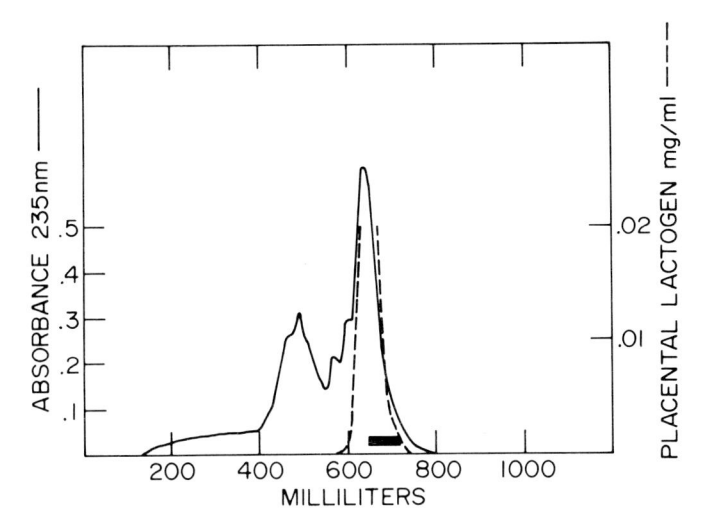

Fig. 3. Chromatography of ovine placental lactogen on CM-cellulose. The active fraction from Figure 2 was applied to a column, 2.5 × 18 cm, in 0.01 M ammonium acetate, pH 5.5, and eluted with a sodium chloride gradient.

TABLE 1

Isolation of ovine placental lactogen

Purification step	Total protein (mg)	Total activity (mg)	Specific activity (mg/mg)	Successive yield (%)	Cumulative yield (%)
NH$_4$HCO$_3$ extract, pH 9.5	7440	67	.009	100	
pH 6.5 supernatant	6580	72	.011	107	100
35-65% (NH$_4$)$_2$SO$_4$ precipitate	3008	15	.005	21	21
Chromatography on Sephadex G-150	329	12	.037	81	17
Chromatography on DEAE-cellulose	25	8.3	.332	68	12
Chromatography on CM-cellulose	4.8	4.7	.984	57	6.5

Starting material was 400 g of surgically obtained placental cotyledons. Specific activity was measured by lactogenic radioreceptor assay.

the precipitate was discarded. The supernatant was brought to 65% ammonium sulfate saturation to precipitate the active hormone which was immediately dissolved and chromatographed on Sephadex G-150 in 0.1 M ammonium bicarbonate, pH 8.3 (Fig. 1). A single peak of activity, determined by the mammary radioreceptor assay (Shiu et al., 1973), was observed in an elution position corresponding to that of an ovine growth hormone (OGH) calibration standard. The active fraction was concentrated and rechromatographed on a column of DEAE-cellulose eluted with 0.01 M Tris-HCl, pH 9.0 and a linear gradient to 0.05 M sodium chloride (Fig. 2). Again a single peak of activity was detected in the mammary radioreceptor assay, as indicated by the broken line. This material was concentrated and chromatographed on a column of CM-cellulose eluted with 0.01 M ammonium acetate, pH 5.5 and a linear gradient to 0.3 M sodium chloride (Fig. 3). Again activity was detected in a single peak in the radioreceptor assay. With adjustment of the buffer gradient to terminate at 0.2 M sodium chloride, this peak, indicated by the broken line, is eluted free of all other proteins.

The results of a typical preparative procedure is summarized in Table I. The final yield is approximately 5 mg of purified OPL from 400 g of wet tissue. Successive yields are adequate except in the ammonium sulfate precipitation step, where there is apparent partial inactivation of OPL with a consequent fall in specific activity. It is our experience that without the final cationic exchange chromatographic step on CM-cellulose, maximum purity of OPL does not exceed 35% to 60%.

CHARACTERIZATION OF OPL

Disc gel electrophoresis of purified OPL monomer at pH 9.0 and a sample concentration of 0.5 mg/ml revealed a single band with mobility slightly greater than that of the major band of OGH, and significantly less than the mobility of ovine prolactin (OPR). At higher concentration (5.0 mg/ml) a second band appeared which migrated more slowly than OGH and represented a stable aggregate of the OPL monomer.

The molecular weight of OPL was estimated by SDS polyacrylamide gel electrophoresis calibrated with 7 reference proteins including OGH and OPR (Fig. 4). These data demonstrate a mobility of OPL intermediate between OPR and OGH, consistent with a monomer molecular weight of slightly less than 23,000.

The amino acid composition of OPL was determined by analysis of duplicate 24-, 48-, and 72-hour hydrolysates (Table 2). From these data, the OPL molecule is estimated to contain 192 amino acid residues, compared with 191 residues for OGH and HPL, and 198 residues for OPR. In general, the amino acid composition of OPL resembles those of OGH and OPR and is somewhat less closely related to that of HPL. The most striking difference is the significantly lower content of the hydrophobic residue leucine in OPL by comparison to its content in the other 3 hormones.

The significant chemical properties of OPL are summarized in Table 3. The isoelectric point of the OPL monomer was determined by gel isoelectric focussing, and found to be 6.7, consistent with its relative mobility on disc gel electrophoresis. The aggregated form of OPL which appears at high concentrations of hormone in solution has a pI of 7.6 or more. While the molecular size of OPL appears to be intermediate between OGH and OPR, it is of particular interest that the molecule closely resembles OGH in total number of residues and in isoelectric point, but is identical with OPR and other prolactins in its content of 2 residues, tryptophan and half-cystine, which are highly conserved in the course of molecular evolution.

Immunological properties of OPL were investigated using high titer rabbit antiserum to purified OPL, bovine placental lactogen (BPL), bovine growth hormone

Subprimate placental lactogens

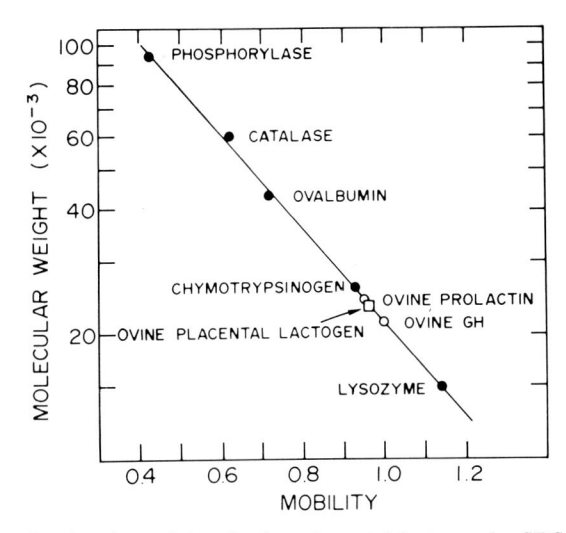

Fig. 4. Estimation of molecular weight of ovine placental lactogen by SDS polyacrylamide gel electrophoresis.

TABLE 2

Amino acid composition of ovine placental lactogen

Amino acid	OPL	OGH	OPR	HPL
Lysine	14.1 (14)	13	9	9
Histidine	4.0 (4)	3	8	7
Arginine	9.6 (10)	13	11	11
Aspartic acid	19.0 (19)	16	22	22
Threonine	9.9 (10)	12	9	12
Serine	15.4 (15)	12	15	18
Glutamic acid	23.6 (24)	25	22	24
Proline	10.1 (10)	8	11	5
Glycine	14.7 (15)	10	11	7
Alanine	12.8 (13)	14	9	6
Half-cystine	6.3 (6)	4	6	4
Valine	11.6 (12)	7	10	7
Methionine	3.9 (4)	4	7	6
Isoleucine	10.2 (10)	7	11	7
Leucine	13.3 (13)	22	22	25
Tyrosine	3.9 (4)	6	7	8
Phenylalanine	6.6 (7)	13	6	11
Tryptophan	2.2 (2)	1	2	1
Number of residues	192	191	198	191

Values are expressed as residues per molecule based on an estimated molecular weight of 22,000. Assumed integral residue numbers are given in parentheses.

TABLE 3

Chemical properties of ovine placental lactogen

Property	OPL	OGH	OPR	HPL
Molecular weight	22,500	21,700	23,300	21,600
Isoelectric point	6.7 (7.6)	6.85	5.73	
Number of residues	192	191	198	191
Number of tryptophan	2	1	2	1
Number of half-cystine	6	4	6	4

TABLE 4

Immunological activities of ovine placental lactogen

Antisera	OPL	OGH	OPR	HPL
anti-OPL	+	(+)	−	−
anti-BPL	+	(+)	−	−
anti-BGH	−	+	−	−
anti-OPR	−	−	+	−

(BGH), and OPR. By double immunodiffusion in agar (Table 4) it was shown that OPL was strongly precipitated by anti-OPL and anti-BPL, but not by anti-BGH or anti-OPR sera. In addition to immunoprecipitation by its own antiserum, BGH cross-reacted very weakly at 36 to 48 hours with both anti-OPL and anti-BPL, suggesting that it has a minor antigenic site similar to a major antigenic site of these placental lactogen molecules. These data provide additional evidence for a greater structural and hence evolutionary similarity between OPL and growth hormone than between the placental lactogen and prolactin, with which it shares its major biological activity.

The biological properties of OPL have been studied in our laboratories in a number of different assay systems (Handwerger et al., 1974, 1975, 1976). In the rabbit mammary radioreceptor assay, OPL is fully as active as OPR or HPL. It also has significant activity in the rabbit mammary intraductal assay and is 25% as potent as OPR in stimulating N-acetyllactosamine synthesis in rabbit mammary explants. In addition OPL has almost 20% of the potency of OGH in the liver radioreceptor assay, where OPR and HPL are considerably less potent. Finally, preliminary data indicate that unlike OPR and HPL it has measurable activity in the rat tibial width assay. Thus OPL, by contrast to HPL, seems to have retained significant amount of growth hormone-like activity in addition to predominant lactogenic properties throughout its evolutionary course.

ISOLATION OF BOVINE PLACENTAL LACTOGEN

The placental lactogen of bovine origin has been isolated from placentas removed surgically from late-pregnant cows in a manner closely similar to that described for the isolation of ovine placental lactogen (Bolander and Fellows, 1976a). A summary of the isolation procedure as applied to 1.5 kg of bovine cotyledons is given in Table 5. After homogenization in absolute ethanol, the powdered residue, representing 15% of the wet weight of cotyledons, was extracted overnight at 4° with 0.1 M ammonium bicarbonate, pH 9.5, containing 1 mM phenylmethylsulfonyl fluoride. The supernatant was fractionated with ammonium sulfate at pH 6.5 and most of the lactogenic receptor activity was obtained in a 45-65% precipitate. This was applied to Sephadex G-150 and eluted with 0.01 M Tris-HCl, pH 9.0, as a single symmetrical peak emerging slightly ahead of the elution position of BGH. By contrast with OPL, 2 peaks of activity were obtained when the active Sephadex pool was rechromatographed on DEAE-cellulose with 0.01 M Tris-HCl, pH 9.0. The first, BPL-1, appeared after a gradient from 0.03 M to 0.05 M sodium chloride, and the second, BPL-2, emerged during a gradient from 0.05 M to 0.065 M sodium chloride. Both active fractions were chromatographed separately on CM-cellulose, in 0.01 M ammonium acetate, pH 5.5 with gradients to 0.2 M sodium chloride. Each yielded single, distinct, symmetrical peaks of activity measured by the mammary radioreceptor assay. As indicated in Table 5, about 60 mg of purified hormone, with BPL-2 constituting the major fraction, was obtained from 1.5 kg of cotyledons. Again by contrast with OPL, the specific activity of the purified BPL is low. At present it is not clear whether this may be due to inactivation in the course of extraction and purification, or perhaps results from inadequacy of the rabbit mammary receptor for the assay of hormone from bovine placenta.

TABLE 5

Isolation of bovine placental lactogen

Purification step	Protein (mg)	Total activity (µg)	Specific activity (ng/mg)	Successive yields (%)	Cumulative yields (%)
NH$_4$HCO$_3$ extract	19,072	610	32		100
pH ammonium sulfate precipitation	5274	451	85	74	74
Sephadex G-150	2046	383	190	85	63
DEAE-cellulose:	355	172		45	41
BPL-1	149	64	430		
BPL-2	205	108	530		
CM-cellulose:	63	82			13
BPL-1	24	32	1310	50	
BPL-2	38	50	1310	46	

Starting material was 1.5 kg of placental cotyledons. Specific activity was measured in the lactogenic radioreceptor assay.

CHARACTERIZATION OF BOVINE PLACENTAL LACTOGEN

When BPL-1 and BPL-2 were subjected to disc gel electrophoresis at pH 9.0, they migrated identically, each as a pair of very closely spaced bands which may, by analogy with growth hormone, represent amide differences within each form. The mobility of BPL was intermediate between that of BGH and BPR, and slightly greater than that of the monomer form of OPL. Within the resolving power of the method there was no evidence of contamination of BPL by either BGH or BPR.

The molecular weight of BPL was determined by chromatography on Sephadex G-200 in 6 M guanidine hydrochloride containing 6.5 mM dithiothreitol (Fig. 5). The column was calibrated with 7 proteins (including BPR), ranging from 14,000 to 70,000 molecular weight. In this system the K_{av} of BPL-1 and BPL-2 were identical and consistent with a molecular weight of 22,150. By contrast, in SDS gel electrophoresis, both forms migrated as though they had a molecular weight of approximately 60,000.

The isoelectric point of BPL was determined by gel isoelectric focussing. Two bands were seen, a major one with a pI of 5.86, and a minor one with pI of 6.1. By disc gel isoelectric focussing, only one band was seen with both BPL-1 and BPL-2; it had a pI of 5.9. This value is consistent with the relatively greater mobility of BPL than OPL observed on analytical disc gel electrophoresis.

The amino acid composition of BPL-1 and BPL-2 were determined by duplicate analysis of 24-, 48- and 72-hour hydrolysates on a Beckman 121 M analyzer (Table 6). BPL differs significantly from either BGH or BPR only in the number of serine, gly-

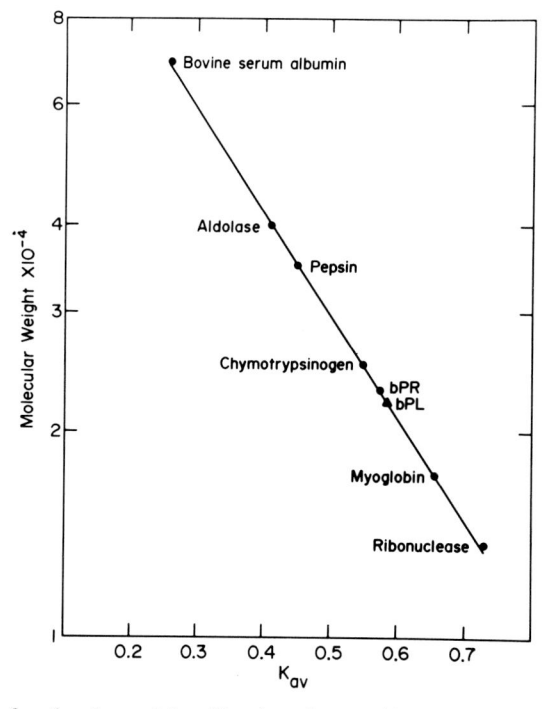

Fig. 5. Estimation of molecular weight of bovine placental lactogen by gel filtration on Sephadex G-200 in 6 M guanidine hydrochloride, 6.5 mM dithiothreitol.

TABLE 6
Amino acid composition of bovine placental lactogen

Amino acid	BPL-1	BPL-2	BGH	BPR
Lysine	11.7 (12)	11.8 (12)	11	9
Histidine	3.9 (4)	4.2 (4)	3	7
Arginine	9.1 (9)	9.0 (9)	13	11
Aspartic acid	21.0 (21)	20.8 (21)	16	22
Threonine	11.0 (11)	11.9 (12)	12	9
Serine	19.8 (20)	19.7 (20)	13	15
Glutamic acid	20.9 (21)	20.8 (21)	24	22
Proline	11.3 (11)	12.1 (12)	6	11
Glycine	18.7 (19)	19.3 (19)	10	11
Alanine	15.8 (16)	16.1 (16)	15	10
Half-cystine	6.1 / 6.1 (6)	6.1 / 6.1 (6)	4	6
Valine	8.9 (9)	10.2 (10)	6.5	9
Methionine	3.1 (3)	2.9 (3)	4	7
Isoleucine	5.1 (5)	4.9 (5)	7	11
Leucine	13.8 (14)	14.0 (14)	26.5	23
Tyrosine	6.1 (6)	5.9 (6)	6	8
Phenylalanine	6.2 (6)	5.9 (6)	13	6
Tryptophan	1.7 / 2.0 (2)	1.7 / 1.9 (2)	1	2
Amides	14.9 (15)	15.2 (15)	17	n.d.
Number of residues	196	199	191	199

Values are expressed as residues per molecule based on an estimated molecular weight of 22,000. Assumed integral residue numbers are given in parentheses.

TABLE 7
Properties of bovine placental lactogen

	BGH	BPL	BPR
Molecular weight	21,500	22,150	22,800
Isoelectric point	6.8	5.9	5.7
Tryptophan content	1	2	2
Cysteine content	4	6	6
COOH-terminus	-Cys · Ala · Phe	-Cys · Ala · Phe	-Asn · Asn · Cys
Net charge (pH 7.4)	+0.5	−6	−6
Cross-reaction to:			
BGH antisera	complete	partial	none
BPR antisera	complete	none	complete
BPL antisera	none	complete	none
OPL antisera	partial	partial	none

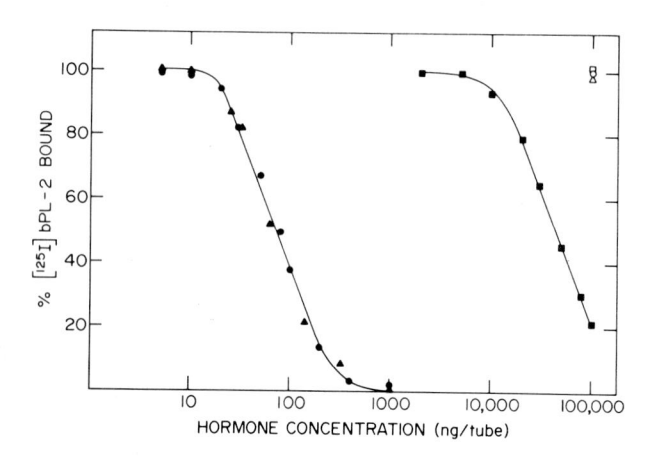

Fig. 6. Radioimmunoassay of bovine placental lactogen. Displacement curves are shown for BPL-1, ▲ ; BPL-2, ● ; and OPL, ■ . Displacement was not observed with BGH, □ ; BPR, ○ ; or HPL, △ .

cine and leucine residues. Changes in these amino acids would not be expected to significantly alter the ionic properties of OPL in a way which would be incompatible with either growth hormone- or prolactin-like activity. The differences between BPL-1 and BPL-2 are limited to single residue differences in threonine, proline and valine. While BPL-1 may represent a cleavage product of BPL-2, these differences are indeed minor and could merely reflect a relative insensitivity of the analytical methodology.

Selected properties of BPL are compared with those of BGH and BPR in Table 7. Like OPL, BPL is intermediate between growth hormone and prolactin in molecular size and isoelectric point. Like OPL, it also shows a very important identity with prolactin in terms of content of tryptophan and cysteine, and its net charge at physiological pH. Digestion with carboxypeptidase A and B, however, has revealed a carboxyl terminal sequence of cysteinyl-alanyl-phenylalanine, identical with the corresponding structure of BGH, but not with BPR. An average yield of 99.7% phenylalanine by carboxypeptidase digestion is convincing proof of the homogeneity of this material. BPL also shares some immunological properties with BGH, since on double immunodiffusion it forms a line of partial identity with BGH against BGH antisera. It also forms a precipitin line of partial identity with OPL against anti-OPL sera, but does not react with anti-OPR. Neither BGH nor BPR form a precipitate with antisera to BPL.

We have recently developed a specific and sensitive radioimmunoassay for bovine placental lactogen utilizing peroxidase-labeled BPL as tracer (Fig. 6). Both BPL-1 and BPL-2 have identical displacement curves which are linear between 20 and 200 ng of hormone. OPL displaces with a curve that is parallel to those of BPL-1 and BPL-2, but is shifted three orders of magnitude to the right. BGH, BPR, and HPR are not detected in this assay, which is currently being used to monitor placental lactogen levels in both dairy and beef cattle throughout pregnancy.

In summary, we have isolated both ovine and bovine placental lactogen in a state of complete homogeneity and have demonstrated that they share significant chemical and immunological properties with both subprimate growth hormone and prolactin.

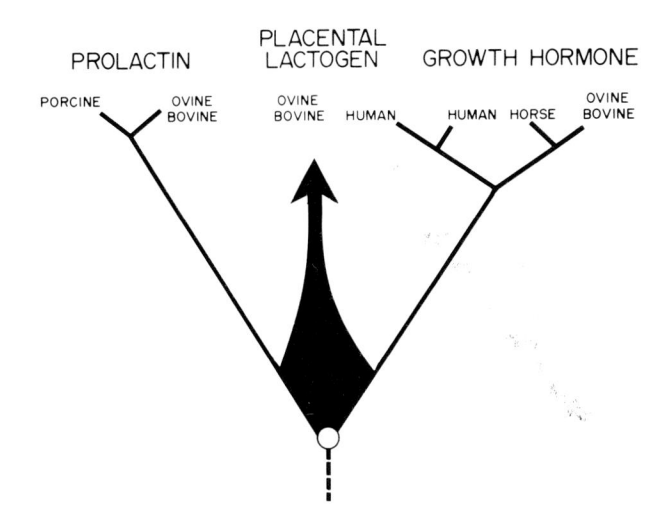

Fig. 7. A hypothetical evolutionary scheme for the growth hormone-prolactin superfamily. (Adapted from Dayhoff et al., 1975.)

Data derived from investigations of primate placental lactogens have led to the view that they have evolved relatively recently as a gene duplication of the growth hormone cistron (Dayhoff et al., 1975). Our studies on ovine and bovine placental lactogens suggest that they are structurally intermediate between growth hormone and prolactin and may represent, as schematically indicated in Figure 7, new evidence for a direct evolutionary line from the primitive precursor which gave rise to both BGH and BPR, rather than being recently divergent from either the growth hormone or prolactin line.

ACKNOWLEDGEMENTS

We wish to thank Dr. Wayne Chamley, Dr. L.C. Ulberg and Dr. M. Carlyle Crenshaw for gifts of placental tissue used in these studies and the Hormone Distribution Program, NIH, for provision of pituitary hormone reference materials.

REFERENCES

Bolander, F.F. and Fellows, R.E. (1976a): *J. biol. Chem.*, in press.
Bolander, F.F. and Fellows, R.E. (1976b): Submitted for publication.
Buttle, H.L., Forsyth, I.A. and Knaggs, G.S. (1972): *J. Endocr.*, 53, 483.
Cerruti, R. and Lyons, W.R. (1960): *Endocrinology*, 67, 884.
Dayhoff, M.O., McLaughlin, P.J., Barker, W.C. and Hunt, L.T. (1975): *Naturwissenschaften*, 62, 154.
Forsyth, I.A. and Buttle, H.L. (1972): In: *Abstracts, IV International Congress of Endocrinology, Washington D.C. 1972*, p. 106. ICS 256, Excerpta Medica, Amsterdam.
Forsyth, I.A. (1974): In: *Lactogenic Hormones, Fetal Nutrition and Lactation*, p. 49. Editor: J.B. Josimovich. John Wiley and Sons, New York.
Gusdon, J.P., Leake, N.H., Van Dyke, A.H. and Atkins, W. (1970): *Amer. J. Obstet. Gynec.*, 107, 441.

Handwerger, S., Maurer, W., Barrett, J., Hurley, T. and Fellows, R.E. (1974): *Endocrine Res. Commun., 1,* 403.
Handwerger, S., Maurer, W., Crenshaw, M.C., Hurley, T. Barrett, J. and Fellows, R.E. (1975): *J. Pediat., 87,* 1139.
Handwerger, S., Fellows, R.E., Crenshaw, M.C., Hurley, T., Barrett, J. and Maurer, W. (1976): *J. Endocr.,* in press.
Hurley, T., Fellows, R.E., Maurer W. and Handwerger, S. (1975): In: *Peptides: Chemistry, Structure and Biology,* pp. 583-588. Editors: R. Walter and J. Meienhofer. Ann Arbor Science Publ.
Josimovich, J.B. and MacLaren, J.A. (1962): *Endocrinology, 71,* 209.
Kelley, P.A., Shiu, R.P.C., Friesen, H.G. and Robertson, H.A. (1973): *Endocrinology, Suppl. 92,* A-233.
Kelley, P.A., Shiu, R.P.C., Robertson, M.C. and Friesen, H.G. (1975): *Endocrinology, 96,* 1187.
Kohomoto, K. and Bern, H.A. (1970): *J. Endocr., 48,* 99.
Lyons, W.R. (1944): *Anat. Rec., 88,* 446.
Matthies, D.C. (1967): *Anat. Rec., 159,* 55.
Shiu, R.P.C., Kelley, P.A. and Friesen, H.G. (1973): *Science, 180,* 968.

CHEMICAL STRUCTURE AND BIOLOGIC AND IMMUNOLOGIC ACTIVITY OF 'BIG' HUMAN PLACENTAL LACTOGEN*

ARTHUR B. SCHNEIDER, KAZIMIERZ KOWALSKI and LOUIS M. SHERWOOD

Department of Medicine, Michael Reese Hospital and Medical Center, and The University of Chicago, Pritzker School of Medicine, Chicago, Ill., U.S.A.

Recent studies from this laboratory identified a higher molecular weight form of human placental lactogen ('big' HPL) in extracts of human placenta and in the sera of pregnant women (Schneider et al., 1975a,b). 'Big' HPL occurs spontaneously and presumably represents a physiologic form of the hormone. Gel exclusion chromatography and polyacrylamide disc gel electrophoresis in sodium dodecyl sulfate (SDS) indicated that it had approximately twice the molecular weight of native monomeric HPL. The 'big' molecule was found to be analogous to corresponding forms of growth hormone and pituitary prolactin which have been identified in other laboratories (Goodman et al., 1972; Gorden et al., 1973a; Suh and Frantz, 1974).

The majority of 'big' HPL was stable to 8 M urea or 4 M NaSCN, indicating that covalent bonds accounted for its higher molecular weight. The nature of the covalent bonds became clearer when it was shown that 'big' HPL could be converted to the molecular weight of the native hormone by reduction with mercaptoethanol under non-denaturing conditions. It was therefore concluded that 'big' HPL was not a prohormone in the sense of proinsulin or proPTH, but a disulfide linked dimer, probably of the native molecule.

Despite these new observations, the biologic role and importance of 'big' HPL and the other 'big' hormones are poorly understood. The present studies were undertaken to provide additional information on the structure of 'big' HPL and to define its biologic and immunologic activity. For these studies, assays of immunoreactive, radioreceptor, and biologic activity of native and 'big' HPL were performed in parallel.

METHODS

Preparation of 'big' HPL

The methods for the purification of 'big' HPL from placenta have been reported in detail (Schneider et al., 1975a,b). Briefly, they consist of: (1) homogenization and extraction in ammonium bicarbonate followed by centrifugation; (2) preliminary precipitation of the supernatant with 20-40% ammonium sulfate; (3) gel filtration of resuspen-

---

* This work was supported by NIH grant No. HD08225, The John A. Hartford Foundation, Inc., and The Michael Reese Research Institute.
Arthur B. Schneider is the recipient of a USPHS Career Development Award No. 1-K04 AM 00103.

ded protein over Sephadex G-100 × 2; (4) anion exchange chromotagraphy on DEAE-cellulose with ammonium bicarbonate; and (5) affinity chromatography of the latter material, followed by a final purification on Sephadex G-100. Homogeneity of the final product was determined by polyacrylamide gel electrophoresis and amino-terminal end group analysis.

Characterization of 'big' HPL

Polyacrylamide disc gel electrophoresis in sodium dodecyl-sulfate (SDS) was carried out according to the methods of Weber et al. (1972). All proteins used to calibrate molecular weight were reduced by heating for 1 minute at 100°C in 1% mercaptoethanol-SDS. The gels were stained with Comassie blue.

The amino-terminal amino acid was determined by reaction with dansyl chloride (Woods and Wang, 1967). Approximately 1 nmole of 'big' HPL (determined by radioimmunoassay) in 20 µl of 0.1 M $NaHCO_3$ was reacted with 20 µl of dansyl chloride (1 mg/ml in acetone). After 16 hours at room temperature the reaction was stopped by evaporating the reagents to dryness and hydrolyzing the residual material in 6 N HCl at 110°C for 16 hours. The resulting sample was applied in acetone:acetic acid (3:2) to a polyamide sheet for 2 dimensional ascending chromatography using H_2O:90% formic acid (200:3) followed by benzene:acetic acid (9:1).

Radioimmunoassay of HPL

The radioimmunoassay of HPL was performed by the double antibody method described earlier (Schneider et al., 1975b). Antibodies to purified HPL were produced in guinea pigs.

Bioassay of HPL

The induction of N-acetyllactosamine synthetase in mammary explants from mid-pregnant mice was used for the bioassay of the hormone preparations. A modification of the method of Loewenstein et al. (1971) was employed. Mammary explants from mid-pregnant mice were incubated in Medium-199 on organ culture grids for 48 hours. Each concentration of HPL or 'big' HPL was assayed in triplicate or quadruplicate. The tissue was homogenized for the determination of enzyme activity which was expressed as DPM ^{14}C-N-acetyllactosamine per mg tissue protein (determined by the method of Lowry et al., 1951).

Radioreceptor (membrane binding) assay of HPL

The displacement assay of ^{125}I-iodo-HPL from mammary gland membranes of cortisol-HPL treated mid-pregnant rabbits was performed by a modification of the procedure of Shiu et al. (1973) and Shiu and Friesen (1974). The iodinated hormone was prepared by the lactoperoxidase method (Thorell and Johansson, 1971). Mammary gland membranes were prepared by differential centrifugation through 0.3 M sucrose, and the 100,000 × g pellet was used at a concentration of 200 µg protein in the final volume of 0.5 ml. Incubation was for 16 hours at room temperature with constant agitation. Average binding in the control tubes was 15%, and the sensitivity of the assay was 2.0 ng/ml.

RESULTS

Structure of 'big' HPL

To perform structural studies, 'big' HPL was purified by ammonium sulfate precipitation, Sephadex gel filtration, ion exchange chromatography, affinity chromatography, and finally Sephadex gel filtration. In order to elute the preparation of 'big' HPL from the affinity column, 4 M NaSCN was required. This is a denaturing solution which resulted in the conversion of a portion of 'big' HPL to the native molecule. Therefore, a final passage over a Sephadex column was used to obtain 2 peaks, i.e., 'stable big' HPL (the majority) and 'little derived from big' HPL. Both of these substances were run on SDS-polyacrylamide disc gel electrophoresis (Fig. 1, gels 1 and 3). After reduction, the positions of both bands were altered (gels 2 and 4). Reduced 'stable big' HPL migrated much more rapidly, indicating a fall in its molecular weight following reduction. 'Little derived from big' HPL migrated slightly slower, reflecting the further unfolding and increase in effective size. Both reduced molecules migrated at the same

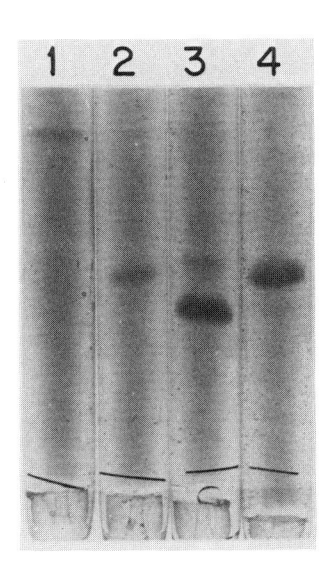

Fig. 1. SDS-polyacrylamide disc gel electrophoresis of unreduced 'stable big' HPL (1), reduced 'stable big' HPL (2), unreduced 'little derived from big' HPL (3) and reduced 'little derived from big' HPL (4). Reduction was carried out by boiling in 1% mercaptoethanol-SDS for 1 minute.

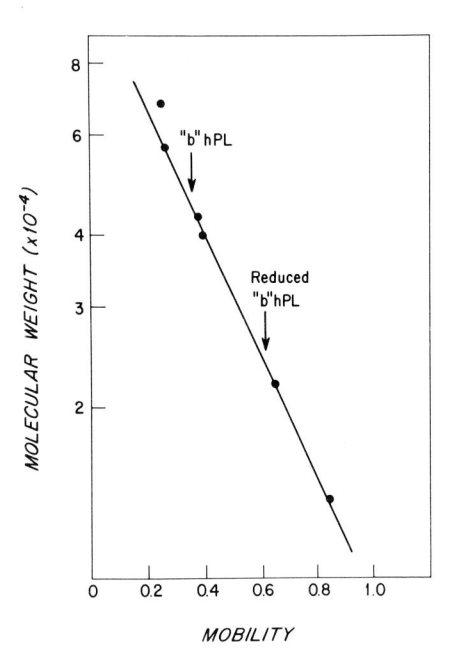

Fig. 2. SDS-polyacrylamide disc gel electrophoresis of unreduced and reduced 'stable big' HPL. Reduction of 'stable big' HPL and the standard proteins ribonuclease, HPL, aldolase, ovalbumin, pyruvate kinase, and human serum albumin was carried out by boiling in 1% mercaptoethanol-SDS for 1 minute. (From Schneider et al., 1975a, by courtesy of the Editors of *Biochemical and Biophysical Research Communications*.)

rate (gels 2 and 4), indicating their probable similarity. The calibration curve of the SDS-polyacrylamide electrophoresis is shown in Figure 2. The molecular weight of 'big' HPL was an estimate since only completely reduced molecules can be analyzed accurately by this method. The estimate indicate that it was approximately twice the size of the native hormone. Following reduction, it migrated to a position very near to, and probably identical with, native HPL.

Reduction of 'big' HPL under non-denaturing conditions (2% mercaptoethanol, 0.2% human albumin, in 0.05 M ammonium carbonate buffer at 24° for 16 hours) resulted in a shift on gel filtration of all of the protein to the elution position of native HPL. Amino-terminal analyses of 'big' and 'little derived from big' HPL were performed with the dansyl procedure. Both molecules showed valine as the only amino-terminal residue available to the dansyl reagent.

Activity of 'big' HPL

Due to the limited supplies of 'big' HPL, the initial activity studies have been performed on material prior to purification by affinity chromatography. As a result, both stable and unstable 'big' HPL were present in the following studies. With the antiserum used in our laboratory, HPL and 'big' HPL had identical displacement curves in the radioimmunoassay. Therefore, the concentration determined in the assay was used as the basis for comparison of biologic activity in the bioassay and radioreceptor assay.

Biologic acitivity was assessed by the induction of N-acetyllactosamine synthetase in explants of mid-pregnant mice maintained in organ culture. The results (Fig. 3) are expressed as the percent stimulation above control (the level found in explants incubated without HPL). Maximum stimulation was shown to be 2-3 times baseline and was achieved by 1000 ng/ml of HPL. The response to 'big' HPL was not distinguishable from that of native HPL, both in terms of the stimulation at any dose and the slope of the dose-response curve (p > 0.05).

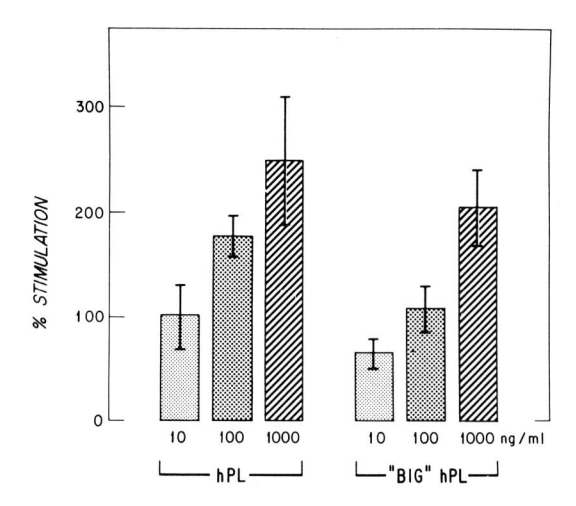

Fig. 3. Bioassay of HPL and 'big' HPL by induction of N-acetyllactosamine synthetase. The results are expressed as the percent stimulation compared to tissue incubated without hormone.

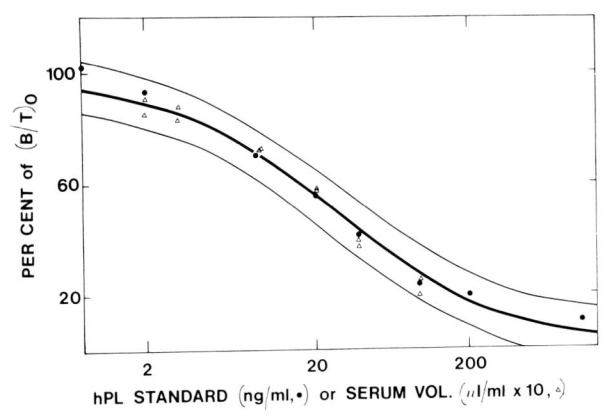

Fig. 4. Radioreceptor (membrane binding) assay of HPL by displacement from rabbit mammary membranes. The data were analyzed by a weighted linear regression through the log-logit transformation of the standard curve. The lines represent the calculated standard curve surrounded by its 95% confidence limits. A third trimester plasma sample (△) is compared with purified hormone (●).

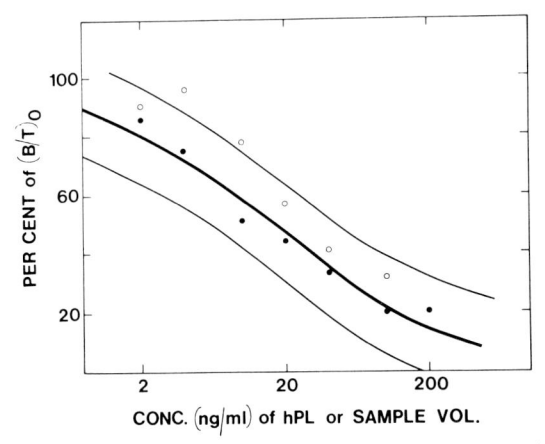

Fig. 5. Radioreceptor (membrane binding) assay of HPL and 'big' HPL. The displacement produced by native HPL (●) is shown with the calculated standard curve and its 95% confidence limits. A solution of 'big' HPL (1 μg/ml, based on immunologic activity) was added in volumes (μl/ml) designed to allow a direct comparison of HPL and 'big' HPL (o).

The binding assay to membranes derived from mammary glands of cortisol-HPL injected mid-pregnant rabbits was also used to test biologic activity. In order to validate the assay, the displacement of ^{125}I-iodo-HPL by a third trimester serum sample was demonstrated (Fig. 4). The displacement produced by the serum sample was parallel to that produced by purified HPL to at least 80% of total displacement. The comparative displacement of 'big' HPL and purified native HPL is shown in Figure 5. The axis and concentrations were selected so that the displacement curves could be compared on the basis of equal activity in the radioimmunoassay. The 'big' HPL curve (open circles) had a slight tendency to be displaced toward the right (i.e., for 'big' HPL to be less potent), but neither the positions nor the slopes of the curves were significantly different from those of native HPL (p > 0.05).

DISCUSSION

The results of our studies have shown that 'big' HPL consists of 2 fractions, a urea-stable fraction (the majority) and a smaller fraction which is urea-labile. The latter reverts to the native form in the presence of denaturing agents and probably represents non-covalent aggregation. The majority of 'big' HPL is a form which is linked covalently by disulfide bonds. Following reduction, under severe (boiling for 1 minute in 1% mercaptoethanol-SDS) as well as mild conditions (2% mercaptoethanol for 16 hours at 24° at mildly alkaline conditions), 'big' HPL is transformed to a molecule of half its molecular weight with characteristics both on gel filtration and polyacrylamide gel electrophoresis similar to native HPL. We suggest, therefore, that 'big' HPL is a disulfide-linked dimer probably of the native hormone. This is further supported by the existence of a single amino-terminal amino acid, valine, in both 'big' and native HPL and the absence of any other peptide noted after reduction.

The location of the disulfide bond or bonds in the dimer is of considerable interest and is currently being investigated by us. Native HPL has 2 intrachain disulfide bonds, a larger one between amino acid residues 53 and 165 and a smaller one closer to the carboxyl-terminal end between residues 182 and 189. It has been reported that selective reduction of 1 disulfide bond in HPL leaves 35% of its immunologic activity intact (Neri et al., 1972), while complete reduction virtually destroys its immunologic activity (Aloj et al., 1972). It is possible that the more easily reduced disulfide bond of HPL (probably carboxyl-terminal) is the one involved in the interchain bond of the 'big' form of the hormone.

The immunologic, radioreceptor (membrane binding) and mammotrophic activity of native and 'big' HPL are strikingly similar. The site or sites responsible for these effects are therefore not likely to be buried in the 3-dimensional conformation of the higher molecular weight form. These relationships will be clarified even further when each of these activities can be expressed on a molar basis for the 2 species of hormone, and the urea-stable and unstable fractions have been tested separately. It will be of considerable interest, for example, to determine whether 'little derived from big' HPL behaves identically to the native hormone.

The observations described here extend further the homologies already shown for HPL and the pituitary peptides, growth hormone and prolactin. Our earlier work showed striking similarities in the amino acid sequences of HPL and human growth hormone (Sherwood et al., 1971), and progress on the sequence of human prolactin to date also suggests that there will be many similarities there as well (Niall et al., 1973). Molecules similar in size to 'big' HPL have been reported both in pituitary extracts and plasma for growth hormone (Goodman et al., 1972; Gorden et al., 1973a) and prolactin (Suh and Frantz, 1974). Likewise recent reports for both growth hormone (Singh et al., 1974; Benveniste et al., 1975) and prolactin (Jacobs and Lee, 1975) suggest that the 'big' forms are disulfide-linked. Of interest is the observation of Gorden et al. (1973b) that 'big' growth hormone may be significantly less active in the radioreceptor assay than the native molecule, and Soman et al. (1975) have described different receptors for 'big' and 'little' growth hormones in the membranes of liver cells.

The similarities in findings for the 3 hormones and the presence of the 'big' forms in plasma suggest that they are of physiologic importance. However, disulfide dimers cannot be viewed as typical hormone precursors in the sense of proinsulin, proparathyroid hormone, proglucagon and others, and their significance remains obscure. Studies of biosynthesis of the larger forms, particularly under in vitro conditions where physiologic stimuli can be applied, may be important in resolving this problem.

REFERENCES

Aloj, S.M., Edelhoch, H., Handwerger, S. and Sherwood, L.M. (1972): *Endocrinology, 91*, 728.
Benveniste, R.B., Stachura, M.E., Szabo, M. and Frohman, L.A. (1975): *J. clin. Endocr., 41*, 422.
Goodman, A.D., Tanenbaum, R. and Rabinowitz, D. (1972): *J. clin. Endocr., 35*, 868.
Gorden, P., Hendriks, C.M. and Roth, J. (1973a): *J. clin. Endocr., 36*, 178.
Gorden, P., Lesniak, M.A., Hendricks, C.M. and Roth, J. (1973b): *Science, 182*, 829.
Jacobs, L.S. and Lee, Y-C. (1975): In: *Program, 57th Annual Meeting of the Endocrine Society, New York*, p. 83.
Loewenstein, J.E., Mariz, I.K., Peake, G.T. and Daughaday, W.H. (1971): *Endocrinology, 33*, 217.
Lowry, O.H., Rosebrough, N.J., Farr, A.L. and Randall, R.J. (1951): *J. biol. Chem., 193*, 265.
Neri, P., Arezzini, C., Canali, G., Cocola, F. and Tarli, P. (1972): In: *Growth and Growth Hormone*, p. 199. Editors: A. Pecile and E.E. Müller. ICS 244, Excerpta Medica, Amsterdam.
Niall, H.D., Hogan, M.L., Tregear, G.W., Segre, G.V., Hwang, P. and Friesen, H. (1973): *Recent Progr. Hormone Res., 29*, 387.
Schneider, A.B., Kowalski, K. and Sherwood, L.M. (1975a): *Biochem. biophys. Res. Commun., 64*, 717.
Schneider, A.B., Kowalski, K. and Sherwood, L.M. (1975b): *Endocrinology, 97*, in press.
Sherwood, L.M., Handwerger, S., McLaurin, W.D. and Lanner, M. (1971): *Nature (Lond.), 233*, 59.
Shiu, R.P.C. and Friesen, H.G. (1974): *Biochem. J., 140*, 301.
Shiu, R.P.C., Kelly, P.A. and Friesen, H.G. (1973): *Science, 180*, 968.
Singh, R.N.P., Seavey, B.K. and Lewis, U.J. (1974): *Endocrine Res. Commun., 1*, 449.
Soman, V., Marsh, P. and Goodman, A.D. (1975): In: *Program, 57th Annual Meeting of the Endocrine Society, New York*, p. 113.
Suh, H.K. and Frantz, A.G. (1974): *J. clin. Endocr., 39*, 928.
Thorell, J.I. and Johansson, B.G. (1971): *Biochim. biophys. Acta (Amst.), 251*, 363.
Weber, K., Pringle, J.R. and Osborn, M. (1972): In: *Methods in Enzymology XXVI. Enzyme Structure. Part C*, p. 3. Editors: C.H.W. Hirs and S.N. Timasheff. Academic Press, New York.
Woods, K.R. and Wang, K-T. (1967): *Biochim. biophys. Acta (Amst.), 133*, 369.

BIOLOGICAL ACTION OF HUMAN CHORIONIC SOMATOMAMMOTROPIN DURING PREGNANCY — ITS LIPOLYTIC ACTION AND FETAL GROWTH*

S. TOJO, M. MOCHIZUKI, H. MORIKAWA and Y. OHGA

Department of Obstetrics and Gynecology, Kobe University, School of Medicine, Kobe, Japan

During the last decade a major impetus to the study of placental protein and polypeptide hormones was provided by Josimovich and MacLaren (1962), who discovered that crude placental extracts cross-react with antiserum to human growth hormone (HGH).

Several different names have been assigned to this protein: placental lactogen, chorionic growth hormone prolactin (CGP), purified placental protein (PPH), and human chorionic somatomammotropin (HCS).

This hormone, HCS, is believed to be synthesized by the syncytiotrophoblast of the placenta and is secreted mainly into the maternal circulation; in the fetus itself the levels are some two orders of magnitude less. The amount synthesized is of the order of 1 g/day in the last trimester and this rate of production is greatly in excess of that of any other placental peptide hormones.

The structure of HCS is very similar to that of HGH (Li et al., 1971). But in contrast to growth hormone, HCS has only weak somatotrophic action, whereas its mammotrophic and lactogenic effects are only a little less than those of prolactin. It has been suggested that it is responsible for the well-known diabetogenic effect of pregnancy, and that this function adjusts glucose and fat metabolism in favor of the fetus. However, the physiological actions of HCS are virtually unknown.

For this reason it seemed desirable to re-examine the biological significance of HCS as a metabolism-regulating hormone during pregnancy.

MATERIAL AND METHODS

This study was carried out in 3 steps. First, HCS was isolated from fresh placenta and purified. Then the biological character of HCS as a metabolism-regulating factor affecting the glucose-fatty acid cycle in rats and their fetal growth was investigated. Finally, we studied the mode of action of HCS on lipolysis in adipose tissue.

No international reference standard is available. A 'Kobe-HCS' was therefore prepared; free fatty acid (FFA) mobilizing activity and cross-reactivity against anti-human growth hormone (HGH) were used as activity indices (Morikawa et al., 1971; Tojo and Mochizuki, 1971). The Kobe-HCS appeared as single band in 7% polyacrylamide disc gel electrophoresis and had a uniform sedimentation pattern in the ultracentrifuge.

* Supported in part by a grant for research on handicapped children from the Ministry of Health and Welfare of Japan.

The effect of HCS on fetal growth was examined as follows. Pregnant Wistar or Sprague-Dawley rats (200-280 g b.w.) were injected daily with 15, 50 and 100 µg of HCS from the 3rd or the 10th to the 21st day of pregnancy, when laparotomy was performed and the fetuses were removed. Body weight, body length, tibial length, liver, heart and kidney weights of the fetus were measured.

At the same time, glucose, FFA, triglyceride (TG) and total nitrogen (TN) in the maternal blood and fetal carcass were estimated by the Hoffman (1937) method for a Technicon Auto Analyzer (1965), Dole's method (1956), Fletcher's acetylacetone method (1968) and Koch and Hanke's method (1948), respectively. The glycogen content of liver and heart was measured by Good's method (Good et al., 1932). In a further experiment, the movement of these substances between mother and fetus in the final stage of pregnancy was examined after administration of HCS (50 to 100 µg per day) to 3-day-fasted mother rats. Subsequently, using oleic acid-1-^{14}C and ^{14}C-D-glucose, the movement of glucose and FFA between mother and fetus and TG content of the fat tissue was examined.

Labeled oleic acid and D-glucose were administered to the pregnant rats prior to HCS treatment, and 4 hours later, ^{14}C-FFA was extracted by Dole's method. ^{14}C-TG was extracted with Bloor's solution (ethanol-ether, 3:1, v/v) and separated by silica gel thin-layer chromatography for counting. ^{14}C-glycogen of the liver was extracted by Good's method and reduced to ^{14}C-glucose. ^{14}C-glucose and ^{14}C-FFA of maternal blood and fetal carcass and ^{14}C-TG of the fat tissue of the mother were estimated by the liquid scintillation counter.

To determine the action of HCS on lipid metabolism, adenylcyclase, adenosine 3',5'-cyclic monophosphate (cAMP), protein kinase and hormone-sensitive lipase (HSL) in rat epididymal adipose tissue were determined by the methods of Salomon et al. (1974), Gilman (1970), Yamamura et al. (1971) and Rizack (1961) with modification by Matsubara (1970), respectively, in experiments in vitro and in vivo. FFA was also determined by the method of Dole.

RESULTS AND DISCUSSION

The effect of HCS on fetal growth and the glucose-fatty acid cycle in normal pregnant rats

First, the effect on the growth of fetuses in normal pregnant rats was observed during continuous administration of HCS from the early or middle to the final stage of pregnancy.

The results are shown in Figure 1. The fetuses of mothers treated with HCS were noted to have increased body, liver and heart weights, longer tibiae, and more glycogen in the liver and heart than the fetuses of the control group. The increases were most marked for body weight and liver glycogen, 25% and 100%, respectively. The mother rats treated with HCS showed significant increases in blood glucose, FFA and total N content (Fig. 2). The fetuses also showed similar increases in parallel.

These interesting results showed that continuous administration of HCS to pregnant rats significantly increased glucose, FFA, TG and total N content in the maternal blood and at the same time promoted growth of the fetus. The fetal body weight, glycogen content in each organ, and glucose, FFA, TG and total N contents of the body had a positive dose-response relationship with the administered HCS. It was noted that all the effects were greater if HCS was administered from the middle to the final stage rather than from the beginning of pregnancy (Mochizuki et al., 1972; Tanaka, 1972).

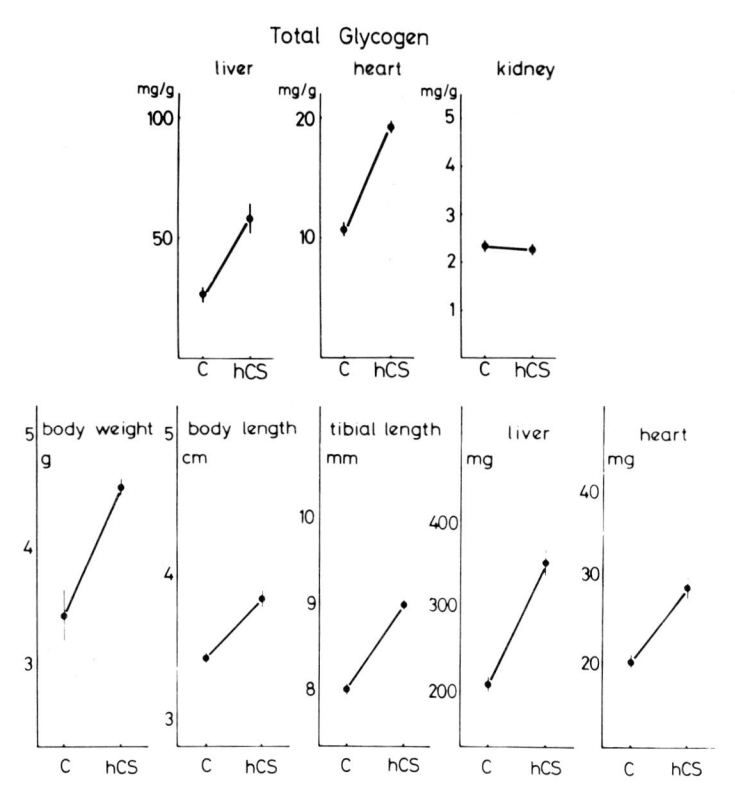

Fig. 1. Effect of HCS on fetal growth and total glycogen in the fetus. Pregnant rats were given intraperitoneally 50 µg of HCS daily from days 10-20 of pregnancy. The value represents the mean ± SE. The standard error is indicated by the vertical bracket. C denotes control group. The body weights and liver glycogen content of fetuses of mothers treated with HCS were significantly higher compared with those of controls ($P < 0.001$).

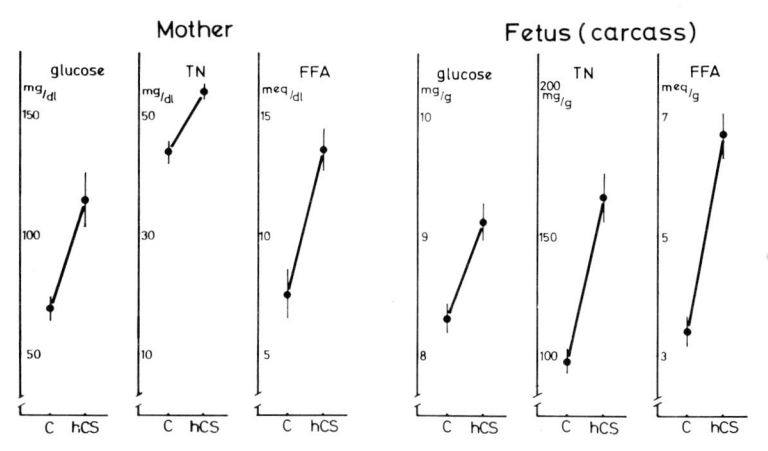

Fig. 2. Effect of HCS on the contents of glucose, total nitrogen and free fatty acid of mother and fetus. Pregnant rats were given intraperitoneally 50 µg of HCS daily from days 10-20 of pregnancy. The value represents the mean ± SE. C denotes control. All substances were significantly increased in the maternal blood and fetal carcass in the mother rats treated with HCS ($P < 0.001$).

The effect of HCS on fetal growth and the glucose-fatty acid cycle in 3-day-fasted mother rats

To investigate in more detail the effect of HCS on the growth of the fetus, the movement of substances between mother and fetus was examined in the final stage of pregnancy after administration of HCS to 3-day-fasted mother rats.

The fetuses of the fasted, HCS-treated group had similar body, liver and heart weights and glycogen contents to the non-fasted group (Fig. 3), although fasted control fetuses displayed dramatic reductions. Maternal blood had significantly increased FFA and decreased glucose after HCS. These changes in the fetuses and mother rats were more obvious in the group which received continuous administration of HCS from days 10-20 of pregnancy.

Thus fat mobilization following HCS administration to pregnant rats is more marked in fasting conditions, indicating that the energy source of the mother depends upon FFA with glucose being conserved for utilization in fetal growth (Mochizuki, 1973).

The effect of HCS on the movement of oleic acid-1-^{14}C and ^{14}C-glucose between mother and fetus

The content of ^{14}C-FFA or cold FFA of blood and ^{14}C-TG of fat tissue increased markedly in the mother rats treated with HCS (Fig. 4). The fetuses from the HCS-treated mother rats also showed ^{14}C-FFA, cold FFA and a marked increase in ^{14}C-TG. Glucose concentrations increased, although ^{14}C-D-glucose decreased in maternal blood after HCS (Fig. 5), while in fetal livers both cold and labeled glucose concentrations were about twice those of controls. It is possible that HCS crosses the placenta to produce the fetal changes.

Therefore, after laparotomy of pregnant rats on the 20th day, HCS was injected into the peritoneal cavity of the fetuses in one uterine horn, but not the other. Four hours

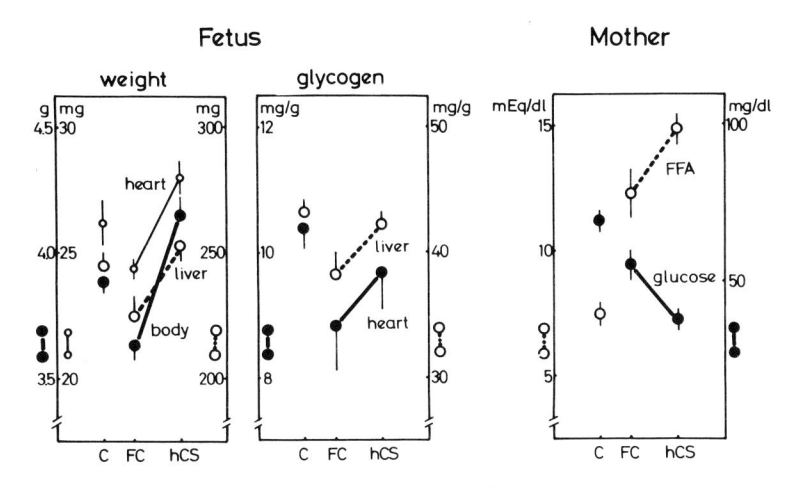

Fig. 3. Effect of HCS on fetal weight and glycogen or FFA content and glucose in 3-day-fasted mother rats. HCS (100 μg per day) was injected into the peritoneal cavity of the fasting mother rats from days 18-20 of pregnancy. The value represents the mean ± SE, C and FC denote non-fasted control and fasting control, respectively. The body weight and liver glycogen content of fetuses from mothers treated with HCS were not decreased compared with those in the non-fasted control.

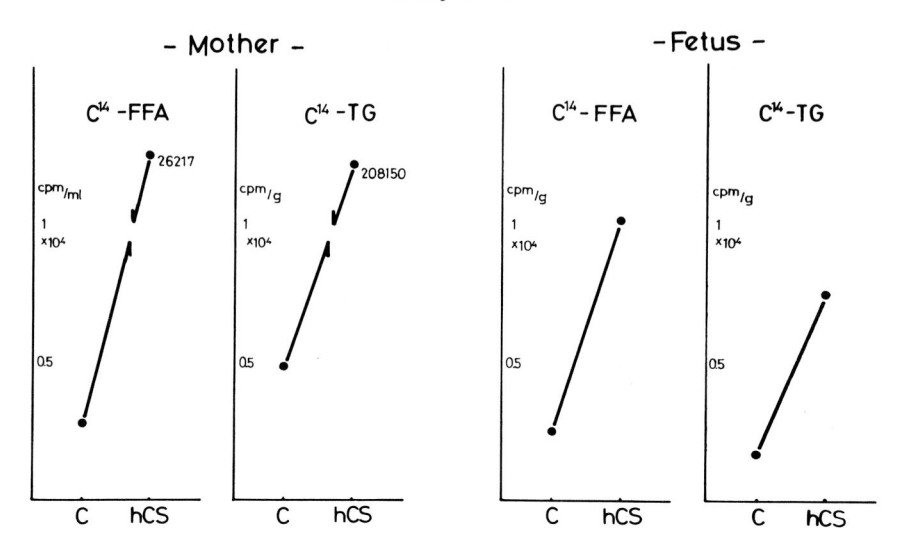

Fig. 4. Effect of HCS on the movement of oleic acid-1-^{14}C between mother and fetus. Emulsion composed of oleic acid-1-^{14}C 100 µCi, cold oleic acid 0.2%, glucose 0.1% and BSA 0.1% was injected intravenously, and thereafter (6 hours) HCS (100 µg) was given to rats intraperitoneally on day 16 of pregnancy. Four hours later, labeled FFA and TG were measured.

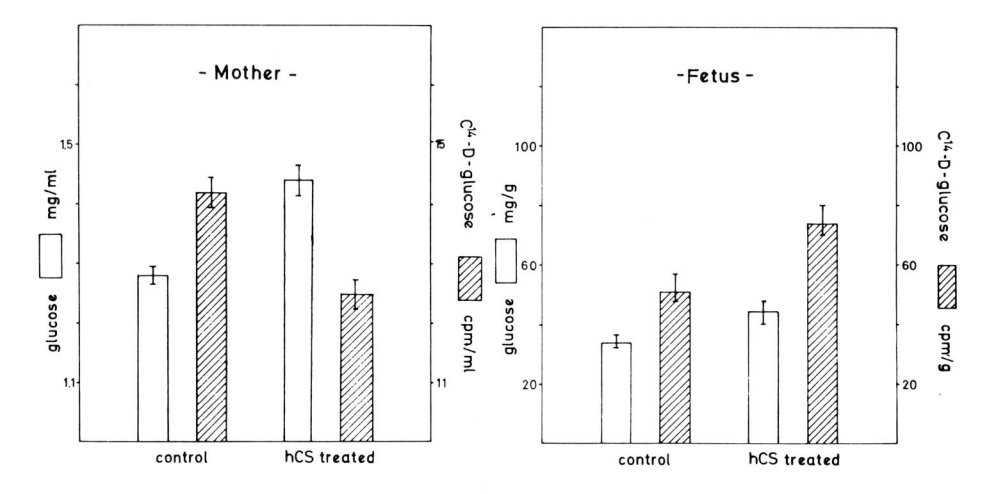

Fig. 5. Effect of HCS on the movement of ^{14}C-D-glucose between mother and fetus. HCS (100 µg) was injected intraperitoneally into rats on day 16 of pregnancy after intravenous administration of an emulsion composed of ^{14}C-D-glucose 17 µCi, cold glucose 0.1%, cold oleic acid 0.2% and BSA 0.1%; 10 hours later labeled glucose and cold glucose were measured. The value represents the mean of 5 animals. The standard error is indicated by the vertical bracket ($P < 0.01$ vs corresponding control).

later, FFA, TG, glucose and total N content of the fetuses in both horns were measured and compared. There were no changes in the amounts of these substances in the mother rats or in the fetuses to which HCS was directly administered. Thus the effects of HCS administration to the mother rats are completely different from those produced by direct administration to the fetuses.

The effect of HCS on the level of adenyl cyclase and cyclic AMP in rat epididymal fat pad in vitro

Administration of HCS (20 µg or 2 µg) increased adenyl cyclase activity in the membrane fractions of epididymal adipose tissue (Fig. 6) though not to the same extent as epinephrine (10^{-4} M). The concentration of cyclic AMP was increased by 35% 5 minutes after addition of 50 µg/ml HCS, but started to decrease at 15 minutes (Fig. 7). When both HCS and caffein (1 mM) were used, the concentrations of cyclic AMP at 5 and 15 minutes were raised by 73% and 36% respectively.

In contrast to the effects of HCS or epinephrine, HGH produced no significant changes in the concent of cyclic AMP (Fig. 8). On the other hand, adipose tissue released FFA into the medium at an accelerated rate for the duration of the incubation period when exposed either to epinephrine or HCS (Mochizuki et al., 1975).

The effect of HCS on protein kinase activity and FFA release of rat epididymal adipose tissue in vitro

Addition of HCS significantly increased the protein kinase activity in adipose tissue after 10 minutes of incubation (Fig. 9). A significant but less pronounced increase in the activity was brought about by HCS after 20 minutes. Without hormones in the medium, FFA was undetectable after 10 minutes, and its concentration was 0.47 ± 0.01 µEq./g at 20 minutes. The respective values were 0.92 and 1.66 µEq./g when HCS was added to the incubation mixture (Mochizuki et al., 1975).

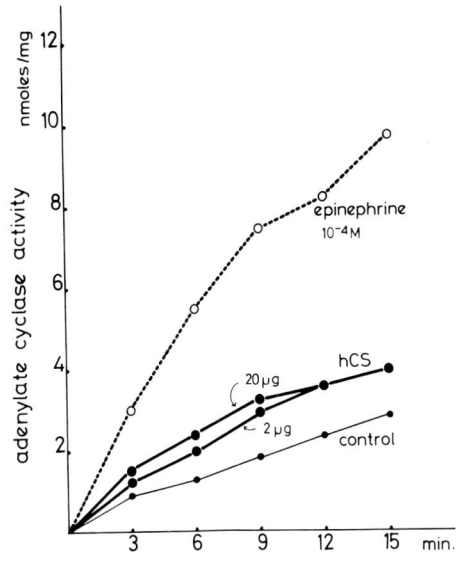

Fig. 6. Time course of adenyl cyclase in the membrane fraction of rat epididymal fat pad exposed to HCS. The value is the mean of 5 samples. Increased adenyl cyclase in the membrane fraction was detectable after addition of 2 µg of HCS per tube.

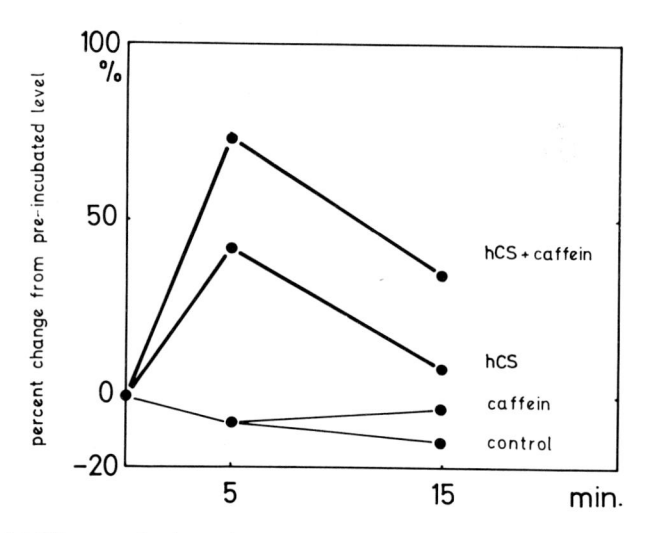

Fig. 7. Effect of HCS on cyclic AMP level in rat epididymal fat pad in vitro. The value is the mean of 3 samples and is expressed as percent increase or decrease from that immediately after the preincubation (0.23 pmole/mg wet tissue). The cyclic AMP level in HCS or HCS + caffein group is significantly higher compared with that of control or caffein group (P < 0.001).

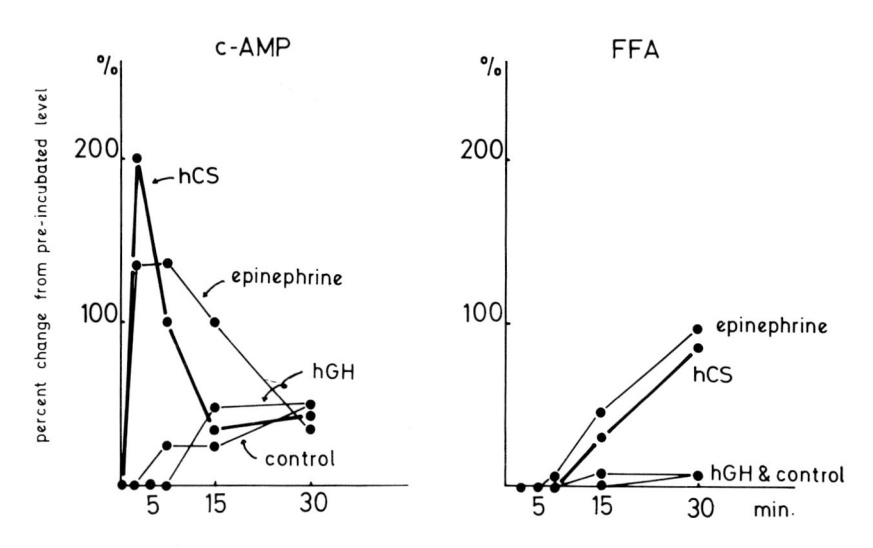

Fig. 8. Time course of cyclic AMP and FFA release from rat epididymal fat pad exposed to HCS, HGH and epinephrine. The value is the mean of 3 samples and is expressed as percent increase from that immediately after the preincubation. The levels of cyclic AMP and FFA release are significantly higher in the HCS group than in the HGH group (P < 0.01).

The effect of HCS on lipid metabolism in vivo
This effect was studied by observing the changes in the cyclic AMP and HSL content in epididymal adipose tissue and the changes in the serum level of FFA in rats after injection of HCS at a dose of 100 μg/100 g.

Cyclic AMP in the epididymal adipose tissue started to increase at 15 minutes, and then rapidly decreased to the pretreatment level by 60 minutes, a state maintained until the end of the observation period of 240 minutes (Fig. 10).

The activity of hormone-sensitive lipase started to rise at 5 minutes, attained a peak at 30 minutes, and began to decrease, first gradually and then rapidly after 90 minutes to about one half of the pretreatment level at 240 minutes. The serum level of FFA began to rise 5 minutes after administration of HCS and continued to increase rapidly until 90 minutes and then more slowly, throughout the observation period (Mochizuki et al., 1975).

Huttunen et al. (1970) reported that cyclic AMP-mediated activation of the lipase involves a phosphorylation step. The results of our experiment appear to indicate that HCS exerts its lipolytic action via the adenyl cyclase-cyclic AMP system both in vitro and in vivo. Grumbach et al. (1966) reported a rise in free fatty acids in hypopituitary subjects after the administration of HCS. Turtle and Kipnis (1967) reported that HCS stimulated lipolysis in isolated fat cells after 60-90 minutes and that this activation was inhibited by actinomycin D or puromycin, suggesting that HCS-activated lipolysis

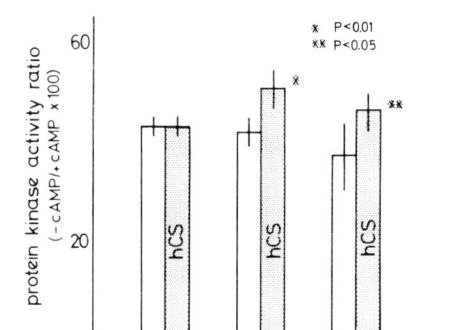

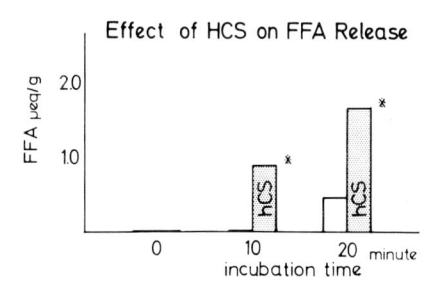

Fig. 9. Effect of HCS on protein kinase activity and FFA release of rat epididymal fat pad in vitro. The value represents the mean of 5 samples. The vertical bracket denotes the standard error.
* $P < 0.01$ vs corresponding control. ** $P < 0.05$ vs corresponding control.

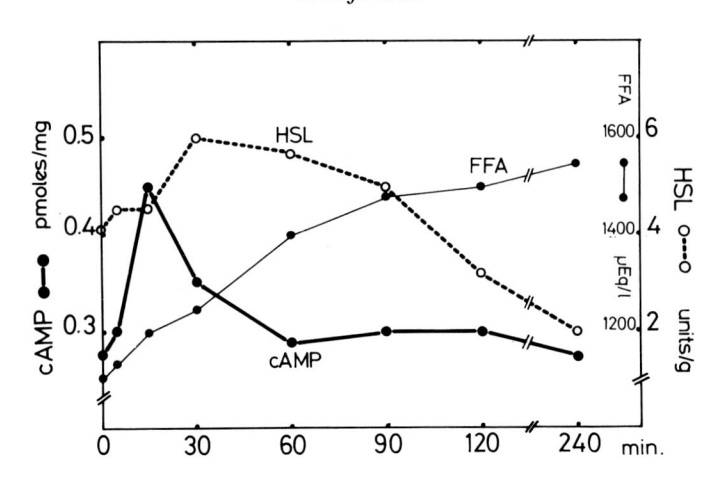

Fig. 10. Time course of cyclic AMP, hormone-sensitive lipase in rat epididymal fat pad and of serum FFA level after administration of HCS in vivo. Male S-D rats were fasted for 12 hours, and thereafter HCS (100 μg/100 g b.w.) was administered intraperitoneally. Each value represents the mean of 5 observations. The levels of cyclic AMP, HSL and FFA in rats treated with HCS increased, but no change was observed in non-treated controls.

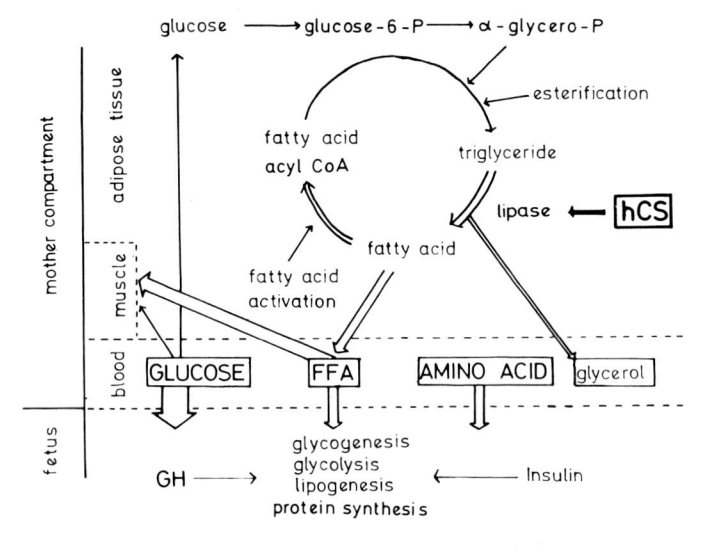

Fig. 11. The suspected physiological role of HCS in materno-fetal substrate supply, and in particular the relationship between HCS and the glucose-fatty acid cycle.

is dependent on synthesis of new protein. Genazzani et al. (1969), and Strange and Swyer (1974) also reported lipolytic activity of HCS shown by the release of glycerol and FFA from isolated epididymal fat cells of rats, although Friesen (1965) did not observe any such FFA releases. Grumbach et al. (1968) postulated that HCS influences the metabolism of the mother to ensure adequate supplies of glucose, amino acids, and minerals for the developing fetus. Our own results support this concept.

In brief, as shown in Figure 11, HCS is secreted principally into the mother's circulation, where it stimulates lipolysis, leading to an increase in serum FFAs. These furnish energy for maternal metabolism and, in addition, the increase in intracellular FFAs in the adipose tissue inhibits glucose utilization and causes hyperglycemia, which influences fetal metabolism via the placenta. The transported glucose may be utilized as an energy source by the fetus, and the transported FFA may be synthesized with α-glycerophosphate into neutral fat. Furthermore, amino acids, transported into the fetus through the placenta, participate in protein synthesis with a hydrogen ion from glycolysis, themselves influenced by HGH from the fetal pituitary and insulin (Mochizuki et al., 1972, 1973).

CONCLUSION

HCS given to the mother increased fetal body weight, total glycogen, total nitrogen, and triglycerides, but had no such effect when given directly to the fetus. HCS has an intrinsic lipolytic action; its effects are primarily on maternal metabolism producing important materials of maternal origin for fetal growth.

ACKNOWLEDGMENTS

The authors wish to express their particular thanks and appreciation to Dr. R.W. Bates, NIAMDD, NIH, Bethesda and Dr. A.E. Wilhelmi, Emory University, Atlanta, Georgia, for the generous supply of human growth hormone.

REFERENCES

Dole, V.F. (1956): *J. clin. Invest., 35,* 150.
Fletcher, M.J. (1968): *Clin. chim. Acta, 22,* 393.
Friesen, H. (1965): *Nature (Lond.), 208,* 1214.
Genazzani, A.R., Benuzzi-Badoni, M. and Felber, J.P. (1969): *Metabolism, 18,* 593.
Gilman, A.G. (1970): *Proc. nat. Acad. Sci. (Wash.), 67,* 305.
Good, C.A., Kramer, H. and Somoggi, M. (1932): *J. biol. Chem., 100,* 485.
Grumbach, M.M., Kaplan, S.L., Abrams, C.L., Bell, J. and Conte, F.A. (1966): *J. clin. Endocr., 26,* 478.
Grumbach, M.M., Kaplan, S.L., Sciarra, J.J. and Burr, I.M. (1968): *Ann. N.Y. Acad. Sci., 148,* 501.
Hoffman, W.S. (1937): *J. biol. Chem., 120,* 51.
Huttunen, J.K., Steinberg, D. and Mayer, S.E. (1970): *Proc. nat. Acad. Sci. (Wash.), 67,* 290.
Josimovich, J.B. and MacLaren J.A. (1962): *Endocrinology, 71,* 209.
Koch, F.C. and Hanke, M.E. (1948): *Practical Methods in Biochemistry.* Williams and Wilkins, Baltimore.
Li, C.H., Dixon, J.S. and Chung, D. (1971): *Science, 173,* 56.
Matsubara, Y. (1970): *Folia endocr. jap., 46,* 163 (in Japanese).

Mochizuki, M., Morikawa, H., Tanaka, Y. and Tojo, S. (1972): In: *Abstracts, IV International Congress of Endocrinology, Washington 1972,* Abstr. Nr. 37. ICS 256, Excerpta Medica, Amsterdam.
Mochizuki, M. (1973): *J. Jap. obstet. gynec. Soc., 25,* 1943 (in Japanese).
Mochizuki, M., Morikawa, H., Ohga, Y. and Tojo, S. (1975): *Endocr. jap., 22,* 123.
Morikawa, H., Mochizuki, M. and Tojo, S. (1971): *Endocr. jap., 18,* 417.
Rizack, M.A. (1961): *J. biol. Chem., 236,* 657.
Salomon, Y., Londos, C. and Rodbell, M. (1974): *Analyt. Biochem., 58,* 541.
Strange, R.C. and Swyer, G.I.M. (1974): *J. Endocr., 61,* 147.
Tanaka, Y. (1972): *Folia endocr. jap., 48,* 698 (in Japanese).
Technicon Auto Analyzer Methodology (1965): *Method file N-9a.*
Tojo, S. and Mochizuki, M. (1971): *Saishin Igaku, 26,* 1148 (in Japanese).
Turtle, J.R. and Kipnis, D.M. (1967): *Biochim. biophys. Acta (Amst.), 144,* 583.
Yamamura, H., Kumon, H., Nishiyama, K., Takeda, M., and Nishizuka, Y. (1971): *Biochem. biophys. Res. Commun., 45,* 1560.

EFFECTS OF HUMAN CHORIONIC SOMATOMAMMOTROPIN ON THE MALE REPRODUCTIVE APPARATUS OF RODENTS AND ON PLACENTAL STEROIDS DURING HUMAN PREGNANCY

P. NERI[1], C. AREZZINI[1], C. FRUSCHELLI[2], E.E. MÜLLER[3], P. FIORETTI[4] and A.R. GENAZZANI[4]

[1] Research Centre, Sclavo Institute, Siena, [2] Institute of Histology, University of Siena, [3] Department of Experimental Endocrinology, University of Milan, and [4] Institute of Obstetrics and Gynecology, University of Cagliari, Italy

The present paper is concerned with the effects of human chorionic somatomammotropin (HCS) on the male reproductive system of rodents, and with the effects of HCS upon threatened abortion and premature labour in humans.

Effects of HCS on the male reproductive system in rodents

Although HCS has closer structural similarities to human growth hormone (HGH) than to ovine prolactin (OPL) (Li, 1972), its somatotropic activity seems less pronounced than its prolactin-like activity (Tarli et al., 1975).

HCS prolactin-like activity does not seem to depend on the complete maintenance of the overall molecular structure. In fact, partial reduction and alkylation (Neri et al., 1971) or oxidation with performic acid (Sherwood et al., 1971; Handwerger et al., 1972) do not interfere with this effect. However, certain chemical and physicochemical properties of HCS, resembling those of OPL, have been described (Bewley and Li, 1974; Tarli et al., 1975) which sustain the biological similarities between these 2 hormones (Li, 1971, 1972).

Recent evidence points to a relationship between the development of the testes and male accessory sex glands, and prolactin secretion, at least in rodents. For example, prolactin affects spermatozoa production in dwarf mice (Bartke, 1966), increases ^{65}Zn uptake by the dorsolateral prostate in the castrated male rat (Moger and Geschwind, 1972), and increases DNA and RNA levels in the prostate of castrated animals (Thomas and Manandhar, 1975). Furthermore, Wuttke (1973) and Negro-Vilar et al. (1973) have clearly shown that plasma levels of prolactin in male rats increase just before puberty, and the authors maintain that prolactin may be involved in testicular and accessory sexual organ development in the male rat.

We thus decided to study whether or not HCS prolactin-like activity might include some biological effects on the testes and accessory sexual organs of male rodents.

MATERIALS AND METHODS

Hormones

The following hormones were used: (1) HCS (Sclavo, batch 42, prepared according to Neri et al. (1970)); (2) ovine prolactin (NIH, batch P-S-10 — 26.4 IU/mg); (3) ovine LH (NIH, batch LH-S-18 — 1.03 NIH-LH-S-1 U/mg); (4) human growth hormone (prepared according to Raben (1957)), and (5) testosterone propionate as a pure commercial preparation.

^{125}I-HCS (60 μCi/μg) labelled by the electrolytic method (Rosa et al., 1964) (CEA-IRE-SORIN, Saluggia, Italy) was used for some in vivo experiments.

Animals

Normal adult male Sprague-Dawley rats (80 days old), from our own colony, were used.

Normal immature (14 days old) and mature (80 days old) male Swiss Webster albino mice, also from our colony, were used intact, and 15 days after castration performed at 36 days of age.

Adult (40 days old) male obese mice (ob/ob) of the C 57BL/6J strain were used. These animals are characterized by hypoplasia of the sex accessory organs, reflecting the reduced endocrine activity of the testes due to deficient gonadotropin secretion (Lidell and Hellman, 1966).

Adult (40 days old) male dwarf mice (dw/dw) of the Snell-Bagg strain, which have a growth hormone and prolactin deficiency (Grüneberg, 1952) with secondary impairment in other organs including the gonads (Bartke, 1966) were also used.

Both strains of animals were descendants of colonies originally obtained from the Jackson Laboratory (Bar Harbour, Maine, U.S.A.). In both experiments, the heterozygous litter mates (ob/+ and dw/+, respectively) were used as control animals. All animals were fed a standard laboratory diet and given tap water ad libitum; the diet of the dwarf mice was supplemented with carrots and sweet biscuits.

Techniques

^{125}I-HCS was administered intravenously at a dose of 0.2 μCi/animal, to 4 groups of 5 normal male rats. The animals were sacrificed by bleeding, respectively 10 minutes, 4, 8 and 16 hours after treatment. Fragments of several tissues (adipose tissue, skeletal muscle, xiphoid and tibial cartilage, and testes) were removed, and weighed samples were counted in an auto-gamma counter. Radioactivity was also determined in blood samples. All the results were expressed as ratios between the specific radioactivity of several organs (c.p.m./g of wet tissue) at various times, and that measured in each organ at 10 minutes.

In the other experiments, at the end of the various treatments with the different hormones (reported in detail below), the animals were decapitated and the testes and accessory sex glands removed, weighed and fixed in Bouin's solution for histological examination. Sections were stained with hematoxylin-eosin and Sudan black B.

Statistical evaluation

The results obtained were examined by Fisher's analysis of variance. The F values were compared with those reported by Snedecor and Cochran (1968).

RESULTS

^{125}I-HCS distribution

As shown in Figure 1 the radioactivity of systemically injected ^{125}I-HCS, while rapidly declining in all the tissues examined, remained high in the testes, with a maximum at 4 hours, and was still present at 16 hours. This indicates an affinity of these glands for the labelled molecule.

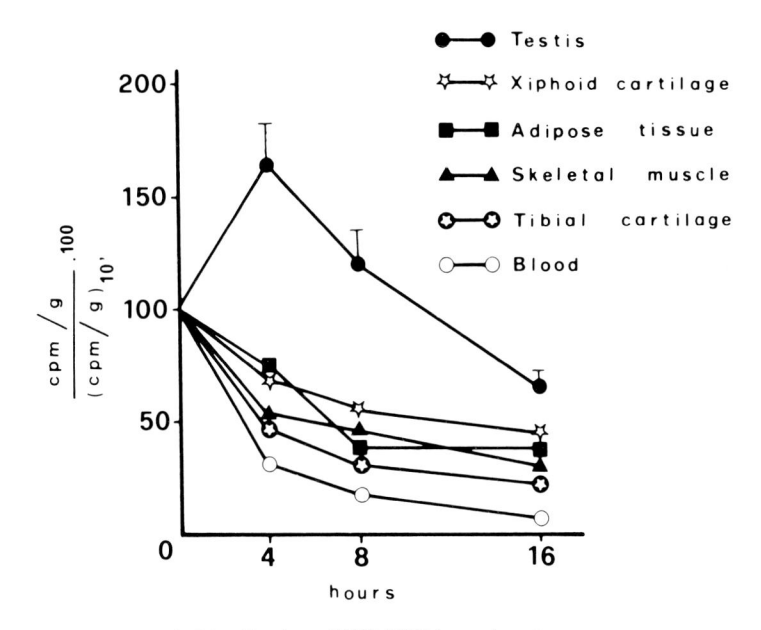

Fig. 1. Distribution of ^{125}I-HCS in male rat organs.

Dwarf mice

The effects of treatment with HCS for 27 or 40 days on the testes and accessory glands of dwarf mice are shown in Table 1. HCS clearly increased the weights of both organs. A similar effect was also obtained after 27 days treatment with OPL, while HGH was less effective.

Saline-treated dwarf mice have small seminiferous tubules with a thin epithelium made up of a few cellular layers (Fig. 2). In the tubular lumen, desquamated cells and occasional multinucleate giant cells occur. Very few Leydig cells are present in the interstitial tissue. Treatment with HCS dramatically stimulated spermatogenesis and activated the interstitial tissue. The thickness of the wall and the lumen of the seminiferous tubules was sharply increased. The increase in tubular size was accompanied by enlargement of the interstitial space with many Leydig cells. Desquamated or multinucleate cells were no longer present in the lumen.

In OPL-treated mice, the tubule size was intermediate between the controls and the HCS-treated mice. Spermatocytogenesis and spermiohistogenesis were active, but Leydig cells were only occasionally evident. On the other hand, HGH treatment

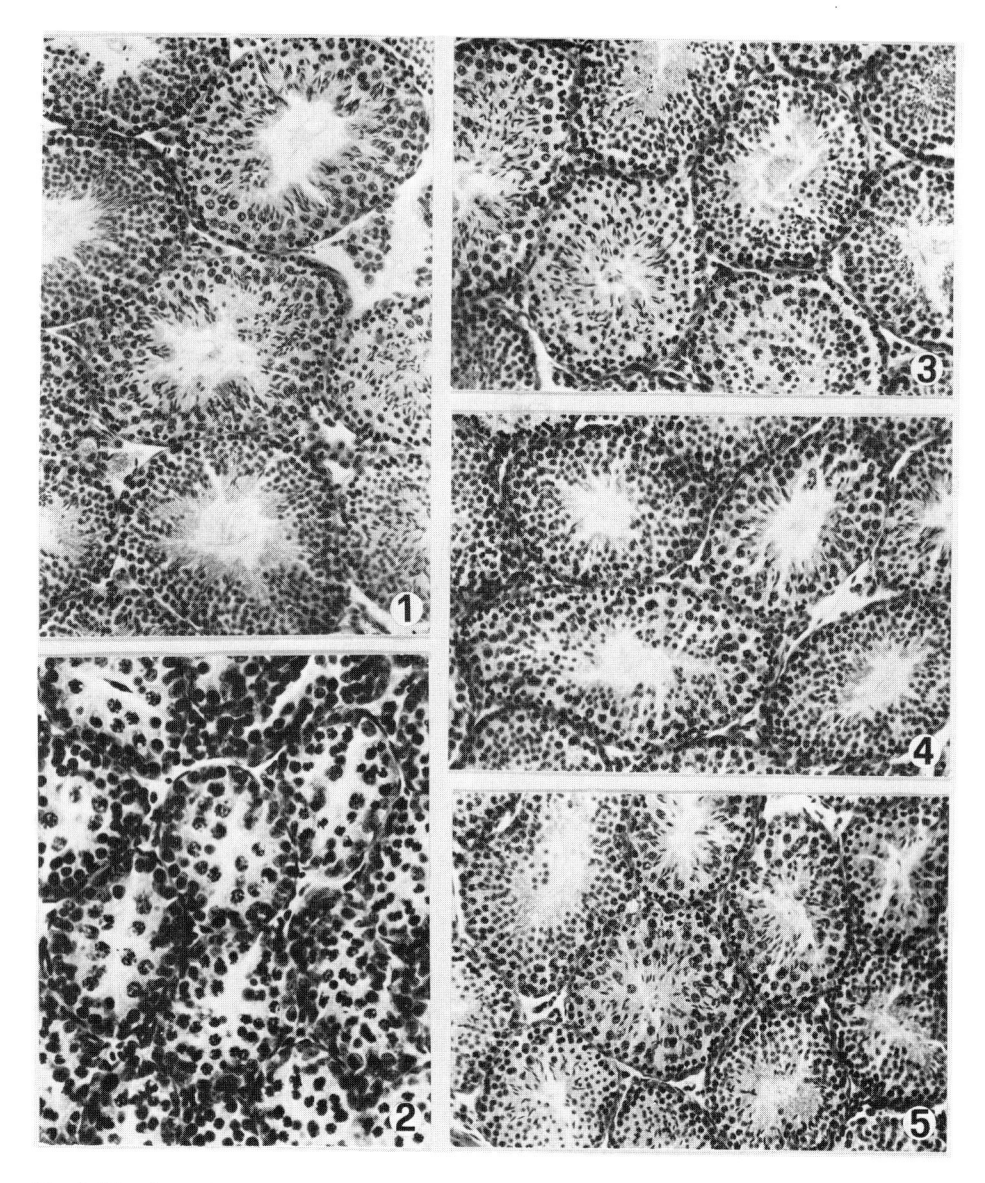

Fig. 2. Histology of mouse testes (60 days old). (1) Normal, 192 ×. (2) Saline-treated dwarf, 300 ×. (3) HCS-treated dwarf, 192 ×. (4) OPL-treated dwarf, 192 ×. (5) HGH-treated dwarf, 192 ×.

induced less striking changes, the tubule size being less than that of HCS or OPL-treated mice and Leydig cells being scarce.

After 40 days' treatment with HCS (Fig. 3), spermatocytogenesis and spermiohistogenesis were active and the interstitial space rich in Leydig cells. Saline-treated animals showed similar histological pictures to those previously described (27-day treatment).

TABLE 1

Effect of HCS, OPL and HGH on weight of testes, seminal vesicles and prostate in the dwarf mouse

	Treatment		Testes	Seminal vesicles	Prostate
			g/100 g body weight ± SE		
27 days	Saline	(10)	0.577 ± 0.194	0.022 ± 0.006	0.004 ± 0.001
	HCS 80 µg/day	(10)	*1.020 ± 0.177*	*0.092 ± 0.012*	*0.015 ± 0.002*
	OPL 80 µg/day	(10)	*0.949 ± 0.073*	*0.094 ± 0.012*	*0.020 ± 0.002*
	HGH 80 µg/day	(10)	0.860 ± 0.182	0.070 ± 0.032	*0.019 ± 0.005*
40 days	Saline	(8)	0.346 ± 0.020	0.017 ± 0.002	0.003 ± 0.000
	HCS 80 µg/day	(8)	*0.847 ± 0.084*	*0.059 ± 0.006*	*0.013 ± 0.001*

Number of animals in brackets. The values in italics indicate a significant difference (analysis of variance) vs. saline.

Whilst in the prostate of saline-treated dwarf mice the tubulo-alveolar glands were small and the interstitium compact, in the HCS-treated animals the alveolar cavities were variable in size and the epithelium was tall.

Obese mice

HCS significantly increased the weight of testes and accessory glands of obese mice, while OPL was ineffective (Table 2). HGH had no effect on testes or seminal vesicles, but significantly increased the prostate weight.

It appears that, in the obese mouse, the tubules are smaller and the interstitial cells fewer than normal. Treatment with HCS seemed to increase tubule size and interstitial cell number (Fig. 4). OPL had similar effects.

Sections of the prostate (Fig. 5) revealed wide tubulo-alveolar glands and scarce interstitial tissue without mononuclear infiltration in HCS- or OPL-treated obese mice, in comparison with saline-treated controls.

Treatment with HCS similarly stimulated the seminal vesicles (Fig. 6). The cuboidal epithelium lacking secretion, typical of the control obese mouse, was transformed into a tall columnar epithelium with intense secretion. A slighter stimulation was induced by OPL.

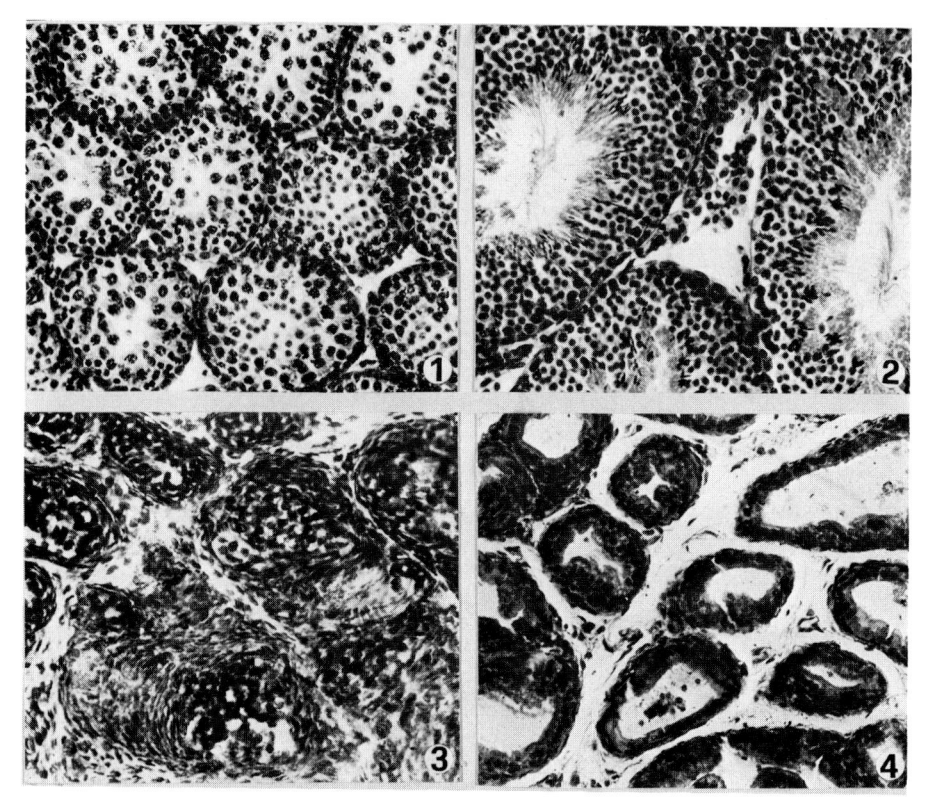

Fig. 3. Histology of testes and prostate of dwarf mouse (73 days old) treated with: (1) Saline; testis section, 175 ×. (2) HCS; testis section, 280 ×. (3) Saline; prostate section, 420 ×. (4) HCS; prostate section, 262.5 ×.

TABLE 2

Effect of HCS, OPL and HGH on weight of testes, seminal vesicles and prostate in the obese mouse

Treatment		Testes	Seminal vesicles	Prostate
		g/100 g body weight ± SE		
Saline	(12)	0.367 ± 0.001	0.022 ± 0.006	0.003 ± 0.001
HCS 80 µg/day	(12)	*0.419 ± 0.016*	*0.049 ± 0.002*	*0.017 ± 0.004*
OPL 80 µg/day	(12)	0.311 ± 0.033	0.025 ± 0.001	0.007 ± 0.002
HGH 80 µg/day	(12)	0.300 ± 0.053	0.021 ± 0.007	*0.018 ± 0.005*

Number of animals in brackets. The values in italics indicate a significant difference (analysis of variance) vs. saline.

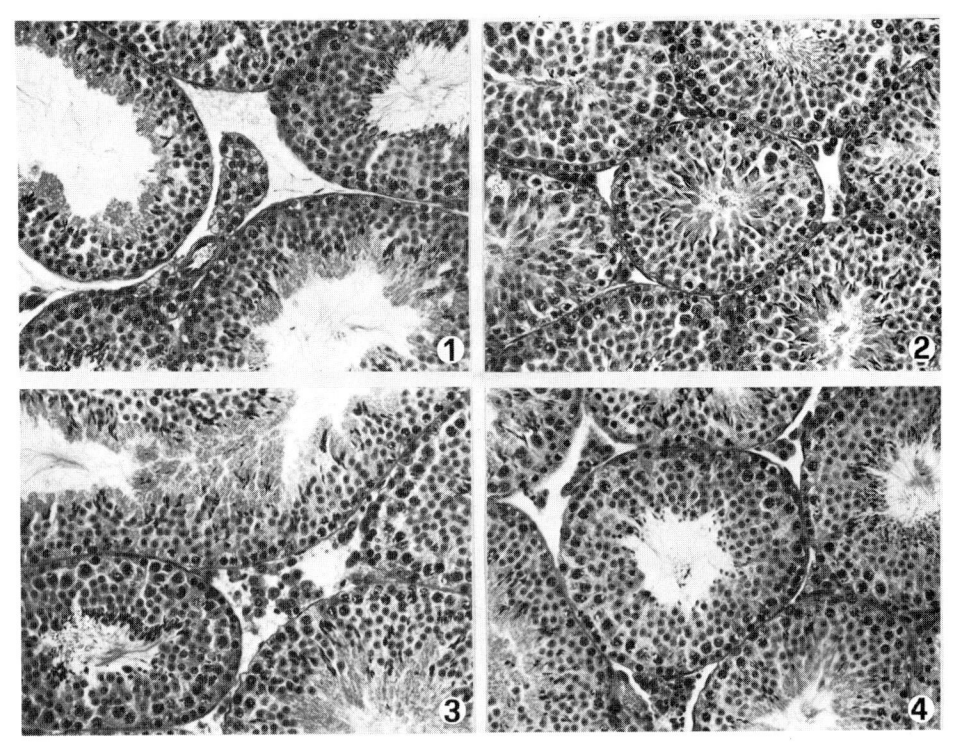

Fig. 4. Histology of mouse testes (70 days old). (1) Normal, 243.7 ×. (2) Saline-treated obese, 243.7 ×. (3) HCS-treated obese, 243.7 ×. (4) OPL-treated obese, 243.7 ×.

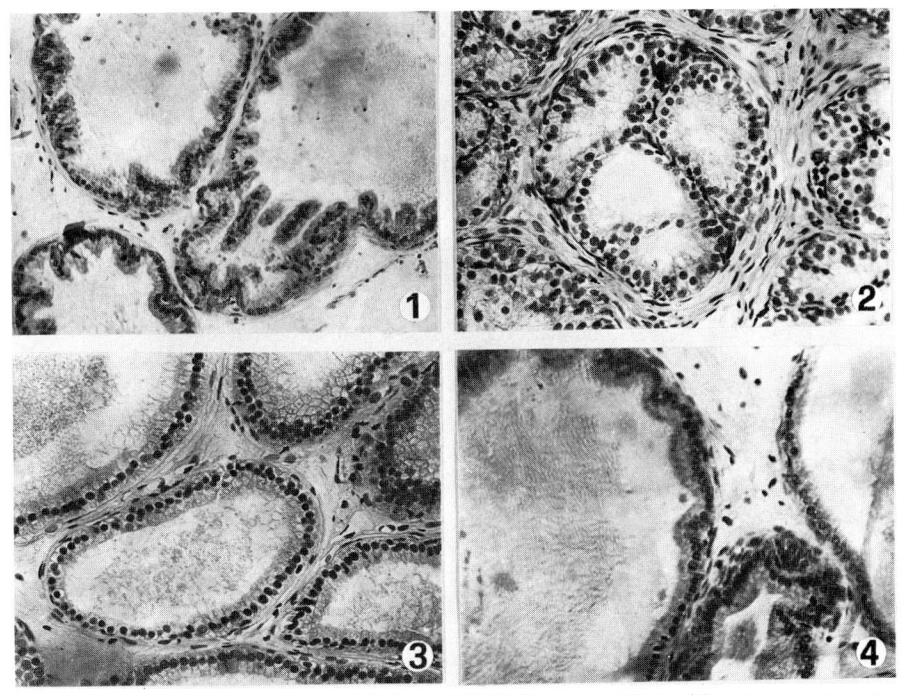

Fig. 5. Histology of mouse prostate (70 days old). (1) Normal, 180 ×. (2) Saline-treated obese, 281.2 ×. (3) HCS-treated obese, 281.2 ×. (4) OPL-treated obese, 281.2 ×.

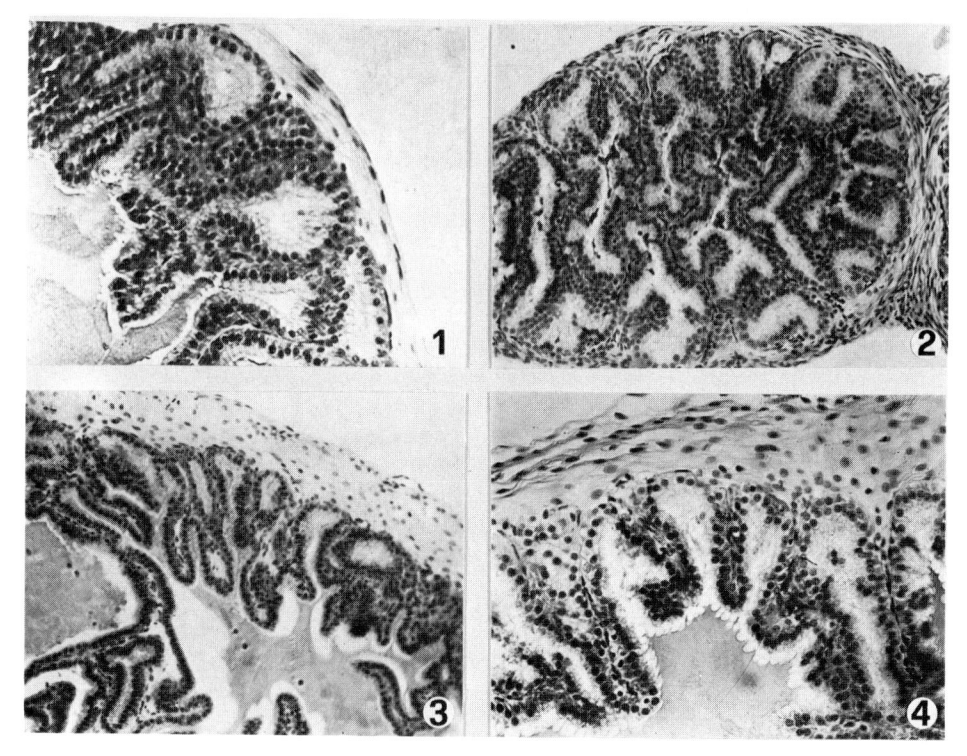

Fig. 6. Histology of mouse seminal vesicles (70 days old). (1) Normal, 281.2 ×. (2) Saline-treated obese, 180 ×. (3) HCS-treated obese, 281.2 ×. (4) OPL-treated obese, 180 ×.

Immature mice

Since previous histological observations of the dwarf and obese mice had shown an increased number of Leydig cells after HCS treatment, immature mice (14 days old) were treated for 9 days with HCS, OPL or HGH to test their effects on maturation of the reproductive apparatus.

These treatments did not influence the weight of the testes or seminal vesicles, but increased the weight of the prostate (Table 3). Testosterone significantly increased the weights of the accessory glands, but not that of the testes.

In the immature, saline-treated mouse, the seminiferous epithelium is not differentiated and the Leydig cells are scarce (Fig. 7); HCS treatment induced moderately wide seminiferous tubules with a highly organized epithelium and spermatocytogenesis without spermiohistogenesis. Numerous Leydig cells were present. The same effects were produced by prolactin. Parallel testosterone treatment did not alter the histological picture of the testes (data not presented).

All the hormonal treatments induced enlargement of tubulo-alveolar glands of the prostate (Fig. 8), producing a tall cuboidal or prismatic epithelium. Testosterone induced complete maturation of the gland (data not presented).

The seminal vesicles (Fig. 9) of the immature saline-treated animal showed a cuboid epithelium with small amounts of intraluminal secretion and simple plicae. After

TABLE 3

Effect of various hormones on weight of testes, seminal vesicles and prostate in the immature mouse

Treatment (9 days)		Testes	Seminal vesicles	Prostate
		\multicolumn{3}{c}{g/100 g body weight ± SE}	
Saline	(23)	0.493 ± 0.012	0.038 ± 0.003	0.010 ± 0.001
HCS 80 µg/day	(19)	0.494 ± 0.022	0.025 ± 0.001	*0.015 ± 0.001*
OPL 80 µg/day	(20)	0.526 ± 0.018	0.033 ± 0.002	*0.018 ± 0.001*
HGH 80 µg/day	(19)	0.453 ± 0.013	0.022 ± 0.002	*0.014 ± 0.002*
Oil	(27)	0.338 ± 0.018	0.035 ± 0.002	0.012 ± 0.001
Testosterone 100 µg/day	(16)	0.284 ± 0.007	*0.287 ± 0.020*	*0.051 ± 0.004*
Testosterone 500 µg/day	(13)	0.366 ± 0.014	*0.302 ± 0.033*	*0.068 ± 0.003*

Number of animals in brackets. The values in italics indicate a significant difference (analysis of variance) vs. control group (saline or oil).

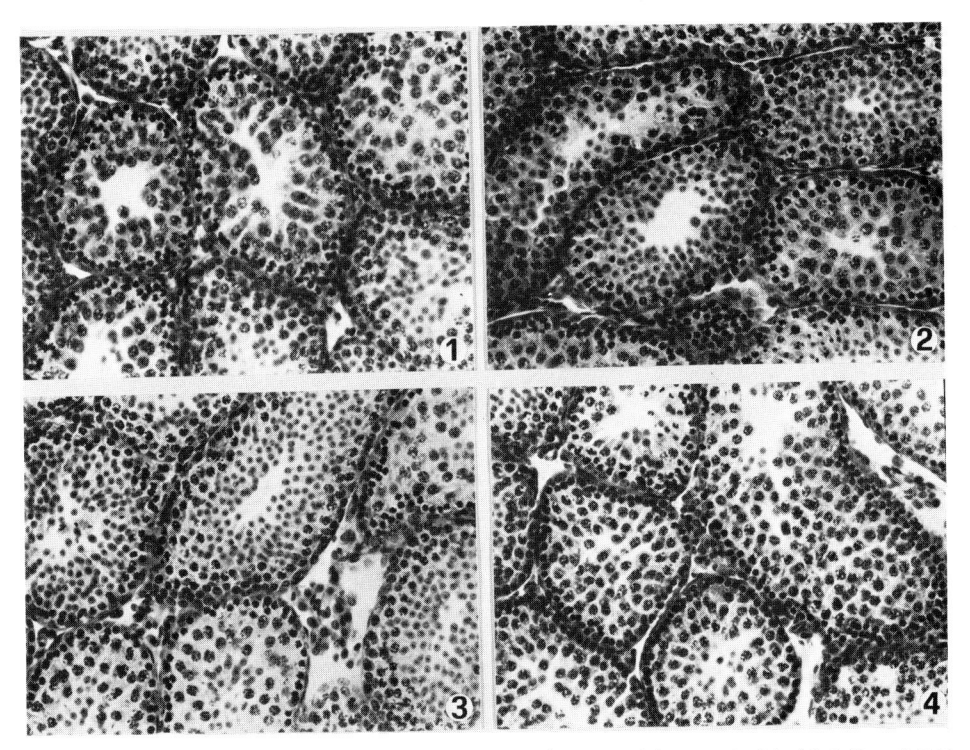

Fig. 7. Histology of testes of normal immature mouse (23 days old) treated with: (1) Saline, 243.7 ×. (2) HCS, 243.7 ×. (3) OPL, 243.7 ×. (4) HGH, 243.7 ×.

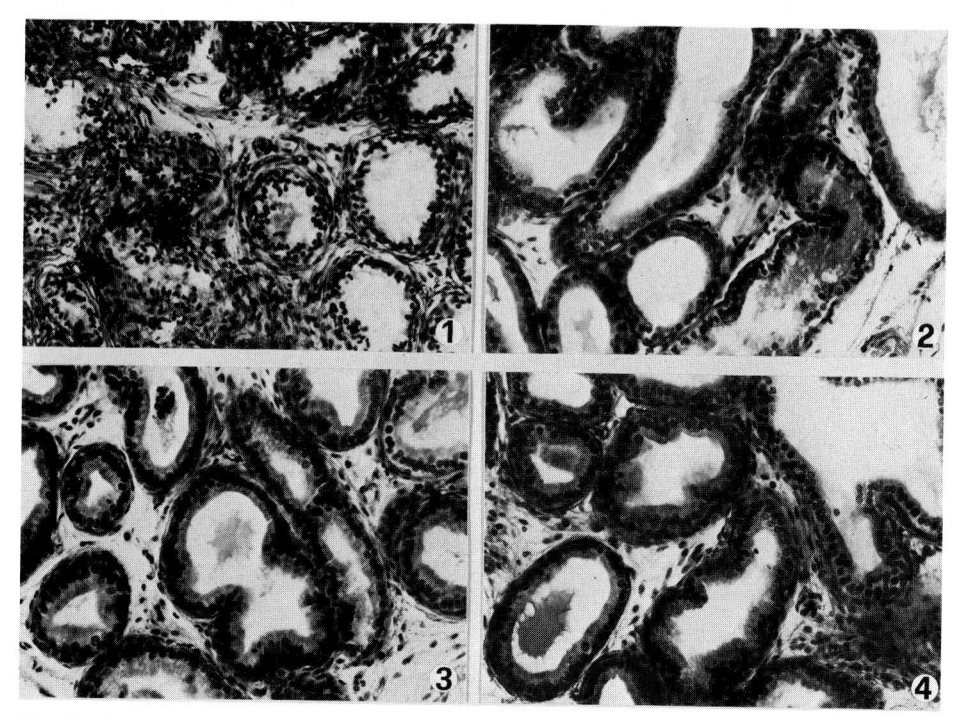

Fig. 8. Histology of prostates of normal immature mouse (23 days old) treated with: (1) Saline, 243.7 ×. (2) HCS, 243.7 ×. (3) OPL, 243.7 ×. (4) HGH, 243.7 ×.

HCS or OPL treatment, the plicae became more complicated, the epithelium was prismatic and intraluminal secretion was evident. HGH treatment had a slighter effect, while testosterone produced complete maturation of the gland (data not presented).

Castrated mice

Castrated mice were used to determine whether HCS was effective after surgical removal of the testes. Under such experimental conditions (Table 4) HCS had no effect on the weight of the accessory glands, while testosterone increases them.

On the other hand, HCS treatment increased the dimensions of the atrophied tubulo-alveolar glands of the prostate (Fig. 10) and markedly reduced the abundant infiltrate of the interstitial tissue. Treatment with testosterone restored the histological picture of the prostate to normal.

The seminal vesicles (Fig. 11), after HCS treatment, showed a prisr increased them. with numerous plicae and abundant secretion, in comparison with a cubic epithelium having desquamated cells and scarce secretory activity in saline-treated, castrated animals.

Testosterone restored the histological picture of the seminal vesicles to normal.

Thus the action of HCS on the accessory glands does not require the presence of the testes.

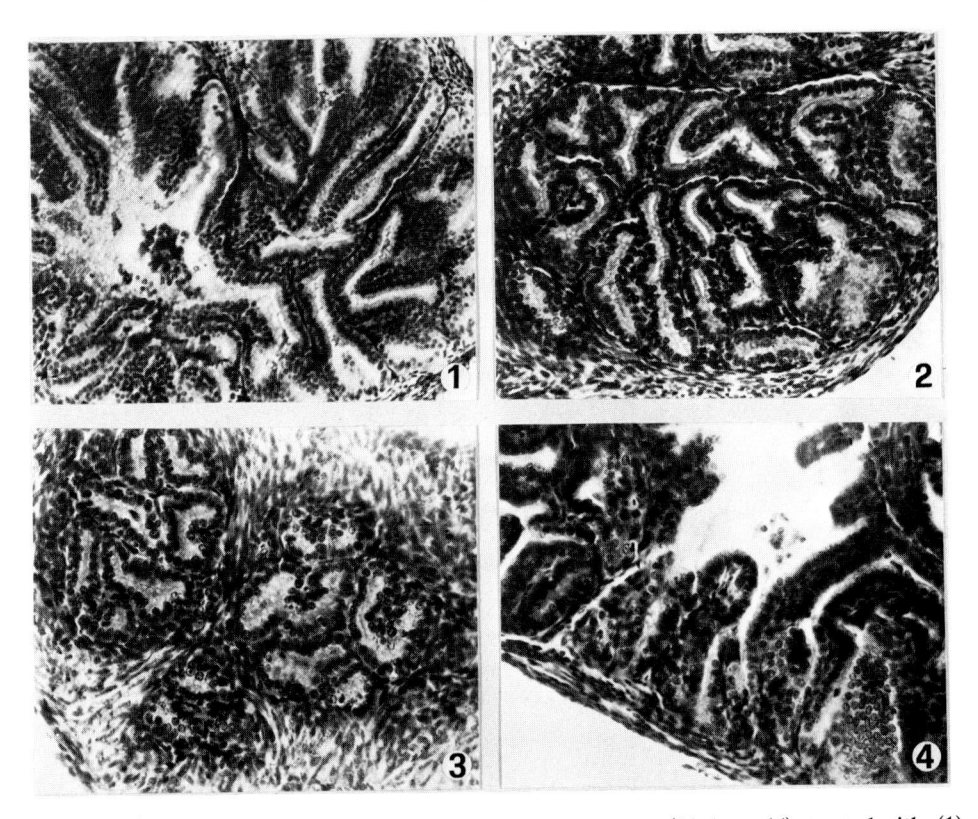

Fig. 9. Histology of seminal vesicles of normal immature mouse (23 days old), treated with: (1) Saline, 262.5 ×. (2) HCS, 262.5 ×. (3) OPL, 168 ×. (4) HGH, 262.5 ×.

TABLE 4

Effect of HCS and testerone on weight of seminal vesicles and prostate in the castrated mouse

Group	Treatment (24 days)		Seminal vesicles	Prostate
			g/100 g body weight ± SE	
Intact animal	—	(15)	0.284 ± 0.036	0.036 ± 0.005
Castrated	Saline	(15)	0.044 ± 0.004	0.012 ± 0.001
Castrated	HCS 80 µg/day	(15)	0.041 ± 0.003	0.014 ± 0.001
Castrated	Testosterone 100 µg/day	(15)	*0.488 ± 0.047*	*0.057 ± 0.003*

Number of animals in brackets. The values in italics indicate a significant difference (analysis of variance) vs. saline.

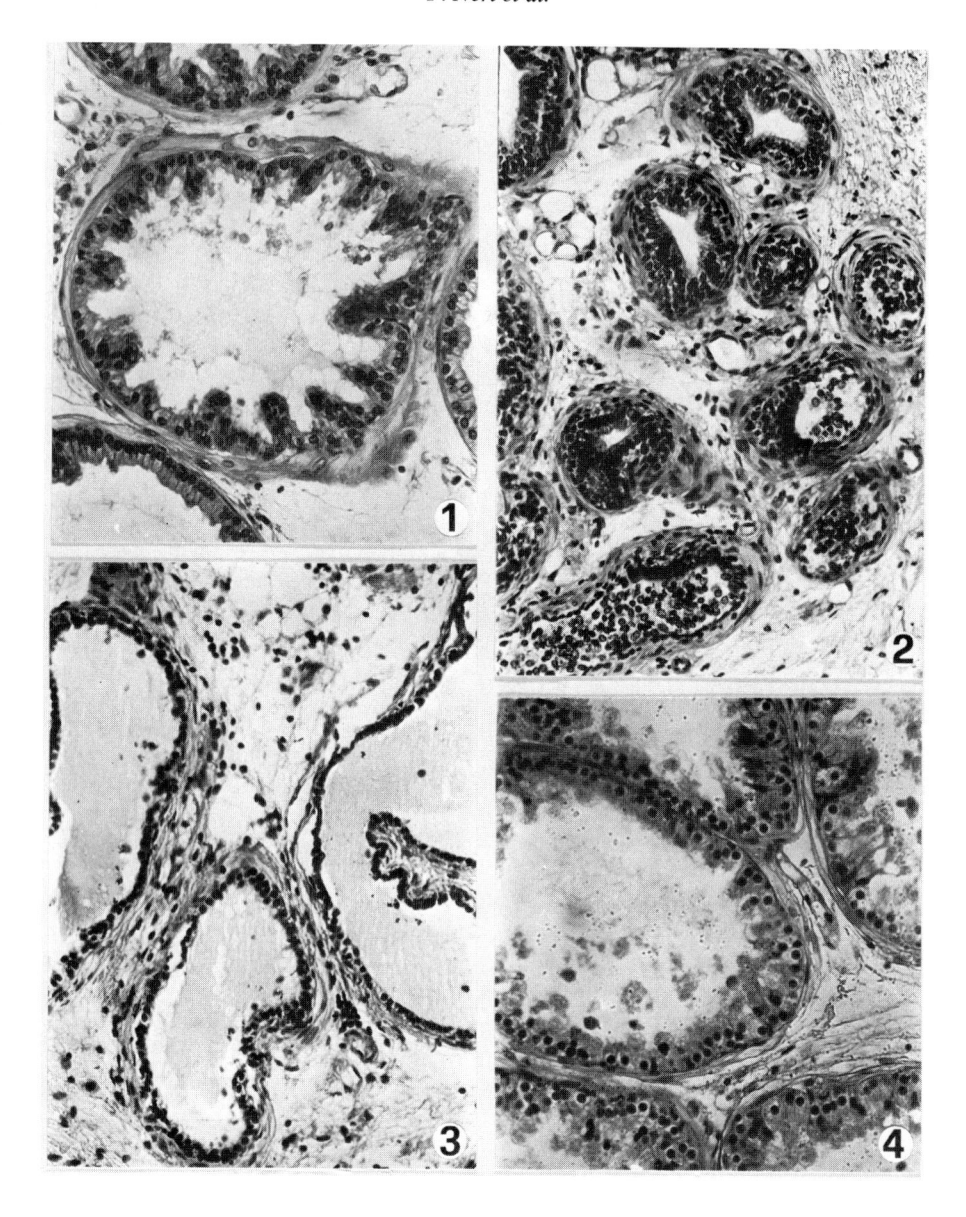

Fig. 10. Histology of mouse prostate (60 days old). (1) Normal, 318.7 ×. (2) Saline-treated castrated, 318.7 ×. (3) HGH-treated castrated, 318.7 ×. (4) Testosterone-treated castrated, 318.7 ×.

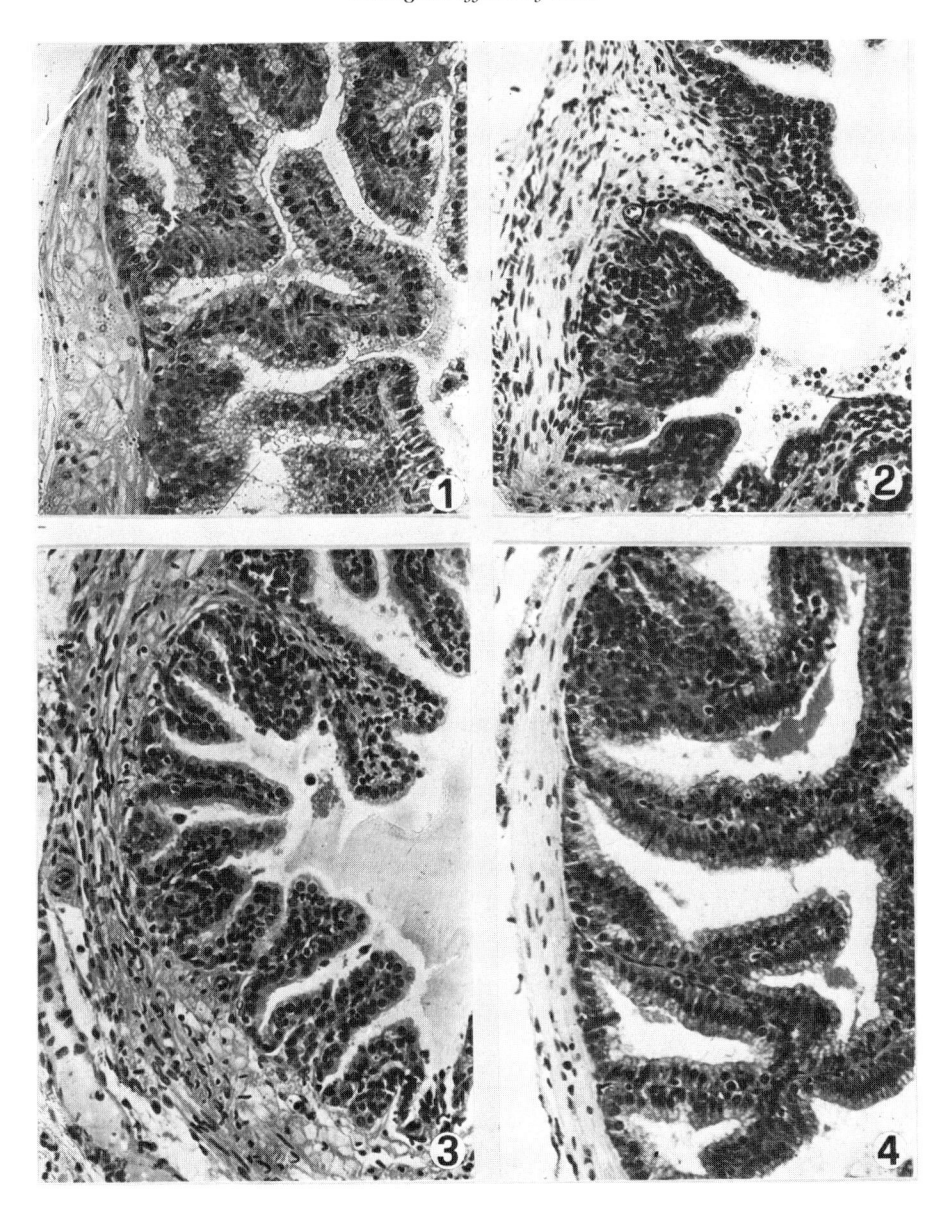

Fig. 11. Histology of mouse seminal vesicles (60 days old). (1) Normal, 318.7 ×. (2) Saline-treated castrated, 318.7 ×. (3) HCS-treated castrated, 318.7 ×. (4) Testosterone-treated castrated, 318.7 ×.

CONCLUSIONS

The present data confirm accumulation of labelled HCS in the testes (Neri et al., 1973). In addition, they strongly suggest that HCS affects the male reproductive system. From a comparison of the effects of equal doses of HCS and OPL on animal models with impaired reproductive functions (genetic, physiologic or surgically induced) it was concluded that

— HCS stimulates the testes and accessory male sex organs of rodents;
— these effects do not seem necessarily to require either the mediation of testosterone, as shown by the histological picture of accessory sex organs of castrated animals, or the presence of pituitary gonadotropins (as shown by the results in obese mice), or prolactin and growth hormone (as shown by the results in dwarf mice).

Collectively, the effects of HCS are clearer than those exerted by OPL. The potential usefulness of this hormone in the treatment of disorders of male reproductive function are readily apparent.

Effects of HCS treatment on oestradiol and progesterone plasma levels in human pregnancy

It is generally agreed that reduced placental function is accompanied by subnormal HCS plasma levels (Genazzani et al., 1972). It has also been shown that a decline in HCS levels in consecutive assays indicates impaired placental function and possibly chronic foetal distress or retarded intrauterine foetal growth (Fioretti et al., 1972; Genazzani et al., 1972, 1974; Letchworth and Chard, 1972; Spellacy et al., 1971).

Although many data are available about the diagnostic and prognostic value of HCS plasma levels, no definitive conclusions can be drawn as to the biological activities of HCS. On the other hand, HCS has been considered one of the most important factors which induces and maintains the switch in maternal metabolism from carbohydrate to lipid utilization (Fioretti et al., 1970; Grumbach et al., 1968). The lipolytic effect of HCS in vitro in the presence of low glucose concentrations has been demonstrated (Genazzani et al., 1969; Turtle and Kipnis, 1967), as also has its liposynthetic activity when high glucose levels are present (Felber et al., 1972; Genazzani et al., 1970). This dual action, mediated by the glucose concentration, is evidence of the major role played by this hormone in sustaining the continual availability of glucose for transfer from mother to foetus. The maternal tissues chiefly utilize the free fatty acids, the reserves of which are always replenished in the hyperglycaemic post-prandial period as a result of the liposynthetic effect of HCS (Fioretti et al., 1970; Genazzani et al., 1970; Grumbach et al., 1968). Moreover, HCS exerts some prolactin-like and luteotropic effects, at least in rodents (Arezzini et al., 1972; Josimovich et al., 1964; Josimovich, 1968). In man, however, Stock et al. (1971) were unable to prolong the luteal phase or to delay the menses with HCS together with, or after, HCG treatment; it was concluded that HCS is not involved in the mechanism which controls maintenance of the corpus luteum in early pregnancy. These data are not in agreement with those for rodents. The reason for this discrepancy could be that whilst the ovary in rodents contains large amounts of 20-progesterone dehydrogenase (Wiest et al., 1963), this enzyme is scarce in human ovaries (Mishell et al., 1963). 20-Progesterone dehydrogenase, the enzyme responsible for conversion of progesterone to the much less active 20-dehydroprogesterone, is inhibited in rodents by prolactin (Armstrong, 1969), and this causes a longer half-life of plasma progesterone and, consequently, increases its biological effects related to the maintenance of pregnancy. A similar role might be proposed for HCS in the placental metabolism of progesterone, especially when it is considered that about half the total amount of progesterone produced daily by the placental tissue is degraded and then reconverted to progesterone (Solomon and Fuchs, 1971). Furthermore, threatened premature labour is accompanied by low progesterone levels in plasma (Csapo et al., 1974; Fioretti et al., 1974) and lower than normal HCS plasma levels (Fioretti et al., 1975; Genazzani et al., 1972). Thus the possibility that intramuscular administration of HCS in cases of threatened abortion or premature labour would alter the clinical evolution of the syndrome was investigated.

MATERIALS AND METHODS

Hormone

Highly purified HCS (Sclavo, batch 42), lyophylized in vials containing 40 mg of hormone, was dissolved at the time of use in 2 ml distilled water and injected intramuscularly. No local or general symptoms of intolerance were found. The treatment pattern varied from case to case, and is presented below.

Patients

All the patients examined were hospitalized in the Department of Obstetrics and Gynecology, University of Cagliari. There were 7 cases of threatened premature labour and 3 cases of threatened abortion. All the subjects were treated daily with 30 mg hysoxysuprine and 3 tablets of Buscopan compositum (Boehringer). In 2 cases of threatened abortion and 3 cases of threatened premature labour, i.m. therapy with HCS was also given; no other hormonal treatment was given in any of the cases. Bed rest was obligatory in all cases for at least 20 hours daily. The clinical data are given in Table 5. Fasting blood samples, taken daily at 8 a.m., were collected in heparinized tubes and immediately centrifuged; the plasma was stored at −20°C until assay. Vaginal smears were also taken from each patient every 5-7 days, to assess vaginal cytological modifications induced by the treatment.

Techniques

HCS radioimmunoassay was performed using a solid phase method (Cocola et al., 1975); the labelled hormone was obtained from CEA-IRE-SORIN (Saluggia, Italy). The various steroids (oestradiol — E_2, oestriol — E_3 and progesterone — P) were measured by radioimmunoassay (RIA) using specific antibodies obtained from CEA-IRE-SORIN, and tritiated molecules from NEN (England) (Fioretti et al., 1974). The plasma was extracted with ethyl ether, and the RIA method included a short incubation and separation of the free labelled hormone from the antibody-bound hormone by charcoal-dextran. In 2 cases, dehydroepiandrosterone sulphate (DHEA-S) was also measured by RIA using an antiserum kindly supplied by Dr. S. Cattaneo and Dr. M. Serio of the Endocrinological Unit of the University of Florence. The tritiated molecule was purchased from NEN (England), and the RIA performed according to Cattaneo et al. (1975) with minor modifications in the antiserum dilution. The vaginal smears were processed according to Papanicolau and Traut (1948).

TABLE 5

Clinical data of the patients submitted to HCS treatment (1bis-5) and those followed as control group (1, 6-10)

Number	Case	Weeks of pregnancy	Diagnosis	Days of observ.	Days of treatm.	HCS total dose (mg)	Result
1	B.S. 74/34/205 25 yr primigravida	23	Threatened abortion	21	—	None	Threatened premature labour
1bis	B.S. 74/34/205 25 yr primigravida	26	Threatened premature labour	35	28	2,320	Favourable
2	C.B. 74/578/231 26 yr primigravida	28	Threatened premature labour	21	12	1,040	Favourable
3	C.M.P. 74/1587/317 25 yr primigravida	16	Threatened abortion	16	12	480	Favourable
4	C.F. 74/1382/319 32 yr primigravida	18	Threatened abortion	23	11	520	Favourable
5	M.M.B. 74/1734/320 27 yr 4 gravida, 3 para	31	Threatened premature labour	21	9	680	Favourable
6	C.C. 74/214 27 yr primigravida	30	Threatened premature labour	10	—	None	Favourable
7	C.T. 74/250 37 yr 6 gravida	31	Threatened premature labour; Cooley's anaemia	12	—	None	Premature delivery
8	S.T. 74/252 22 yr primigravida	30	Threatened premature labour	13	—	None	Favourable
9	P.R. 74/323 43 yr 3 gravida, 2 para	36	Threatened premature labour	21	—	None	Retarded foetal growth
10	M.A. 74/342 38 yr primigravida	26	Threatened abortion	21	—	None	Threatened premature labour

RESULTS

Clinical evaluation

All 5 cases, treated with HCS (Figs. 12-16), were characterized by resolution of the syndrome within the first 3-5 days of therapy. They were then treated for at least 7 more days and controlled within the next 5-10 days, prior to discharge from hospital. All these cases delivered normal, full-term foetuses.

The control group included case no. 1 during the first 2 weeks of hospitalization: during this period, spontaneous uterine contractions became even more marked, and disappeared only after HCS treatment was started. The syndrome resolved in cases 6, 8 and 10 and all delivered normal, full-term foetuses. Case no. 7, followed for 12 days (30th to 32nd week) delivered prematurely at the 37th week; case no. 9, followed for 21 days, had a caesarean section at the 39th week for acute foetal distress, and delivered a foetus which was small chronologically (2350 g).

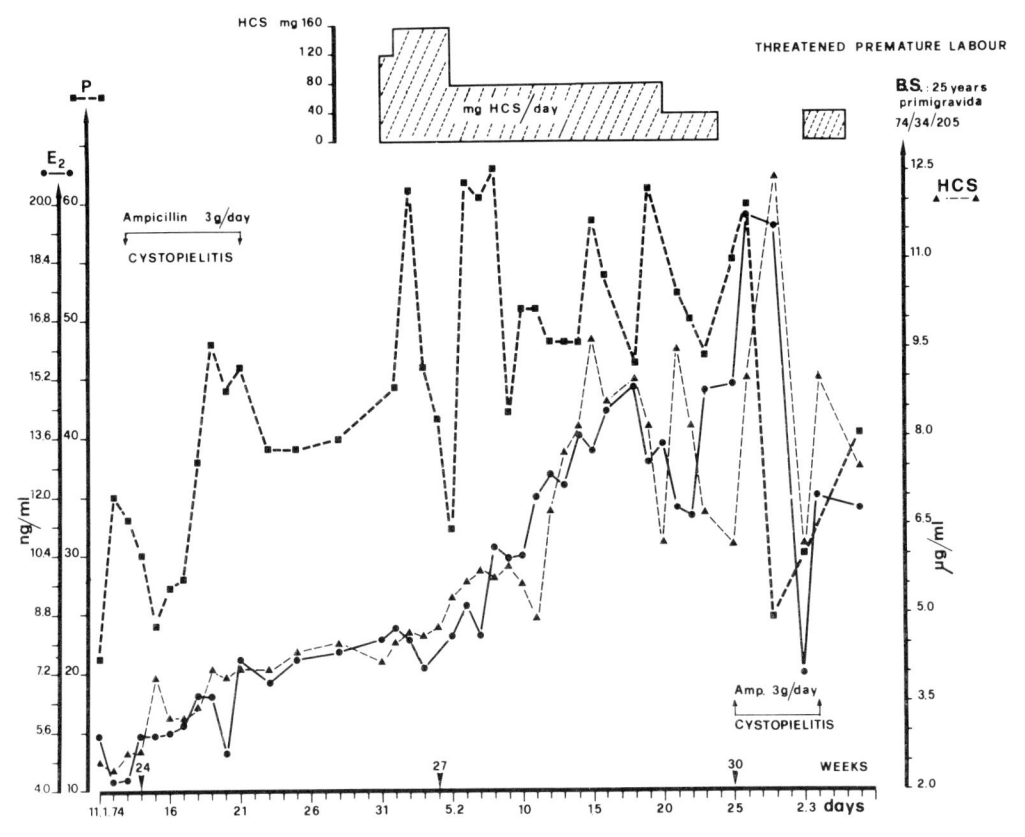

Fig. 12. Pattern of oestradiol, HCS and progesterone plasma levels in case number 74/34/205, prior, during and after i.m. HCS treatment. The daily HCS dose is given in the upper part of the figure.

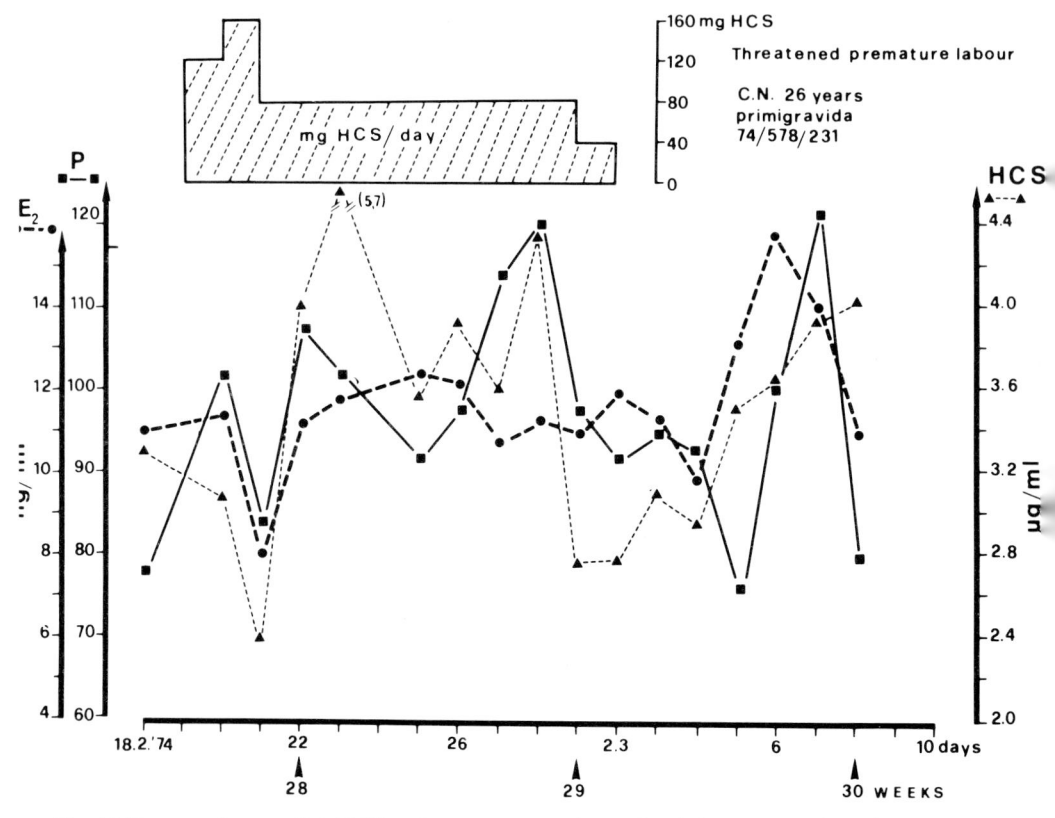

Fig. 13. Pattern of oestradiol, HCS and progesterone plasma levels in case number 74/578/231, prior, during and after i.m. HCS treatment. The daily dose is given in the upper part of the figure.

Hormonal evaluation

The results of the various hormone assays in cases treated with HCS are presented in Figures 12 to 16. Under treatment, the HCS plasma levels increased progressively in case 1 and 4; in the others (2, 3 and 5) slight increases were found in subsequent assays, but the differences were not statistically significant from the control subjects.

The progesterone levels in plasma did not show major alterations under HCS therapy, and were similar to those of the control group. However, progesterone levels showed greater day-to-day variations than did HCS and oestradiol, both during treatment, and compared with the controls.

After 2-4 days of HCS therapy, the oestradiol plasma levels progressively increased in all cases. This increase was moderate in case 2, progressive and constant reaching double the original values in case 1, and very marked in cases 3, 4 and 5. In these cases, the oestradiol levels increased 3-5 times over 5-8 days and remained at these values during the week after ending treatment. Case 1 (Fig. 12) had a second episode of cystopyelitic fever after ending HCS treatment; thereafter, during a second short course of HCS treatment, the HCS, oestradiol and progesterone levels fell steadily.

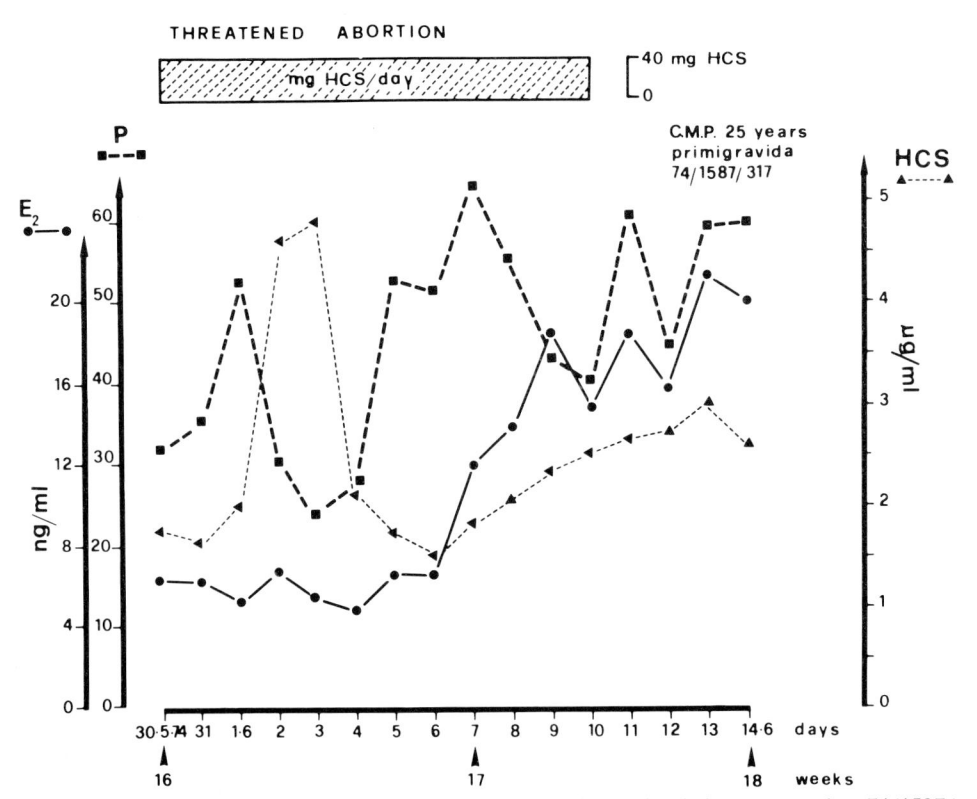

Fig. 14. Pattern of oestradiol, HCS and progesterone plasma levels in case number 74/1587/317, prior, during and after i.m. HCS treatment. The daily dose is given in the upper part of the figure.

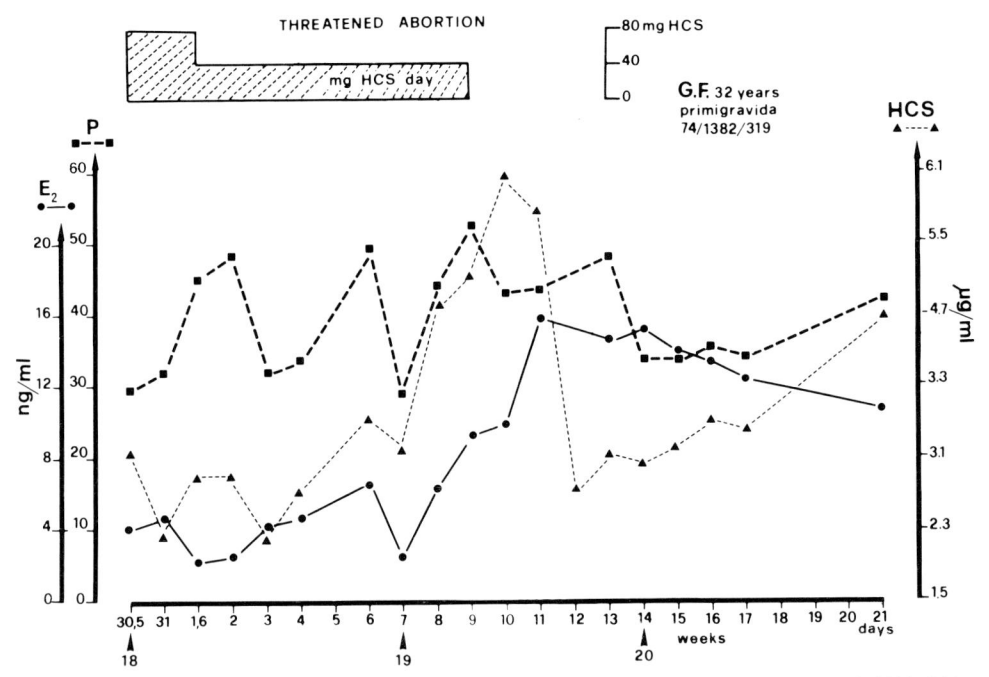

Fig. 15. Pattern of oestradiol, HCS and progesterone plasma levels in case number 74/1382/319, prior, during and after i.m. HCS treatment. The daily dose is given in the upper part of the figure.

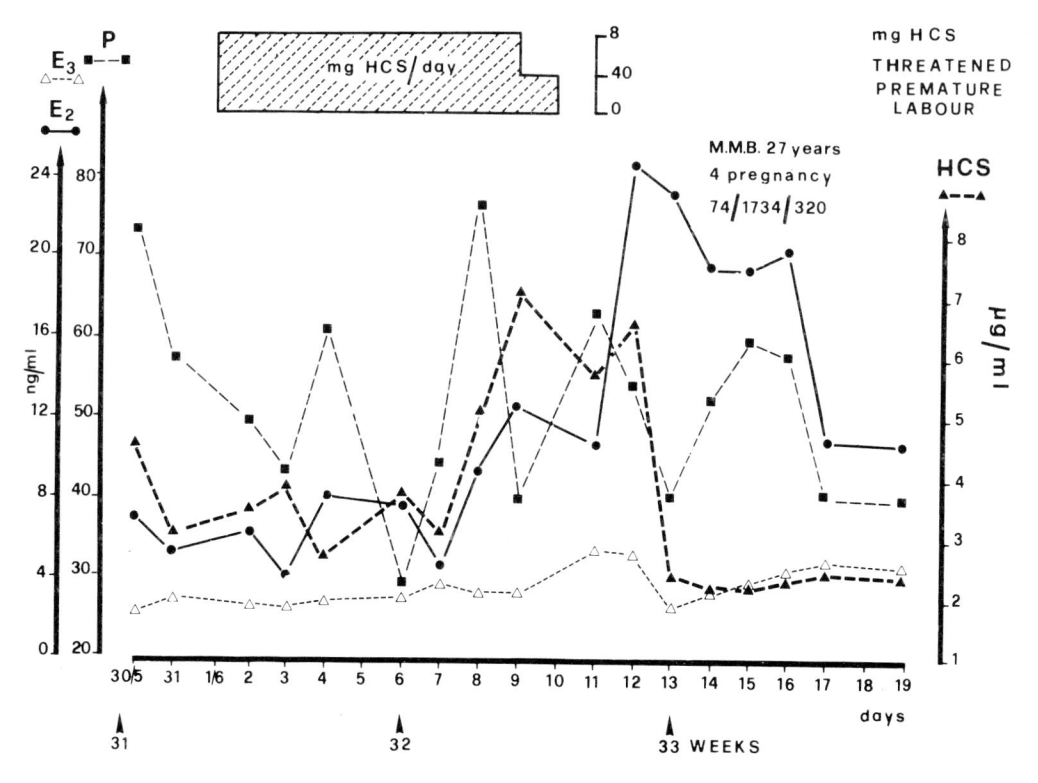

Fig. 16. Pattern of oestradiol, oestriol, HCS and progesterone plasma levels in case number 74/1734/320, prior, during and after i.m. HCS treatment. The daily dose is given in the upper part of the figure.

Oestriol determined in case 3 (Fig. 14) showed little variation during or after HCS treatment.

To obtain a more clear evaluation of the effects of HCS on the various plasma hormonal levels, and to compare them with the spontaneous modifications found in the control group, the percentage variation of each hormonal value was calculated; the values measured on the first day of observation (control group) or the day prior to HCS were taken as 100%.

No significant difference was detected in the percent variations of HCS and progesterone plasma levels in treated and control groups (Fig. 17). The oestradiol levels proved to be significantly higher on the 8th-10th day of therapy (P <0.05) and this difference became even greater (P < 0.001) thereafter. Moreover, the percent variation in oestradiol values remained constantly high throughout the observation period after ending treatment.

Preliminary data on DHEA-S behaviour under HCS therapy, only studied in cases no. 1 and 2, showed that in case 1 the DHEA-S levels decreased progressively from an initial 102 µg/100 ml to 64 µg/100 ml at the end of treatment; in case no. 2 they decreased from 92 to 55 µg/100 ml. The vaginal cytology is summarized in Table 6. HCS treatment induced a marked improvement in all cases; in the control group the bed rest and standard non-hormonal treatment did not, in most cases, induce any change in the vaginal cell appearance.

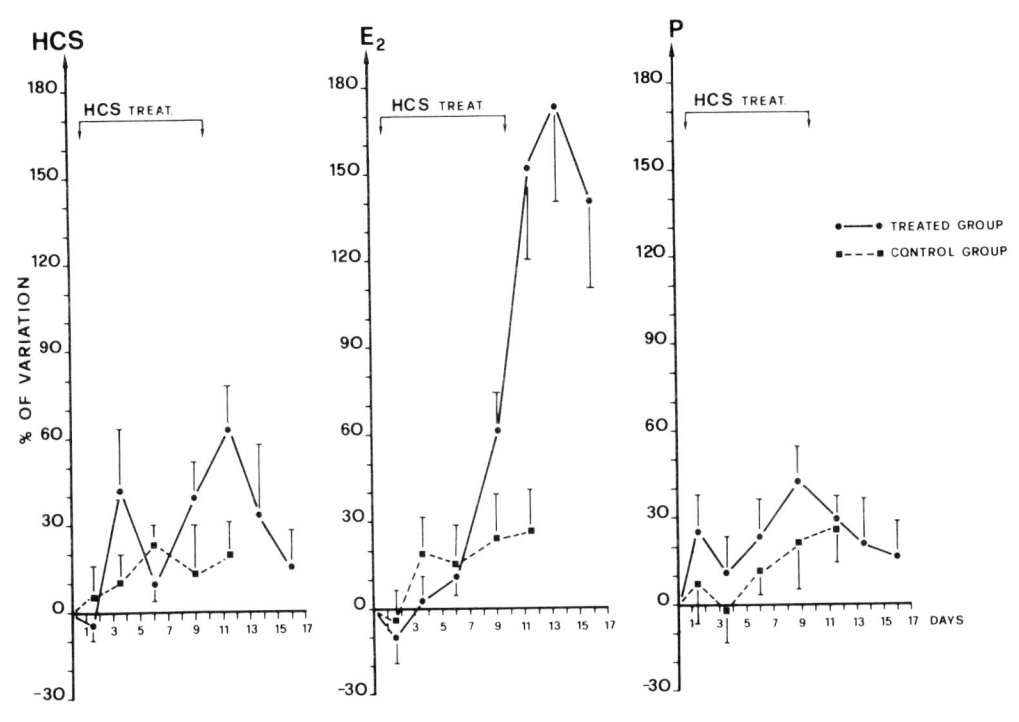

Fig. 17. Mean ± S.E. of the percentage variation in HCS, oestradiol and progesterone plasma levels, in control and HCS-treated groups.

CONCLUSIONS

From the clinical and endocrine data, it is concluded that
— the treatment with i.m. HCS in cases of threatened abortion or premature labour removes the clinical symptoms 3 to 5 days after the start of therapy;
— all treated patients delivered normal foetuses at term;
— no significant variations were found in plasma HCS and progesterone levels during HCS treatment compared with the control group who showed similar symptoms and were followed for the same period;
— on the other hand, the oestradiol plasma levels increased significantly during and after HCS treatment, reaching values 2 to 5 times higher than the original values after 10-15 days;
— the oestradiol increase was accompanied by a simultaneous, progressive decrease in plasma DHEA-S levels, in the 2 cases studied;
— the hormonal changes occuring during and after HCS treatment concur with an improved vaginal cytology.

The present data suggest that HCS stimulates placental conversion of DHEA-S to oestradiol, an hypothesis strongly supported by the production of oestradiol associated

TABLE 6

Modifications in vaginal cytology with and without HCS treatment

Number	Case	Weeks of pregnancy	Diagnosis	Treatment total dose HCS (mg)	Vaginal cytology Before	Vaginal cytology After
1	B.S. 74/34/205 25 yr primigravida	23	Threatened abortion	None	Medium deficiency	Medium deficiency
1 bis	B.S. 74/34/205 25 yr primigravida	26	Threatened premature labour	2,320	Medium deficiency	Good evolution
2	C.N. 74/578/231 26 yr primigravida	28	Threatened premature labour	1,040	Grave deficiency	Good evolution
3	C.M.P. 74/1587/317 25 yr primigravida	16	Threatened abortion	480	Slight deficiency	Good evolution
4	G.F. 74/1382/319 32 yr primigravida	18	Threatened abortion	520	Medium deficiency	Slight deficiency
5	M.M.B. 74/1734/320 27 yr 4 gravida, 3 para	31	Threatened premature labour	680	Grave deficiency	Medium deficiency
6	C.C. 74/214 27 yr primigravida	30	Threatened premature labour	None	Grave deficiency	Grave deficiency
7	C.T. 74/250 37 yr 6 gravida, 5 para	31	Threatened premature labour; Cooley's anaemia	None	Medium deficiency	Medium deficiency
8	S.T. 74/252 22 yr primigravida	30	Threatened premature labour	None	Grave deficiency	Medium deficiency
9	P.R. 74/323 43 yr 3 gravida, 2 para	36	Threatened premature labour	None	Medium deficiency	Medium deficiency
10	M.A. 74/342 38 yr primigravida	26	Threatened premature labour	None	Grave deficiency	Medium deficiency

with decreasing DHEA-S levels found in the 2 cases studied. A possible stimulatory effect of HCS on the production of DHEA-S and a subsequent increase in plasma oestradiol, is, however, contrary to Genazzani et al. (1975) who found that HCS failed to stimulate the in vivo production of adrenal steroids in rodents. This possible stimulatory effect of HCS on the conversion of DHEA-S to oestradiol has clinical support in the observations of Magrini et al. (*This Volume*, p. 369), who found subnormal HCS and oestradiol levels alongside raised DHEA-S values in cases of retarded intrauterine foetal growth. Further studies must be performed to elucidate the effect of HCS on oestradiol levels in plasma.

ACKNOWLEDGEMENTS

We thank the National Institutes of Health, Bethesda, for donations of OPL and OLH. We also thank L. Villa and C. Alessandrini for preparation of the histological sections, and G. Canali, A. Ruspetti and F. Zappalorto for technical assistance.

REFERENCES

Arezzini, C., De Gori, V., Tarli, P. and Neri, P. (1972): *Proc. Soc. exp. Biol. (N.Y.), 141,* 98.

Armstrong, D.T. (1969): In: *Progress in Endocrinology,* p. 89. Editor: C. Gual. ICS 184, Excerpta Medica, Amsterdam.

Bartke, A. (1966): *J. Endocr., 35,* 419.

Bewley, T.A. and Li, C.H. (1974): In: *Lactogenic Hormones, Fetal Nutrition and Lactation,* p. 19. Editors: J.B. Josimovich, M. Reynolds and E. Cobo. J. Wiley and Sons, New York.

Cattaneo, S., Forti, G., Fiorelli, G., Barbieri, U. and Serio, M. (1975): *Clin. Endocr., 4,* 505.

Cocola, F., Nasi, A., Genazzani, A.R. and Neri, P. (1975): *J. nucl. Biol. Med.,* in press.

Csapo, A., Pohanka, O. and Kaihola, H. (1974): *Brit. med. J., 1,* 137.

Felber, J.P., Zaragoza, M., Benuzzi-Badoni, M. and Genazzani, A.R. (1972): *Hormone metab. Res., 4,* 293.

Fioretti, P., Genazzani, A.R., Aubert, M.L., Gragnoli, G. and Pupillo, A. (1970): *J. Obstet. Gynaec. Brit. Cwlth, 77,* 745.

Fioretti, P., Genazzani, A.R., Cocola, F., Scarselli, G. and Mello, G. (1972): *Minerva ginec., 24,* 609.

Fioretti, P., Genazzani, A.R., Facchinetti, F., Nasi, A., Melis, G.B., Paoletti, A. (1974): In: *Atti, 56° Congresso Nazionale della Società Ostetrica e Ginecologica, Padova 1974,* p. 637.

Fioretti, P., Genazzani, A.R., Facchinetti, F., Nasi, A., Melis, G.B., Paoletti, A. and Medda, F. (1975): *Acta endocr. (Kbh.), Suppl. 199,* 383 (abstract).

Genazzani, A.R., Benuzzi-Badoni, M. and Felber, J.P. (1969): *Metabolism, 18,* 593.

Genazzani, A.R., Cocola, F., Nasi, A., Neri, O. and Fioretti, P. (1974): *J. nucl. Biol. Med., 18,* 60.

Genazzani, A.R., Cocola, F., Neri, P. and Fioretti, P. (1972): *Acta endocr. (Kbh,), 71, Suppl. 167.*

Genazzani, A.R., Fraioli, F., Hurliman, J., Fioretti, P. and Felber, J.P. (1975): *Clin. Endocr., 4,* 1.

Genazzani, A.R., Zaragoza, M., Neri, P. and Felber, J.P. (1970): *Ital. J. Biochem., 19,* 230.

Grumbach, M.M., Kaplan, S., Sciarra, S.S. and Burt, I.M.(1968): *Ann. N.Y. Acad. Sci., 148,* 501.

Grüneberg, H. (1952): In: *The Genetics of the Mouse,* p. 122. Editor: H. Grüneberg. Nijhoff, The Hague.

Handwerger, S., Pang, E., Aloj, S. and Sherwood, L. (1972): *Endocrinology, 91,* 721.

Josimovich, J.B. (1968): *Endocrinology, 83,* 530.

Josimovich, J.B., Atwood, B.L. and Goss, D.A. (1964): *Endocrinology, 73,* 410.

Letchworth, A.T. and Chard, T. (1972): *Lancet, 1,* 704.

Li, C.H. (1971): In: *Growth and Growth Hormone*, p. 17. Editors: A. Pecile and E.E. Müller. ICS 244, Excerpta Medica, Amsterdam.
Li, C.H. (1972): In: *Lactogenic Hormones*, p. 7. Editors: G.E.W. Wolstenholme and J. Knight. Churchill/Livingstone, Edinburgh and London.
Lidell, C. and Hellman, B. (1966): *Metabolism, 15,* 444.
Mishell Jr, D.R., Wide, L. and Gemzell, C.A. (1963): *J. clin. Endocr., 23,* 125.
Moger, W.H. and Geschwind, I.I. (1972): *Proc. Soc. exp. Biol. (N.Y.), 141,* 1017.
Negro-Vilar, A., Krulick, L. and McCann, S. (1973): *Endocrinology, 93,* 660.
Neri, P., Arezzini, C., Canali, G., Cocola, F. and Tarli, P. (1971): In: *Growth and Growth Hormone*, p. 199. Editors: A. Pecile and E.E. Müller. ICS 244, Excerpta Medica, Amsterdam.
Neri, P., Arezzini, C., Cocola, F. and De Gori, V. (1973): *IRCS med. Sci.,* (73-6) 3-5-10.
Neri, P., Tarli, P., Arezzini, C., Cocola, F., Pallini, V. and Ricci, C. (1970): *Ann. Sclavo, 12,* 663.
Papanicolau, G.N. and Traut, H.F. (1948): *The Diagnosis of Uterine Cancer by Vaginal Smears.* The Commonwealth Foundation, New York.
Raben, M.S. (1957): *Science, 125,* 883.
Rosa, U., Scassellati, G.A. and Pennisi, F. (1964): *Biochim. biophys. Acta (Amst.), 86,* 519.
Sherwood, L.M., Handwerger, S., McLaurin, W.D. and Pang, E. (1971): In: *Growth and Growth Hormone*, p. 209. Editors: A. Pecile and E.E. Müller. ICS 244, Excerpta Medica, Amsterdam.
Snedecor, G.W. and Cochran, W.G. (1968): In: *Statistical Methods,* p. 116. Iowa State University Press, Ames, Iowa.
Solomon, S. and Fuchs, F. (1971): In: *Endocrinology of Pregnancy,* p. 66. Editors: F. Fuchs and A. Klopper. Harper and Row, New York.
Spellacy, W.N., Teoh, E.S., Buhi, W.C., Birk, S.A. and McCreary, S.A. (1971): *Amer. J. Obstet. Gynec., 109,* 588.
Stock, R.J., Josimovich, J.B., Kosor, B., Klopper, A. and Wilson, G.R. (1971): *J. Obstet. Gynaec. Brit. Cwlth, 78,* 549.
Tarli, P., Arezzini, C., Neri, P., Fruschelli, C. and Ricci, C. (1975): In: *Atti delle Giornate Endocrinologiche Senesi,* p. 3. Editors: E.E. Müller and S. Piazzi. Edizione I.S.V.T. Sclavo, Siena, Italy.
Thomas, J.A. and Manandhar, M. (1975): *J. Endocr., 65,* 149.
Turtle, J.R. and Kipnis, D.M. (1967): *Biochim. biophys. Acta (Amst.), 144,* 583.
Wiest, W.B., Wilcox, R.B. and Kirschbaum, T.H. (1963): *Endocrinology, 73,* 588.
Wuttke, W. (1973): In: *Human Prolactin,* p. 143. Editors: J.L. Pasteels and C. Robyn. ICS 308, Excerpta Medica, Amsterdam.

CORRELATION BETWEEN PLASMA LEVELS OF HUMAN CHORIONIC SOMATOMAMMOTROPHIN (HCS), SEXUAL STEROIDS AND THEIR PRECURSORS IN NORMAL AND PATHOLOGICAL PREGNANCIES

G. MAGRINI, J.P. FELBER, F. MÉAN and A. CURCHOD

Division de Biochimie Clinique, Département de Médecine et Service d'Obstétrique et de Gynécologie, Centre Hospitalier Universitaire Vaudois*, Lausanne, Switzerland

Several authors (Saxena et al., 1969; Genazzani et al., 1969, 1972; Haoùr et al., 1971; Spellacy et al., 1971, 1974, and others) have shown the importance of HCS as a guide to the outcome of threatened abortion. For this reason, the hormone was called 'the watchdog of fetal distress' (England et al., 1974).

However, from the numerous studies published on the subject, it appears that a rather high percentage of complicated pregnancies with fetal distress has remained undetected by low HCS values or by descending values when successive measurements were performed. This can be understood from the knowledge that the hormone originates in the syncytiotrophoblast cells of the placenta and is secreted almost entirely into the maternal circulation. HCS reflects the placenta, not the fetus; it is a sensitive indicator of placental function, not of fetal well-being. Low or descending values of plasma HCS are therefore indicative of placental insufficiency rather than of fetal distress.

Plasma steroids, also, have been used by several workers (Tulchinsky and Korenman, 1971; Townsley et al., 1973; Lindberg et al., 1974, and others) to estimate high-risk pregnancies.

Differing from HCS, which originates entirely in the placenta, plasma and urinary steroids have 3 sources: the fetus, the placenta and the mother. The placenta does not, like the ovary, possess a whole enzymatic system for steroid synthesis. It synthesizes estrone and 17β-estradiol mainly from dehydroepiandrosterone sulfate (DHEA-S), of fetal and, to a lesser degree, maternal origin (Lauritzen, 1971). Estriol does not derive directly from estrone and 17β-estradiol, but it is synthesized by the placenta from the 16α-hydroxy derivatives of DHEA-S and estrone sulfate, which are almost solely of fetal origin. In late pregnancy, the 17β-estradiol precursors have been shown (De Hertogh et al., 1975) to originate in about equal quantities from the fetal and maternal compartments, whereas more than 90% of the estriol precursors are of fetal origin. In the fetus, as in the mother, the precursors are synthesized in the adrenals, and sulfation takes place in the liver. The mechanism that regulates the maternal and fetal adrenal functions indirectly controls the secretion of placental estrogens into the maternal blood where they are measured.

In normal pregnancies, Loriaux et al. (1972) observed that the increase in the plasma concentration of estrone and estriol and their sulfates correlates better with

---

* Formerly named: Hôpital Cantonal Universitaire.

the fetal than with the placental weight, and De Hertogh et al. (1975) found that HCS fluctuations were not parallel to those of unconjugated plasma estrone, 17β-estradiol and estriol.

In the hope of obtaining better prognostic information about fetal distress, the levels of HCS and steroids and their precursors in the plasma were compared in obstetrical diseases in which the fetus is directly involved: intrauterine fetal growth retardation and toxemic pregnancy.

MATERIALS AND METHODS

Fifteen normal women were followed throughout pregnancy, and blood samples were taken monthly from the 16th week until delivery. All women had normal pregnancies with healthy normal babies. Aliquots from the same sample were used for HCS and steroid determinations.

Measurement of HCS

HCS was measured by means of a rapid radioimmunoassay technique, with ethanol precipitation (Phadebas, Pharmacia, Uppsala, Sweden).

Measurement of steroids

Unconjugated 17β-estradiol (E$_2$), estriol (E$_3$) and progesterone (P) were measured by radioimmunological methods after selective solvent extractions. Unconjugated dehydroepiandrosterone (DHEA) and pregnenolone were determined by radioimmuno-

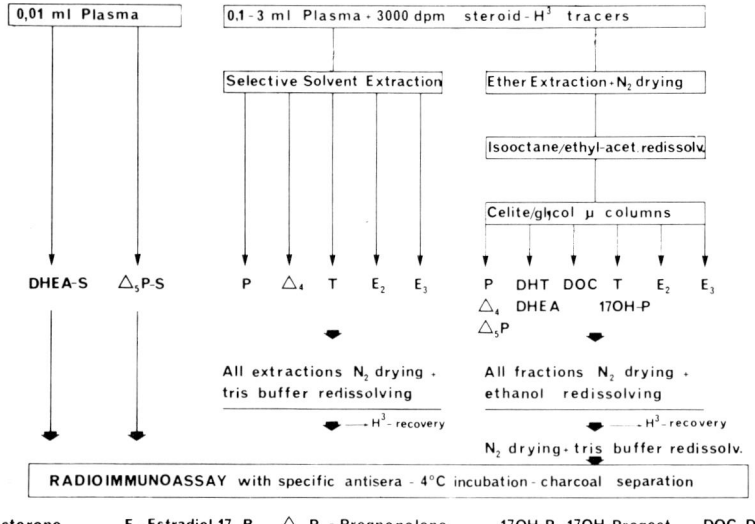

Fig. 1. Scheme of the methods used for the radioimmunological determination of the various steroids.

assay after solvent extraction and chromatographic separation on a celite microcolumn. DHEA-S and pregnenolone sulfate were measured in plasma by radioimmunological methods, essentially as described by Buster and Abraham (1972, 1973). Highly specific antibodies were used for E_2, E_3, P and DHEA radioimmunoassays. The radioimmunoassay methods for E_2, E_3 and P were checked by means of chromatographic procedures on celite columns, essentially as previously described (Magrini and Felber, 1974). They are summarized in Figure 1 and will be published in detail elsewhere.

RESULTS

Study of normal pregnancies

The results of HCS and steroid measurements are given in Figures 2A and 2B.

Study of retarded fetal growth

In a first series of cases of intrauterine fetal growth retardation (Fig. 3), HCS values were compared, using the means and 99% confidence limits, during the last phase of pregnancy. In most cases, HCS tended to remain at the lower limit of normal, but any differences from normal were seldom clear-cut; only a few cases had values markedly below normal levels.

HCS and steroid measurements were compared in 4 established and 2 suspected cases of retarded fetal growth (Fig. 4). Three of the former cases ended with fetal death.

Case 1 presented severe intrauterine fetal growth retardation with fetal death occurring 1 hour after delivery at term. The HCS pattern started slightly below normal, falling to clearly pathological levels. Plasma E_2 levels were within normal limits at the beginning; thereafter, they markedly decreased to pathological levels. E_3, although remaining within normal limits, did not follow the physiological rise after the 26th week, and ended at the lower limit of the normal range. DHEA-S was normal until the 30th week. It increased moderately thereafter. Progesterone was constantly below normal after the 25th week, and pregnenolone sulfate remained at the lower limit of normal.

Cases 2, 3 and 4 presented severe intrauterine retardation of fetal growth. HCS, E_2 and pregnenolone sulfate levels were noticeably low, whereas DHEA-S was increased. E_3 levels remained normal in cases 2 and 3, and were low at death in case 4.

Case 5 presented suspected fetal growth retardation. This was not confirmed at delivery, although the newborn suffered malformation of the lower limbs. The low HCS pattern, and the low levels, except for one value of E_2, as well as the high levels of DHEA-S, monitored fetal distress better than did E_3, which never left the normal range.

Case 6 presented suspected fetal growth retardation, which was not, however, confirmed. Delivery had to be induced. At the beginning, plasma levels of HCS, E_2 and E_3 were normal, progesterone and pregnenolone sulfate were lower and DHEA-S higher than the normal range. Thereafter, HCS progressively decreased to pathological values, but returned to the lower limits of normal before term. E_3 decreased to the lower limit of normal, but later than did HCS. Only E_2 remained normal in all the assays. DHEA-S remained above normal, and progesterone below normal limits. Pregnenolone sulfate fluctuated around the lower limit of normal. The last HCS

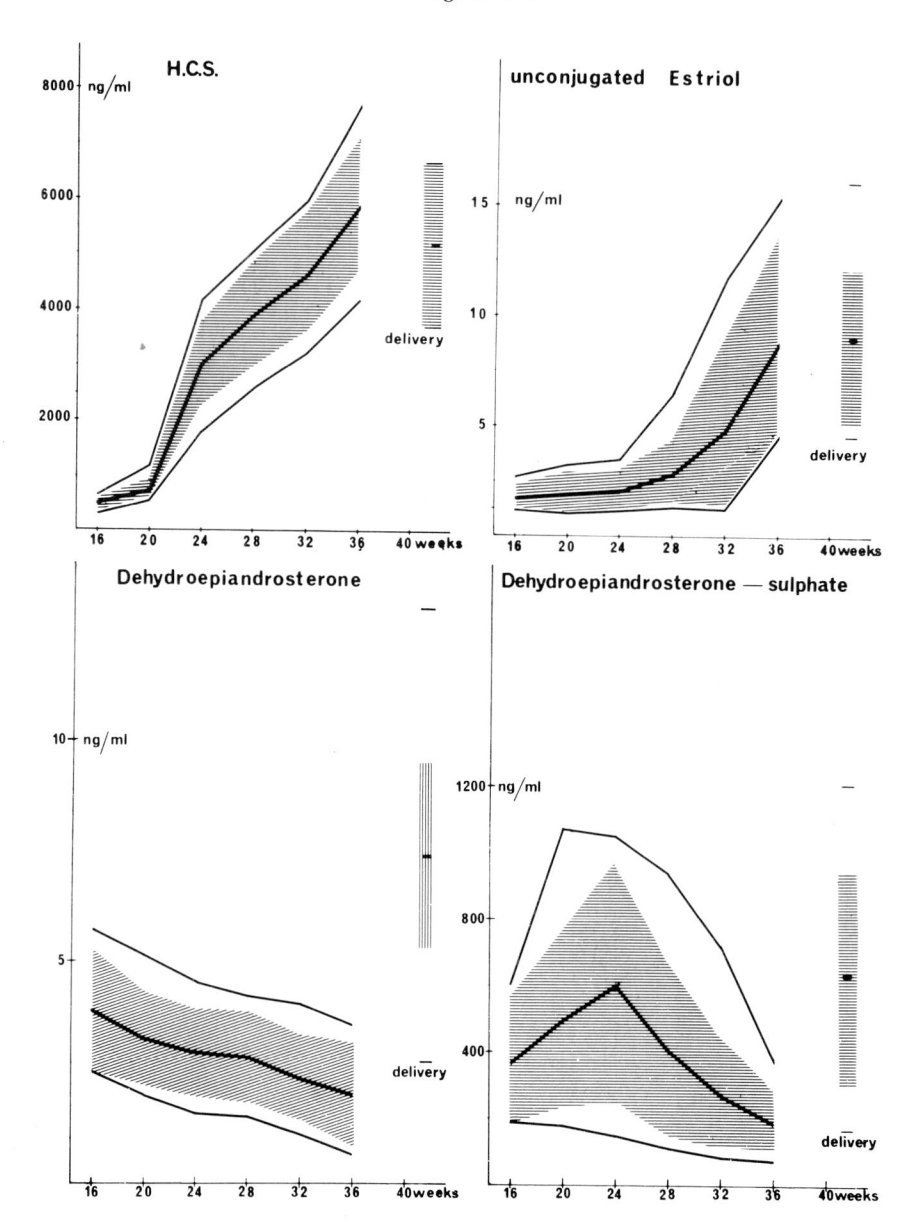

Fig. 2A. Ranges with confidence limits (shaded area) and dispersion limits in the measurement of HCS, unconjugated estriol, dehydroepiandrosterone and dehydroepiandrosterone sulfate in a group of 15 normal pregnancies, from the 16th-36th week and at delivery.

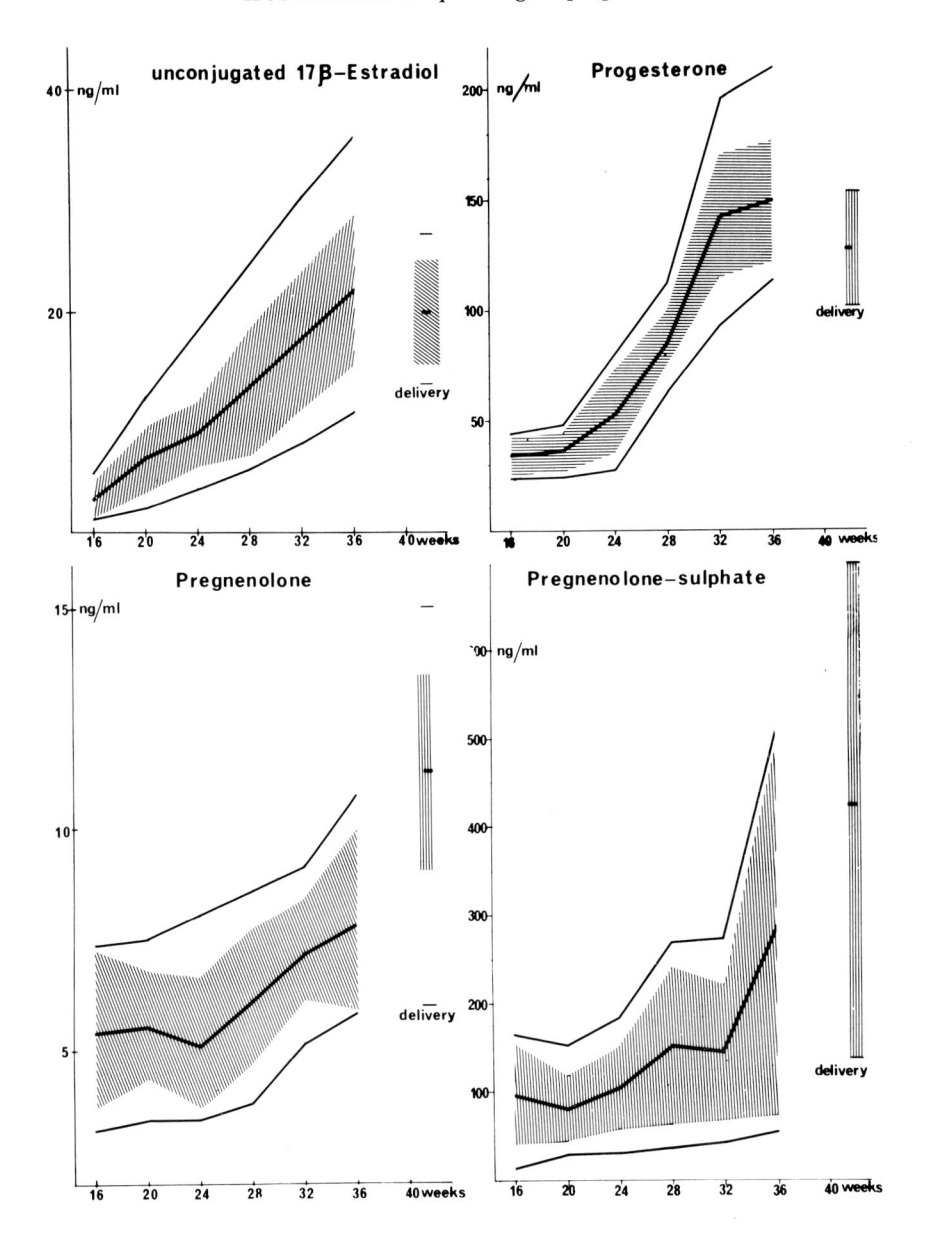

Fig. 2B. Ranges with confidence limits (shaded area) and dispersion limits in the measurement of 17 β-estradiol, progesterone, pregnenolone and pregnenolone sulfate in a group of 15 normal pregnancies, from the 16th-36th week and at delivery.

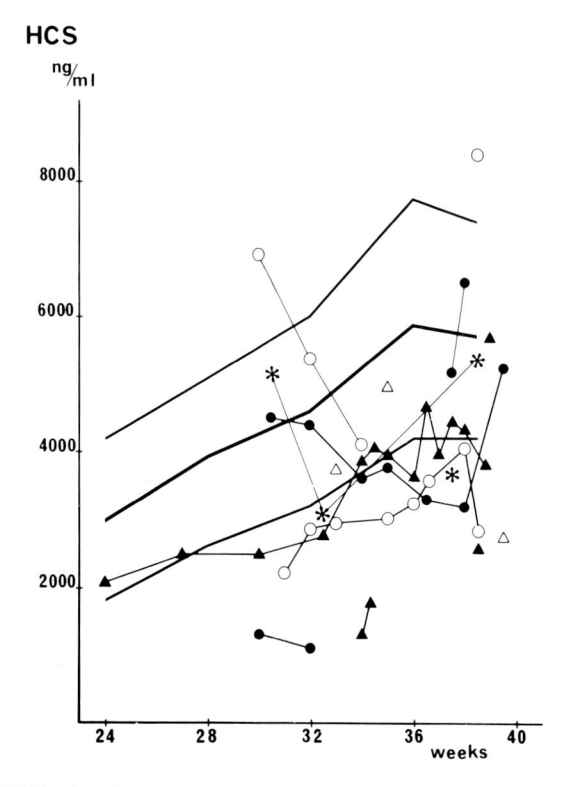

Fig. 3. Plasma HCS values in a group of cases of intrauterine fetal growth retardation.

value, within the normal range, correlated with the E_2 values which remained within the normal range.

Study of toxemic pregnancy

In a first series of toxemic pregnancies with edema, proteinuria and hypertension (Fig. 5), the HCS levels did not seem to correlate with the disease except for 3 cases in which they were clearly below normal, and 2 others in which they fell late. HCS and steroid measurements were compared in 1 case of toxemic pregnancy and 3 cases of toxemic pregnancy + intrauterine fetal growth retardation (Fig. 6). Fetal death occurred in one case.

Case 1 presented with toxemia without intrauterine fetal growth retardation. The parameters were all normal, except for slightly elevated values of DHEA-S.

Case 2 showed severe toxemia with intrauterine fetal growth retardation. It showed wide HCS fluctuations, from normal to pathological values. Plasma E_2 displayed similar, though reduced, fluctuations with a gradual decrease to pathological values. Plasma E_3 became pathological only at delivery. DHEA-S was slightly above normal. Progesterone decreased from normal to low values, and pregnenolone sulfate remained at the lower limit of the normal range.

Case 3 similarly showed severe toxemia with intrauterine fetal growth retardation. The patterns of HCS, E_2, E_3 and progesterone were clearly pathological except at the

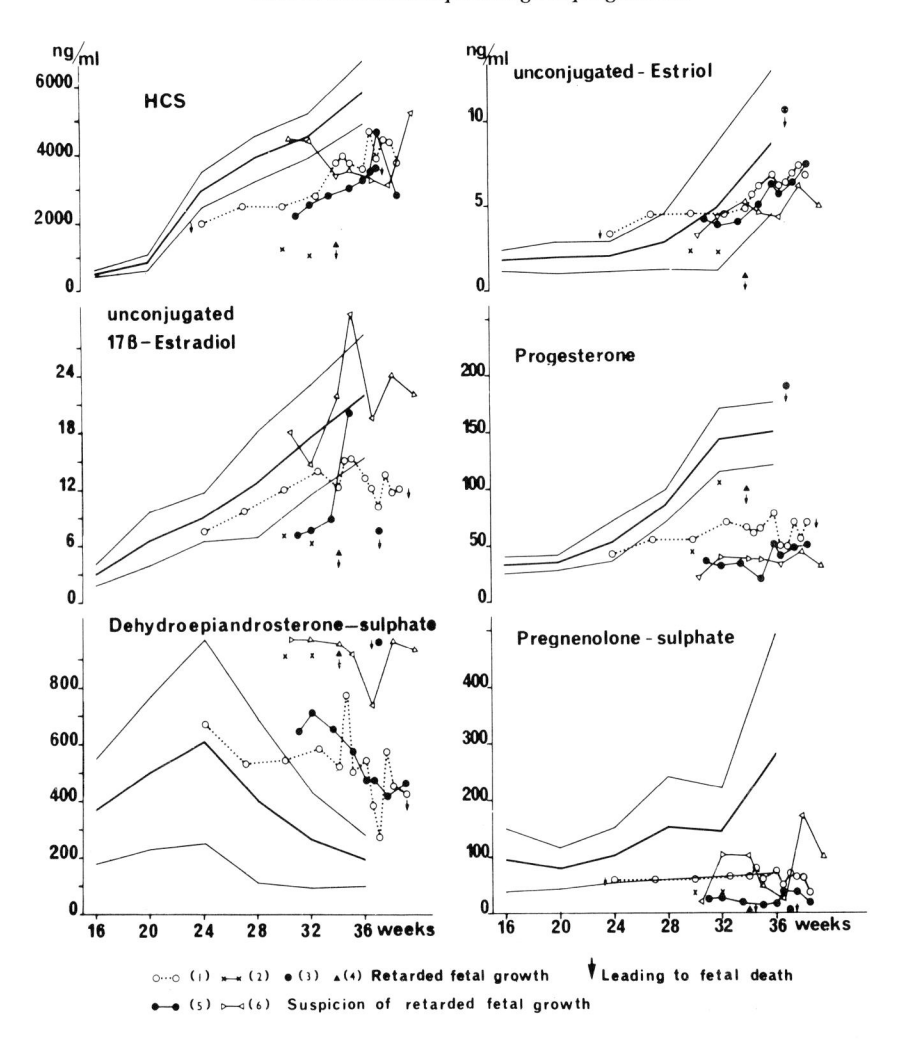

Fig. 4. Plasma HCS, unconjugated 17β-estradiol, dehydroepiandrosterone sulfate, unconjugated estriol, progesterone and pregnenolone sulfate in 4 cases of intrauterine fetal growth retardation, and 2 suspected cases of intrauterine fetal growth retardation.

beginning of the observations when only HCS and DHEA-S were outside normal limits, HCS below and DHEA-S above. The case ended with fetal death.

Case 4 again showed toxemia with intrauterine fetal growth retardation. HCS values were within normal limits at the beginning, but decreased during the following weeks. E_2 fluctuated around the lower limit of the normal range, E_3 was normal, DHEA-S went from normal to above-normal levels. Progesterone and pregnenolone sulfate decreased in parallel with HCS.

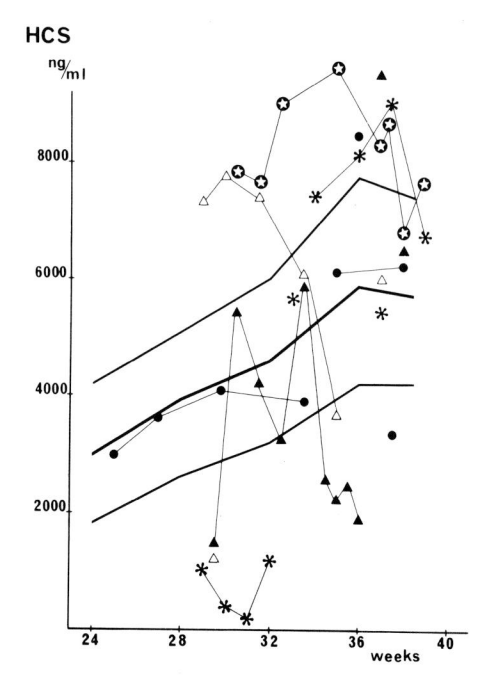

Fig. 5. Plasma HCS values in a group of cases of toxemic pregnancy.

DISCUSSION

The values and shapes of the curves for normal pregnancies are similar to those of other authors (De Hertogh et al., 1975). HCS increased gradually during pregnancy, but decreased before delivery. The rise of E_3 after the 24th week can be explained by an increased synthesis of the steroid from fetal adrenal precursors, whereas the decrease in DHEA-S, also after the 24th week, may perhaps be explained by an increased extraction, by the placenta, of circulating maternal DHEA-S for its conversion into E_2 and estrone (Gant et al., 1971; Buster et al., 1974a). Plasma levels of DHEA-S increased again at delivery. Unconjugated E_2 increased gradually during pregnancy, to decrease at delivery, like HCS, confirming the observations of Munson et al. (1970), Loriaux et al. (1972), Tulchinsky and Korenman (1971), and De Hertogh et al. (1975), but differing from those of Townsley et al. (1973).

The study of plasma HCS levels in cases of intrauterine fetal growth retardation (Fig. 3) and toxemia (Fig. 5) confirms the frequent overlap of the values between pathological and normal cases described by other authors. It justifies the search for better criteria of fetal distress.

Although, in this study, the origin of the various steroids measured in maternal plasma is related to 3 sources: i.e. maternal, placental and fetal, a tendency of the steroids to fluctuate in parallel was sometimes observed. An inverse correlation was, however, found in most cases of intrauterine fetal growth retardation and toxemia between decreased HCS and E_2 levels and increased DHEA-S levels. It must also be realized that, like cortisol, DHEA-S depends on ACTH and can be increased by stress.

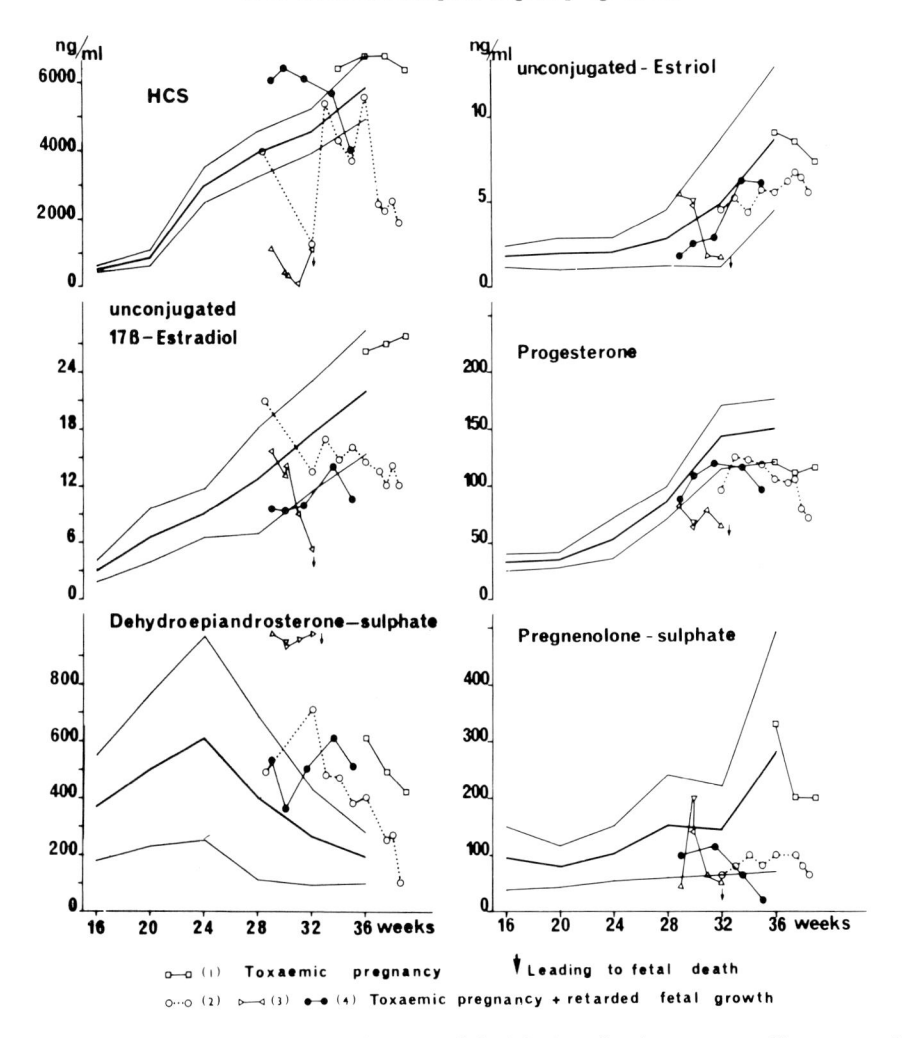

Fig. 6. Plasma HCS, unconjugated 17β-estradiol, dehydroepiandrosterone sulfate, unconjugated estriol, progesterone and pregnenolone sulfate in 1 case of toxemic pregnancy and 3 cases of toxemic pregnancy with retarded fetal growth.

On the other hand, Townsley et al. (1970) reported that cortisol suppresses placental sulfatase, thus lowering the placental estrogen production. Rahman et al. (1975) pointed out that the various enzymatic activities responsible for the conversion of precursors into E_2 are markedly lowered in toxemic pregnancies.

The information from the rather limited number of cases presented here does not entirely fulfil the hope of finding a method to give information on the condition of the fetus rather than of the placenta. In particular, the results do not confirm the value of unconjugated E_3 determinations which might be expected to mirror the activity of the fetal adrenals. This confirms the work of Lindberg et al. (1974). Klopper et al. (1975) also observed a good deal of overlap between individuals, although they demonstrated the advantage of measuring total rather than unconjugated E_3 in intrauterine fetal growth retardation. The measurement of steroids, or their precursors, increases the prognostic value of HCS measurements, but does not dispense with them. Steroids do not seem to offer a greatly improved indication of the state of the fetus itself. In fact, like HCS, steroids in maternal blood mainly reflect placental function.

A lowered or decreasing pattern of progesterone and pregnenolone sulfate also correlates with HCS in most of the cases studied, in keeping with the postulated role of pregnenolone sulfate as a maternal precursor of placental progesterone. Scommegna et al. (1972) suggested that, in normal pregnancies, in contrast to estrogen synthesis in the placenta, maternal cholesterol seems to play an exclusive role in the production of progesterone.

Like HCS, plasma steroids require serial determination to yield information on the clinical situation. Other authors (Lauritzen and Schneider, 1968; Buster et al., 1974a,b) obtained good tests of feto-placental function by administering steroid hormone precursors (DHEA-S or pregnenolone sulfate) to the mother and then measuring urinary steroid hormone excretion. Lauritzen (1968) described a change in the metabolic rate of DHEA-S in some pathological pregnancies.

In the present limited study, the steroids that correlate best with fetal distress are E_2 and possibly its precursor, DHEA-S. They often vary inversely, the mean value of the precursor decreasing in normal pregnancy with an increased synthesis of estrogen. The E_2/DHEA-S ratio can be a useful index of feto-placental pathology, especially if HCS measurements are used in addition. This was evident particularly in case 4 (Fig. 6) with toxemic intrauterine fetal growth retardation: the patterns of E_2, at the lower limit of normal, and of DHEA-S above normal, yielded additional information on feto-placental function, whereas HCS remained within the normal range until the 34th week. In case 6 (Fig. 4), the E_2 values remained within the normal range although, in this patient with suspected but unconfirmed intrauterine fetal growth retardation, the HCS pattern decreased, except for the final value.

ACKNOWLEDGEMENTS

The authors wish to thank G. Chiodoni, P. Inaudi, A. Bonaventura and Miss L. Bianchi for their skilful technical help, and Miss F. Tapernoux for the preparation of the manuscript.

REFERENCES

Buster, J.E. and Abraham, G.E. (1972): *Analyt. Letters, 5/8,* 543.
Buster, J.E. and Abraham, G.E. (1973): *Analyt. Letters, 6/2,* 147.
Buster, J.E., Abraham, G.E., Kyle, F.W. and Marshall, J.R. (1974a): *J. clin. Endocr., 38,* 1031.
Buster, J.E., Abraham, G.E., Kyle, F.W. and Marshall, J.R. (1974b): *J. clin. Endocr., 38,* 1038.
De Hertogh, R., Thomas, K., Bietlot, Y., Vanderheyden, I. and Ferrin, J. (1975): *J. clin. Endocr., 40,* 93.
England, P., Fergusson, J.C., Lorrimer, D., Moffatt, A.M. and Kelly, A.M. (1974): *Lancet, 1,* 5.
Gant, N.F., Hutchinson, H.T., Siiteri, P.K. and McDonald, P.C. (1971): *Amer. J. Obstet. Gynec., 111,* 555.
Genazzani, A.R., Aubert, M.L., Casoli, M., Fioretti, P. and Felber, J.P. (1969): *Lancet, 2,* 1385.
Genazzani, A.R., Cocola, F., Neri, P. and Fioretti, P. (1972): *Acta endocr. (Kbh.), Suppl. 167.*
Haoùr, F., Cohen, M. and Bertrand, J. (1971): *Europ. J. clin. biol. Res., 16,* 124.
Klopper, A., Jandial, V. and Wilson, G. (1975): *J. Steroid Biochem., 6,* 651.
Lauritzen, C. (1968): In: *Verhandlungen, 13. Symposion der Deutschen Gesellschaft für Endokrinologie, Köln,* p. 313. Editor: J. Kracht. Springer-Verlag, Berlin-Heidelberg-New York.

Lauritzen, C. (1971): In: *Proceedings, III International Congress on Hormonal Steroids, Hamburg 1970*, p. 37. Editors: V.H.T. James and L. Martin. ICS 219, Excerpta Medica, Amsterdam.
Lauritzen, C. and Schneider, H.P.G. (1968): *Geburtsh. u. Frauenheilk., 28*, 396.
Lindberg, B.S., Nilsson, B.A. and Johansson, E.D.B. (1974): *Acta obstet. gynec. scand., 53*, 329.
Loriaux, D.L., Ruder, H.J., Knab, D.R. and Lipsett, M.B. (1972): *J. clin. Endocr., 35*, 887.
Magrini, G. and Felber, J.P. (1974): In: *Clinical Biochemistry — Principles and Methods. Vol. 1*, p. 784. Editors: H.Ch. Curtius and M. Roth. W. de Gruyter, Berlin.
Munson, A.K., Mueller, R.J. and Yannone, M.E. (1970): *Amer. J. Obstet. Gynec., 108*, 340.
Rahman, S.A., Hingorani, V. and Laumas, K.R. (1975): *Clin. Endocr., 4*, 333.
Saxena, B.N., Emerson Jr, K. and Selenkow, H.A. (1969): *New Engl. J. Med., 281*, 225.
Scommegna, A., Burd, L., Seals, C. and Wineman, C. (1972): *Amer. J. Obstet. Gynec., 113*, 60.
Spellacy, W.N., Buhi, W.C., Birk, S.A. and McCreary, S.A. (1974): *Amer. J. Obstet. Gynec., 120*, 214.
Spellacy, W.N., Teoh, E.S., Buhi, W.C., Birk, S.A. and McCreary, S.A. (1971): *Amer. J. Obstet. Gynec., 109*, 588.
Townsley, J.D., Gartman, L.J. and Crystle, C.D. (1973): *Amer. J. Obstet. Gynec., 115*, 830.
Townsley, J.D., Scheel, D.A. and Rubin, E.J. (1970): *J. clin. Endocr., 31*, 670.
Tulchinsky, D. and Korenman, S.G. (1971): *J. clin. Invest., 50*, 1490.

CLINICAL APPLICATION OF HUMAN CHORIONIC SOMATOMAMMOTROPHIN

A.T. LETCHWORTH

Department of Obstetrics and Gynaecology, Royal Hampshire County Hospital, Winchester, and Department of Human Reproduction and Obstetrics, Southampton University, Southampton, United Kingdom

There has been considerable controversy over the value of human chorionic somatomammotrophin (HCS) estimations in the management of pregnancy. There are a number of factors, apart from the inherent problem of the evaluation of any test of placental function, that might explain this state of affairs. Many of the reports from earlier workers are based on relatively small numbers of cases without close attention to the complication of pregnancy under study. The assay techniques used often took days to complete and were associated with varying results from different laboratories. However, recent reports include several large series from a number of centres and these have suggested that HCS levels in pregnancy can play a valuable role in the management of both normal and abnormal pregnancy.

ADVANTAGES OF HCS AS A TEST OF PLACENTAL FUNCTION

Theoretical

Synthesis is related to placental mass, as a significant correlation between HCS levels and placental mass has been reported by a number of workers (Sciarra et al., 1968; Saxena et al., 1969; Seppala and Ruoslahti, 1970; Spellacy et al., 1971a; Varma et al., 1971). The lipolytic action of HCS may be the factor responsible for the breakdown of maternal fat to glucose. The foetus under normal circumstances depends almost entirely on glucose as a primary energy substrate. If this is lacking there is an increased incidence of pathological syndromes of the neonate and these have important implications in the long-term development of the child. Thus low levels of HCS might mean this action is impaired and the amount of glucose being available for the foetus is reduced.

From these considerations it would seem that HCS levels will reflect placental function as well as being a guide to the availability of the primary energy substrate of the foetus.

Practical advantages

The immunological activity of HCS in serum is retained at room temperature for at least 3 days and is, therefore, easily transported. When frozen its activity remains for at least 6 months (Letchworth et al., 1971). The half life is short and has been shown to be between 12-25 minutes (Samaan et al., 1966; Spellacy et al., 1966; Beck and Daughaday, 1967; Grumbach et al., 1968). Changes in production, therefore, will be rapidly reflected in maternal levels. Although some subjects show fluctuation in levels, the day-to-day variation is considerably less than that of urinary oestriol. These occasional fluctuations show no rhythm and samples may be taken at any time in the knowledge that the results will be equally valid (Pavlou et al., 1972; Lindberg and Nilsson, 1973a), but this data stresses the need for serial sampling.

Levels of HCS have been examined following a variety of physiological stimuli. The levels are not consistently influenced by posture, exercise, bed rest or smoking (Pavlou et al., 1973; Ylikorkala et al., 1973; Letchworth et al., 1974). The levels remain unaltered following a protein meal or the intravenous infusion of prostaglandin or oxytocin (Tyson et al., 1971). There are conflicting reports on levels following infusion of glucose and insulin (Samaan et al., 1966; Grumbach et al., 1968; Spellacy, 1971; Pavlou et al., 1973).

The presence of high circulating levels enables measurement to be made with an assay that does not require high sensitivity. This has the advantage of avoiding an initial dilution step, rapid equilibrium, improved precision and the ability to process large numbers with ease (Letchworth et al., 1971).

NORMAL RANGE

The definition of the normal range is an essential step in assessment of placental function for without it, no meaningful statements can be made about the abnormal. It is, therefore, surprising that this aspect has been neglected in so many publications. Furthermore, comparisons between centres have been of limited value as different assay techniques and standards have been used.

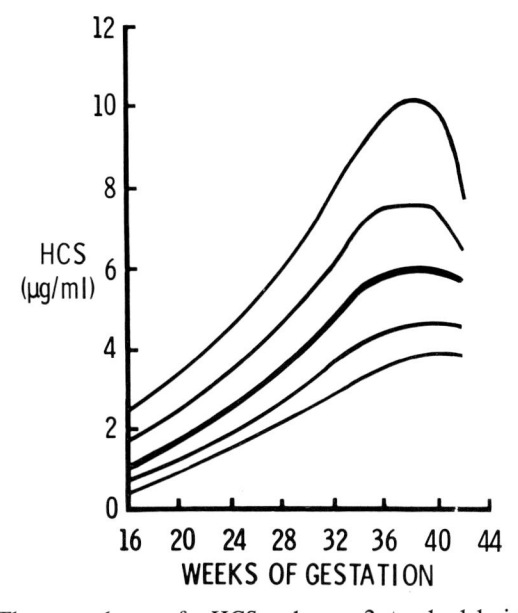

Fig. 1. The normal range for HCS and mean 2 standard deviations.

The normal range developed from 200 well-defined normal patients from which over 2000 samples were taken is shown in Figure 1. The deviation either side of the mean is unequal. This is the result of the correction for a skewed distribution by logarithmic transformation of the data.

ABNORMAL PREGNANCY

The clinical value of HCS levels for assessing placental function in abnormal pregnancy will depend upon the abnormality under review. The problems in early pregnancy will be examined, followed by examination in late pregnancy.

Early pregnancy

The clinical value of HCS levels in early pregnancy complicated by bleeding has been reported by a number of groups. Those who have examined large numbers (Genazzani et al., 1971; Niven et al., 1972; Ylikorkala et al., 1973; Garoff and Seppala, 1974; Zuckerman et al., 1974) reach similar conclusions, namely that a single HCS level is of little guidance in predicting the outcome of the pregnancy before the 10th week. Thereafter, prediction can be made with a fair degree of certainty, particularly after the 14th week. It is considered that between the 10th and 14th week it is best to rely on serial samples. The levels taken from patients before bleeding and subsequent abortion are no different to the normal population (Niven et al., 1972). This substantiates our knowledge that most abortions are not due to poor placentation, but to other factors.

In conclusion, the value of HCS levels in early pregnancy is confined to assessing the prognosis in threatened abortion. In the absence of contra-indications such as maternal age or infertility, low levels would constitute grounds for evacuation of the uterus which could reduce the in-patient stay with all its benefits.

Late pregnancy

Hypertension

The value of HCS levels in pregnancies complicated by hypertension has been confused by the wide variation in results reported by early workers. Those who have studied large numbers (Genazzani et al., 1971; Keller et al., 1971; Spellacy et al., 1971b; Letchworth and Chard, 1972a; Lindberg and Nilsson, 1973b; Ylikorkala, 1973; Crosignani et al., 1974; Spellacy et al., 1974) reach similar conclusions that the levels are either normal or low and that HCS levels are a valuable guide in the management of this complication. Spellacy's (1971b) early reports on a large group of hypertensives demonstrated that those subjects with levels of less than 4 μg/ml after the 30th week showed a foetal mortality of 24%. A further report on 949 hypertensive patients showed that the lowest levels of HCS were present in those pregnancies complicated by chronic vascular problems (Spellacy et al., 1974). Lindberg and Nilsson (1973b) examined 98 patients with pre-eclampsia and found that HCS levels of less than 4 μg/ml were associated with a high perinatal mortality and a high incidence of low birth weights and low Apgar scores.

Thus, most authors who have reported on extensive series find reduced values in this complication. The lowest levels are found in the severe group. Where levels are below 4 μg/ml, there is a high incidence of placental malfunction and this is a further indication to interrupt the pregnancy.

Rhesus iso-immunisation

The levels of HCS have been considered to be of limited value in pregnancies complicated by rhesus iso-immunisation. Most authors (Spellacy et al., 1971b; Lindberg and Nilsson, 1973b; Crosignani et al., 1974; Spellacy et al., 1974; Ward et al. 1974) find raised levels which can be attributed to an enlarged placenta. Some have noted that the levels are highest in those pregnancies with the most severe disease (Ward et al., 1974). It has also been observed that there is a sharp rise in levels prior to foetal death and that serial determinations might be of help in management, particularly if intrauterine transfusions rendered amniotic fluid examinations useless (Lindberg and Nilsson, 1973b; Crosignani et al., 1974; Ward et al., 1974). The results might dictate the timing of repeat intra-uterine transfusions or delivery. Ward and his co-workers (1974) used the finding of elevated levels in early pregnancy to identify those patients who required treatment in centres specialising in this problem.

Diabetes mellitus

There is little agreement about how levels are affected in this condition and the interpretation of results is also disputed. Most authors (Selenkow et al., 1971; Spellacy et al., 1971b; Spona and Janisch, 1971; Lindberg and Nilsson, 1973b; Persson et al., 1973; Ursell et al., 1973; Ylikorkala, 1973) who have examined large numbers have found levels to be high, as one would expect because of the increase in placental mass. There are several factors which could explain these findings. Few authors have given details of the severity, duration, complications or the control of the diabetes when reporting on HCS estimations in this condition. Foetal death can occur in the presence of satisfactory HCS levels. Interpretation of the results is complicated by the higher levels found in the diabetic as a level which is normal for the non-diabetic may in this condition be low and of prognostic significance.

Selenkow and colleagues (1971) studied 84 diabetics and noted generally high levels, but in patients with small babies and those with placental malfunction, the levels were low. These findings have been supported by other groups (Genazzani et al., 1971; Haour et al., 1971; Selenkow et al., 1971; Ursell et al., 1973; Ylikorkala, 1973).

Thus, there are wide differences in the results obtained and in their interpretation. It is important to recognise that HCS levels in this condition must be judged in relation to the range for normal diabetics and not the normal population. Most workers conclude that HCS levels are low in the presence of placental malfunction and the test had a potential usefulness in diabetic pregnancy.

Intra-uterine growth retardation

The evaluation of HCS levels as a guide to foetal welfare in this complication is incomplete. The main problem is that in the absence of other pathology, this is a relatively rare complication of pregnancy and authors who report on this condition have studied very limited numbers. In general, the levels are low, particularly in the severe cases (Saxena et al., 1969; Josimovich et al., 1970; Lindberg and Nilsson, 1973b; Ylikorkala, 1973).

Foetal distress

Several authors have examined the levels of HCS and the appearance of the signs of foetal distress in the perinatal period (Seppala and Ruoslahti, 1970; Lindberg and Nilsson, 1973b; Yates, 1973; England et al., 1974). In spite of the variable causes of these clinical phenomena, it has been shown by several groups that low levels of HCS are associated with the appearance of foetal distress. Letchworth and Chard (1972b) found that in a clinically normal group of patients, those whose only abnormality was perinatal distress had a higher percentage of low HCS levels. The prognostic significance of a level below 4 µg/ml after the 35th week of pregnancy was examined. For 1 or more low levels the risk was 30%, while for 3 or more it was 71%. The converse was also of value in that for 3 levels above 5, the risk of these complications occurring was 4%. This finding is an extremely important one in the clinical situation as it is of great value to have the reassurance of satisfactory HPL levels, particularly in cases with high risk factors.

England and colleagues (1974) found a 56% chance of perinatal complications if HCS levels were 2 standard deviations below normal in 547 normal patients. Spellacy and his group (1972) could find no relation to the foetal heart pattern and the 5-minute Apgar score, but found significantly lower levels in cases with meconium staining of the amniotic fluid and in a low 1-minute Apgar score.

In view of these findings, a good case can be made for screening all pregnancies using HCS levels.

CONCLUSIONS

The value of estimating HCS in apparently normal pregnancies is the detection of certain at risk cases which could not be identified clinically. A good case can be produced for screening all pregnancies.

In abnormal pregnancies, the merits of HCS measurements will depend upon the abnormality. In pregnancies complicated by rhesus iso-immunisation and diabetes they are of limited value, while in those with hypertension or intra-uterine growth retardation, the levels are of considerable value. The strongest evidence comes from a random treatment series by Spellacy's group on 2733 high risk cases. They found that in the group in which the levels were known and treatment was instituted, the foetal death rate was 2.6%, while in the control group the rate was 14%. The neonatal death rate was the same for both groups which suggests that early delivery was not giving rise to increased loss from prematurity.

In conclusion, HCS levels provide a convenient screening method for the assessment of placental function. Together with other parameters of placental function, they play an important role in modern day obstetrics.

REFERENCES

Beck, P. and Daughaday, W.H. (1967): *J. clin. Invest., 46,* 103.
Crosignani, P.G., Trojsi, L., Attanasio, A.E.M. and Lombroso, G.C. (1974): *Obstet. and Gynec., 44,* 673.
England, P., Lorrimer, D., Fergusson, J.C., Moffatt, A.M. and Kelly, A.M. (1974): *Lancet, 1,* 5.
Garoff, L. and Seppala, M. (1974): *Amer. J. Obstet. Gynec., 121,* 257.
Genazzani, A.R., Cocola, F., Casoli, M., Mello, G., Scarselli, G., Neri, P. and Floretti, P. (1971): *J. Obstet. Gynaec. Brit. Cwlth, 78,* 577.
Grumbach, M.M., Kaplan, S.L., Sciarra, J.J. and Burr, I.M. (1968): *Ann. N.Y. Acad. Sci., 148,* 501.
Haour, F., Cohen, M. and Bertrand, J. (1971): *Europ. J. clin. biol. Res., 16,* 124.
Josimovich, J.B., Kosor, B., Bocella, L., Mintz, D.H. and Hutchinson, D.L. (1970): *Obstet. and Gynec., 36,* 244.
Keller, P.J., Baertschi, U., Bader, P., Gerber, C., Schmid, J., Soltermann, R. and Kopper, E. (1971): *Lancet, 2,* 729.
Letchworth, A.T., Boardman, R., Bristow, C., Landon, J. and Chard, T. (1971): *J. Obstet. Gynaec. Brit. Cwlth, 78,* 535.

Letchworth, A.T. and Chard, T. (1972a): *J. Obstet. Gynaec. Brit. Cwlth, 79,* 680.
Letchworth, A.T. and Chard, T. (1972b): *Lancet, 1,* 704.
Letchworth, A.T., Howard, L. and Chard, T. (1974): *Obstet and Gynec., 43,* 702.
Lindberg, B.S. and Nilsson, B.A. (1973a): *J. Obstet. Gynaec. Brit. Cwlth, 80,* 619.
Lindberg, B.S. and Nilsson, B.A. (1973b): *J. Obstet. Gynaec. Brit. Cwlth, 80,* 1046.
Niven, P.A.R., Landon, J. and Chard, T. (1972): *Brit. med. J., 3,* 799.
Pavlou, C., Chard, T., Landon, J. and Letchworth, A.T. (1973): *Europ. J. Obstet. Gynec. reprod. Biol., 3/2,* 45.
Pavlou, C., Chard, T. and Letchworth, A.T. (1972): *J. Obstet. Gynaec. Brit. Cwlth, 79,* 629.
Persson, B., Lunell, N.O., Aubert, M.L., Carlstrom, K. and Felber, J.P. (1973): *Acta obstet. gynec., scand., 52,* 63.
Samaan, N., Yen, S.C.C., Friesen, H. and Pearson, O.H. (1966): *J. clin. Endocr., 26,* 1303.
Saxena, B.N., Emerson, K. and Selenkow, H.A. (1969): *New Engl. J. Med., 281,* 225.
Sciarra, J.J., Sherwood, L.M., Varma, A.A. and Lundberg, W.B. (1968): *Amer. J. Obstet. Gynec., 101,* 413.
Selenkow, H.A., Varma, K., Younger, D., White, P. and Emerson Jr, K. (1971): *Diabetes, 20,* 696.
Seppala, M. and Ruoslahti, E. (1970): *Acta obstet. gynec. scand., 49,* 143.
Spellacy, W.N. (1971): *Acta endocr. (Kbh.), 67, Suppl. 155,* 82 (abstract).
Spellacy, W.N., Buhi, W.C. and Birk, S.A. (1975): *Amer. J. Obstet. Gynec., 121,* 835.
Spellacy, W.N., Buhi, W.C., Birk, S.A. and McCreary, S.A. (1974): *Amer. J. Obstet. Gynec., 120,* 214.
Spellacy, W.N., Buhi, W.C., Birk, S.A. and Holsinger, R.N. (1972): *Amer. J. Obstet. Gynec., 114,* 803.
Spellacy, W.N., Buhi, W.C., Schram, J.D., Birk, S.A. and McCreary, S.A. (1971a): *Obstet. and Gynec., 37,* 567.
Spellacy, W.N., Carlson, K.L. and Birk, S.A. (1966): *Amer. J. Obstet. Gynec., 96,* 1164.
Spellacy, W.N., Teoh, E.S., Buhi, W.C., Birk, S.A. and McCreary, S.A. (1971b): *Amer. J. Obstet. Gynec., 109,* 588.
Spona, J. and Janisch, H. (1971): *Acta endocr. (Kbh.), 68,* 401.
Tyson, J.E., Jones, G.S., Huth, J. and Thomas, P. (1971): *Amer. J. Obstet. Gynec. 110,* 934.
Ursell, W., Brudenell, M. and Chard, T. (1973): *Brit. med. J., 2,* 80.
Varma, K., Driscoll, S.G., Emerson Jr, K and Selenkow, H.A. (1971): *Obstet. and Gynec., 38,* 487.
Ward, R.H.T., Letchworth, A.T., Niven, P.A.R. and Chard, T. (1974): *Brit. med. J., 1,* 347.
Yates, M.J. (1973): *Lancet, 2,* 1092.
Ylikorkala, O. (1973): *Acta obstet. gynec. scand., 26,* 1.
Ylikorkala, O., Haapalahti, J. and Jarvinen, P.A. (1973): *J. Obstet. Gynaec. Brit. Cwlth, 80,* 546.
Zuckerman, H., Gendler, L., Schwarz, M. and Harpaz, S. (1974): *Israel J. med. Sci., 10,* 490.

VIII. Prolactin

POLYPEPTIDE HORMONE PRODUCTION BY CELLS CULTURED ON ARTIFICIAL CAPILLARIES

RICHARD A. KNAZEK and JAY S. SKYLER*

Laboratory of Pathophysiology, National Cancer Institute; and Hypertension-Endocrine Branch, National Heart and Lung Institute, National Institutes of Health, Bethesda, Md., U.S.A.

Human pituitary hormones are used in both clinical and laboratory settings. The heavy demand for these hormones is largely satisfied, in the United States, by the National Pituitary Agency. The supply to meet the increasing needs, however, appears to be remaining at a constant level — approximately 80,000 pituitaries for biochemical extraction per year (P.G. Condliffe, personal communication). Alternative sources must be developed. Although synthesis of several pituitary hormones has been accomplished, the techniques required are extremely complex and not practical for commercial application. An alternative is to produce these hormones in vitro using cell cultures. This would necessitate that hormone-producing cells be readily available for use in equipment that is relatively inexpensive and easily maintained.

Rodent cell lines exist that produce pituitary hormones (Yasumura and Sato, 1966; Tashjian et al., 1968, 1973) but no human cell lines have, as yet, been established to produce such factors. Several investigators (Kohler et al., 1969; Gailani et al., 1970) have shown that dispersed human pituitary cells or explants may produce microgram amounts of hormone over several months while maintained in vitro. Although it would be ideal if human pituitary cells could be established as cell lines, primary cultures from normal or neoplastic human pituitaries may serve as alternative sources of hormones. If large numbers of cells can be maintained in vitro, and the method of maintenance is straightforward, then this may serve as a viable alternative to biochemical extraction techniques.

There are many techniques that have been developed to increase both the numbers of cells and the ease with which they can be cultured in vitro. Modifications of surface support such as helices (McCoy, 1965) and stacked plates (Weiss and Schleicher, 1968) increase the surface area for cell attachment and growth. Roller bottles (Kruse and Miedema, 1965) accomplish this while improving nutrient supply by perfusion, while gas-permeable support membranes increase oxygen transport (Jensen et al., 1974). Sephadex beads (Van Wezel, 1967) offer the advantages of suspension culture to cells that can only be grown on solid surfaces; and the spin-filter device (Himmelfarb et al., 1969) combines the properties of suspension cultures and perfusion systems. Each method possesses certain advantages to meet specific needs of the investigator.

One technique, which is described below, has unique properties that make it particularly well suited to the production of pituitary hormones and for the study of secre-

---

* Present address: Department of Medicine, Duke University Medical Center, Durham, N.C., U.S.A.

tory dynamics. This technique of cell culture on artificial capillaries overcomes some of the problems of maintaining cells in vitro and harvesting their secretory products.

Most established cell lines form multilayers in monolayer culture if permitted to persist in a confluent state for extended periods of time. Nutrients in the culture media then must diffuse from the media to the cell layers to feed the innermost cells. This becomes increasingly difficult as the number of cell layers increases. Metabolic products present the same problem in reverse: these substances must be removed before toxic or inhibitory concentrations are reached. The problems of diffusion are even more severe when the demands of oxygen or high molecular weight serum factors are considered. If the nutrient medium is continuously or periodically replaced to enhance the diffusion gradients, then sudden changes occur in the fluid composition immediately adjacent to the cells. Such changes of medium also prevent the accumulation of macromolecular cell products which modify cell growth or function (Eagle et al., 1960; Smith and Temin, 1974; Weiss et al., 1975). These changes of the pericellular microenvironment may alter cellular controls of behavior or those controls needed for the formation of dense tissue masses. Such problems are overcome, at least in part, by the technique of cell culture on artificial capillaries. Semi-permeable membranes in the shape of fine hollow tubes having outer diameters of approximately 250-350 microns with walls 25-75 microns thick and numbering 100-320 are bundled randomly together and pulled into a polycarbonate shell. The ends of the bundle are temporarily crimped and then dipped into a catalyzed silicone rubber monomer. After polymerization, the excess rubber is trimmed flush to the shell ends, exposing the patent lumens of the capillaries. Fluid applied to one end of this capillary culture unit (CCU) thus flows only through the capillaries but not within the extracapillary space (Fig. 1).

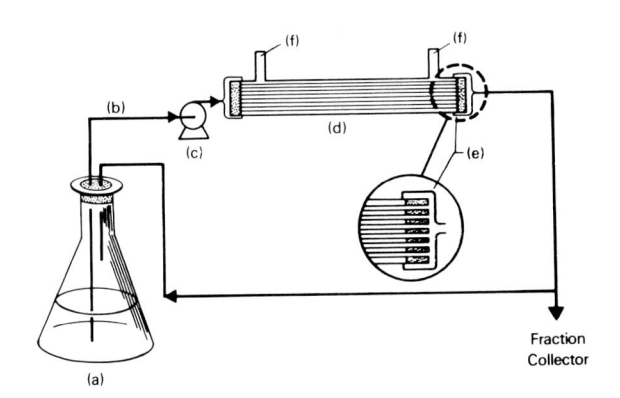

Fig. 1. The perfusion circuit consists of a capillary culture unit (CCU) (d) fed by a roller pump (c) with medium from reservoir (a). All components were connected by Silastic tubing (b) which also serves as an oxygenator and pCO_2 control. The flow pattern may be changed from recirculating to the single pass mode by clamping the Silastic tubing at the appropriate points. The capillary bundle consists of an admixture of silicone polycarbonate and nutrient capillaries in a 1 cm × 8 cm polycarbonate shell. Both ends are potted with silicone rubber (insert) so fluid applied at one end passes through the capillaries without causing bulk flow within the extracapillary space. Ground glass female tapered fittings allow insertion of the CCU into the perfusion circuit. Cells injected into the shell ports (f) settle onto the capillaries and receive nutritional support from the perfusate by diffusion through the capillary walls. The entire circuit is operated in a humidified 5% CO_2 air incubator at 37°C. (Reproduced from Knazek and Skyler, 1975, by courtesy of the *Proceedings of the Society of Experimental Biology and Medicine.*)

Several types of capillaries are now available for cell culture, each type having a spectrum of permeabilities. Cellulose acetate capillaries (CA) from Dow Chemical Company (Walnut Creek, Calif.), and polysulfone (PM) or polyvinyl chloride-acrylic copolymer (XM) from Amicon Corporation (Lexington, Mass.), are permeable to substances of up to 100,000 molecular weight. Silicone polycarbonate capillaries (SPC), manufactured by Dow Chemical, are permeable only to gases and are usually incorporated in a ratio of 3 SPC to 1 nutrient capillary to increase oxygen transport from the perfusion medium into the extracapillary space. Each CCU shell has side ports through which cells are injected into the extracapillary space. The shell ends are tapered to accept complementary adapters that are inserted into the perfusion circuit (Fig. 1). Medium is drawn from a 250-ml reservoir by a roller pump (911, Extracorporeal, King of Prussia, Penn.) through 2-3 meters of Silastic rubber tubing (Dow Corning, Midland, Mich.) at a rate of 0.2-10.0 ml/minute. The connecting tubing is permeable to the ambient gas within a humidified 5% CO_2 air incubator and therefore serves to maintain oxygenation and pH control. Thus stabilized, the nutrient medium passes through the capillary bundle and returns to the reservoir. Cells injected into the extracapillary space settle on and attach to the outer surface of the capillaries. Nutrients diffuse through the capillary walls to the cells while the products of cellular metabolism are removed in the opposite direction (Knazek et al., 1972).

Although primary explants appear to be maintained in a functional form, they do not visibly increase in mass. However, when established cell lines are injected into the extracapillary space, the cells attach and divide, forming multilayers on the supporting capillaries. They eventually reach a point when the outermost cells can no longer be adequately nourished by that capillary. At that time, however, the cells have grown into the sphere of influence of an adjacent capillary which supplies the required nutrients, thus allowing continued cell division and function. The cells form a well-nourished and functioning solid tissue within the extracapillary space. If the cells used have the capacity to secrete hormones, this tissue mass acts as a source of the hormone. Initial studies of the culture technique delineated the nutritional environment in which the cells would best grow and function. Many varied cell types have been cultured successfully using the artificial capillary technique and some have been of endocrine origin (Chick et al., 1974; Knazek, 1974; Wolf and Munkelt, 1975). This report will discuss cultures of rat and human pituitary cells using the artificial capillary technique to study prolactin and growth hormone secretion in vitro.

RAT PITUITARY CELLS

An established rat pituitary tumor cell line, GH_1 (Yasumura and Sato, 1966), grows rapidly and secretes large amounts of growth hormone (RGH) and prolactin (RPRL). Monolayers of these cells were trypsinized and washed twice in complete Ham's F-10 medium consisting of 13.1% horse serum and 2.1% fetal calf serum (Gibco, Grand Island, N.Y.) supplemented with 50 units penicillin and 50 µg streptomycin per ml. Cells suspended at densities of 3×10^6 per ml complete medium were then injected into the extracapillary space of various types of capillary culture units.

Initially, the only evidence of cellular presence was the acidic appearance of the extracapillary fluid when the perfusion was temporarily halted. However, after 2 weeks with media changes every 1-3 days, cells became grossly visible. A histological cross section through such a capillary bundle is shown in Figure 2. Note that the extracapillary volume is nearly filled with cells, providing a cell density approaching that of solid tissue. These cells produce RPRL (Fig. 3), initially increasing logarith-

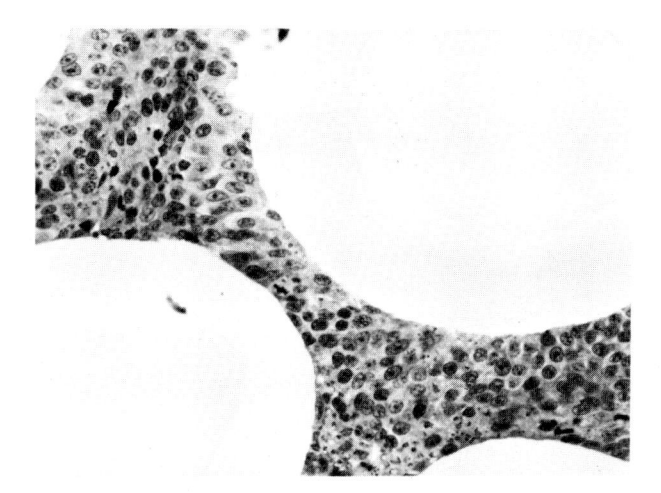

Fig. 2. Histologic cross section through a CA-SPC capillary culture unit in which GH₁ rat pituitary cells had been cultured for 23 days. Note the cells completely fill the space between the capillaries, forming a solid tissue mass.

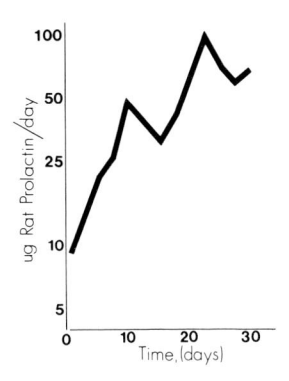

Fig. 3. Rat pituitary tumor cells, GH₁, were grown within a CA-SPC capillary culture unit. Prolactin secreted by the cells diffused across the capillary wall and accumulated in the perfusate which was replaced periodically.

mically and then slowing after 2 weeks. The maximum amounts of hormones range between 10 and 100 μg per day, with some concentrations exceeding 1 μg RPRL per ml medium.

In general, bundles composed of admixtures of SPC and nutrient capillaries were superior to those containing only nutrient capillaries. The time period between inoculation and significant levels of RPRL secretion was shorter and the ultimate secretory rates were higher. This is probably a result of improved oxygen transport from the perfusate into the extracapillary space.

HUMAN ENDOCRINE CELLS

An obvious extrapolation of the above observation is to the production of human pituitary hormones. Unfortunately, no established human cell lines exist that secrete these hormones. However, JEG-7 human choriocarcinoma, established in monolayer by Kohler et al. (1971), secretes several hormones. One, human chorionic gonadotropin (HCG), is a glycoprotein with a molecular weight of 37,000, considerably larger than the anterior pituitary hormones, growth hormone (HGH) or prolactin (HPRL).

These cells grow well when injected into artificial capillary culture units composed of 30 XM and 90 SPC capillaries. The nominal permeability of the nutrient capillaries limited diffusion to molecules 50,000 MW or less and, therefore, allowed HCG to diffuse readily into the perfusate for harvest when the nutrient medium was replaced every 1-4 days. The rate of HCG secretion increased exponentially during the month of culture reaching 10 IU per day. When terminated the extracapillary contents were assayed for DNA revealing that the cells within the CCU were secreting 11-fold more hormone per cell than the same cells in monolayer (Knazek et al., 1974). This is a reflection of the improved nutritional environment of the perfused cells since the confluent monolayers, when fed 20 ml per flask per day depleted the glucose within the medium during that period. Attempts to increase the nutrient supply to monolayers by simply increasing the volume of nutrient medium aggravates the anoxic state that probably exists at the surface of the confluent monolayer (Werrlein and Glinos, 1974). The glucose concentrations in the perfusates of the above CCU experiments did not fall below 70 mg%.

Efforts have been made to establish cell lines that produce human pituitary hormones. Notable examples are those efforts directed toward HGH. Kohler and his coworkers (1969) were able to maintain normal or neoplastic pituitary explants in vitro for 4 or 12 months, respectively. During these periods, the tissues synthesized and secreted approximately 45 μg of HGH per culture flask. The long doubling times together with the propensity for fibroblastic overgrowth made it impossible to clone and establish a secretory cell line. Gailani and coworkers (1970) studied human fetal pituitaries showing a maximum secretion of 10 μg HGH per day per culture flask. It was apparent from these studies that primary tissue explants might be used to produce significant quantities of pituitary hormones.

To obtain tissue for culturing, we studied the tumor of a 51-year-old male with an eroded sella, high circulating levels of HPRL and complaints of visual field cuts who was found to have a chromophobe adenoma at surgery. Pieces of the pituitary tumor were immediately placed in complete Ham's F-10 medium to which crude collagenase (Worthington Biochemical Corp., Freehold, N.J.) had been added to yield a concentration of 100 mg%. The suspension was stirred for half an hour at 37°C in a 5% CO_2 air incubator, centrifuged at 100 $\times$ g for 5 minutes, washed, recentrifuged, and then injected into a sterile capillary culture unit composed of 90 SPC and 84 CA capillaries. The unit was maintained as noted previously, the perfusate replaced every 1-4 days with fresh complete medium and frozen at $-20°C$ for subsequent analysis. Radioimmunoassay (RIA) (Rogol and Rosen, 1974) of the perfusate showed that the 50 mg of tissue maintained within a single CCU produced more than 3 mg of HPRL during the first 4 months of culture (Knazek and Skyler, 1975). This indicates that large quantities of a human pituitary hormone can be provided by a technique utilizing a simple, relatively inexpensive apparatus that requires minimal technical expertise.

Several attempts to duplicate this experiment with an acromegalic tumor have been only partially successful since the initial rates of HGH secretion dropped rapidly from 1 μg per day to non-detectable levels. The reasons for the lack of success are uncertain but efforts are continuing to study HGH secretion using this technique.

STUDIES OF SECRETORY DYNAMICS

If the perfusion circuit is modified to a single-pass mode (Fig. 1), nutrient medium can be collected in fractions immediately after passage through the CCU. Cells cultured on artificial capillaries can be viewed as a tissue mass grown within a vascular network having a single artery and vein and perfused by solutions under tightly controlled conditions. When secretory cells are used, the characteristics of the single-pass system lend themselves to the study of the secretory dynamics of the cells. This arrangement was used to observe the effect of thyrotropic releasing hormone (TRH, Abbott Laboratories, N. Chicago, Ill.), using another human pituitary tumor that produced prolactin in vivo. After 3 days of culture in the CCU in the recirculating mode, the medium was replaced with fresh medium and the single-pass mode was employed. After allowing several hours for equilibration, TRH was added to the perfusate in the amount of 100 ng/ml medium. Tritiated leucine was also added to mark the presence of the TRH stimulus. Figure 4 shows that the tissue responded to the stimulus within 8 minutes by releasing a massive amount of HPRL, raising the rate of secretion from 15 ng/minute to more than 125 ng/minute. This was followed by a gradual decline in secretion. Subsequent stimulations of the same culture did not cause the rapid HPRL release. This implies that other factors, as yet unknown, are necessary for storage in the forms available for rapid release by appropriate stimuli. This exemplifies the technique by which transient secretory responses to pharmacologic and physiologic stimuli may be followed in vitro (Knazek and Skyler, 1975).

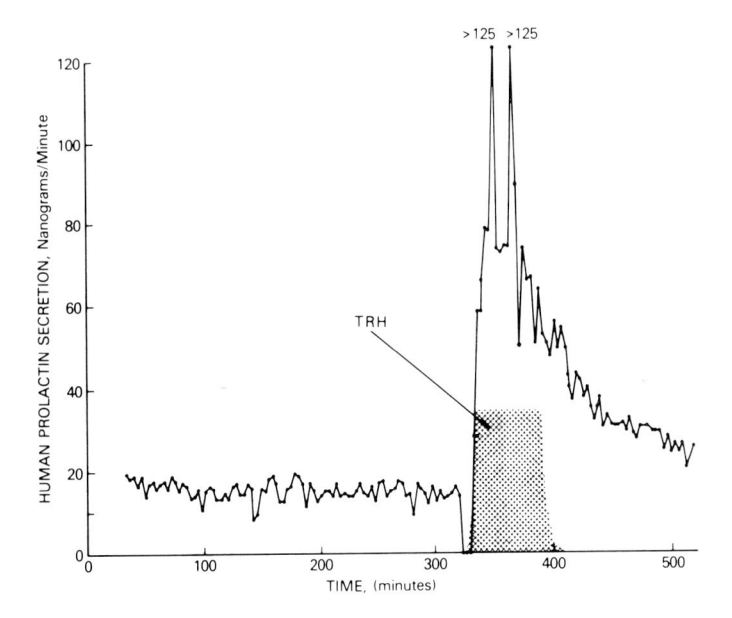

Fig. 4. Human pituitary tumor explants were injected into a CA-SPC CCU. When changed to single-pass mode, they secreted 15 ng HPRL into the perfusate per minute. Within 8 minutes after exposure to TRH, a massive amount of HPRL was released followed by a gradual decline toward the basal secretion rate. Tritiated leucine was mixed with the TRH and used as a tracer.

CHARACTERIZATION OF HORMONES

Should cell lines be established that produce the desired hormones, it will be incumbent upon the investigators to show that (1) the hormone produced is indistinguishable from the native hormone both biochemically, and biologically in its effect, and (2) the hormone produced is safe to use in a clinical setting. To this end, 2 hormones produced in tissue culture were studied to determine if they differed from the native species extracted from normal pituitaries obtained at autopsy.

Several methods of chemical characterization have been used. All compare unfractionated hormone in tissue culture medium with native hormone extracted from pituitary glands and obtained from the Hormone Distribution Program, National Institute of Arthritis, Metabolism and Digestive Diseases. The comparison to the pituitary-extracted hormone assumes that neither hormone has undergone artifactual changes prior to analysis.

Immunochemical comparison is made using specific antibody generated against pituitary hormone. The technique is based on the assumption that immunochemical differences in antigens will produce non-parallel dose-response curves in antibody-binding experiments. These are performed in a manner analogous to RIA, using displacement of ^{125}I-labeled hormone. Data are analysed, slopes calculated, and parallelism tested by parallel line dose-response bioassay statistics using the methods of Rodbard (1974). Dose-response curves of HGH obtained from a monolayer culture of an acromegalic pituitary and HGH from pituitary extraction were parallel (Fig. 5) (Skyler et al., 1975). This indicates that the anti-HGH antibody reacts with both tissue culture-produced HGH and pituitary-extracted HGH in the same way. The results have been quite different for HPRL. We have used 3 pituitary tumors grown in culture (SK-1, RUS, MCG) and 3 anti-HPRL antisera generated in different laboratories (Becker RB-4, supplied by Drs. Robert Becker and Judith Vaitukaitus, V-L-S-2 supplied by Dr. U.J. Lewis, and Freisen 65-5 supplied by Dr. Henry Freisen). Only 1 of the

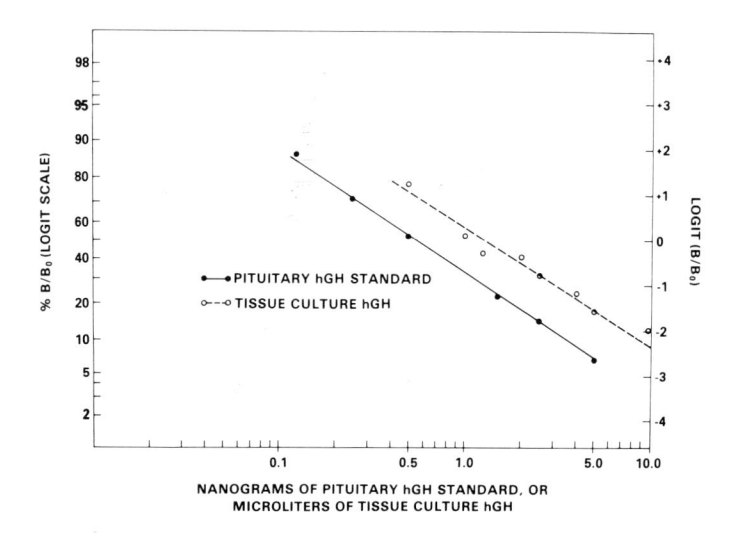

Fig. 5. Dose-response curves comparing the binding of pituitary-extracted HGH and tissue culture-produced HGH to a single specific antiserum. The lines are statistically parallel, indicating that the hormone recognition site of the antibody fails to distinguish any differences between these hormones.

9 combinations, SK-1 prolactin with RB-4 antisera, yielded a dose-response curve parallel to that of the HPRL extracted from normal pituitary glands. The non-parallel curves seen in the other 8 combinations indicate that the HPRL produced in culture has immunochemical differences from the standard pituitary preparation. The significance of these differences is not clear. They may represent changes in the extraction process of HPRL, changes during storage, or real differences between the hormones from the 2 sources. It is of interest to note that with a given antiserum, prolactin from the 3 cultures studied yielded dose-response curves that were parallel to each other. The only exception was noted above.

In an analogous way, hormones binding to a specific cellular binding receptor in radioreceptor assay (RRA) can yield dose-response curves, the slopes of which can be compared by the same techniques as in RIA (Rodbard, 1974). Here, the assumption is made that parallel dose-response curves demonstrate the inability of the cellular binding receptor to distinguish structural differences between the hormone preparations. This implies that the RRA is a better test of structural differences which would be physiologically significant since the cellular binding receptor presumably represents biological specificity as opposed to the immunological specificity. Human growth hormone generated in monolayer culture and growth hormone derived by biochemical extraction of autopsied pituitaries have been compared in the RRA of Lesniak et al. (1974) using cultured IM-9 lymphocytes. The dose-response curves were parallel, indicating that the growth hormone receptor on these cultured cells did not distinguish between the tissue culture and pituitary-extracted growth hormones.

Elution patterns of human growth hormone and of human prolactin on a Sephadex G-100 column revealed that tissue culture-produced hormones and the pituitary-extracted hormones had identical partition coefficients, K_{av}, and band widths, $\sigma_{K_{av}}$. The HGH and HPRL from both sources eluted with $K_{av} = 0.41$ for the major peaks. Both hormones also showed smaller peaks at the void volume and $K_{av} = 0.20$-0.22, corresponding to 'big-big' and 'big' hormones respectively. The tissue culture-produced HGH and HPRL, therefore, appear to be predominantly the monomeric species with smaller amounts of the larger forms. Since the nominal permeability of the artificial capillaries is only 50,000 daltons, the presence of material in the void volume would suggest that this represents aggregated forms.

The hormones were also examined in quantitative polyacrylamide gel electrophoresis (PAGE). Multiphasic buffer systems were optimized for each hormone (Chrambach and Skyler, 1975). The resolving gel pH was set at 7.50 to study HGH with bis-tris as the common counter ion (system 1954.5.0) (Chrambach and Skyler, 1975). The system optimized for characterization of human prolactin had an operative pH of 8.16 in the resolving gel, with a common counter ion of N-ethylmorpholine (system 2333.0.VII) (Rogol and Chrambach, 1975).

The technique of quantitative PAGE allows one to distinguish proteins and polypeptides on the basis of molecular net charge and molecular size by electrophoresis at several acrylamide gel concentrations. A plot of the logarithm of relative mobility, R_F, vs total gel concentration plot, is known as the Ferguson plot (Ferguson, 1964). The slope of this straight line is equal to the retardation coefficient, K_R, which is a parameter of molecular size. Extrapolation of the line to the zero gel concentration yields the y-intercept, termed the relative free mobility, Y_0, a parameter of molecular net charge. This linear regression analysis can be performed using the method and programs of Rodbard and Chrambach (1974) permitting the calculation of K_R and Y_0 with 95% confidence limits. The computer programs used also yield the joint 95% confidence envelopes of the relationship of K_R to Y_0. This can be graphically shown as 'K_R-Y_0 ellipses' when plotted on semi-logarithmic paper using the logarithmic function for

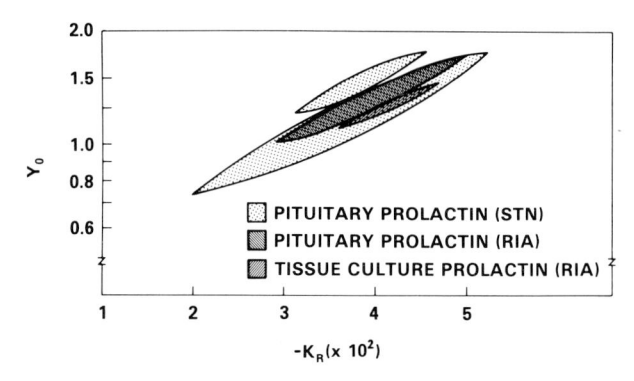

Fig. 6. Depicted are joint 95% confidence envelopes of K_R and Y_0 for HPRL in system 2333.0.VII. The stippled areas are the major and minor components of human pituitary prolactin, defined by staining. Both the pituitary-extracted hormone and the tissue culture-derived hormone, detected by RIA, are completely included within the staining ellipse of the major pituitary component and are indistinguishable by this technique.

Y_0. Chrambach and Rodbard (1971) have shown that the K_R-Y_0 ellipse derived from such experiments in a given buffer milieu allows determination of a relatively precise relationship between molecular size and molecular net charge. This technique can also be used to test the identity of 2 species on the basis of their joint 95% confidence envelopes: when the 95% confidence envelopes for 2 proteins do not overlap, the proteins are significantly different. When ellipses partially overlap, an F-test may be used to indicate the likelihood of identity. This test only allows the investigator to say that the 2 species are not distinguishable from one another by these criteria, and clearly does not prove identity. Such criteria have been applied to the comparison of tissue culture-produced prolactin and pituitary-extracted prolactin (Fig. 6) and to the comparison of the tissue culture-derived HGH and pituitary-extracted HGH. The tissue culture-produced materials were indistinguishable from the pituitary-extracted counterpart.

No biological comparisons have been made, to date, of the tissue culture-produced and pituitary-extracted hormones. Such comparisons would be a prerequisite to the utilization of tissue culture-produced hormones in place of pituitary-extracted hormones. It is further noted that no attempts have as yet been made to purify significant quantities of tissue culture-produced hormone. All comparisons have been made on unfractionated tissue culture media. The purified materials would have to be shown to be unchanged by the purification process and free of contamination by other substances, e.g. virus and foreign protein. The demonstration that growth hormone produced in culture is indistinguishable from HGH extracted from pituitaries by the 4 criteria listed above gives impetus to answering these additional questions.

CONCLUSIONS

The technique of cell culture on artificial capillaries has the potential of providing large quantities of hormones and other cell products for clinical and laboratory study. Before this can be practical, however, there will have to be both an increase in the rates of hormone secretion per cell in culture for prolonged periods of time and a relatively inexpensive purification process to separate the hormones from the tissue culture media. When these problems are solved, cell culture on artificial capillaries could become a standard source of polypeptide hormones.

The technique of artificial capillary cell culture also offers a simple method for the study of the secretory dynamics and transient biological responses of cells stimulated under controlled physiological conditions. Such studies should provide new insights and clearer understandings of these physiologic processes.

REFERENCES

Chick, W.L., Like, A.A. and Lauris, V. (1974): *Science, 187,* 1184.
Chrambach, A. and Rodbard, D. (1971): *Science, 172,* 440.
Chrambach, A. and Skyler, J. (1975): *Protides of the Biological Fluids, 22nd Colloquium, 22,* 701. Pergamon Press.
Eagle, H., Agranoff, B.W., and Snell, E.E. (1960): *J. biol. Chem., 235,* 1891.
Ferguson, K.A. (1964): *Metabolism, 13,* 985.
Gailani, S.D., Nussbaum, A., McDougall, W.J. and McLimans, W.F. (1970): *Proc. Soc. exp. Biol. (N.Y.), 134,* 27.
Himmelfarb, P., Thayer, P.S. and Martin, H.E. (1969): *Science, 164,* 555.
Jensen, M.D., Wallach, D.F.H. and Liu, P.-S. (1974): *Exp. Cell Res., 84,* 271.
Knazek, R.A. (1974): *Fed. Proc., 33,* 1978.
Knazek, R.A., Gullino, P.M., Kohler, P.O. and Dedrick, R.L. (1972): *Science, 178,* 65.
Knazek, R.A., Kohler, P.O. and Gullino, P.M. (1974): *Exp. Cell Res., 84,* 251.
Knazek, R.A. and Skyler, J.S. (1975): *Proc. Soc. exp. Biol. (N.Y.),* in press.
Kohler, P.O., Bridson, W.E., Hammond, J.M., Weintraub, B., Kirschner, M.A. and Van Theil, D.H. (1971): *Acta endocr. (Kbh.), Suppl., 153,* 137.
Kohler, P.O., Bridson, W.E., Rayford, P.L. and Kohler, S.E. (1969): *Metabolism, 18,* 782.
Kruse, P.F. and Miedema, E. (1965): *J. Cell Biol., 27,* 273.
Lesniak, M.A., Gorden, P., Roth, J. and Gavin III, J.R. (1974): *J. biol. Chem., 249,* 1661.
McCoy, T.A. (1965): In: *Tissue Culture,* p. 108. Editor: C.V. Ramakrishnan. W. Junk, The Hague.
Rodbard, D. (1974): *Clin. Chem., 20/10,* 1225.
Rodbard, D. and Chrambach, A. (1974): In: *Electrophoresis and Isoelective Focusing in Polyacrylamide Gel,* p. 28. Editors: R.C. Allen and H.R. Maurer. deGroyter, Berlin-New York.
Rogol, A.D. and Chrambach, A. (1975): *Endocrinology, 97,* 406.
Rogol, A.D. and Rosen, S.W. (1974): *J. clin. Endocr., 39,* 359.
Tashjian Jr, A.H., Hinkle, P.M. and Dannies, P.S. (1973): In: *Proceedings, IV International Congress of Endocrinology, Washington, 1972,* p. 648. Editor: R.O. Scow. ICS 273, Excerpta Medica, Amsterdam.
Tashjian Jr, A.H., Yasumura, Y., Levine, L., Sato, G.H. and Parker, M.L. (1968): *Endocrinology, 82,* 342.
Skyler, J.S., Rogol, A.D., Lovenberg, W. and Knazek, R.A. (1975): Submitted for publication.
Smith, G.L. and Temin, H.M. (1974): *J. Cell Physiol., 84,* 181.
Van Wezel, A.L. (1967): *Nature (Lond.), 216,* 64.
Weiss, L., Poste, G., MacKearnin, A. and Willett, K. (1975): *J. Cell Biol., 64,* 135.
Weiss, R.E. and Schleicher, J.B. (1968): *Biotechnol. Bioeng., 10,* 601.
Werrlein, R.J. and Glinos, A.D. (1974): *Nature (Lond.), 25,* 317.
Wolf, C.F.W. and Munkelt, B.E. (1975): *Trans. Amer. Soc. artif. intern. Org., 21,* 16.
Yasumura, Y. and Sato, G. (1966): *Science, 154,* 1184.

PROLACTIN AND FERTILITY CONTROL IN WOMEN

C. ROBYN, M. VEKEMANS*, P. DELVOYE, V. JOOSTENS-DEFLEUR, A. CAUFRIEZ and M. L'HERMITE

Human Reproduction Research Unit, Hôpital Saint-Pierre, Université Libre de Bruxelles, Brussels, Belgium

When human prolactin was firmly established in 1971, it first appeared that this hormone played no significant role in regulating reproductive processes in women. In contrast to other animal species prolactin secretion in man was not thought to be influenced by endogenous nor exogenous estrogens (Hwang et al., 1971; Midgley and Jaffe, 1973). Human prolactin was not specifically bound to sections of human ovaries while the labeled hormone bound extensively and exclusively to mammary alveolar epithelium (Midgley and Jaffe, 1973).

These preliminary conclusions seem erroneous. Already in 1852 Chiari and co-workers and Frommel described the persistence of amenorrhea and lactation for long periods of time after delivery. Such an amenorrhea-galactorrhea syndrome also occurs in patients with pituitary tumors (Argonz and Del Castillo, 1953; Forbes et al., 1954). Occasionally the amenorrhea-galactorrhea syndrome is associated with primary hypothyroidism: in most of the cases where circulating prolactin could be assayed, the basal levels were supranormal (e.g., Robyn, 1976). Prior to sensitive and specific assays for human prolactin, galactorrhea was considered to indicate prolactin hypersecretion: circulating levels of growth hormone were normal. With the advent of such assays, it soon became clear that galactorrhea only occurs in 30-50% of hyperprolactinemic women. The incidence of galactorrhea depends for the greater part on the clinician's obstinacy and endeavor to establish non-puerperal lactation. The discharge of milk may have been noticed by the patient herself and then appears among her complaints or it may be a casual discovery during a physical examination. If hyperprolactinemia without galactorrhea is frequent, galactorrhea with normal serum prolactin is also common (e.g., Robyn, 1976). Hyperprolactinemia in women is frequently, if not systematically, associated with menstrual cycle disorders: amenorrhea, oligomenorrhea, short luteal phases. It is obvious, however, that menstrual cycle disorders are more frequently due to endocrine disturbances other than hypersecretion of prolactin. Nevertheless, all women with anomalies of the menstrual cycle deserve an evaluation of their prolactin secretion. The present communication reviews our own data on the relationship between prolactin secretion and the endocrine control of fertility in women, and compares them with those reported by others.

---

* Aspirant at the 'Fonds National de la Recherche Scientifique' of Belgium.

PROLACTIN AND ESTROGENS

There are several indirect indications that prolactin secretion is enhanced by endogenous estrogens. In pregnant women circulating levels of prolactin increase in parallel with estrogens (Robyn, 1976). In the rhesus monkey, there is no massive increase in circulating estrogens, and circulating prolactin remains low during gestation (Friesen et al., 1972).

No consistent changes in serum prolactin concentrations were seen to occur during the menstrual cycle (Hwang et al., 1971; Midgley and Jaffe, 1973; McNeilly et al., 1973; Yuen et al., 1974). Vekemans et al. (1972) using an homologous ovine system for the radioimmunoassay of human prolactin, first demonstrated significant changes: mean serum prolactin was higher at midcycle and during the luteal phase than during the early follicular phase. Robyn et al. (1973) confirmed this biphasic pattern and showed that it closely paralleled serum estrogens. Guyda and Friesen (1973) also reported higher values during the luteal phase and Lequin (1973), Tamura and Igarashi (1973) and Schmidt-Gollwitzer et al. (1975, personal communication) found similar serum prolactin patterns. It should be emphasized that serum prolactin concentration varies during the day; in addition to a circadian variation, irregular bursts also occur (e.g., Robyn et al., 1973). These variations often have greater amplitudes than those occurring at midcycle and during the luteal phase. The blood samples should be taken late in the morning or during the afternoon rather than early in the morning at a time when the night surge is frequently incomplete. Even then, consistent changes in basal levels of serum prolactin are not always seen in individual cycles. However, as shown in Figure 1, even when 4 cycles are pooled, the rises at midcycle and during the luteal phase become apparent. With an increasing number of cycles, the biphasic pattern is more and more definite (Figs. 2 and 3).

Recently, Vekemans and Robyn (1975a) reported that serum prolactin concentration in women decreased with age, between 15 and 65 years (Fig. 4), a decline which parallels that of urinary and plasma estrogens (Pincus et al., 1954; Heusghem, 1956; Furuhjelm, 1966; Longcope, 1971). In men, serum prolactin does not fall with age; there is even a trend toward higher levels at the age of 65. Finally, in amenorrheic women with normal prolactin secretion and with normal gonadotrophin-releasing hormone tests, serum prolactin is found lower than in normal cycling women (Robyn et al., 1973; Alexander et al., 1975).

These data suggest indirectly that endogenous estrogens physiologically affect prolactin secretion in women.

High doses of ethinylestradiol (2 mg/day) stimulate prolactin secretion in women with recurrent breast cancer (L'Hermite et al., 1972, 1974). 400 μg acted similarly in regularly menstruating women (Robyn et al., 1973). More recently, the sensitivity of prolactin secretion to estrogens has been further investigated in regularly menstruating and post-menopausal women. Oral doses of 50, 200 and 400 μg ethinylestradiol were administered daily to groups of 4 regularly menstruating women for 20 days. Blood samples were taken daily during a control cycle and during the period of treatment. Prolactin was measured by an homologous ovine radioimmunoassay (L'Hermite et al., 1972); the results were expressed with reference to a pool of sera collected during the early post-partum period and used as laboratory standard: ·1.0 U of this pool is the amount of immunoreactive prolactin contained in 1.0 ml of this pool. One unit was found equivalent to 2.3 ampoule of the serum prolactin Research Standard 71/167 (Medical Research Council, National Institute for Biological Standards and Control, London). In women receiving 50 μg ethinylestradiol, the basal levels of serum prolactin were within the limits of those found during the control cycles (Fig. 1): the

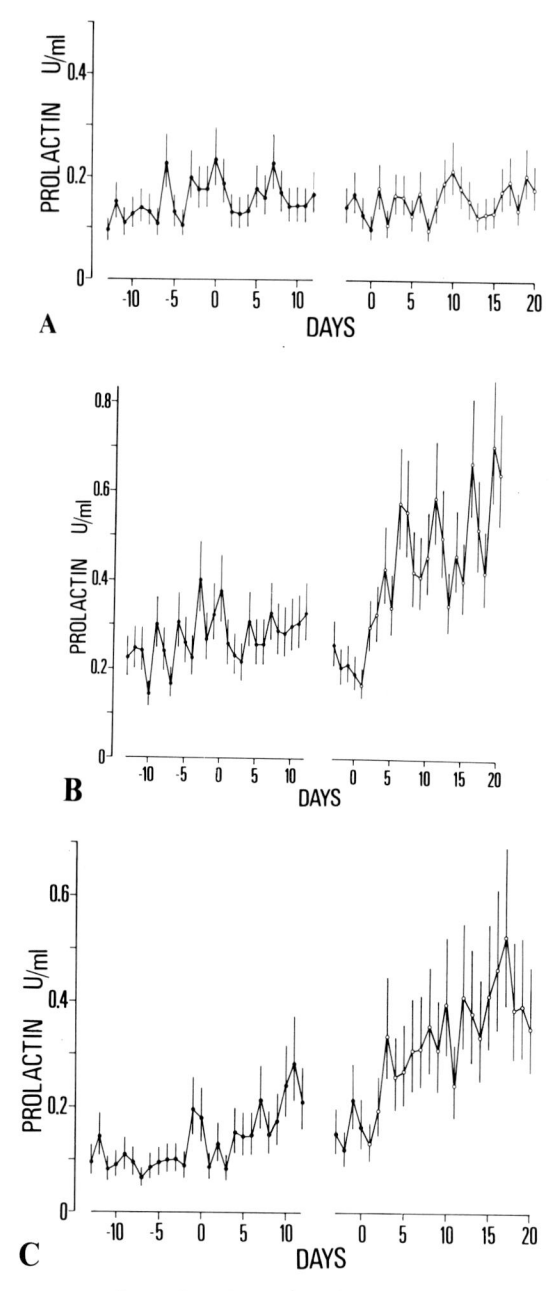

Fig. 1. Mean serum concentrations of prolactin (PRL) in 4 control cycles (left) and in 4 cycles under treatment with 50 (A), 200 (B) and 400 µg (C) ethinylestradiol (right). Four different women are included in each of the 3 groups. Oral administration of the estrogen starts on day 0 of the second cycle. Vertical bars represent one standard error of the means. The PRL results are expressed in units (U): 1 U is the amount of immunoreactive prolactin contained in 1 ml of a pool of sera collected during the early post-partum and used as laboratory standard.

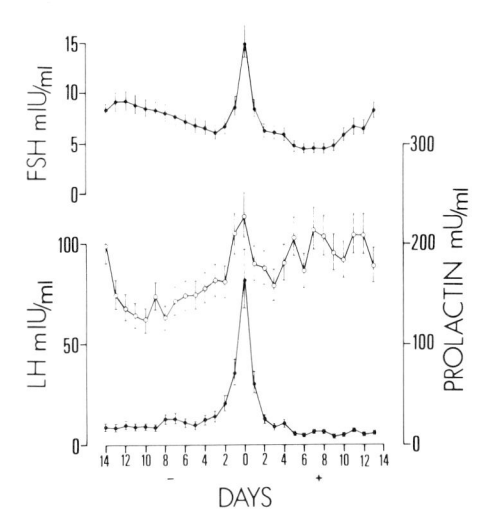

Fig. 2. Changes in mean serum prolactin (PRL), LH and FSH concentrations based on 20 ovulatory cycles from 20 different women. The PRL results are expressed in milli-units of the laboratory standard (see Fig. 1). Vertical bars represent one standard error of the means.

ratio between the mean value during estrogen administration and the mean control value was 1.05. However, both 200 and 400 µg significantly increased the basal levels in a dose-related manner: the ratio between the mean value during estrogen administration and the mean control value was 1.70 with 200 µg and 2.50 with 400 µg. It has been shown recently (Vekemans and Robyn, 1975b) with an homologous human assay for prolactin (distributed by the National Pituitary Agency USA) (Aubert et al., 1974) that ethinylestradiol administration not only enhances prolactin secretion but also modifies the 24-hour pattern of the circulating hormone: under estrogen treatment the increase in prolactin starts earlier in the afternoon and exhibits a biphasic pattern with peak levels at 10 p.m. and 6 a.m. Robyn and Vekemans (1974, 1975) measured serum LH, FSH and prolactin in 5 post-menopausal women on a daily oral dose of 25 µg ethinylestradiol for 28 days. The time elapsed from surgical or spontaneous menopause varied from 2-28 years. Serum prolactin (homologous human assay kit; National Pituitary Agency) increases within the first week of treatment (Fig. 5). During estrogen administration, the mean basal level is 2.5 times higher than the mean control level. Within 2 weeks after interruption of treatment, the serum levels of prolactin return to the control range. It thus appears that 50 µg ethinylestradiol stimulates prolactin secretion to an equivalent extent as endogenous estrogens during an ovulatory cycle. However, after suppression of ovarian steroidogenesis at the menopause, prolactin secretion is much more sensitive to exogenous estrogens: 25 µg ethinylestradiol are then already effective and significantly increase serum prolactin concentration. It is important to know whether elevated prolactin levels are maintained in post-menopausal women with estrogens administered for periods of time much longer than those applied in the present study. It should also be established whether elevated circulating prolactin, resulting from chronic treatment with estrogens would increase breast cancer risk in post-menopausal women (Robyn, 1975). Conjugated estrogens appear less effective than free estrogens in stimulating prolactin secre-

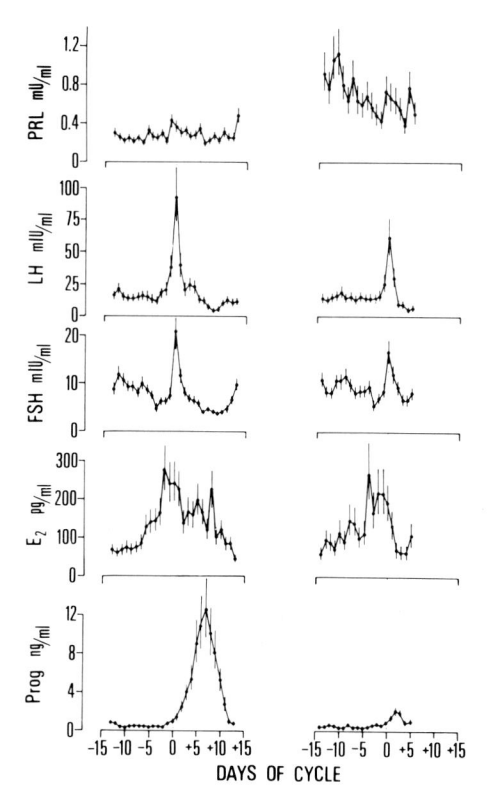

Fig. 3. Changes in mean serum prolactin (PRL), LH, FSH, estradiol (E$_2$) and progesterone (Prog) concentrations based on 7 control cycles (on the left) and on 6 cycles with short luteal phase (on the right) observed during or just after hyperprolactinemia induced with sulpiride (Dogmatil®, Delagrange Laboratories, Paris). Sulpiride administration occurred during the follicular phase. Vertical bars represent one standard error of the means. Prolactin results are expressed in milli-units of the Research Standard 71/222 (Cotes, 1973) distributed by the Medical Research Council (London). (From Robyn et al., 1976.)

tion: in post-menopausal women daily oral administration of a combination of 0.8 mg estradiol sulfate and 1.2 mg estriol sulfate did not increase circulating prolactin (Delvoye et al., 1972). However, here both circulating LH and FSH were much less reduced than with 25 µg ethinylestradiol (Robyn and Vekemans, 1975).

EFFECTS OF EXPERIMENTAL HYPERPROLACTINEMIA ON THE MENSTRUAL CYCLE

The supply of human prolactin is insufficient for studies of its biological effects in man. An indirect approach to assess the physiological significance of prolactin is to use drugs that inhibit or stimulate prolactin secretion. Prerequisites for the selection of such drugs are that they have a long-lasting action, that they have no direct influence on the secretion of other hormones and that they have minimal side effects (Robyn, 1976).

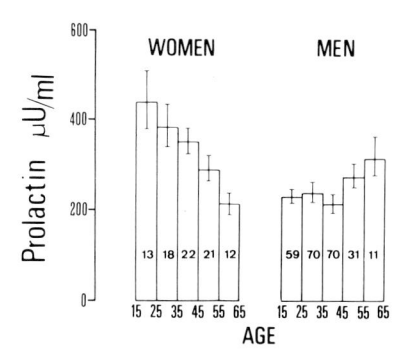

Fig. 4. Evolution with age of mean serum prolactin (PRL) concentration expressed in micro-units (µU) of the Research Standard 71/222 (see Fig. 3), in men and women. The number of individual determinations are indicated in the columns. Vertical bars represent one standard error of the means. (From Vekemans and Robyn, 1975a, by courtesy of the Editors of the *British Medical Journal*.)

The thyrotrophin-releasing hormone (TRH) has been used to induce experimental hyperprolactinemia during menstrual cycles in primates and in women. Stevens et al. (1973) reported anti-LH effects in the baboon: midcycle LH surges were occasionally suppressed and, when ovulation occurred, a luteal deficiency was observed. Such preliminary data were not further confirmed. In similar experiments conducted in women, Jewelewicz et al. (1974) and Zarate et al. (1974) could not find changes either in length or in hormone production during the menstrual cycle in women receiving repeated oral doses of TRH. However, TRH stimulated both TSH and prolactin secretions and its hyperprolactinemic effect is short (1 or 2 hours).

Sulpiride (Dogmatil®, Delagrange Laboratories, Paris) exerts a much more spe-

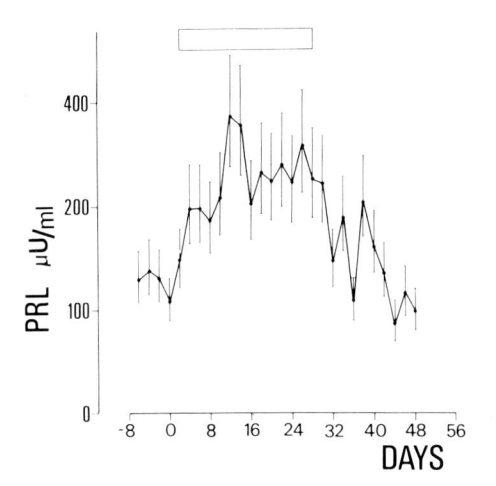

Fig. 5. Mean serum prolactin (n = 6) concentration expressed in micro-units (µU) of the Research Standard 71/222 before, during and after administration of a daily oral dose of 25 µg ethinylestradiol (open rectangle) during 28 days. Vertical bars represent one standard error of the means. (From Robyn and Vekemans, 1976, by courtesy of the Editors of *Acta endocrinologica*.)

cific and a much longer stimulating effect on prolactin secretion (L'Hermite et al., 1975). This drug appears to be a potent dopamine antagonist. Delvoye (1973) and Delvoye et al. (1974) induced experimental hyperprolactinemia in 10 regularly menstruating women, starting oral administration of 3 × 50 mg sulpiride daily at midcycle. Blood samples were collected daily during a control cycle and during the cycle with sulpiride administration. In 2 cases, ovulation did not occur: the LH peak was suppressed and progesterone did not increase. In 4 out of the 8 ovulatory cycles, the increase in serum prolactin concentration started before the occurrence of the LH peak. The luteal phases under hyperprolactinemia were shortened by 1-4 days, when compared to the control cycles in the same women. In addition, under sulpiride treatment, LH decreased by some 50%, FSH by some 40% and progesterone by some 30% (Fig. 6).

In another 7 regularly menstruating women, 3 × 50 mg sulpiride was administered daily per os starting the first day of menstruation (Robyn et al., 1975). The treatment was maintained for 2 consecutive cycles or for at least 43 days in 6 women. Blood samples were collected daily during a control cycle and during the entire period of induced hyperprolactinemia. In 3 cases, blood sampling continued after the interruption of the sulpiride administration until the next menses. Prolactin, LH, FSH, estradiol and progesterone were measured by radioimmunoassays. Such experimental hyperprolactinemia has a deleterious effect, primarily on the granulosa cells: in the group of 6 women investigated for periods varying from 43-76 days, 6 cycles were observed with luteal phase of less than 8 days and with a progesterone peak below 5.0 ng/ml. Examples of cycles with short luteal phases are shown in Figure 7. Considering the evolution of mean serum prolactin, LH, FSH, estradiol and progesterone in these 6 cycles, compared with those observed during the control cycles studied in the same group of women, it appears that the basal levels of both LH and FSH during the follicular phases are almost identical. At midcycle, the LH peak is lower under hyperprolactinemia (60 mIU/ml) than in the control cycles (90 mIU/ml). The reduction in amplitude of the FSH peak at midcycle is less apparent. The rises in serum estrogens

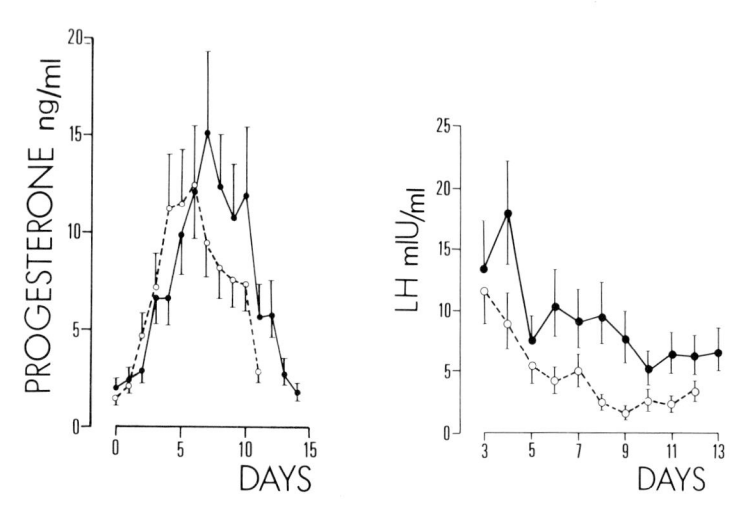

Fig. 6. Changes in LH and progesterone serum concentrations during 6 control cycles (●—●) and 6 luteal cycles under hyperprolactinemia (o---o) induced by sulpiride (see Fig. 3), from the same 10 women. Sulpiride administration started at midcycle. Vertical bars represent one standard error of the means.

during the follicular phases under hyperprolactinemia and in the control cycles are almost similar; the mean values on day −1 are at 215 pg/ml and 240 pg/ml, respectively. Progesterone levels are markedly reduced under hyperprolactinemia compared with those found in the control cycles; the mean peak values are at 2.0 ng/ml and 12.6 ng/ml, respectively. During the luteal phase, the estradiol concentration also falls within 2 days to levels equivalent to those of the early follicular phase. Such a decline occurs in the control cycle too, but only after day +10. Hyperprolactinemia thus exerts more deleterious effects on the corpus luteum function, when increased PRL is initiated during follicular growth and maturation than when it is initiated during the luteal phase. Beside the occurrence of short luteal phase cycles, episodes of increased circulating estrogen concentration followed by, or concomitant with increased LH secretion also take place, but without significant progesterone secretion. Such signs of follicular growth are preceded by increased serum FSH. In these conditions it seems that the circulating estrogens are unable to trigger an LH surge to induce ovulation.

The apparent deleterious effect of high circulating levels of prolactin on corpus luteum function is confirmed clinically in that increased basal serum prolactin is found in women with short luteal phase menstrual cycles (Del Pozo et al., 1975). In such patients, treatment with CB-154, an inhibitor of prolactin secretion, apparently acting

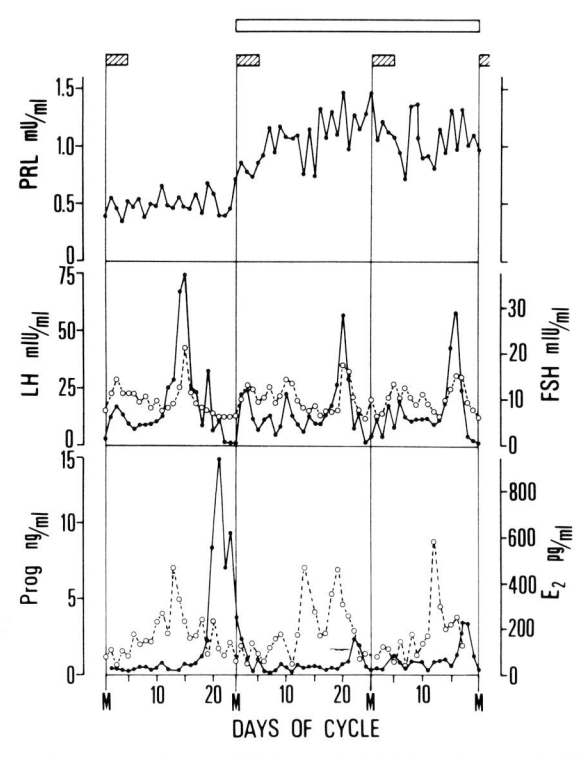

Fig. 7. Development of serum prolactin, LH (●—●), FSH (o---o), estradiol (E$_2$; o---o) and progesterone (Prog; ●—●) concentrations in a woman before and during hyperprolactinemia induced by sulpiride (see Fig. 3). Prolactin results are expressed in milli-units of the Research Standard 71/222. Open rectangle indicates the period of sulpiride administration. M indicates the first day of menstruation (hatched area).

by stimulation of dopamine receptors, restores luteal phases of normal duration and increases progesterone secretion. CB-154 administration to amenorrhea-galactorrheic women with hyperprolactinemia, re-establishes ovulatory cycles, pregnancies and term deliveries of normal children (Lutterbeck et al., 1971; Besser et al., 1972; Varga et al., 1973). In such cases, when the dose of CB-154 is inappropriate, episodes of menstrual cycles appear first with short luteal phases. With maintained or reinforced treatment, normal ovulatory cycles occur. Marked luteal deficiency has also been found during the first menstrual cycles reappearing post-partum during resolution of the hyperprolactinemia characterizing pregnancy and lactation (Reyes et al., 1972; Rolland et al., 1975).

Our findings also concur with the in vitro data (McNatty et al., 1974): physiological concentrations of prolactin are required for the production of progesterone by granulosa cells in vitro. Suppression of prolactin by specific antisera or prolactin excess results in suppression of progesterone secretion by these granulosa cells in vitro. It seems that in vivo a moderate increase in prolactin secretion, when occurring from the first day of the menstrual cycle, is sufficient to alter corpus luteal function: in our study levels of some 500-1000 μU/ml (Research Standard 71/222; Medical Research Council, London) were associated with luteal insufficiency. Our data also suggest that hyperprolactinemia affects other endocrine mechanisms controlling fertility in women. Although FSH secretion is probably unaffected since follicular growth and estrogen production by theca cells are not grossly impaired, the feedback relationships between gonadotrophins and estrogens seem to be altered such that the LH level is inadequate for ovulation or at least luteinization. This is an explanation for the occurrence of amenorrhea in hyperprolactinemic patients. Recently Glass et al. (1975) reported that 13 out of 14 patients with amenorrhea and hyperprolactinemia but no evidence of pituitary tumors failed to release LH in response to estrogen. This abnormality might explain the anovulation in these patients since positive feedback to estrogens is necessary for the ovulatory surge of gonadotrophins (Monroe et al., 1972). These clinical data and our own observations account for the frequently negative response to clomiphene in patients with the amenorrhea-galactorrhea syndrome, and for the need to inject human chorionic gonadotrophin in addition to achieve apparent ovulation (Gambrell et al., 1971; Thorner et al., 1974; Glass et al., 1975). It has been shown that ovaries of patients with the amenorrhea-galactorrhea syndrome are refractory to gonadotrophins (McNeilly, 1974). It has been proposed that in vivo steroid feedback to the hypothalamic-pituitary unit is reduced, with subsequent suppression of ovulation. Such refractoriness of ovaries to gonadotrophins has also been observed during the amenorrhea of the post-partum period (Reyes et a., 1972; Zarate et al., 1972; Rolland et al., 1975). Our data do not negate such possible antigonadotrophic action of increased circulating prolactin on estrogen production by the ovary. Indeed, such alteration may appear after long periods of hyperprolactinemia than those achieved in the present study.

Our results confirmed earlier observations of normal basal gonadotrophins and estrogens in women with hyperprolactinemia (Mortimer et al., 1973; Thorner et al., 1974; Glass et al., 1975).

When prolactin secretion is stimulated by a drug, it may be that at least some of the effects seen are inherent to the drug rather than to the induced hyperprolactinemia. Even if entirely justified, this criticism should be considered alongside the fact that our data on experimental hyperprolactinemia obtained with sulpiride agree closely with those obtained from clinical studies of hyperprolactinemic patients and from the treatment of such patients with prolactin inhibitors. The etiology of prolactin hypersecretion in pathological conditions may involve an abnormal biogenic amine fuction at the hypothalamo-hypophysial level and include abnormal dopamine receptors and/or metabolism. Such an abnormality at the hypothalamo-hypophysial level may affect other endocrine mechanisms than prolactin secretion. In these conditions, hyperprolactinemia would be more a symptom of the disease, than the factor responsible for the disorders of the menstrual cycle. A drug such as sulpiride acting at the hypothalamo-hypophysial level, may then be considered to mimic the disease first described as amenorrhea-galactorrhea syndrome. Pharmacological agents, stimulating dopamine receptors, may suppress the hypothalamo-hypophysial abnormality responsible for the menstrual cycle disturbances. Considering the possible role of dopamine as a neurotransmitter in the peripheral autonomic nervous system (Thorner, 1975), it is tempting to postulate that, in some cases at least, the abnormality may involve dopamine receptors or its metabolism outside of the central nervous system.

ACKNOWLEDGMENTS

We are indebted to the National Pituitary Agency (University of Maryland, School of Medicine), National Institute of Arthritis, Metabolism and Digestive Diseases for the generous gift of human prolactin radioimmunoassay kits (V-L-S).

The expenses of the investigations presented in this report were defrayed by grants to Prof. P.O. Hubinont from the Ford Foundation (U.S.A.) and from the Fonds de la Recherche Scientifique Médicale (Belgium).

REFERENCES

Alexander, S., Robyn, C. and Schwers, J. (1975): *Aménorrhées Secondaires et Spanioménorrhées*. Presses Universitaires de Bruxelles. In press.
Argonz, J. and Del Castillo, E.B. (1953): *J. clin. Endocr., 13*, 79.
Aubert, M.L., Becker, R.L., Saxena, B.B. and Raiti, S. (1974): *J. clin. Endocr., 38*, 1115.
Besser, G.M., Parke, L., Edwards, C.R.W., Forsyth, I.A. and McNeilly, A.S. (1972): *Brit. med. J., 3,* 669.
Chiari, J.B.V.L., Braun, C. and Spaeth, J. (1852): *Klinik der Geburtshilfe und Gynäkologie.* Enke, Erlangen.
Cotes, P.M. (1973): In: *Human Prolactin,* pp. 97-101. Editors: J.L. Pasteels and C. Robyn. ICS 308, Excerpta Medica, Amsterdam.
Del Pozo, E., Wyss, H., Varga, L. and Obolensky, W. (1975): In *Abstracts, Workshop Human Prolactin, Amsterdam, 1975.*
Delvoye, P. (1973): In: *Human Prolactin,* Discussion. Editors: J.L. Pasteels and C. Robyn. ICS 308, Excerpta Medica, Amsterdam.
Delvoye, P., Taubert, H.-D., Jürgensen, O., L'Hermite, M., Delogne, J. and Robyn, C. (1974): *C.R. Acad. Sci. (Paris), Sér. D, 279,* 1463.
Delvoye, P., Vekemans, M., Calaf, J., L'Hermite, M. and Robyn, C. (1972): *J. Gynec. obstet. Biol. Reprod., Suppl. I,* 357.
Delvoye, P., Vekemans, M., L'Hermite, M. and Robyn, C. (1975): Effects of graded doses of ethinyl-oestradiol on serum prolactin, LH and FSH in regularly menstruating women. In preparation.
Forbes, A.P., Henneman, P.H., Griswold, G.L. and Albright, F. (1954): *J. clin. Endocr., 14,* 265.
Friesen, H., Hwang, P., Guyda, H., Tolis, G., Tyson, J. and Myers, R. (1972): In: *Prolactin and Carcinogenesis,* pp. 64-97. Editors: A.R. Boyns and K. Griffiths. Alpha Omega Alpha, Cardiff.
Frommel, R. (1852): *Z. Geburtsh. Gynäk., 7,* 305.
Furuhjelm, M. (1966): *Acta obstet. gynec. scand., 45,* 352.
Gambrell, R.D., Greenblatt, R.B. and Mahesh, V.B. (1971): *Amer. J. Obstet. Gynec., 110,* 838.
Glass, M.R., Shaw, R.W., Butt, W.R., Logan Edwards, R. and London, D.R. (1975): *Brit. med. J., 3,* 274.

Guyda, H.J. and Friesen, H.G. (1973): *Pediat. Res., 7,* 534.
Heusghem, C. (1956): *Contributions à l'Etude Analytique et Biochimique des Oestrogènes Naturels.* Thone, Liège.
Hwang, P., Guyda, H. and Friesen, H. (1971): *Proc. nat. Acad. Sci. (Wash.), 68,* 1902.
Jewelewicz, R., Dyrenfurth, I., Worren, M., Frantz, A.G. and Vande Wiele, R. (1974): *J. clin. Endocr., 39,* 387.
Lequin, R.M. (1973): In: *Human Prolactin,* pp. 221-222. Editors: J.L. Pasteels and C. Robyn. ICS 308, Excerpta Medica, Amsterdam.
L'Hermite, M., Delvoye, P., Nokin, J., Vekemans, M. and Robyn, C. (1972): In: *Prolactin and Carcinogenesis,* pp. 81-97. Editors: A.R. Boyns and K. Griffiths. Alpha Omega Alpha, Cardiff.
L'Hermite, M., Heuson, J.C. and Robyn, C. (1973): *Louvain méd., 93,* 161.
L'Hermite, M., Virasoro, E., Golstein, J., Copinschi, G., Van Haelst, L. and Robyn, C. (1975): Profile of the effects of Sulpiride on luteinizing hormone, follicle stimulating hormone, growth hormone and thyrotropin secretion in men and women. In preparation.
Longcope, C. (1971): *Amer. J. Obstet. Gynec., 111,* 778.
Lutterbeck, P.M., Pryor, J.S., Varga, L. and Wenner, R. (1971): *Brit. med. J., 3,* 228.
Midgley, A.R. and Jaffe, R.B. (1973): In: *Endocrinology,* pp. 629-635. Editor: R.O. Scow. ICS 273, Excerpta Medica, Amsterdam.
Monroe, S.E., Jaffe, R.B. and Midgely, A.R. (1972): *J. clin. Endocr., 34,* 342.
Mortimer, C.H., Besser, G.M., McNeilly, A.S., Marshall, J.C., Harsoulis, P., Tunbridge, W.M.G., Gomez-Pan, A. and Hall, R. (1973): *Brit. med. J., 4,* 73.
McNatty, K.P., Sawers, R.S. and McNeilly, A.S. (1974): *Nature (Lond.), 250,* 653.
McNeilly, A.S. (1974): *Brit. J. Hosp. Med., 12,* 57.
McNeilly, A.S., Evans, G.E. and Chard, T. (1973): In: *Human Prolactin,* pp. 231-232. Editors: J.L. Pasteels and C. Robyn. ICS 308, Excerpta Medica, Amsterdam.
Pincus, G., Romanoff, L.P. and Carlo, J. (1954): *J. Geront., 9,* 113.
Reyes, F.I., Winter, J.S.D. and Faiman, C. (1972): *Amer. J. Obstet. Gynec., 114,* 589.
Robyn, C. (1975): *Path. et Biol., 23,* 783.
Robyn, C. (1976): In: *Basic Applications and Clinical Uses of Hypothalamic Hormones.* Editors: A.L. Charro Salgado, R. Fernández Durango and J.G. Lopez del Campo. ICS 374, Excerpta Medica, Amsterdam. In press.
Robyn, C., Delvoye, P., Nokin, J., Vekemans, M., Badawi, M., Perez-Lopez, F.P. and L'Hermite, M. (1973): In: *Human Prolactin,* pp. 167-188. Editors: J.L. Pasteels and C. Robyn. ICS 308, Excerpta Medica, Amsterdam.
Robyn, C. and Vekemans, M. (1974): In: *International Workshop: Estrogens in the Post Menopause.*
Robyn, C. and Vekemans, M. (1976): *Acta endocr. (Kbh.),* in press.
Robyn, C., Vekemans, M., Caufriez, A. and L'Hermite, M. (1976): *Int. Res. Commun. Syst., 4,* 14.
Rolland, R., Lequin, R.M. and Schellekens, L.A. (1975): *Clin. Endocr., 4,* 15.
Stevens, V.C., Powell, J.E. and Sparks, S.J. (1973): In: *Abstracts, 55th Annual Meeting of the Endocrine Society, Chicago,* A 64-31.
Tamura, S. and Igarashi, M. (1973): *Endocr. jap., 20,* 483.
Thorner, M.O., McNeilly, A.S., Hagan, C. and Besser, G.M. (1974): *Brit. med. J., 2,* 419.
Thorner, M.O. (1975): *Lancet, 1,* 662.
Varga, L., Wenner, R. and Del Pozo, E. (1973): *Amer. J. Obstet. Gynec., 117,* 75.
Vekemans, M., Delvoye, P., L'Hermite, M. and Robyn, C. (1972): *C.R. Acad. Sci. (Paris), Sér. D., 275,* 2247.
Vekemans, M. and Robyn, C. (1975a): *Brit. med. J., 4,* 738.
Vekemans, M. and Robyn, C. (1975b): *J. clin. Endocr., 40,* 886.
Yuen, B.H., Keye Jr, W.R. and Jaffe, R.B. (1973): *Obstet. gynec. Surv., 28,* 527.
Zarate, A., Canales, E.S., Soria, J., Riuz, F. and MacGregor, C. (1972): *Amer. J. Obstet. Gynec., 112,* 1130.
Zarate, A., Schally, A.V., Soria, J., Jacobs, L.S. and Canales, E.S. (1974): *Obstet. and Gynec., 43,* 487.

THE RELATIONSHIP BETWEEN RELAXIN AND PROLACTIN IMMUNOACTIVITIES IN VARIOUS REPRODUCTIVE STATES: RADIOIMMUNOASSAY USING PORCINE RELAXIN*

WAYNE A. CHAMLEY,[1] ROGER D. HOOLEY[1] and GILLIAN D. BRYANT[2]

[1]Reproductive Research Section, University of Melbourne, S.S. Cameron Laboratory Werribee, Australia; and [2]Department of Anatomy and Reproductive Biology, John A. Burns School of Medicine, University of Hawaii, Honolulu, Hawaii, U.S.A.

Shortly before Stricker and Greuter showed the existence of a lactogenic factor in hypophyseal extracts in 1928, Hisaw described relaxin, a non-steroidal factor derived from pregnant sow ovaries which caused the relaxation of the interpubic ligament in the estrogen-primed guinea pig. Although our knowledge of the physiology of prolactin was by no means insignificant prior to the advent of specific radioimmunoassays, once these were developed for the sheep, rat and human great advances in the depth of our understanding were accomplished in a relatively short period of time. However, a prerequisite of these radioimmunoassays was in each case the availability of a purified preparation of the hormone. It has been the lack of a purified, well characterized relaxin preparation which is largely responsible for our lack of knowledge today of the physiological role of relaxin in any species.

The biological activities of relaxin have been reviewed by Hall (1960). In summary they are: (1) The estrogen-primed mouse responds to single injections of relaxin with dose-proportional increases in the length of the interpubic ligament (Steinetz et al., 1960). (2) The spontaneous motility of the estrogen-dominated uterus is inhibited in vitro (Wiquist and Paul, 1958) in several small mammals. This inhibition has been ascribed to an uterine-relaxing factor which may be an intrinsic bioactivity of relaxin (Griss et al., 1967) or that of a related peptide.

In addition, there is also good evidence that the mammary gland is a target organ for ovarian relaxin (Zarrow and McClintock, 1966) and that relaxin may act synergistically with estrogen and progesterone to develop the mammary apparatus during pregnancy in the rat (Harness and Anderson, 1975).

From histochemical, biochemical and electron microscopic studies it appears that the granulosa cells are probably the site of ovarian relaxin production. In addition, however, it may also be synthesized by the uterus and/or placenta. Hence in a non-pregnant female the total relaxin activity in blood may arise from 2 sources, the ovaries and uterus, whilst in the pregnant female it may arise from the ovaries, uterus and/or placenta.

---

* This work was supported by grant number HD 06633 from the National Institute of Child Health and Human Development, a Ford Foundation grant number 66202 and the Cancer Center of Hawaii (grant number CA 15655). One of us (GDB) was supported by a Research Career Development Award HD 70516 and (WAC) and (RDH) were supported by the Australian Wool Corporation.

In 1971 we attempted to develop a specific radioimmunoassay for relaxin in unextracted plasma without recourse to the bioassay or to purified relaxin (Bryant, 1972). The radioimmunoassay developed was then used to examine its own specificity in terms of ovarian and uterine relaxins, their metabolism and heterogeneity in plasma and tissue extracts (Bryant and Stelmasiak, 1974). The radioimmunoassay has also been applied to the study of relaxin physiology in the sheep and human, to seek further evidence of its specificity (Chamley et al., 1975; Bryant et al., 1975). The results of those studies relating relaxin and prolactin secretion will be described in this paper.

Sherwood and O'Byrne (1974) took a more classical approach to the problem: isolation, bioassay, radioimmunoassay, application. Relaxin was purified by column chromatography using the mouse interpubic ligament bioassay; physico-chemical studies were then undertaken before proceeding to the development and limited application of a radioimmunoassay (Sherwood et al., 1975).

We have repeated their purification and compared their major bioactive fractions with our major immunoactive component. The detailed results shown in the paper by Kwok et al. (*This Volume*, p. 414) suggest that the radioimmunoassay used in the study here is measuring in plasma a larger molecular size precursor relaxin, which is related to the biologically active molecule as precursor:product and that they are secreted in phase.

METHODS

Radioimmunoassay of relaxin

The technique of Bryant (1972) was used exactly as described, except that separation of bound and free hormone was effected by a conventional double antibody procedure (Bryant and Stelmasiak, 1974). Levels are expressed as ng porcine relaxin NIH-R-P1/ml since the immunoactivity of sheep and human plasma paralleled that of the standard. The sensitivity of the assay is > 6.3 ng relaxin/ml of plasma which is measured with a maximum percentage error ($P = 0.05$) of $\pm 50\%$ when estimates at 3 concentrations are performed.

Radioimmunoassay of ovine prolactin

The method of Bryant and Greenwood (1968) using a charcoal-dextran separation of bound and free hormone was employed. The sensitivity of the assay was >5.9 ng/ml of plasma. At this concentration the maximum percentage error ($P = 0.05$) would be $\pm 50\%$ for determinations performed at 2 concentrations.

Radioimmunoassay of human prolactin

Human prolactin was measured using human prolactin supplied by the National Pituitary Agency and antiserum to human prolactin was kindly obtained from Dr. U.J. Lewis. The radioimmunoassay was a modification of that published by Dr. Lewis and his colleagues (Sinha et al., 1973). The sensitivity of the assay was > 3.1 ng/ml plasma.

There was no cross-reaction in either of the prolactin assays with relaxin at 1 µg/ml or any cross-reaction in the relaxin radioimmunoassay by any of the known anterior and posterior pituitary hormones at levels of 1 µg/ml.

RESULTS

Human breast feeding

Whilst studying the effect of breast feeding on human prolactin secretion, relaxin immunoactivity was concomitantly measured. It was observed that relaxin immunoactivity in plasma increased as acutely as prolactin but followed it by 5 minutes when first one breast was suckled and then the other (Bryant, 1973). On the basis of this serendipitous result suckling experiments were carried out in sheep.

Suckling and milking in the sheep

In these experiments plasma prolactin and relaxin immunoactivities rose between 3-10 minutes after the start of the suckling stimulus (Bryant and Chamley, 1976) and the concentrations showed a highly significant correlation in 6 animals.

Likewise in a second series of experiments in lactating ewes undergoing machine milking, there was a consistent release of both prolactin and relaxin upon milking; one such experiment is shown in Figure 1. It can be seen that at 1.5 minutes after the beginning of the machine milking, levels of both hormones had increased. By 2 minutes, relaxin immunoactivity was almost back to pre-milking levels, whereas prolactin decreased and then showed a steady increase over the following 3 minutes. A third series of experiments were carried out on teat stimulation in lactating ewes. Figure 2 shows the results of one such experiment. In this experiment relaxin immunoactivity rose faster than the prolactin activity, and by 2 minutes after the commencement of teat stimulation had risen to its peak level of 135 ng/ml, whereas prolactin reached a peak of 185 ng/ml by 6 minutes and stayed at this level for a further 3 minutes.

An attempt has been made to dissect the mechanism of the relaxin release, to see whether the ovarian/uterine relaxin response is a direct result of prolactin release or

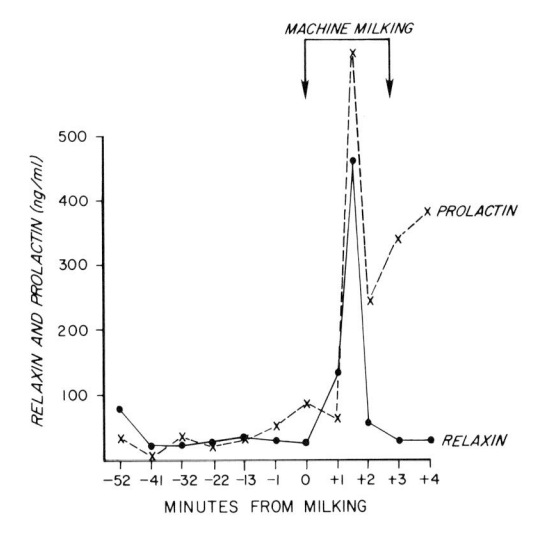

Fig. 1. Prolactin (x----x) and relaxin (●——●) immunoactivities are shown over machine milking in a lactating ewe.

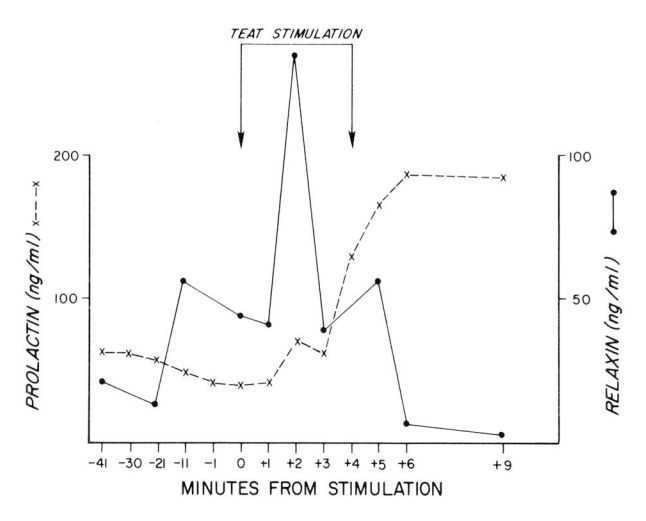

Fig. 2. Prolactin (x----x) and relaxin (●——●) immunoactivities are shown over a 4-minute period of teat stimulation in a lactating ewe.

whether the concomitant release of oxytocin acts upon the uterus alone to stimulate the release of uterine relaxin. A series of oxytocin infusion studies were carried out on intact anestrous ewes pretreated with progesterone and estradiol-17β. Blood samples were collected using an indwelling catheter at 5-minute intervals for 25 minutes prior to the start of the infusion. Oxytocin (60 IU total dose) was then infused over a 90-minute period with blood samples collected every 5 minutes. Figure 3 shows the prolactin and relaxin levels in the control animal which received no infusion. There were 2 small spontaneous peaks of relaxin immunoactivity of 200 and 100 ng/ml respectively during the course of the experiment, whereas prolactin levels remained low throughout. One test animal showed a rise of relaxin immunoactivity from 5 ng/ml at 5 minutes after the start of the infusion to 1115 ng/ml by 20 minutes (Fig. 4). Prolac-

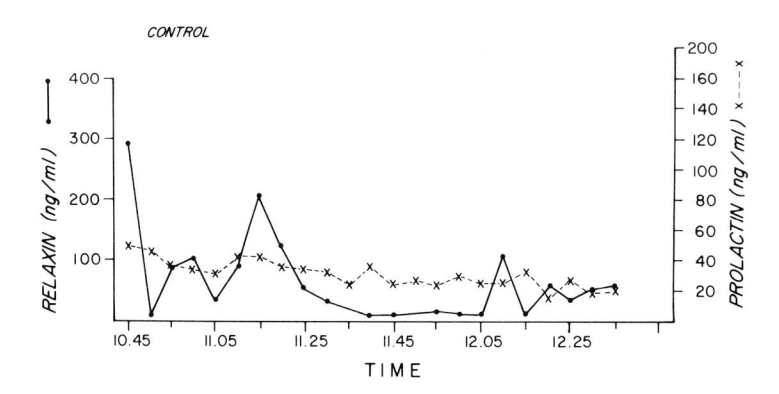

Fig. 3. A control ewe was pretreated with 20 mg progesterone every second day for 7 days and given 100 mg estradiol-17β, (i.m.) 1 day prior to the experiment. Plasma prolactin (x---x) and relaxin (●——●) immunoactivities were measured at 5-min. time intervals for 1 hour 50 min.

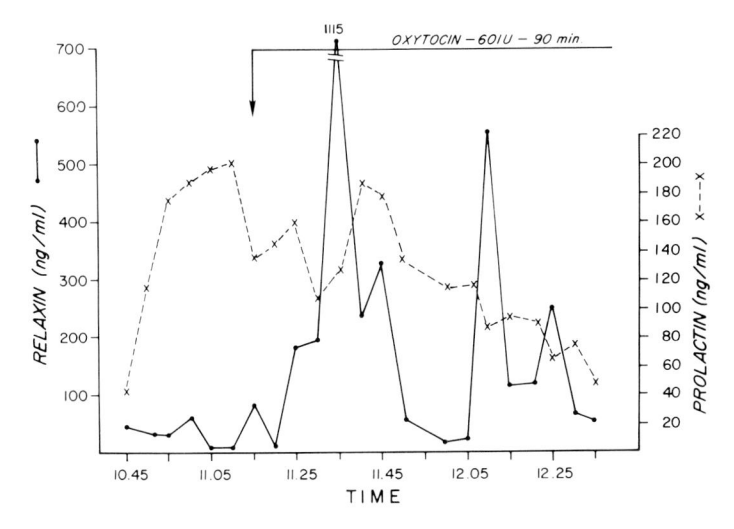

Fig. 4. A test animal was pretreated with estrogen and progesterone in the same manner as the control ewe. Oxytocin (60 IU total dose) was infused over a 90-minute period from 11:15 am. Blood samples were collected at 5-minute intervals and prolactin (x----x) and relaxin (●——●) immunoactivities measured.

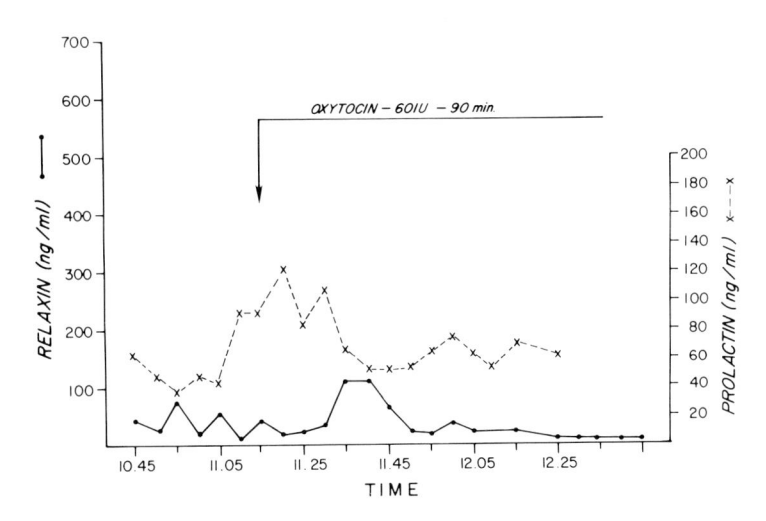

Fig. 5. A second test animal was pretreated in the same manner as the control ewe with estrogen and progesterone. Oxytocin (60 IU total dose) was infused over a 90-minute period from 11:15 am. Blood samples were collected at 5-minute intervals and prolactin (x----x) and relaxin (●——●) immunoactivities measured.

tin levels were higher both prior to and during the infusion than in the control animal. Another test animal showed a modest rise in relaxin immunoactivity during the infusion of up to 105 ng/ml, but this was no greater than the 2 spontaneous peaks recorded in the control study (Fig. 5).

Studies have been carried out in the sheep in order to determine the broader reproductive role of relaxin, in some experiments prolactin has also been measured in order to establish whether a relationship exists between these hormones.

Ovine estrous cycle, human menstrual cycle

In a study of the ovine estrous cycle relaxin immunoactivity was measured every 2 hours during 4-day periods in a series of sheep to cover the 17 days of the estrous cycle. In one animal over the estrous period both prolactin and relaxin immunoactivities were measured and there was no significant correlation found between them (Chamley et al., 1975). Likewise, a study during the human menstrual cycle was carried out with samples collected from 9 women on a selected bleeding schedule with prolactin, relaxin, LH, FSH, estradiol-17β and progesterone being measured in the same samples, and the data combined according to the day of the LH peak. No correlation was found between the relaxin and prolactin immunoactivities in this study (Bryant et al., 1975).

Pregnancy and parturition

Prolactin and relaxin immunoactivities have been studied in late pregnancy and parturition in the ewe. Three sheep were bled every 2 hours for 24 hours at 45 and 110 days of gestation. Mean prolactin levels were significantly higher for each animal at 110 days than at 45 days, however, only 1 animal had a significantly higher mean plasma relaxin level at the later date. In late pregnancy there was no correlation between the 2 hormone levels but when studied acutely over the period of expulsion of the lamb both prolactin and relaxin immunoactivities rose and were significantly correlated.

DISCUSSION

These results are surprising: prolactin and relaxin secreted are correlated but only in one physiological circumstance — after a suckling or milking stimulus. The correlation even in lactation is not precise, the relaxin response may just precede or follow or rise simultaneously with prolactin. The latter imprecision may result from differing half lives of these hormones and the less than continuous sampling.

Attempts to extend these correlations in lactation to the pregnant or non-pregnant, non-lactating animal were uniformly negative and it seemed possible that correlation was associated only with a concomitant and acute release of oxytocin. Parenthetically it may be noted that prolactin increases on mating (Bryant et al., 1970), a situation in which oxytocin is also released (Fox and Knaggs, 1969) but one not studied for both prolactin and relaxin. The results of oxytocin infusion in the intact, non-pregnant ewe were inconclusive despite priming with estrogen and progesterone. Neither prolactin nor relaxin were conclusively elevated by oxytocin infusion in these animals, and it is tempting to speculate that such infusions are inappropriate physiologically and that our experimental animal is not a model for a suckled ewe in lactation.

The constancy of the relaxin response to suckling, milking or teat stimulation in the lactating ewe strongly suggests a physiological role for relaxin at this time. More studies need to be done to determine whether the relaxin is released from the ovary or the uterus and whether this is mediated through oxytocin or a nerve pathway to the uterus. Receptors for relaxin in the mammary gland may be inferred from the work of Harness and Anderson (1975) and a search for such receptors has been initiated in order to develop a radioreceptor assay. Similarly the identification of uterine receptors for ovarian relaxin would suggest that one of the consequences of the suckling stimulus would be uterine contraction, from oxytocin, and uterine relaxation from ovarian relaxin. What seems clear at this time is that prolactin, oxytocin and relaxin are released on suckling but that this triad is not always released in concert, a further example of the orchestration of hormone responses to a stimulus.

REFERENCES

Bryant, G.D. (1972): *Endocrinology, 91,* 1113.
Bryant, G.D. (1973): Comment in discussion. In: *Human Prolactin,* p. 92. Editors: J.L. Pasteels and C. Robyn. ICS 308, Excerpta Medica, Amsterdam.
Bryant, G.D. and Chamley, W.A. (1976): *J. Reprod. Fertil.,* in press.
Bryant, G.D. and Greenwood, F.C. (1968): *Biochem. J., 109,* 831.
Bryant, G.D., Linzell, J.L. and Greenwood, F.C. (1970): *Hormones, 1,* 26.
Bryant, G.D., Panter, M.E.A. and Stelmasiak, T. (1975): *J. clin. Endocr., 41,* 1065.
Bryant, G.D., Sassin, J.F., Weitzman, E.D., Kapen, S. and Frantz, A (1976): *J. clin. Endocr.,* submitted for publication.
Bryant, G.D. and Stelmasiak, T. (1974): *Endocrine Res. Commun., 1,* 415.
Chamley, W.A., Stelmasiak, T. and Bryant, G.D. (1975): *J. Reprod. Fertil., 45,* 455.
Fox, C.A. and Knaggs, G.S. (1969): *J. Endocr., 45,* 145.
Griss, G., Keck, G., Englehorn, R. and Tuppy, H. (1967): *Biochim. biophys. Acta (Amst.), 140,* 45.
Hall, K. (1960): *J. Reprod. Fertil., 1,* 368.
Harness, J.R. and Anderson, R.R. (1975): *Proc. Soc. exp. Biol. (N.Y.), 148,* 933.
Sherwood, O.D., Rosentreter, K.R. and Birkhimer, M.L. (1975): *Endocrinology, 96,* 1106.
Sherwood, O.D. and O'Byrne, E.M. (1974): *Arch. Biochem., 160,* 185.
Sinha, Y.N., Selby, F.W., Lewis, U.J. and Vanderlaan, W.P. (1973): *J. clin. Endocr., 36,* 509.
Steinetz, B.G., Beach, V.L., Kroc, R.L., Stasilli, N., Nussbaum, R.E., Nemith, P.J. and Dun, R.K. (1960): *Endocrinology, 67,* 102.
Stricker, P. and Greuter, F. (1928): *C.R. Soc. Biol. (Paris), 99,* 1978.
Wiquist, N. and Paul, K.G. (1958): *Acta endocr. (Kbh.), 29,* 135.
Zarrow, M.X. and McClintock, J.A. (1966): *J. Endocr., 36,* 377.

THE RELATIONSHIP BETWEEN RELAXIN AND PROLACTIN IMMUNOACTIVITIES IN VARIOUS REPRODUCTIVE STATES: PHYSICAL-CHEMICAL AND IMMUNO-BIOLOGICAL STUDIES*

SIMON C.M. KWOK[1], JOHN P. McMURTRY[2] and GILLIAN D. BRYANT[2]

[1]Department of Biochemistry and Biophysics, [2]Department of Anatomy and Reproductive Biology, John A. Burns School of Medicine, University of Hawaii, Honolulu, Hawaii, U.S.A.

A radioimmunoassay for immunoreactive relaxin has been applied to the study of plasma immunoactivities in different physiological states in man and sheep and correlated only in the lactation state to prolactin (Chamley et al., *This Volume*, p. 407). We have also undertaken a parallel study of the immunoactivity and biological activities obtained from pregnant porcine ovaries. These studies were undertaken to adduce further evidence or not for the specificity and physiological significance of measurements of relaxin obtained by radioimmunoassay. We have isolated the 3 biologically active components identified by Sherwood and O'Byrne (1974) and an additional component of higher molecular weight with some biological activity. The latter component is measured by the radioimmunoassay (Bryant, 1972) and the 3 bioactive components identified by Sherwood and O'Byrne (1974) show cross-reactions in this assay. An additional radioimmunoassay has been developed specifically for the major bioactive components. These studies and some preliminary experiments of the uptake in vivo of labeled relaxin fractions by ovary, uterus, cervix and the interpubic ligament have encouraged us to speculate that the 2 radioimmunoassays are measuring a prohormone and hormone respectively.

In addition, a study has been made of the immunological and biologically active moieties of the NIH-R-P1 preparation of porcine relaxin which was used for the development of the original radioimmunoassay of Bryant (1972).

METHODS

The radioimmunoassay for plasma relaxin has been used exactly as described by Bryant (1972) with the bound and free hormone separated by conventional antibody technique.

Porcine relaxins termed CM-B, CM-a and CM-a' by Sherwood and O'Byrne (1974) were isolated as described by these authors and the isolation monitored by the mouse interpubic ligament assay of Steinetz et al. (1960).

---

* This work was supported by grant number HD 06633 from the National Institute of Child Health and Human Development, a Ford Foundation grant number 66202 and the Cancer Center of Hawaii (grant number CA 15655). One of us (GDB) was supported by a Research Career Development Award number HD 70516.

Those biologically active components identified by Sherwood and O'Byrne (1974), CM-B, CM-a and CM-a', lack tyrosine and histidine and cannot be labeled by the chloramine-T method. Component CM-a' was iodinated using the following modification of the acylating technique of Bolton and Hunter (1973). These authors have described a method in which a succinimide ester is iodinated by the chloramine-T method, prior to attaching the labeled ester to a protein. This method was modified by first reacting relaxin with the succinimide ester and then reacting with iodine. 1 μg of crystalline N-succinimidyl 3-(hydroxyphenyl) propionate was reacted with 25 μg protein. The iodination of the acylated protein was carried out using the same concentrations of reagents as described by Bolton and Hunter (1973). Separation of radiolabeled relaxin from the unreacted succinimide ester and free ^{125}I was accomplished by passage through a Sephadex G-25 column.

Uptake studies were performed using the selected label from NIH-R-P1 (Bryant, 1972); 500 ng was injected (i.v.) into 16 estrogen-primed mice which were then killed at 15, 60, 120, and 240 minutes post-injection. The following tissues were removed, weighed and counted: uterus, ovary, cervix, interpubic ligament, heart, kidney and liver. To serve as a control, the same number of estrogen-primed mice were injected with ^{125}I-bovine serum albumin. Tissue uptake was expressed as c.p.m./mg tissue. Similar studies were performed using the acylation label CM-a'. Seventeen estrogen-primed mice were injected (i.v.) with 100 ng of Cm-a'-^{125}I relaxin and killed at 15, 30, 60, 120 and 240 minutes post-injection. The following tissues were removed, weighed and counted: interpubic ligament, uterus, cervix, ovary, adrenal, kidney, heart, spleen and lung.

RESULTS

Figure 1 shows the elution pattern of 500 mg of crude extract obtained from 100 grams of pregnant sow ovaries, when chromatographed on a Sephadex G-50 (fine) column (6.5 × 114 cm). The regions of major immuno- and biological activities are marked and it can be seen that they are quite distinct and without overlap. Nevertheless, the major immunoreactive area has some biological activity but this assay was qualitative and a full assay is awaited. Essentially similar results can be obtained by Sephadex chromatography of NIH-R-P1 with the exception that the large retarded peak of Figure 1 is not seen. This peak is inactive biologically and immunologically. The major immunoactive area is larger in molecular size (approximately 18,000) than the bioactivity as described previously (Bryant and Stelmasiak, 1974); the bioactivity has a molecular size of 6,000 (Sherwood and O'Byrne, 1974). The bioactivity (peak G-2) was rechromatographed on a CMC column and 3 peaks of biological activity were obtained and designated CM-B, CM-a, and CM-a' in accordance with the nomenclature of Sherwood and O'Byrne (1974). They reported that these fractions were almost equipotent biologically, with the same electrophoretic pattern on polyacrylamide gel electrophoresis and with similar amino acids composition. Table 1 shows the relative bioactivities of these peaks in terms of the standard NIH-R-P1 preparation. This Table also shows the relatively low immunoactivity of the 3 bioactive peaks in a labeled NIH-R-P1/anti-NIH-R-P1 system against an NIH-R-P1 standard in the radioimmunoassay of Bryant (1972). These inhibitions showed the typical characteristics of cross-reactions, non-parallel to the standard and the amounts required for 30% inhibition were used to calculate the relative immunoactivities. These results show that the radioimmunoassay of Bryant (1972) would be sensitive to these bioactive components in plasma but only if they were present at the μg level. These components

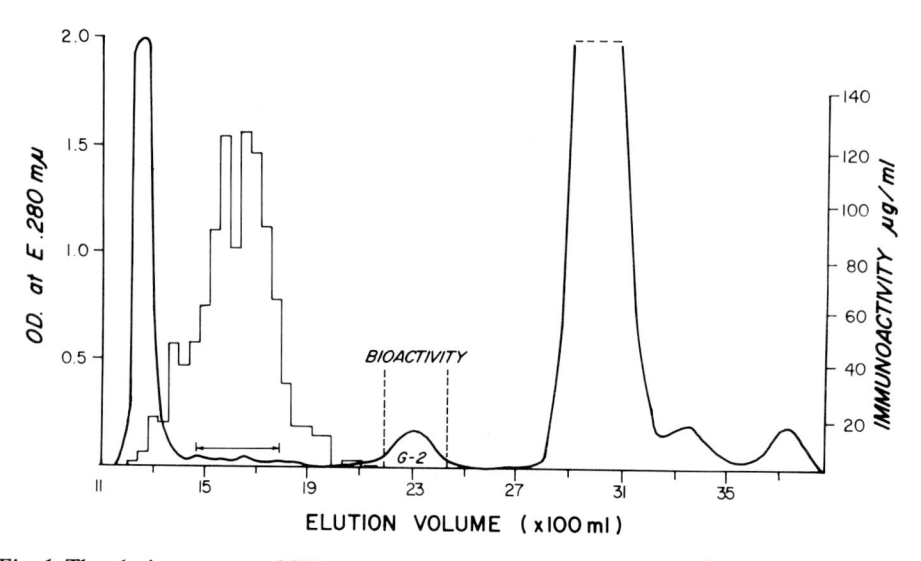

Fig. 1. The elution pattern of 500 mg of a crude extract obtained from 100 grams of pregnant sow ovaries is shown when chromatographed on a Sephadex G-50 (fine) column (6.5 × 114 cm). The biologically active area is indicated and the immunologically active area is shown as an histogram.

were reported not to contain tyrosine (Sherwood and O'Byrne, 1974) and hence cannot be part of the labeled fraction obtained by Bryant (1972) from NIH-R-P1 using chloramine-T. We have confirmed this for 2 of the bioactive fractions CM-a and CM-a' but did achieve labeling with chloramine-T using bioactive fraction CM-B (Table 1). It is assumed at this stage of our knowledge that this labeling with chloramine-T was to a contaminant of a tyrosine-containing peptide. Labeling of fraction CM-a' was successful using the modified acylation technique of Bolton and Hunter (1973) and labeled fraction CM-a' was bound avidly to our antiserum, raised against NIH-R-P1. Fraction

TABLE 1

A comparison of the bio- and immuno-properties of relaxin

	CM-B*	CM-a*	CM-a'*	NIH-R-P1
Biological activity units/mg	1547	2210	1547	442
Relative immunoactivity	4.2%	1.6%	0.8%	100%
Iodination:				
(1) Greenwood et al. (1963)	++	---	---	++
(2) Bolton and Hunter (1973)			++	
Binding to antiserum (NIH-R-P1)	---		++++	+++
Uptake of label by pubic ligament			++	---
Uptake of label by uterus, ovary and cervix			---	++

* Method of Sherwood and O'Byrne (1974)

CM-a has not yet been tested for binding to this antiserum nor has fraction CM-B been so tested when similarly labeled by acylation. That fraction of component CM-B labeling with cloramine-T does not bind to our antiserum. In summary, therefore, our antiserum generated to the mix of peptides in the NIH material contains antibodies to a high molecular weight component (Fig. 1) of high immunopotency, some bioactivity and also contains antibodies to a low molecular weight biologically active component cross-reacting in the original radioimmunoassay where it competes with labeled NIH-R-P1. The antiserum was generated to NIH-R-P1 and used to develop a radioimmunoassay to the bioactive relaxin component CM-a', the latter being labeled by acylation. This assay, a priori, would be sensitive primarily to circulating CM-a' material in plasma. This system was appropriately inhibited by the bioactive components CM-B, CM-a and CM-a' with approximately equal potencies, a result in agreement with their relative bioactivities, and by NIH-R-P1 at a level more or less consistent with its contents of these components (Fig. 2). Thence, the inhibitions obtained by plasma, parallel to and measured in terms of the CM-a' material, may be equated with bioactive relaxin material in plasma, released in this instance in response to suckling (Fig. 3). The results obtained for immunoreactive relaxins by the labeled NIH-R-P1/anti-NIH-R-P1 assay are shown and evidently there is a concomitant release of immunoactive relaxin-like material of high molecular weight. The secretions are not completely in phase since the immunoactive components appear to decline more rapidly than the biological components.

Tissue uptake studies using the NIH-R-P1 label showed that the uterus, ovary and cervix exhibited a linear uptake, with time, of approximately 6 c.p.m./mg tissue at 15 minutes to over 30 c.p.m./mg at 240 minutes. No uptake was noted in the pubic liga-

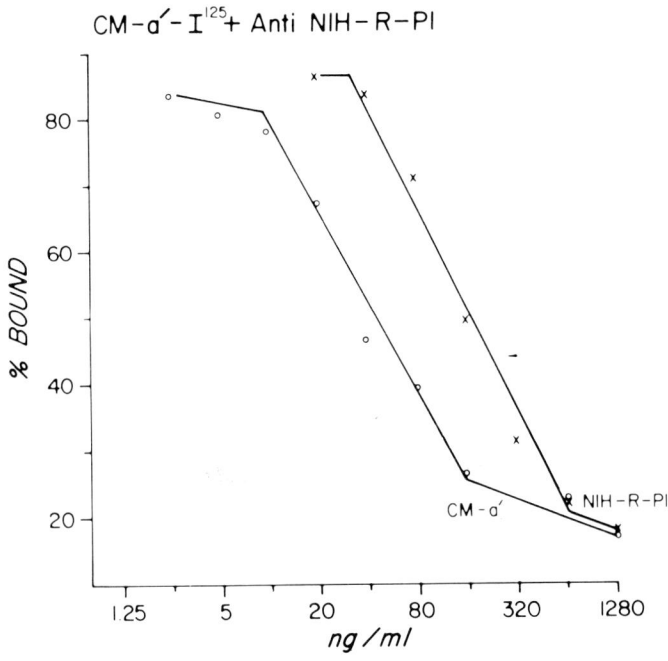

Fig. 2. Inhibition is shown by NIH-R-P1 (x——x) and by Cm-a' (o——o) in the labeled CM-a'/anti NIH-R-P1 radioimmunoassay system.

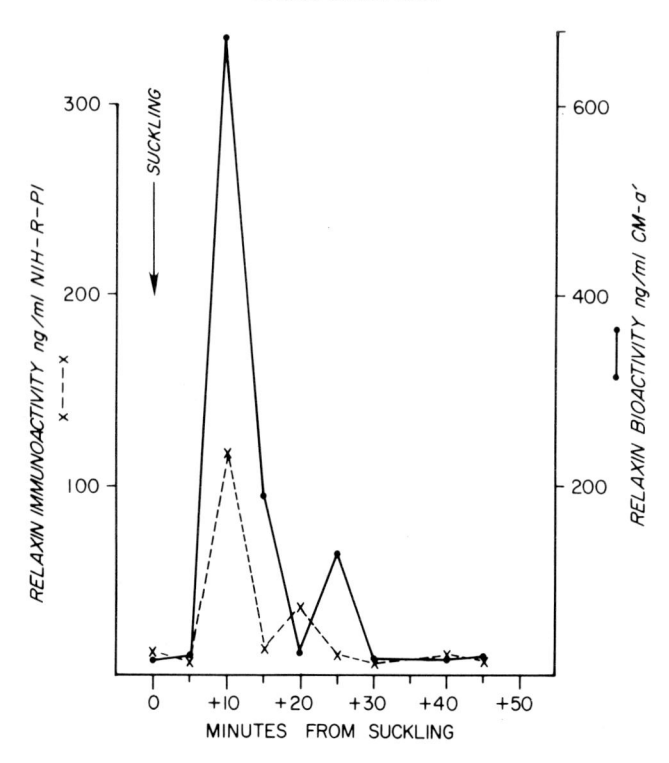

Fig. 3. The relaxin immunoactivity as measured in the labeled NIH-R-P1/anti NIH-R-P1 radioimmunoassay system is shown (x-----x) as well as the bioactivity as measured in the labeled CM-a′/anti NIH-R-P1 radioimmunoassay system (●——●) after a suckling stimulus in a lactating ewe.

ment. Compared to the other tissues, the kidney contained most of the radioactivity, exceeding 100 c.p.m./mg tissue at 60 minutes. The radioactivity in the heart and liver never exceeded 10 c.p.m./mg tissue. No discernible pattern of ^{125}I-BSA uptake was noted in any of the tissues.

A different pattern of tissue uptake was observed with the CM-a′ relaxin label. The radioactivity in the pubic ligament exceeded that noted in any of the tissues except for the kidney; 25 c.p.m./mg tissue at 30 minutes and then reaching a peak uptake of 35 c.p.m. at 120 and 240 minutes. A value of 80 c.p.m. was observed in the kidney at 30 minutes which then declined to values less than 10 c.p.m./mg of tissue at 60 minutes. No other tissue exceeded 10 c.p.m./mg tissue during the time span studied.

It has been noted that the NIH-R-P1 material chromatographs on Sephadex G-50 (fine) in all essence as if it were a crude extract of pregnant porcine ovarian tissue. It was interesting, therefore, to study this material in some detail since it has been deduced that it contained components with immunological and with biological activity and ability to bind to specific tissues. Fractionation of 750 mg of NIH-R-P1 monitored by bioassay and by immunoassay gave peaks equivalent to those of Figure 1 except as previously noted. The major immunoactive area has been pooled and rechromatographed on Sephadex G-100. The immunologically active material is not homogeneous as yet but significant biological activity is an intrinsic activity of this large molecular weight material. A quantitative bioassay against the CM-B, CM-a and CM-a′ peaks has to be performed, but it is already apparent that the activity is very much less than these components.

The selection of label after radioiodination for use in the radioimmunoassay is important since a particular label may select specific antibodies and be inhibited by different immunoreactive moieties in plasma. This following experiment was carried out during an early purification of NIH-R-P1 on a Sephadex G-200 column, which yielded several immunoreactive fractions. These were lyophilized and radioiodinated and used with the same antiserum in radioimmunoassays for the assay of plasmas from sheep and human subjects. One fraction, 22, was labeled and used in a radioimmunoassay for measurement of plasma relaxin immunoactivity in a woman over breast feeding. The same plasmas were also measured using a labeled relaxin prepared normally on Sephadex G-50 (Bryant, 1972). Results were calculated using fraction 22 as the standard with the labeled fraction 22 assay and the usual NIH-R-P1 as the standard using the normal assay procedure. The results (Fig. 4) show that both assays detected bursts of plasma immunoreactivity, different in magnitude, being expressed in terms of different standards but the bursts are not in a phase. One interpretation of

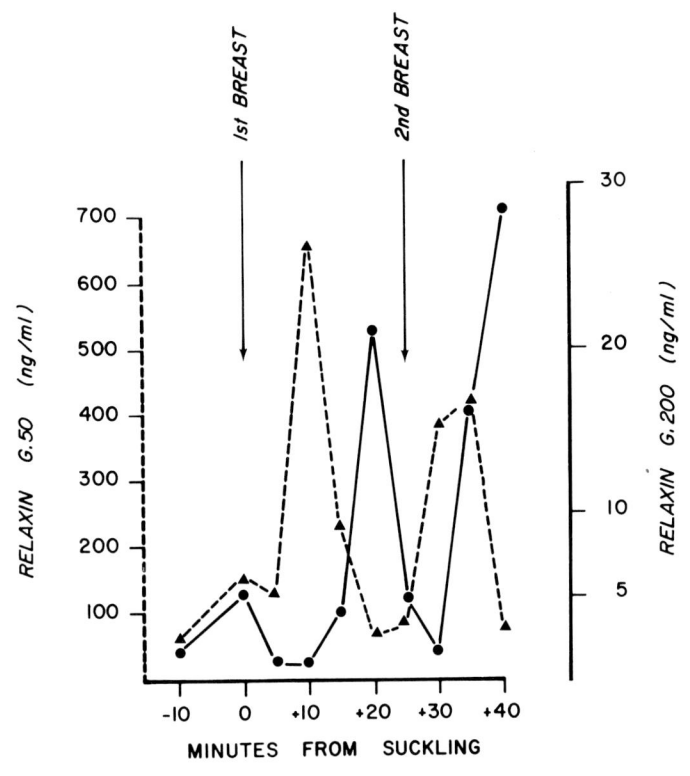

Fig. 4. Relaxin immunoactivity was assayed in human plasma at 5-minute intervals over breast feeding using the chloramine-T labeled NIH-R-P1/anti NIH-R-P1 radioimmunoassay (▲-----▲). The samples were also assayed using a label (fraction 22) prepared by passage of 10 mg NIH-R-P1 through a Sephadex G-200 column which was then lyophilized and labeled by the chloramine-T method (●——●). Each was used as its own standard in the respective assays.

the results is that the highly selected label, fraction 22, selects only some of the populations of antibodies in the antiserum and that this radioimmunoassay partially discriminates in favor of relaxin immunoactivity arising quantitatively in time from secreted relaxin. It is interesting to speculate that one label might discriminate in favor of the more biologically active components or might be discriminating in favor of metabolites of relaxin.

DISCUSSION

Work on the physiology and chemistry of relaxin and related peptides has been reactivated by the development of a radioimmunoassay and of more up-to-date procedures for the fractionation of pregnant sow ovaries. Reactivation must be clearly distinguished from clarification: relaxin is in the confusion before the dawn. A number of peptides have been identified in corpora lutea and pregnant sow ovarian tissue. Major biological activity is associated with 3 small molecular weight components with a subunit structure, devoid of tyrosine (Sherwood and O'Byrne, 1974), but able to be labeled by techniques other than chloramine-T, and detected with high sensitivity by a radioimmunoassay recently developed in our laboratory. A high molecular weight prohormone for this material is postulated (Bryant and Stelmasiak, 1974). Plasma immunoactivity is heterogeneous (Bryant and Stelmasiak, 1974) but further studies are needed using the 2 radioimmunoassays now available.

The present evidence for a prohormone is circumstantial and based on fractionation, physiological, immunological and biological results. The apparent prohormone is measured by our original radioimmunoassay in high amounts in a granule fraction from porcine corpora lutea of pregnancy, kindly supplied by Dr. H. Tyndale-Biscoe of the Australian National University, Canberra. He showed that the granule fraction had biological activity and we found that it had high immunological activity, indicating that both activities reside simultaneously within the granules of the granulosa lutein cells of pregnancy, but are of differing molecular size.

Using the two radioimmunoassays to detect both immuno- and bioactivities it was observed that both an immuno-component and a bioactive component are released after suckling in a lactating ewe. There are ample precedents for the simultaneous release of prohormone and hormone in the endocrine system but one stimulus may also release 2 distinct hormones. Hence, 2 assays are now being applied to obtain concurrent measurement in other physiological situations, to determine whether they are always secreted simultaneously, which would provide evidence for a prohormone-product relationship. In addition, plasma fractionations will be carried out to complement these studies since the activities differ in molecular size.

The biologically active peptide CM-a' shows a cross-reaction in the NIH-R-P1/anti-NIH-R-P1 radioimmunoassay system. The reactivity of the immunoactive fractions obtained from pregnant sow ovaries has not been tried in the labeled CM-a'/anti-NIH-R-P1 radioimmunoassay system. If a prohormone exists it would be expected that it would inhibit to some unknown degree in a labeled hormone/antihormone system.

The major biologically active components of pregnant sow ovarian tissue are the small molecular weight CM-B, CM-a and CM-a'. However, we have shown that the larger immunologically active peptide has at least 1:250 of the biological activity of the small peptide. This has not been fully quantitated as yet but the assay is highly specific for relaxin and it would seem unlikely that it is due to anything other than a biologically active sequence of relaxin.

Studies in vivo have been carried out using the selected label from NIH-R-P1 and the acylated labeled biologically active CM-a'. It has been shown that the latter is actively taken up by the mouse interpubic ligament and by the ovaries, uterus and cervix. Conversely significant uptake by the labeled NIH relaxin was observed in these tissues and none by the interpubic ligament. There may be a relationship therefore between the prohormone and the very elusive uterine-relaxing factor (URF), first postulated by Krantz et al. (1950). Ms. Marlene Bagoyo in our laboratory is in the process of setting up this assay to see whether URF bioactivity is present in the prohormone, the bioactive relaxin fractions and the crude ovarian extracts.

It should be apparent that further studies are required in order to clarify the physiology and biochemistry of relaxin. This hormone has not yet earned a firm place in endocrinology despite its isolation nearly 50 years ago. It is a problem which promises to add dramatically to our knowledge of the endocrinology of reproduction.

In conclusion, the first radioimmunoassay applied to the measurement of relaxin immunoactivity in plasma (Bryant, 1972) in a number of physiological situations is a specific index of relaxin secretion. The application of this and the new edition of the radioimmunoassay reported here to the same plasmas should yield data on the levels of the postulated prohormone and hormone in plasma, their survival and their metabolism and provide a more refined index of relaxin secretion.

ACKNOWLEDGEMENTS

We would like to acknowledge the most valuable help given with the bioassays by many people working in our respective departments.

REFERENCES

Bolton, A.E. and Hunter, W.M. (1973): *Biochem. J., 133,* 529.
Bryant, G.D. (1972): *Endocrinology, 91,* 1113.
Bryant, G.D. and Stelmasiak, T. (1974): *Endocrine Res. Commun., 1,* 415.
Greenwood, F.C., Hunter, W.M. and Glover, J.S. (1963): *Biochem. J., 89,* 114.
Krantz, J.D., Bryant, H.H. and Carr, C.J. (1950): *Surg. Gynec. Obstet., 90,* 372.
Sherwood, O.D. and O'Byrne, E.M. (1974): *Arch. Biochem., 160,* 185.
Steinetz, B.G., Beach, V.L., Kroc, R.L., Stasilli, N., Nussbaum, R.E., Nemith, P.J. and Dun, R.K. (1960): *Endocrinology, 67,* 102.

MAMMOTROPHIC HORMONES IN RUMINANTS

ISABEL A. FORSYTH and I.C. HART

National Institute for Research in Dairying, Shinfield, Reading, United Kingdom

There have been important developments in ruminant endocrinology in recent years, such as the discovery of placental lactogen in ruminants (Buttle et al., 1972; Forsyth, 1973), the application of competitive binding assays to the measurement of protein and steroid hormones (Convey, 1974) and the specific manipulation of endogenous hormone levels by the use of drugs and hypothalamic-inhibiting and -releasing factors. Two further developments are now in progress: the extension to ruminants of mammary gland organ culture techniques and the study of hormone receptors in the mammary gland. This short review is concerned mainly with protein hormones and their relation to udder growth and function. Anatomical, biochemical and ultrastructural changes in the developing and secretory mammary gland and their endocrine control have been reviewed recently by Cowie (1971), Cowie and Tindal (1971), Ceriani (1974) and by several authors in Larson and Smith (1974).

GROWTH OF THE UDDER

Information on postnatal udder development in ruminants is far from complete. Sinha and Tucker (1969) observed allometric duct growth in Holstein heifers between 3 and 9 months of age, beginning some 3 months before behavioural oestrus, but little is known about its hormonal control. The gross size of the udder may vary widely in unmated females, but is a poor guide to development because of the high and variable proportion of stroma to parenchyma.

In primiparous goatlings there were only limited changes in the structure of the udder during the first half of pregnancy (Cowie, 1971). A period of rapid change with the establishment of lobules of alveoli occurs in goats between days 70 and 80 of pregnancy (gestation length 140 days; Cowie, 1971) and in cows between days 110 and 140 (gestation length 280 days; Hammond, 1927; Kwang, 1940; Turner, 1952). Lobulo-alveolar growth continues during the second half of pregnancy and the stroma is reduced to narrow bands of connective tissue. Although in several laboratory species an important component of mammary gland growth occurs in early lactation, in the sheep (Denamur, 1965; Anderson, 1975) and the cow (Baldwin, 1966) udder growth as measured by DNA content is essentially complete at parturition.

Secretory activity begins during the last third of pregnancy in ruminants as shown by histology, the appearance of lactose or increases in mammary RNA content (cow: Hammond, 1927; goat: Cowie, 1971 and E.A. Jones, unpublished observations; sheep: Denamur, 1965). Its extent can be judged by the considerable milk yields resulting from pre-partum milking in late pregnancy in cows and goats. Even in animals milked pre-partum a further increase in the yield of milk constituents occurs at about the time of parturition. Changes in biosynthetic and secretory capacity in late pregnancy and early lactation have recently been studied in some detail in cows in their first or subsequent pregnancy (Hartmann, 1973; Mellenberger et al., 1973) and in ewes (Hartmann et al., 1973).

HORMONAL INDUCTION OF UDDER GROWTH AND LACTATION

In vivo

In ovariectomized, hypophysectomized goatlings, oestrogen + progesterone had no mammogenic effect (Cowie et al., 1966). The further addition of prolactin + GH + adrenocorticotrophin gave limited lobulo-alveolar growth. The restoration of milk yields in lactating animals following hypophysectomy has been studied in goats and sheep (Cowie and Tindal, 1971; Denamur, 1971). In both species a combination of prolactin, GH, thyroid hormones and adrenal steroids restored milk yields to preoperative levels. If prolactin was then withdrawn, milk secretion continued for several weeks in the goat (Cowie and Tindal, 1971). The importance of prolactin for the initiation and growth hormone for the maintenance of lactation in ruminants is further discussed under 'Hormone levels in lactation'. These experiments help to define the minimal hormonal requirements for lobulo-alveolar growth and milk secretion in ruminants.

There has recently been a revival of interest in regimes for the induction of lactation in unmated or infertile ruminants, using oestrogen + progesterone, and sometimes also dexamethasone (see Smith and Schanbacher, 1973; Fulkerson and McDowell, 1974). However, histological studies (Cowie et al., 1968; Howe et al., 1975) suggest that oestradiol + progesterone in themselves elicit only limited parenchymal development even in the presence of the pituitary and that it is only after application of the milking stimulus and thereby release of pituitary hormones that full lobulo-alveolar growth occurs. Steroids may act to sensitize the mammary gland or to stimulate hormone output from the pituitary. Further work is needed to elucidate the hormonal events in induced lactogenesis and it should be remembered that in the primigravid animal the first establishment of a lobulo-alveolar structure is a rather rapid event.

In vitro

The in vitro technique of organ culture has been very successful in studying the effect of hormones on mammary development and the initiation of secretory activity, especially in the mouse, rat and rabbit (see reviews by Forsyth, 1971, 1975; Topper and Oka, 1974). This technique is now being extended to ruminants.

Using biopsy samples containing mainly ductal tissue from 2-year-old heifers, Djiane et al. (1975) initiated secretory activity after 5 days in culture with insulin + cortisol + prolactin. This hormone combination is also effective in most of the other mammals so far tested (Denamur, 1971; Forsyth, 1971), although the precise role of insulin remains controversial (see Forsyth, 1975). The mammary gland of ewes between 80 and 90 days of pregnancy reacted similarly (Delouis, 1975). However, at 30 days of pregnancy in primigravid ewes, no secretory response was obtained unless the tissue was first exposed in vitro to oestrogen + progesterone + insulin + cortisol + prolactin + GH. When the steroids were withdrawn after 2 days and culture then continued for a further 8 days, lobulo-alveolar development and initiation of secretion were observed. Progesterone was not obligatory but improved the response (Jeulin-Bailly et al., 1973; Delouis, 1975).

This difference between virgin heifers and early pregnant ewes is not readily explained. In preliminary experiments (Jones and Forsyth, unpublished data) we have found heifer mammary tissue generally unresponsive to prolactin even after pretreatment with oestrogen + progesterone. However, the results of Delouis and co-workers are in broad general agreement with in vivo studies. They raise the interesting possibility that at certain developmental stages priming by steroids may be required before a lactogenic reponse can occur.

PLACENTAL LACTOGEN IN RUMINANTS

Using co-culture techniques, we have shown that the cotyledonary placenta secretes a lactogenic hormone in goats, cows, sheep and fallow deer (Forsyth, 1973). Placental production of the hormone could be detected in vitro from 30 days of pregnancy in goats and sheep (Forsyth, 1973 and unpublished observations) and from day 36 in cows (Buttle and Forsyth, 1975). Purification of ovine placental lactogen has been reported (Handwerger et al., 1974; Martal and Djiane, 1975).

A specific radioimmunoassay for ruminant placental lactogen has not yet been developed, but total lactogenic activity in plasma can be measured by bioassay (Buttle et al., 1972) and with more sensitivity and precision by radioreceptor assay (Shiu et al., 1973; Parke and Forsyth, 1975). Such assays detect prolactin as well as placental lactogen, but specific measurement of prolactin is possible by radioimmunoassay, since there is no cross-reaction between ruminant prolactin and placental lactogen (Buttle et al., 1972; Handwerger et al., 1974; Buttle and Forsyth, 1975; Djiane and Kann, 1975; Martal and Djiane, 1975). The activity of goat and cow placental extracts in a radioreceptor assay is shown in Figure 1.

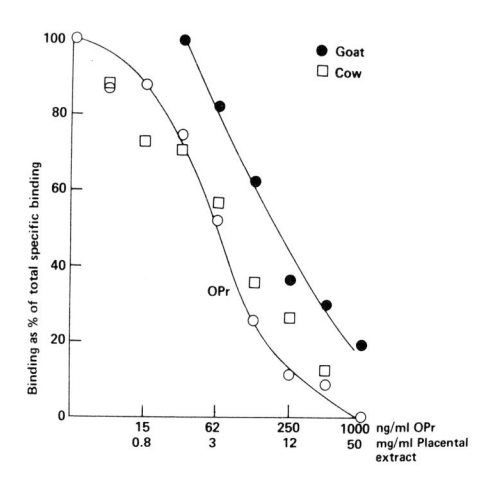

Fig. 1. Activity of extracts of the placenta of a 138-day pregnant goat (●) and of a cow at term (□) in a rabbit mammary gland receptor assay for lactogenic activity based on the method of Shiu et al. (1973). The standard was ovine prolactin (OPr) (NIH-P-S 9).

HORMONE LEVELS DURING PREGNANCY AND PARTURITION

Prolactin and placental lactogen

In primiparous goats (Buttle et al., 1972) and in sheep (Kann and Denamur, 1974; Kelly et al., 1974; Djiane and Kann, 1975) prolactin levels remain low until a few days before parturition (see also Convey, 1974). Total lactogenic activity in pregnancy is largely accounted for by placental lactogen, and has been measured in goats by in vitro bioassay (Buttle et al., 1972; Forsyth, 1973) and radioreceptor assay (Fig. 2; C.R. Thomas, unpublished results) and in sheep by radioreceptor assay (Kelly et al., 1974; Djiane and Kann, 1975). Levels of total lactogenic activity show considerable individual variation, but all these studies show a similar pattern of increase from about mid-pregnancy to reach peak values, which may be in excess of 1 µg/ml, between 100 and 140 days. Levels then fall again before parturition. Maximum levels appear somewhat lower and are reached slightly later in the goat than in the sheep. In a small series of goats, multiple pregnancies were associated with higher levels of placental lactogen (Fig. 2).

In the cow, although in vitro production rates and cotyledonary content of placental lactogen seem to be rather similar to those found in the goat and sheep, plasma levels are very considerably lower. By bioassay, we have been unable to detect placental lactogen in 7 primiparous heifers (Buttle and Forsyth, 1975). However, these heifers were inseminated early in the year. They showed the usual day-length and temperature related seasonal variation in prolactin levels (Schams, 1972) which were therefore relatively high during much of the second half of pregnancy. Animals pregnant during the winter would be expected to have low prolactin levels, and it will be of interest to

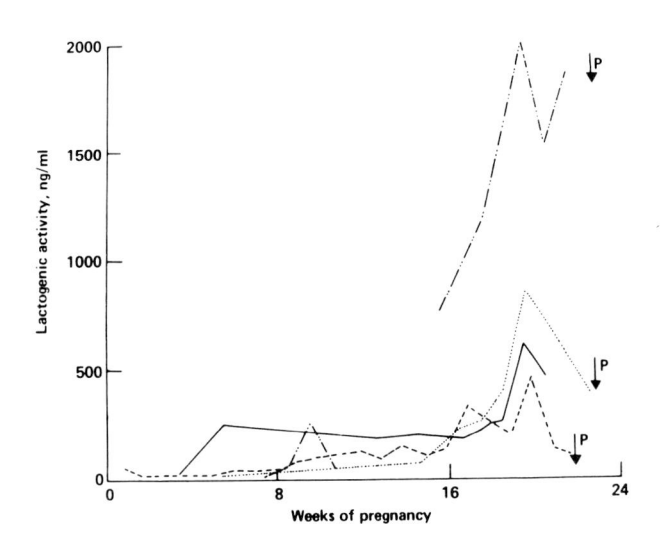

Fig. 2. Total lactogenic activity in the plasma of 4 pregnant goats measured by the method of Parke and Forsyth (1975) against ovine prolactin (NIH-P-S6) as standard: Valetta (---) primigravid, 1 ♀; Winifred (····) primigravid, 2 ♀; Tamara (—·—) multiparous, 1 ♀ 2 ♂; Wistful (———) primigravid, 2 ♀, 1 ♂. ↓P indicates the day of parturition. Wistful died of pregnancy toxaemia in late gestation. Lactogenic activity remained less than 50 ng/ml in a non-pregnant goat sampled over the same time period (C.R. Thomas, unpublished work).

examine these for any evidence of a reciprocal relationship between the 2 hormones.

In sheep and goats the close association between rising levels of placental lactogen and a rapid phase of morphological and secretory udder development in the second half of pregnancy, while prolactin levels remain low, does suggest an important role for placental lactogen in this development.

During labour in the goat, large increases in plasma prolactin are often associated with intense uterine contractions and expulsion of the foetus (Hart, 1972). These prolactin peaks, possibly evoked by stress, were unrelated to increases in oxytocin which normally occur at this time (McNeilly and Hart, 1973) thus tending to disprove the hypothesis of Benson and Folley (1956). The pre-partum rise in prolactin in cows can be suppressed without affecting the normal delivery of the foetus (Hoffmann et al., 1973). However, consistent with observations in hypophysectomized lactating ruminants (see page 423) high blood concentrations of prolactin shortly before and after parturition are essential for the onset of full milk secretion in the cow (Schams et al., 1972). Injections of thyrotrophin-releasing hormone (TRH) a potent stimulator of prolactin release, for 20 days starting 15 days before parturition in cows, actually increased the level of milk production over the first month of lactation (Karg and Schams, 1974). However, TRH also stimulates the release of GH and TSH from the anterior pituitary in cows (Convey et al., 1973; Kelly et al., 1973). In sheep, exogenous prolactin administration between days 120 and 140 of pregnancy did not accelerate the onset of copious milk secretion suggesting that levels of prolactin activity are not a limiting factor (Delouis and Denamur, 1967).

Growth hormone

Replacement experiments in endocrinectomized ruminants (see page 423) have implicated GH in udder development and function. However, the limited information available indicates that, like prolactin, circulating GH remains low in ruminants throughout gestation, rising 1-10 days before parturition (cow: Oxender et al., 1972; Olsen et al., 1974; sheep: Bassett et al., 1970). In none of these studies was regular sampling carried out throughout the whole of pregnancy. Olsen and collaborators (1974) found that during parturition plasma GH remained high in 2 of 3 cows. Whether the pre-partum increase in GH is significant in terms of the onset of lactation is not known.

Placental lactogen in primates is known to have GH-like properties (see Kaplan and Grumbach, 1974) but whether the same is true in ruminants remains to be convincingly demonstrated. Little is known of the factors controlling the rate of secretion and metabolic clearance of prolactin and GH in pregnant ruminants. In women, high levels of placental lactogen are associated with suppression of GH in late pregnancy (Kaplan and Grumbach, 1974) and it is tempting to suggest that placental lactogen might partially inhibit the release of prolactin and GH in ruminants until shortly before parturition when its influence is removed.

Some authors have suggested that the rise in prolactin and GH might occur in response to the stress of delivery. A firm relationship between stress and the release of prolactin has been established, but evidence for a similar relationship for GH in ruminants is less convincing as conflicting results have been obtained (Eaton et al., 1968; Tucker, 1971). Furthermore, in some cases the hormonal increases precede by several days the onset of uterine contractions associated with delivery of the foetus.

Pre-partum increases in prolactin and GH might also be affected by the sudden absence of the inhibitory effect of progesterone in the presence of sharply increasing concentrations of oestrogen as outlined by Hart (1973a). In addition to steroid effects on pituitary hormones, there is convincing evidence for stimulatory actions of glucocorticoids on mammary cells (Oka and Perry, 1974) and inhibitory effects of progesterone on the initiation of secretory activity (see Delouis, 1975) but not in established lactation. The direct effects of oestrogen are more uncertain but it has been claimed to increase subsequent milk yield independent of an effect on prolactin when given prepartum to ewes (Delouis and Terqui, 1974).

Insulin

As with monogastric animals, insulin plays an important role in the regulation of ruminant metabolism, blood concentrations of the hormone being markedly affected by the metabolic status of the animal at the time of sampling (Blair, 1974). It is unwise, therefore, to lay too much emphasis on results obtained in a given physiological situation using small numbers of animals or infrequent sampling techniques. None the less the small amount of research carried out in cows indicates that circulating insulin decreases as pregnancy progresses and is inversely related to blood-free fatty acids, a fact consistent with the anti-lipolytic character of the hormone (Koprowski and Tucker, 1973b; Grigsby et al., 1974). Blum and co-workers (1973) found plasma insulin to be elevated during the last 3 weeks of gestation, but found no significant differences between pre- and post-partum concentrations of the hormone.

HORMONE LEVELS IN LACTATION

Prolactin

Although ruminants release large quantities of prolactin in response to the milking/suckling stimulus (Johke, 1970; Hart, 1975a), which some workers have positively correlated with the stage of lactation (Johke, 1970; Koprowski and Tucker, 1973a), it is becoming increasingly evident that once lactation has been established high circulating concentrations of prolactin are unnecessary to maintain the level of milk secretion. In non-pregnant lactating cows both the quantity of prolactin released at milking and the basal level of the hormone are related to the season of the year and not the stage of lactation (Schams, 1972). These seasonal fluctuations are positively correlated with, and influenced by, day-length and temperature (goat: Buttle, 1973; Hart, 1975b; cow: Schams and Reinhardt, 1974; Wettemann and Tucker, 1974). The quantity of prolactin released at milking in goats can be maintained at the midsummer level, by housing lactating goats in summer light conditions, without preventing the normal autumn decline in milk yield (Hart, 1975b). Unpublished results obtained in this laboratory indicate further that throughout the whole of lactation there is no significant difference in circulating prolactin between high yielding and low yielding cows (see Fig. 4).

Even more convincing are the results in lactating ruminants using the ergot alkaloid, 2-Br-α-ergocryptine methane sulphonate (bromocryptine). Prolonged use of the drug during established lactation in the cow (Karg et al., 1972) and goat (Hart, 1973b) reduced the plasma concentration of prolactin to sub-basal levels without significantly affecting the milk yield, thereby indicating that the mammary gland functions normally in the absence of high circulating levels of the hormone. In none of these experiments was prolactin entirely eliminated from the circulation; low blood levels of the hormone might still play a part in maintaining milk secretion by binding to hormone receptor sites (see below) at the mammary gland. Immunoreactive prolactin is found in goat's and cow's milk (Malven and McMurtry, 1972) and the high metabolic clearance rate of prolactin in lactating sheep may be due to its excretion in milk (Davis and Borger, 1973). Preliminary results in the goat, however, indicate that prolactin in milk is probably directly related, via a concentration gradient, to the level of prolactin in the blood and that the hormone is not removed from the circulation by active transport. Nicoll and co-workers (1973) have evidence to suggest that some radioimmunoassays for prolactin and GH measure precursor forms of the hormones and not the biologically active entity. Bromocryptine might therefore inhibit immunoreactive but not biologically active prolactin. However, as treatment with bromocryptine inhibits milk secretion during early lactation in cows and has been shown to inhibit both prolactin release and milk secretion in monogastric animals, it seems likely that the drug inhibits biologically active prolactin in ruminants.

Growth hormone

Unlike prolactin, GH is essential for the maintenance of established lactation in the hypophysectomized lactating goat (Cowie and Tindal, 1971), but as a specific inhibitor for GH release has not been discovered (somatostatin is not specific for GH release in ruminants; Davis, 1975) less is known about the role of the hormone in the maintenance of ruminant lactation. Although workers have failed to elicit a GH response to milking and suckling in the cow (Tucker, 1971; Reynaert et al., 1972) others have demonstrated variable increases in GH in the goat (Hart and Flux, 1973) and sheep (Martal, 1975); blood sampling at milking, however, showed no consistent trend in the release of GH during early and late lactation in the goat (Hart and Flux, 1973). Neither was a significant difference found between basal levels of GH in lactating and non-lactating goats (Hart et al., 1975b). Hart and Buttle (1975) were unable to detect any effect of season on plasma GH in samples taken from male goats in which the seasonal effect on prolactin concentration (see above) had already been demonstrated.

A comparison of plasma GH in samples taken at hourly intervals throughout a 48-hour period from lactating beef (low milk yield) and dairy (high milk yield) cattle which had been matched for diet, age and date of parturition demonstrated significantly higher levels of the hormone in the high yielding cows (which had lost body weight during lactation) than in the low yielding cows (which had gained body weight

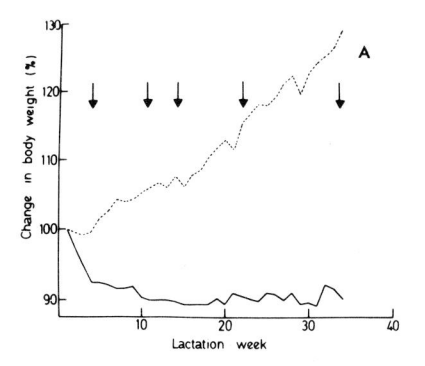

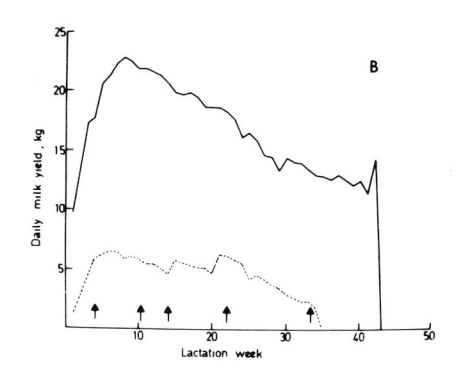

Fig. 3. (A) The average percentage liveweight change and (B) the average daily milk yield throughout lactation in 3 beef (----) and 4 dairy (———) heifers matched for age, diet and stage of lactation. ↑ = dates of blood sampling.

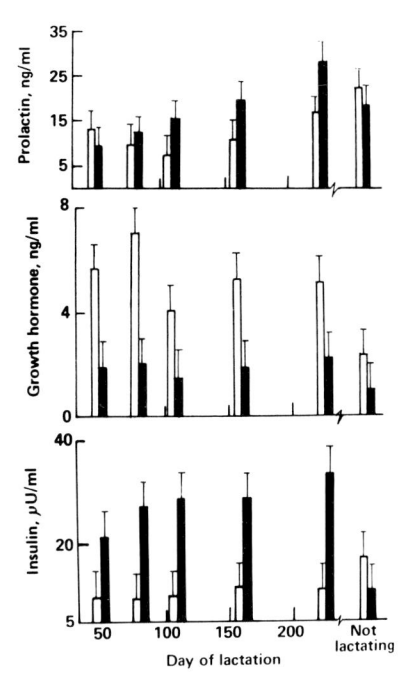

Fig. 4. The mean concentrations (± SEM) of prolactin, GH and insulin measured in hourly blood samples taken throughout selected 48-hour periods from 3 beef (■) and 4 dairy (☐) heifers at intervals throughout lactation and the dry period. Animals were matched for age, diet and stage of lactation.

during lactation) (Hart et al., 1975a). This difference was maintained throughout lactation (Figs. 3 and 4) and is of particular interest as it has been known for many years that administration of GH to lactating cows will increase the milk yield (Cotes et al., 1949; Machlin, 1973). There is considerable evidence implicating GH in ruminant carbohydrate metabolism, free fatty acid mobilization and protein balance (Blair, 1974); it seems likely therefore that the elevated levels of GH in the high yielding dairy cows exert an effect by making available to the mammary gland increased concentrations of milk precursors as outlined by Hart et al. (1975a). The possibility that GH might exert a direct effect on the mammary gland should not however be overlooked.

Insulin

There has been considerable interest in the effects of insulin administration on milk secretion and composition in cows; the general conclusions being that the hormone decreases milk production, lactose and glucose, while increasing milk fat and protein (Kronfeld et al., 1963; Schmidt, 1966). The limited amount of research carried out on blood levels of insulin in cows indicates that concentrations of the hormone are inversely related to the milk yield and that the rate of secretion of the hormone is probably controlled by the demand of the mammary gland on dietary energy and body reserves. Koprowski and Tucker (1973b) have demonstrated a negative association between serum insulin and milk yield in the cow and claim that levels of the hormone are increased at milking. The results, however, were complicated both by feeding and concurrent pregnancy. Levels of insulin were significantly higher in low yielding beef cows than in high yielding dairy cows (Hart et al., 1975a). This difference was maintained throughout lactation but was no longer apparent in the dry period (Fig. 4). Plasma insulin was negatively correlated with milk yield in the dairy cows. It would appear therefore that insulin levels are high in the low yielding cow where there is less demand on sources of energy, and in this situation the hormone might assume its anabolic effect on glycogen, fat and protein synthesis as evidenced by an increase in body weight (Fig. 3). In the high yielding cow where the demand on metabolites is greater than that which can be met by dietary energy, circulating insulin is low, possibly to allow the utilization of body tissues as evidenced by a fall in body weight (Fig. 3).

MEASUREMENT OF HORMONE RECEPTORS

Recognition by hormones relates to the presence of specific receptors in target cells. Steroid hormones enter cells and are bound by receptors in the cytoplasm for transfer to the nucleus. Protein hormones are bound by receptors on the cell surface. Following the initial interaction between hormones and their receptors, specific biochemical events are triggered in the cell.

Experiments in this area on ruminants are still very limited. Posner and co-workers (1974) reported specific binding of more than 1% but less than 3% of added ^{125}I-labelled ovine prolactin to 150 µg udder membrane protein from a 105-day pregnant sheep. Less than 1% of ^{125}I-insulin was specifically bound to the same tissue. In preliminary experiments on udders from 3 heifers and 2 cows using a similar technique for the preparation of membranes we have found specific binding of 1-4% to 800 µg membrane protein. These levels of binding are rather similar to those reported for normal rat mammary tissue similarly prepared (see Kelly et al., 1974). It seems possible that the high binding capacity of rabbit mammary gland membranes for prolactin is related to the remarkable response of this species to prolactin alone as a lactogenic hormone (Cowie and Tindal, 1971). In ruminants where a hormone complex is required it may be that the binding of several hormones to the cell membrane must occur before intracellular events are triggered leading to milk biosynthesis and secretion.

Experiments on rat liver indicate that prolactin receptors may be induced by prolactin itself (Posner et al., 1975a), by thyroxine, and by oestrogen (Gelato et al., 1975; Posner et al., 1975b) acting either directly or via stimulation of prolactin secretion (Posner et al., 1975a). This important area of fundamental research therefore suggests that rising levels of hormones may produce their effects by inducing or unmasking receptors as well as by interacting with them. On the basis of in vitro experiments on mouse mammary gland Vanderhaar and Topper (1974) suggested that mammary cells become hormone-sensitive in terms of the induction of milk synthesis only during a part of the G_1 phase of the cell cycle, but whether this relates in any way to the induction of hormone receptors is not known.

REFERENCES

Anderson, R.R. (1974): In: *Lactation. A Comprehensive Treatise, Vol. 1*, pp. 97-140. Editors: B.L. Larson and V.R. Smith. Academic Press, New York.

Anderson, R.R. (1975): *J. Animal Sci., 41*, 118.

Baldwin, R.L. (1966): *J. Dairy Sci., 49,* 1533.
Bassett, J.M., Thorburn, G.D. and Wallace, A.L.C. (1970): *J. Endocr., 48,* 251.
Benson, G.K. and Folley, S.J. (1956): *Nature (Lond.), 177,* 700.
Blair, T. (1974): *Hormonal Influences on Metabolism and Milk Secretion in Small Ruminants.* Thesis, University of Leeds.
Blum, J.W., Wilson, R.B. and Kronfeld, D.S. (1973): *J. Dairy Sci., 56,* 459.
Buttle, H.L. (1973): *J. Reprod. Fertil., 37,* 95.
Buttle, H.L. and Forsyth, I.A. (1975): *J. Endocr.,* in press.
Buttle, H.L., Forsyth, I.A. and Knaggs, G.S. (1972): *J. Endocr., 53,* 483.
Ceriani, R.L. (1974): *J. invest. Derm., 63,* 93.
Convey, E.M. (1974): *J. Dairy Sci., 57,* 905.
Convey, E.M., Tucker, H.A., Smith, V.G. and Zolman, J. (1973): *Endocrinology, 92,* 471.
Cotes, P.M., Crichton, J.A., Folley, S.J. and Young, F.G. (1949): *Nature (Lond.), 164,* 992.
Cowie, A.T. (1971): In: *Lactation,* pp. 123-140. Editor: I.R. Falconer. Butterworths, London.
Cowie, A.T., Knaggs, G.S., Tindal, J.S. and Turvey, A. (1968): *J. Endocr., 40,* 243.
Cowie, A.T. and Tindal, J.S. (1971): *The Physiology of Lactation.* Edward Arnold, London.
Cowie, A.T., Tindal, J.S. and Yokoyama, A. (1966): *J. Endocr., 34,* 185.
Davis, S.L. (1975): *J. Animal Sci., 40,* 911.
Davis, S.L. and Borger, M.L. (1973): *Endocrinology, 92,* 1414.
Delouis, C. (1975): *Mod. Probl. Pädiat., 15,* 16.
Delouis, C. and Denamur, R. (1967): *C.R. Acad. Sci. (Paris), 264D,* 2493.
Delouis, C. and Terqui, M. (1974): *C.R. Acad. Sci. (Paris), 278D,* 307.
Denamur, R. (1965): In: *Proceedings, II International Congress of Endocrinology,* pp. 434-462. Editor: S. Taylor. ICS 83, Excerpta Medica, Amsterdam.
Denamur, R. (1971): *J. Dairy Res., 38,* 237.
Djiane, J., Delouis, C. and Denamur, R. (1975): *J. Endocr., 65,* 453.
Djiane, J. and Kann, G. (1975): *C.R. Acad. Sci. (Paris), 280D,* 2785.
Eaton, L.W., Klosterman, E.W. and Johnson, R.R. (1968): *J. Animal Sci., 27,* 1785.
Forsyth, I.A. (1971): *J. Dairy Res., 38,* 419.
Forsyth, I.A. (1973): In: *Le Corps Jaune,* pp. 239-255. Editors: R. Denamur and A. Netter. Masson et Cie, Paris.
Forsyth, I.A. (1975): In: *Organ Culture in Biomedical Research,* p. 000-000. Editors: M. Balls and M.A. Monnickendam. Cambridge University Press, Cambridge. In press.
Fulkerson, W.J. and McDowell, G.H. (1974): *J. Endocr., 63,* 167.
Gelato, M., Marshall, S., Boudreau, M., Bruni, J., Campbell, G.A. and Meites, J. (1975): *Endocrinology, 96,* 1292.
Grigsby, J.S., Oxender, W.D., Hafs, H.D., Britt, D.G. and Merkel, R.A. (1974): *Proc. Soc. exp. Biol. (N.Y.), 147,* 830.
Hammond, J. (1927): *The Physiology of Reproduction in the Cow.* Cambridge University Press, Cambridge.
Handwerger, S., Maurer, W., Barrett, J., Hurley, T. and Fellows, R.E. (1974): *Endocrine Res. Commun., 1,* 403.
Hart, I.C. (1972): *J. Endocr., 55,* 51.
Hart, I.C. (1973a): *Hormonal Studies on the Control of Lactation in the Goat with Special Reference to Prolactin.* Thesis, University of Reading.
Hart, I.C. (1973b): *J. Endocr., 57,* 179.
Hart, I.C. (1975a): *J. Endocr., 64,* 305.
Hart, I.C. (1975b): *J. Endocr., 64,* 313.
Hart, I.C., Bines, J.A., Balch, C.C. and Cowie, A.T. (1975a): *Life Sci., 16,* 1285.
Hart, I.C. and Buttle, H.L. (1975): *J. Endocr.,* in press.
Hart, I.C. and Flux, D.S. (1973): *J. Endocr., 57,* 177.
Hart, I.C., Flux, D.S., Andrews, P. and McNeilly, A.S. (1975b): *Hormone metab. Res., 7,* 35.
Hartmann, P.E. (1973): *J. Endocr., 59,* 231.
Hartmann, P.E., Trevethan, P. and Shelton, J.N. (1973): *J. Endocr., 59,* 249.
Hoffmann, B., Schams, D., Gimenez, T., Ender, M.L., Herrmann, C. and Karg, H. (1973): *Acta endocr. (Kbh.), 73,* 385.
Howe, J.E., Heald, C.W. and Bibb, T.L. (1975): *J. Dairy Sci., 58,* 853.

Jeulin-Bailly, C., Delouis, C. and Denamur, R. (1973): *C.R. Acad. Sci. (Paris), 277D,* 2525.
Johke, T. (1970): *Endocr. jap., 17,* 393.
Kann, G. and Denamur, R. (1974): *J. Reprod. Fertil., 39,* 473.
Kaplan, S.L. and Grumbach, M.M. (1974): In: *Lactogenic Hormones, Fetal Nutrition, and Lactation,* pp. 183-191. Editors: J.B. Josimovich, M. Reynolds and E. Cobo. John Wiley & Sons, New York.
Karg, H. and Schams, D. (1974): *J. Reprod. Fertil., 39,* 463.
Karg, H., Schams, D. and Reinhardt, V. (1972): *Experientia (Basel), 28,* 574.
Kelly, P.A., Bedirian, K.N., Baker, R.D. and Friesen, H.G. (1973): *Endocrinology, 92,* 1289.
Kelly, P.A., Bradley, C., Shiu, R.P.C., Meites, J. and Friesen, H.G. (1974): *Proc. Soc. exp. Biol. (N.Y.), 146,* 816.
Kelly, P.A., Robertson, H.A. and Friesen, H.G. (1974): *Nature (Lond.),. 248,* 435.
Koprowski, J.A. and Tucker, H.A. (1973a): *Endocrinology, 92,* 1480.
Koprowski, J.A. and Tucker, H.A. (1973b): *Endocrinology, 93,* 645.
Kronfeld, D.S., Mayer, G.P., Robertson, J.M. and Raggi, F. (1963): *J. Dairy Sci., 46,* 559.
Kwang, F.J. (1940): *J. Amer. vet. med. Ass., 46,* 36.
Larson, B.L. and Smith, V.R. (Eds.) (1974): *Lactation: A Comprehensive Treatise,* Vols. 1 and 2. Academic Press, New York.
McMurtry, J.P. and Malven, P.V. (1974): *Endocrinology, 95,* 559.
McNeilly, A.S. and Hart, I.C. (1973): *J. Endocr., 56,* 159.
Machlin, L.J. (1973): *J. Dairy Sci., 56,* 575.
Malven, P.V. and McMurtry, J.P. (1972): *J. Dairy Sci., 55,* 715.
Martal, J. (1975): *C.R. Acad. Sci. (Paris), 280D,* 197.
Martal, J. and Djiane, J. (1975): *Biochem. biophys. Res. Commun., 65,* 770.
Mellenberger, R.W., Bauman, D.E. and Nelson, D.R. (1973): *Biochem. J., 136,* 741.
Nicoll, S.C., Mena, F., Sauguannoi, H., Tai, M. and Green, S. (1973): In: *Abstracts, The Endocrine Society,* p. 250.
Oka, T. and Perry, J.W. (1974): *J. biol. Chem., 249,* 3586.
Olsen, J.D., Trenkle, A., Witzel, D.A. and McDonald, J.S. (1974): *Amer. J. vet. Res., 35,* 1131.
Oxender, W.D., Hafs, H.D. and Edgerton, L.A. (1972): *J. Animal Sci., 35,* 51.
Parke, L. and Forsyth, I.A. (1975): *Endocr. Res. Commun., 2,* 137.
Posner, B.I., Kelly, P.A. and Friesen, H.G. (1975a): *Science, 188,* 57.
Posner, B.I., Kelly, P.A. and Friesen, H.G. (1975b): *Proc. nat. Acad. Sci. (Wash.), 71,* 2407.
Posner, B.I., Kelly, P.A., Shiu, R.P.C. and Friesen, H.G. (1974): *Endocrinology, 95,* 521.
Reynaert, R., De Paape, M. and Peeters, G. (1972): *Ann. Endocr. (Paris), 33,* 541.
Schams, D. (1972): *Acta endocr. (Kbh.), 71,* 684.
Schams, D. and Reinhardt, V. (1974): *Hormone Res., 5,* 217.
Schams, D., Reinhardt, V. and Karg, H. (1972): *Experientia (Basel), 28,* 697.
Schmidt, G.H. (1966): *J. Dairy Sci., 49,* 381.
Shiu, R.P.C., Kelly, P.A. and Friesen, H.G. (1973): *Science, 180,* 968.
Sinha, Y.N. and Tucker, H.A. (1969): *J. Dairy Sci., 52,* 507.
Smith, R.L. and Schanbacher, F.L. (1973): *J. Dairy Sci., 56,* 738.
Topper, Y.J. and Oka, T. (1974): In: *Lactation. A Comprehensive Treatise,* Vol. 1, pp. 327-348. Editors: B.L. Larson and V.R. Smith. Academic Press, New York.
Tucker, H.A. (1971): *J. Animal Sci., 32, Suppl. 1,* 137.
Turner, C. (1952): *The Mammary Gland. 1. The Anatomy of the Udder of Cattle and Domestic Animals.* Lucas Brothers, Columbia, Mo.
Vanderhaar, B.K. and Topper, Y.J. (1974): *J. Cell Biol., 63,* 707.
Wettemann, R.P. and Tucker, H.A. (1974): *Proc. Soc. exp. Biol. (N.Y.), 146,* 908.

PROLACTIN BINDING TO PLASMA MEMBRANES, AND ITS EFFECT ON MONOVALENT CATION TRANSPORT IN MAMMARY ALVEOLAR TISSUE — A POSSIBLE MECHANISM OF ACTION*

IAN R. FALCONER

Department of Biochemistry and Nutrition, University of New England, Armidale, N.S.W., Australia

PROLACTIN UPTAKE AND BINDING

Prolactin, like other polypeptide and glycoprotein hormones, does not appear to enter the cells of its target tissue. Evidence is rapidly accumulating for hormone receptors on the plasma membranes of mammary alveolar and other cells, which show specific binding properties towards prolactin and closely related polypeptides.

To reach these receptors, pituitary prolactin is distributed through the bloodstream, in which it has a short half-life. In the rabbit we showed a half-life of 16 minutes for ^{125}I-labelled prolactin (Birkinshaw and Falconer, 1972). The majority of prolactin is metabolised by the kidneys, and appears as low molecular-weight products in the urine. The distribution of radioactivity in tissues and secretions obtained 30 minutes after intravenous administration of ^{125}I-prolactin to a rabbit is illustrated in Table 1, clearly showing the importance of the urinary route of excretion. Our in vitro studies of degradation of ^{125}I-prolactin by slices of kidney, liver and mammary gland (into 5% trichloroacetic acid soluble products) showed that over 2 hours of incubation, kidney was twice as effective as liver, and no significant degradation occurred in the presence of mammary gland.

The small proportion of circulating prolactin which is taken up by mammary tissue, however, exhibits a long half-life, about 45-50 hours. Detailed autoradiographic studies of the localization of this prolactin have shown that the alveolar cell

TABLE 1

Radioactivity of tissues and secretions obtained from a lactating rabbit 30 minutes after intravenous injection of ^{125}I-labelled prolactin

Tissue	^{125}I (c.p.m./g ± S.E.)	Secretion	^{125}I (c.p.m./ml ± S.E.)
Liver	667 ± 50	Bile	997 ± 97
Kidney	3093 ± 110	Urine	4812 ± 148
Mammary gland	597 ± 45	Milk	1923 ± 29
Blood	1380 ± 18		

* These studies were supported by the Nuffield Foundation, the National Health and Medical Research Council of Australia and the University of New England.

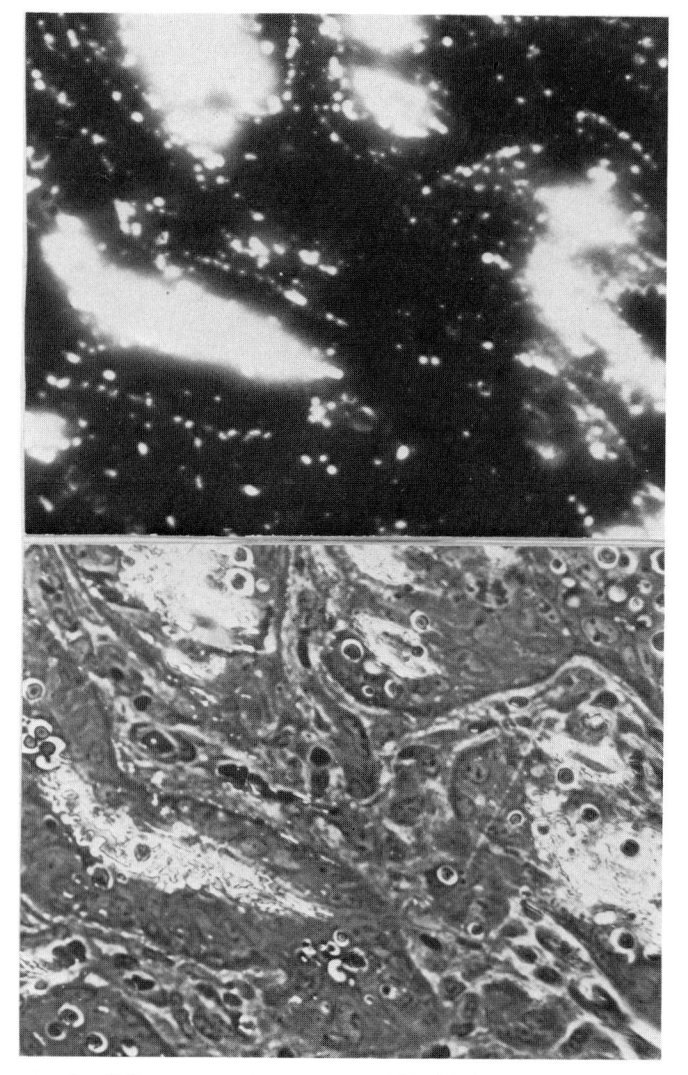

Fig. 1. Mammary alveoli from a pseudopregnant rabbit 1.5 hours after the intravenous injection of ^{125}I-labelled prolactin. Upper image photographed by dark-field illumination to show silver grains (small bright spots), lower image by phase contrast to show structure ($\times$ 350).

surface adjacent to the vascular supply is uniquely labelled (Fig. 1) irrespective of the route of administration of the ^{125}I-labelled hormone (Birkinshaw and Falconer, 1972).

We have also studied the localization of prolactin receptors through the cellular fractionation approach. The initial experiments were carried out by incubating slices of lactating mammary tissue in solutions containing ^{125}I-prolactin. The slices were then homogenised and subcellular fractions collected by differential centrifugation (Falconer, 1972).

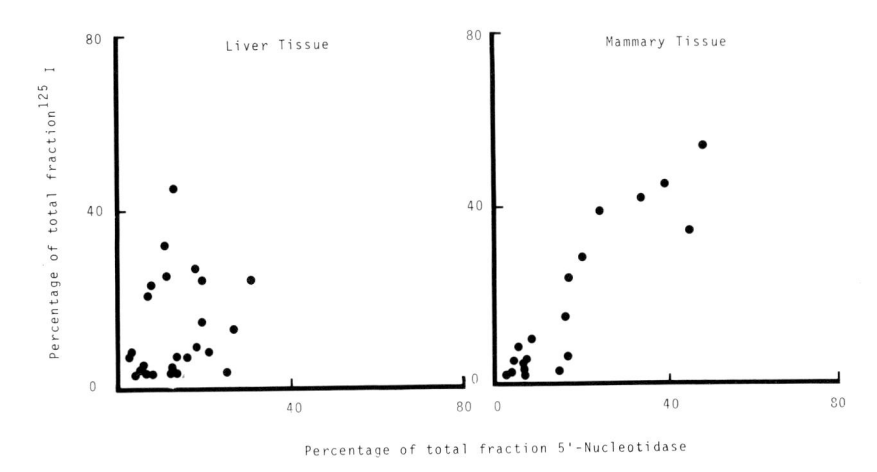

Fig. 2. Correlation of the distribution of ^{125}I-prolactin radioactivity with 5′-nucleotidase in subcellular fractions of mammary gland and liver after in vivo administration of prolactin.

More recently Stephen Turley and I administered ^{125}I-prolactin intravenously to lactating rabbits, and then obtained mammary tissue post mortem at intervals from 1-24 hours after injection. The tissue was chopped finely, washed and homogenised in 0.25 M sucrose, 0.05 mM CaCl$_2$, pH 7.4. The homogenate was layered on top of a continuous sucrose density gradient from 24% to 59% sucrose, and spun in a swing-out rotor at 3000 × g.av for 3 hours. Five fractions were collected from each tube, and assayed using standard techniques for 'marker' enzymes and DNA. The marker enzymes were lactate dehydrogenase — a cytoplasmic enzyme of mammary alveolar cells (Gul and Dils, 1969); 5′-nucleotidase — a plasma membrane enzyme (Huang and Keenan, 1972); inosine diphosphatase — an endoplasmic reticulum enzyme (Novikoff and Heus, 1963); succinate dehydrogenase — a mitochondrial enzyme (Allmann et al., 1966). The ^{125}I radioactivity of all fractions was also measured. Samples of liver were collected at the same time, and processed identically for purposes of comparison. Figure 2 shows the results for radioactivity measurement plotted against 5′-nucleotidase activity for fractions from liver tissue and mammary tissue.

Statistical evaluation of the data showed no significant changes in radioactivity distribution in fractions obtained from 1-24 hours after injection. The data were therefore pooled for the calculation of correlation coefficients between the proportions of radioactivity and marker enzymes in different fractions. The results of these calculations are shown in Table 2.

Clear correlations between ^{125}I radioactivity and enzymes of mammary gland plasma membranes, mammary gland cytoplasm and liver endoplasmic reticulum can be seen. The fraction containing the cytoplasm was collected from the top of the sucrose gradient, and hence would contain radioactive ^{125}I not bound to cellular components as well as protein-bound ^{125}I-prolactin. To examine the nature of the labelled constituents of the 'cytoplasm' fraction, mammary tissue was collected from a lactating rabbit 12 hours after intravenous injection of ^{125}I-prolactin. The tissue was chopped, blotted free of exuded milk, homogenised in isotonic sucrose solution and centrifuged at 100,000 × g for 30 minutes. The supernatant was removed and passed down a Sephadex G-200 column with an EDTA/Tris/sucrose buffer, pH 8.0. The

TABLE 2

Statistical analysis of the correlation between 'marker' enzymes and ^{125}I radioactivity in mammary gland fractions collected after separation by sucrose density gradient centrifugation

Subcellular fraction	Enzyme or component	Correlation coefficient Mammary gland	Liver
Cytoplasm	Lactate dehydrogenase	0.90*	0.43
Plasma membranes	5'-nucleotidase	0.92*	0.22
Endoplasmic reticulum	Inosine diphosphatase	0.56	0.80*
Mitochondria	Succinate dehydrogenase	−0.28	0.18
Nuclei	DNA	−0.38	0.15

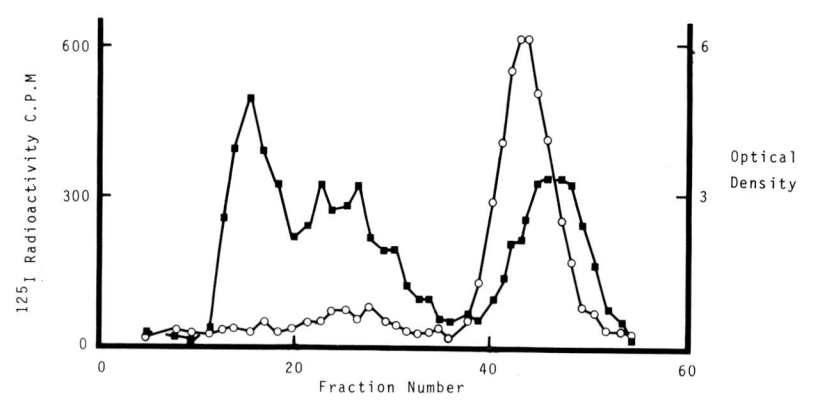

*$P < 0.001$.

Fig. 3. Separation on Sephadex G-200 of the 100,000 × g supernatant obtained from mammary tissue from a lactating rabbit intravenously injected with ^{125}I-labelled prolactin 12 hours earlier. —o—, ^{125}I radioactivity; — ■ —, optical density.

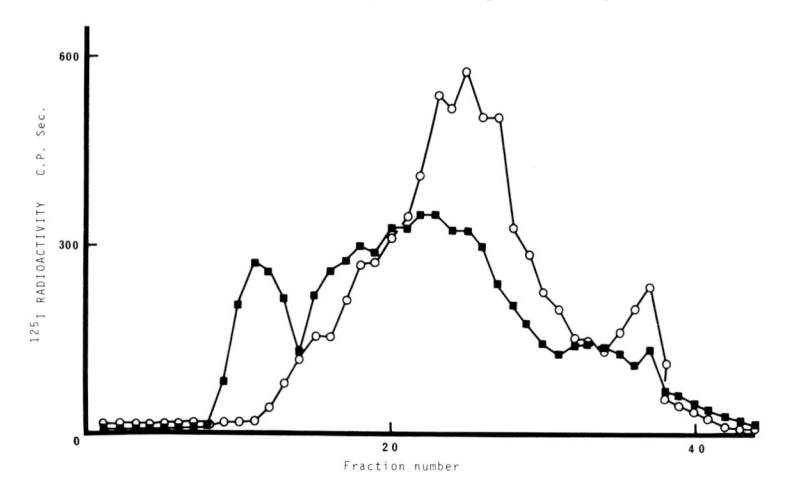

Fig. 4. Separation on Sephadex G-200 of ^{125}I-labelled prolactin alone (—o—), and after incubation with a 100,000 × g supernatant fraction from a mammary gland homogenate (— ■ —).

results of the separation are shown in Figure 3, which shows the majority of the ^{125}I in a low molecular weight fraction, which could not be precipitated with 5% trichloroacetic acid. The statistical correlation between lactate dehydrogenase and ^{125}I (Table 2) therefore does not reflect prolactin binding, but the presence of degradation products in the mammary glands. To clarify any possibility of prolactin binding to cytoplasmic proteins, ^{125}I-prolactin was incubated in vitro for 30 minutes at 37°C with a 100,000 × g supernatant fraction from a mammary gland homogenate. The supernatant + labelled prolactin was passed down a Sephadex G-200 column as before; the results of this separation are shown in Figure 4, together with the separation of ^{125}I-prolactin alone. The only clear evidence of binding of labelled prolactin to constituents of the mammary gland supernatant appeared in the high molecular weight fraction excluded from the Sephadex, which would contain membrane fragments and proteins above 200,000 daltons molecular weight.

Since the high-speed supernatant fraction from lactating mammary gland homogenates inevitably contains some residual milk, it was of interest to examine ^{125}I radioactivity in milk. Milk always contains iodide, concentrated from the blood by the mammary gland (see Falconer, 1963) and a proportion of this iodide becomes incorporated into milk proteins. To investigate this a sample of milk was taken from a rabbit injected with ^{125}I-prolactin intravenously 12 hours earlier. Casein-like proteins were precipitated at pH 4.0, and the remaining soluble proteins precipitated by 5% trichloroacetic (TCA). The majority (60%) of the milk radioactivity appeared in the casein-like protein precipitate, with a further proportion (20%) in the TCA precipitate.

We concluded from these experiments that the observed correlation between ^{125}I radioactivity and lactate dehydrogenase did not represent cytoplasmic binding of prolactin, but was due to the presence of ^{125}I-iodide and ^{125}I-prolactin degradation products in the tissue and milk of the gland appearing in the supernatant fraction.

To verify the in vivo binding of ^{125}I-prolactin to plasma membranes of mammary alveolar cells, lactating rabbits were given intravenous ^{125}I-prolactin. After 12 hours membrane fractions were obtained by differential centrifugation of homogenates of the alveolar tissue. These fractions were layered onto tubes containing continuous sucrose density gradients, and centrifuged as described before. Fractions collected down the sucrose gradient were assayed for 5'-nucleotidase and ^{125}I radioactivity; statistical analysis of the distribution of activities showed a highly significant correlation. The majority of both activities were found in the 1.103-1.116 g/ml region of sucrose density, corresponding to the light subfraction of plasma membrane (Evans, 1970).

On the basis of these studies we consider that the in vivo binding of prolactin in rabbit mammary tissue is to the plasma membrane of alveolar cells, on the side adjacent to the vascular supply.

PROLACTIN RECEPTORS

Recent in vitro studies by Turkington, Friesen and their co-workers have done much to quantify and characterise prolactin receptors (Franz et al., 1974; Shiu and Friesen, 1974a,b). Both groups have published dissociation constants of the order of 1×10^{-9} M, and have determined the saturation binding capacity of their membrane fractions.

We have attempted to measure these characteristics for suspensions of isolated mammary alveolar cells, to obtain a figure for prolactin receptors per cell. Cells were obtained from lactating rabbits pre-treated for 2 days with 1 mg/kg of bromocriptin, to deplete prolactin from receptor sites. The tissue was sliced, incubated with collagenase (Sigma type II) at 25 mg/50 ml Tyrode's solution at 37°C for 1 hour and then dispersed through nylon gauze. The cells were washed twice by centrifuging at 600 × g for 10 minutes followed by resuspension in fresh solution, the final resuspension being in Tyrode's solution pH 7.4 containing 1% rabbit serum.

The prolactin used was NIH PS9 (ovine) and ^{125}I-labelling carried out by the method of free chloride oxidation (Redshaw and Lynch, 1974). The technique for determining prolactin binding that we have used is to add varying quantities of normal prolactin to a fixed amount of high specific activity ^{125}I-prolactin. Dissociation constants and saturation capacity have been determined from Scatchard plots (Scatchard, 1949). Our measurements of the dissociation constant (Kd) average approximately 6×10^{-10} M, which is between the figures already published (Franz et al., 1974; Shiu and Friesen, 1974a). Determinations of the saturation capacity per cell for bound prolactin fall in the range 2×10^3 to 1.5×10^4, with a mean of 7.6×10^3 molecules/cell. We are currently trying to improve the precision of these measurements.

Other workers have published data for other hormones including a saturation capacity of 1×10^4 molecules of insulin per fat cell (Cuatrecasas, 1971) and 500 molecules of TSH per cultured thyroid cell (Lissitzky et al., 1973). These measurements may over-estimate the actual number of sites needed for maximal stimulation, since recent work by Birnbaumer and Pohl (1973) demonstrated maximum activation of adenyl cyclase by glucogen on liver membranes with only 10-20% of the binding sites occupied.

POSSIBLE MECHANISMS FOR PROLACTIN ACTION

Investigations by Turkington and colleagues showed that prolactin does not activate adenyl cyclase in mammary cell membranes, nor could cyclic AMP added to mammary explant cultures mimic the effects of prolactin (Turkington, 1972). Recently a detailed study of cyclic AMP concentration and enzyme activity in mammary tissue during pregnancy and lactation in rats has shown rising cyclic AMP concentrations during pregnancy, and an abrupt fall in lactation (Sapag-Hagar and Greenbaum, 1974). Subsequent in vitro work using mammary explants cultured with very high concentrations of cyclic AMP (10^{-3} M) showed reductions in biosynthetic enzyme activities, especially fatty acid synthetase (Sapag-Hagar et al., 1974). It was suggested that the decrease in cyclic AMP during the onset of lactation may therefore be a component of the control mechanism. It is difficult to interpret this, since insulin is one of the most effective hormonal inhibitors of adenyl cyclase, but does not trigger lactation in culture or in vivo.

Our own approach to the problem of the initial mechanism by which prolactin will initiate, and then maintain, lactation has been directly derived from the well-described osmotic role of prolactin (see Ensor and Ball, 1972). Since prolactin has direct effects on salt transport and Na$^+$/K$^+$ activated ATPase in renal tissues of a variety of vertebrate classes, including man (Horrobin et al., 1971), it was of interest to investigate any comparable effects on mammary tissue. The major tissues responsive to prolactin are of epithelial origin, including the nasal salt glands of ducks (Peaker et al., 1970), fish gills (Epstein et al., 1967) and mammary alveolar tissue (Cowie, 1974). Prolactin appears to exert its effects on osmo-regulation in teleost fish through modulation of Na$^+$ permeability across membranes, and control of Na$^+$/K$^+$-ATPase activity (Ensor and Ball, 1972). We therefore started by investigating the influence of prolactin on ^{22}Na$^+$ transport into and out of mammary alveolar tissue, in order to differentiate these alternatives.

PROLACTIN AND MONOVALENT CATION TRANSPORT

This experimental approach used slices of mammary gland from lactating rabbits, incubated at 37°C in Tyrode's solution (pH 7.4) containing 1% rabbit serum. To deplete endogenous prolactin, bromocriptin (1 mg/kg) was given subcutaneously to the rabbits for 3 days before killing. $^{22}Na^+$ was added to the incubation medium and in the first series of experiments the slices were incubated for 2 hours. Prolactin, bovine growth hormone or serum albumin were added to the incubation at concentrations of 0.5 µg/mg tissue slices (for prolactin this corresponded to a concentration of 4.4 × 10^{-7} M). The results are shown in Table 3, which demonstrated a significant reduction in $^{22}Na^+$ content in the presence of prolactin, with a lesser reduction due to growth hormone. Inclusion of 5×10^{-5} M ouabain in these experiments abolished the decreased $^{22}Na^+$ content due to prolactin, and resulted in an approximately 10% increase in tissue $^{22}Na^+$, whether prolactin was present or not (Table 4).

In these experiments the content of $^{22}Na^+$ in the tissue after incubation would be due to entry by diffusion, opposed by active extrusion through the operation of the Na^+ pump (Na^+/K^+-activated ATPase). The abolition of the prolactin-induced decrease in $^{22}Na^+$ content by ouabain indicates that prolactin has activated the Na^+ pump, rather than decreased membrane permeability to Na^+.

To confirm this a second series of experiments were carried out, in which slices of lactating mammary gland were pre-incubated for 1 or 16 hours at 3°C in K^+-free Tyrode's solution containing $^{22}Na^+$ to 'load' cells with $^{22}Na^+$. After 'loading' the slices were rapidly washed, counted for $^{22}Na^+$ radioactivity, then incubated for 30 minutes at 37°C in normal Tyrode's solution + 1% rabbit serum. After this the slices were drained, blotted and $^{22}Na^+$ radioactivity measured. The final radioactivity of individual

TABLE 3

Effect of prolactin, growth hormone or albumin on $^{22}Na^+$ content of mammary gland slices after 2 hours incubation at 37°C with 0.1 µCi of ^{22}Na

Treatment	Concentration (µg protein per mg tissue)	$^{22}Na^+$ content of 100 mg-slices (c.p.m. ± S.E.) (3 experiments, 6 replicates in each)
Control	—	376.2 ± 16.5
Prolactin	0.5	326.8 ± 13.3†
Growth hormone	0.5	342.6 ± 13.8*
Albumin	0.5	381.1 ± 18.5

*P ≃ 0.05; †P < 0.01.
(From Falconer and Rowe (1975), by courtesy of the Editors of *Nature*.)

TABLE 4

Effect of ouabain (5×10^{-5} M) on the $^{22}Na^+$ content of mammary gland slices in the presence or absence of prolactin (4.4×10^{-7} M)

Treatment	$^{22}Na^+$ content of 100 mg slices (c.p.m. ± S.E.)
Control	355.4 ± 20.9
Ouabain	382.4 ± 22.8
Ouabain + prolactin	402.0 ± 34.3

TABLE 5

^{22}Na ions retained in mammary alveolar tissue (0.1 g of 0.7 mm slices) after incubation at 37°C for 30 minutes in the presence of prolactin (4.4 × 10^{-8} M) or ouabain (5 × 10^{-5} M) or both. Tissue 'loaded' with ^{22}Na for 16 hours at 3°C in K$^+$-free buffer. Six samples per treatment

Treatment	^{22}Na$^+$ in tissue (mean ± S.E. c.p.m./100 mg)
Control	238 ± 10
Ouabain	272 ± 13*
Prolactin	198 ± 28
Ouabain + prolactin	262 ± 25*

* Significant differences (P < 0.01) between treatments with and without ouabain.

incubations was corrected for variations in the initial activity of the sample. These experiments measured ion outflow alone, and were carried out with varying concentrations of prolactin, growth hormone, albumin and ouabain present in the final incubation medium. In 10 experiments in which prolactin was added to the incubation medium, 9 showed a lower tissue retention of ^{22}Na$^+$ after incubation, averaging a

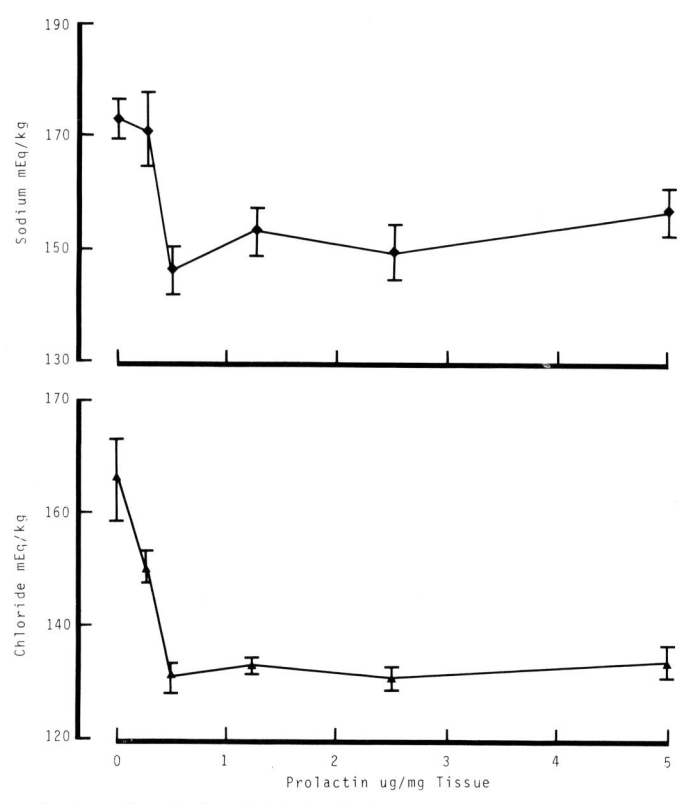

Fig. 5. Changes in the sodium (♦) and chloride (▲) content of guinea-pig mammary tissue after incubation for 1 hour at 37°C in the presence of increasing amounts of prolactin.

reduction of 13.5% (P < 0.01) compared with control incubations. No effects were seen due to the addition of growth hormone or albumin. The addition of ouabain increased tissue ^{22}Na$^+$ retention significantly, and abolished the reduction due to prolactin (Table 5). As the tissue ^{22}Na$^+$ content in these experiments was solely due to variations in ion outflow, the results show that prolactin increases sodium outflow, and that ouabain both reduces outflow and abolishes the effect of prolactin. Together with the data in Tables 3 and 4, these results strongly indicate a direct action of prolactin on Na$^+$ extrusion by the active pump, since this is specifically blocked by ouabain (Skou, 1965).

To examine whether these in vitro responses to prolactin could also be detected by other techniques, John Rowe and I carried out a series of experiments measuring the monovalent ion content of incubated slices of lactating mammary tissue. These studies employed a range of prolactin concentrations from 0 to 17.4×10^{-8} M, and measured Na$^+$, K$^+$ and Cl$^-$ concentrations after 1 hour of incubation at 37°C in Krebs bicarbonate buffer (pH 7.4) with continuously bubbled 95% O$_2$, and 5% CO$_2$. The results of one experiment are shown in Figure 5, the maximal response in this example being at 0.5 µg/mg, a concentration in the medium of 1.7×10^{-7} M. Subsequent experiments have been carried out with lower concentrations of prolactin, and with speedier handling of tissue kept at 37°C between excision and incubation. The result was that sodium content of control incubations was lower (100-120 mEq/kg), and that significant decreases ($0.05 > P > 0.01$) in tissue sodium and chloride were observed at 4.4×10^{-8} M prolactin (Falconer and Rowe, 1975).

We have also studied the effect of addition of ouabain to the incubation medium. This series of experiments are still in progress, but preliminary results show an abolition of the effect of prolactin by ouabain, which results in elevated tissue Na$^+$ concentration, and depressed K$^+$ concentration.

CONCLUSION

From the in vitro studies described, I conclude that a rapid response of mammary alveolar tissue to prolactin is the activation of the sodium pump of the cell membrane, resulting in decreased intracellular concentration of sodium and chloride. Evidence that this may occur in lactation is provided by the observation that sodium concentration in milk rises during late lactation, and this rise can be reversed by prolactin treatment (Gachev, 1963; Taylor et al., 1975). The ionic content of milk is considered due to two components, passive diffusion of intracellular ions from secretory cells (Linzell and Peaker, 1971) and 'leaking' of ions between cells (Taylor et al., 1975). The data of this paper support the hypothesis that intracellular sodium and chloride concentrations in secretory cells are modulated by prolactin, and these modulations may be reflected in changes in milk composition.

A potential role for the modulation of intracellular Na$^+$ and K$^+$ concentrations by prolactin during lactogenesis is worthy of investigation. Since small changes in intracellular Na$^+$ will have comparatively large effects on the Na$^+$/K$^+$ ratio, this presents a possible mechanism for modulation of enzyme activity and nuclear function in alveolar cells. Studies with both lymphocytes (Quastel and Kaplan, 1970) and cells in tissue culture (Lubin, 1967) have shown marked changes in DNA, RNA and protein synthesis on modification of the intracellular Na$^+$/K$^+$ ratio. In particular, activation of lymphocytes by phytohaemagglutinin could be blocked by very low ouabain concentrations ($2\text{-}5 \times 10^{-7}$ M), which elevate cell Na$^+$ and deplete cell K$^+$ (Quastel and Kaplan, 1970).

It is therefore possible that further investigation of the role of prolactin in modulating ion transport in mammary alveolar cells may cast valuable light on the mechanism of action of prolactin in lactogenesis and in the maintenance of lactation.

ACKNOWLEDGEMENT

I thank John Rowe, Paul Keys, Steven Turley, Tom Buckley, Kam Tadros and Rhyl McLeod for their participation in this project, the Endocrinology Study Section of NIAMDD for gifts of prolactin, and Prof. E. Flükiger and Dr. H. Friedle of SANDOZ, Basle for gifts of bromocriptin (CB 154).

REFERENCES

Allmann, D.W., Bachmann, E. and Green, D.E. (1966): *Arch. Biochem., 115,* 165.
Birkinshaw, M. and Falconer, I.R. (1972): *J. Endocr., 55,* 323.
Birnbaumer, G.L. and Pohl, S.L. (1973): *J. biol. Chem., 248,* 2056.
Cuatrecasas, P. (1971): *Proc. nat. Acad. Sci. (Wash.), 68,* 1264.
Cowie, A.T. (1969): *J. Endocr., 44,* 437.
Cowie, A.T. (1974): *J. invest. Derm., 63,* 2.
Ensor, D.M. and Ball, J.M. (1972): *Fed. Proc., 31,* 1615.
Epstein, F.M., Katz, A.I. and Pickford, G.E. (1967): *Science, 156,* 1245.
Evans, W.H. (1970): *Biochem. J., 116,* 833.
Falconer, I.R. (1963): *J. Endocr., 25,* 533.
Falconer, I.R. (1972): *Biochem. J., 126,* 8P.
Falconer, I.R. and Rowe, J.M. (1975): *Nature (Lond.), 256,* 327.
Franz, W.L., MacIndoe, J.H. and Turkington, R.W. (1974): *J. Endocr., 60,* 485.
Gachev, E. (1963): *Zh. obshchei Biol., 24,* 382.
Gul, B. and Dils, R. (1969): *Biochem. J., 112,* 293.
Horrobin, D.F., Lloyd, I.J., Lipton, A., Burstyn, P.G., Durkin, N. and Muiruri, K.L. (1971): *Lancet, 2,* 352.
Huang, C.M. and Keenan, T.W. (1972): *J. Dairy Sci., 55,* 862.
Kroeger, H. (1966): *Exp. Cell Res., 41,* 64.
Kroeger, H. (1967): In: *Endocrine Genetics,* pp. 55-66. Editor: S.G. Spickett. Endocrine Society Symposium, 15. Cambridge University Press, Cambridge.
Linzell, J.L. and Peaker, M. (1971): *Physiol. Rev., 51,* 564.
Lissitzky, S., Fayet, G. and Verrier, B. (1973): *FEBS Letters, 29,* 20.
Lubin, M. (1967): *Nature (Lond.), 213,* 451.
Novikoff, A.B. and Heus, M. (1963): *J. biol. Chem., 238,* 710.
Peaker, M., Phillips, J.G. and Wright, A. (1970): *J. Endocr., 47,* 123.
Quastel, M.R. and Kaplan, J.G. (1970): *Exp. Cell Res., 62,* 407.
Redshaw, M.R. and Lynch, S.S. (1974): *J. Endocr., 60,* 527.
Sapag-Hagar, M. and Greenbaum, A.L. (1974): *Europ. J. Biochem., 47,* 303.
Sapag-Hagar, M., Greenbaum, A.L., Lewis, D.J. and Hallowes, R.C. (1974): *Biochem. biophys. Res. Commun., 59,* 261.
Scatchard, G. (1949): *Ann. N.Y. Acad. Sci., 51,* 660.
Skou, J.C. (1965): *Physiol. Rev., 45,* 596.
Shiu, R.P.C. and Friesen, H.G. (1974a): *Biochem. J., 140,* 301.
Shiu, R.P.C. and Friesen, H.G. (1974b): *J. biol. Chem., 249,* 7902.
Taylor, J.C., Peaker, M. and Linzell, J.L. (1975): *J. Endocr., 65,* 26P.
Turkington, R.W. (1972): In: *Lactogenic Hormones,* pp. 111-127. Editors: G.E.W. Wolstenholme and Julie Knight. Churchill-Livingstone, Edinburgh.

INHIBITION OF PROLACTIN SECRETION BY DOPAMINE AND PIRIBEDIL (ET-495)*

ROBERT M. MACLEOD, HIROKO KIMURA and IVAN LOGIN

Departments of Medicine and Neurology, University of Virginia School of Medicine, Charlottesville, Va., U.S.A.

The inhibitory influence that the hypothalamus exerts on prolactin secretion has been recognized since Pasteels (1961) and Talwalker et al. (1963) observed that tissue extracts inhibited the in vitro secretion of prolactin. Since that time many efforts have been made to define the active prolactin-inhibiting component of hypothalamic extracts, and some early inconclusive investigations were conducted with the catecholamines. When Kanematsu et al. (1963) observed that the administration of reserpine caused lactation in normal ovariectomized rabbits but not in ovariectomized animals with hypothalamic lesions, the role of catecholamines became better established. More recently, Hökfelt (1967) observed that the tubero-infundibular dopaminergic neurons terminate in the hypothalamus adjacent to the portal blood vessels which perfuse the pituitary gland. Thus the possibility that the hypothalamic catecholamines may influence pituitary function was realized by Van Maanen and Smelik (1968) and they postulated that the monoaminergic tubero-infundibular neural system may inhibit prolactin secretion by a direct action at the pituitary gland.

The first evidence supporting the hypothesis that catecholamines directly inhibit prolactin secretion was reported by MacLeod (1969) and Birge et al. (1970). Koch et al. (1970) and Shaar and Clemens (1974) subsequently confirmed many of these findings. This topic has been recently reviewed (MacLeod, 1976).

RESULTS

Dopamine and dopaminergic blocking agents

Our results are based upon the measurement of prolactin by radioimmunoassay and the in vitro synthesis and release of the pituitary hormones through the use of radioactive amino acids (MacLeod and Abad, 1968; MacLeod and Lehmeyer, 1974; Augustine and MacLeod, 1975). Incorporation of ^{3}H-leucine into prolactin and growth hormone is readily observed when the pituitary gland is incubated in tissue culture Medium-199 for several hours. Aliquots of the incubation medium and pituitary homogenates were subjected to polyacrylamide gel electrophoresis, and the stained radioactive hormone bands were identified by conventional techniques. These radioactive bands of prolactin and growth hormone were counted by liquid scintillation

* This research was supported by U.S.P.H.S. Grant CA07535 from the National Cancer Institute.

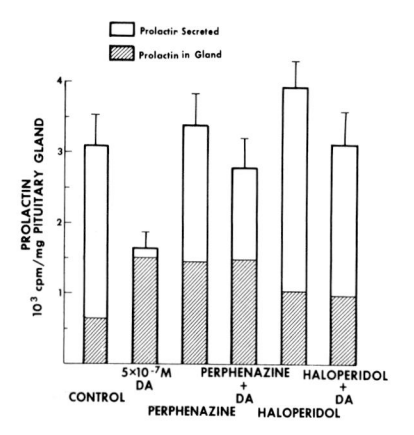

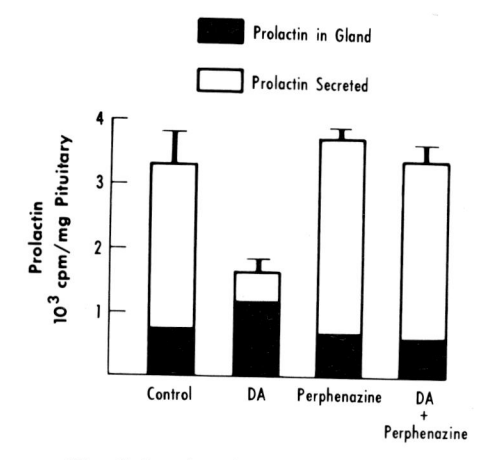

Fig. 1. Effect of perphenazine and haloperidol injection on the dopamine-mediated inhibition of prolactin secretion. Perphenazine and haloperidol (1.0 mg and 0.1 mg) were injected daily for 4 days. Four hemipituitary glands from different rats were incubated in the absence and presence of 5×10^{-7} M dopamine in 1 ml tissue culture medium-199 containing 10 μCi ^{3}H-leucine for 6 hours in 95% O_2—5% CO_2 (From MacLeod and Lehmeyer, *Endocrinology*, 1974, by courtesy of Charles C Thomas, Publisher.)

Fig. 2. In vitro blockade by perphenazine of the dopamine-induced inhibition of prolactin secretion. Four hemipituitary glands were incubated with 5×10^{-7} M dopamine (DA), 1.25×10^{-5} M perphenazine, or both substances.

methods. Most of the radioactive prolactin is released into the incubation medium, and little is retained in the pituitary glands of control rats. The introduction of 5×10^{-7} M dopamine into the incubation medium caused a decrease of 80-90% in labeled prolactin released into the incubation medium, and caused a concomitant increase in radioactive prolactin retained by the pituitary gland. Incubation in the presence of 5×10^{-8} M dopamine caused an approximate 50% decrease in radioactive prolactin released into the incubation medium.

It has long been known that injection of rats with reserpine, an agent which decreases brain catecholamine levels, or various neuroleptic agents such as perphenazine, haloperidol and pimozide causes an increase in serum prolactin levels (Ben-David et al., 1970; Koch et al., 1970; Lu et al., 1970; Dickerman et al., 1974; MacLeod and Lehmeyer, 1974). The results of previous investigations demonstrated that the injection of rats with these neuroleptic agents causes the pituitary gland to be refractory to the inhibitory effect that dopamine exerts on prolactin secretion. In the presence of only dopamine, in a concentration of 5×10^{-7} M, there was a 95% inhibition in prolactin secretion in vitro (Fig. 1). However, this inhibition was dramatically decreased when the rats were injected with perphenazine, haloperidol or pimozide prior to in vitro incubation of the pituitary gland with dopamine. Although the effect of these drugs to increase prolactin secretion could possibly be due to their action in the hypothalamus, additional data demonstrated conclusively that the drugs have a direct and potent action on the pituitary gland.

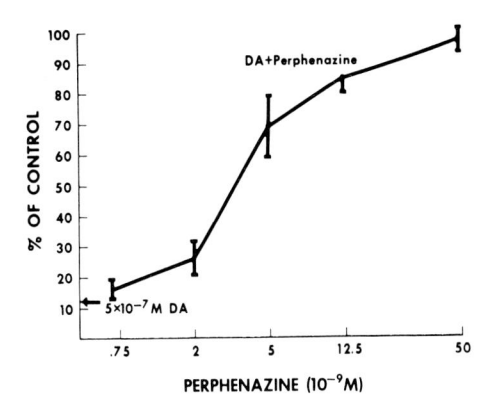

Fig. 3. Titration by perphenazine of the dopamine-mediated inhibition of prolactin secretion.

Fig. 4. Blockade by haloperidol of the dopamine-mediated inhibition of prolactin secretion.

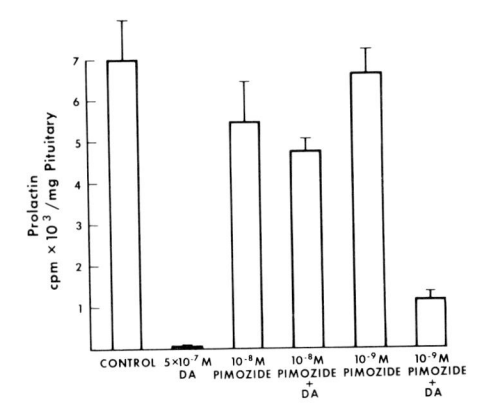

Fig. 5. Effect of pimozide and dopamine on the in vitro secretion of prolactin.

The data presented in Figure 2 show that, in vitro, dopamine caused a decrease in prolactin secretion and that perphenazine, in a concentration of 1.25×10^{-5} M, had no direct effect on the secretion of the hormone. However, concomitant incubation of perphenazine with dopamine completely blocked the normal inhibitory effect of the catecholamine and prolactin secretion. That perphenazine is extremely potent in overcoming the in vitro effects of dopamine is demonstrated by the results presented in Figure 3. An 88% decrease in prolactin secretion was produced by the presence of 5×10^{-7} M dopamine. Incubation of the pituitary with both 5×10^{-7} M dopamine and 2×10^{-9} M perphenazine significantly increased the amount of prolactin released, and 5×10^{-8} M perphenazine restored prolactin secretion to normal levels. Similar direct effects on the pituitary gland were observed with haloperidol (Fig. 4). Thus, although 5×10^{-9} M haloperidol alone had no significant effect on prolactin secretion, it completely blocked the inhibitory action that 5×10^{-7} M dopamine exerts on the pituitary gland. Similar results were also produced with pimozide (Fig. 5). The data show that 10^{-8} M pimozide almost completely overcame the inhibitory action of 5×10^{-7} M dopamine. At 10^{-9} M pimozide this agent was not as effective in overcoming the inhibitory effects of the catecholamine. The results of these studies conclusively demonstrate that either α-adrenergic or dopaminergic blocking agents successfully compete with dopamine for the biological sites on the pituitary gland which govern the secretion of prolactin.

Dopamine agonists

Apomorphine is known to have some of the pharmacological activities of dopamine and can stimulate dopamine receptors. Thus its effects on prolactin secretion have been studied by many investigators. It was originally thought to cause a decrease in prolactin secretion by acting at the hypothalamic level (Ojeda et al., 1974; Smalstig et al., 1974), but more recent evidence demonstrates that the alkaloid can act directly on the pituitary gland (MacLeod and Lehmeyer, 1974). Prior injection of rats with perphenazine, haloperidol or another dopaminergic blocking agent completely protects the prolactin-secreting cells from the in vitro inhibitory effects of apomorphine (Table 1). The data presented in Figure 6 demonstrate that a low concentration of apomorphine caused a 60% decrease in the amount of labeled prolactin released into the incubation medium. These data also demonstrate that low concentrations of perphenazine added to the incubation medium almost completely reversed the direct inhibitory action that apomorphine exerted on the pituitary gland.

One of the most thoroughly investigated series of pharmacological agents used to inhibit prolactin secretion are the derivatives of ergotamine (Meites and Clemens, 1972). The data presented in Figure 7 demonstrate that ergotamine tartrate, an α-adrenergic stimulating agent, significantly decreased the amount of prolactin released in vitro, although the drug had no direct effect on the secretion of growth hormone. Two synthetic derivatives of ergotamine, ergocryptine and ergocornine, were more potent inhibitors of prolactin secretion. It is of interest that at 10^{-5} M, the agents also cause a significant inhibition in the biosynthesis of growth hormone but had little direct effect on growth hormone release. Because these agents were potent inhibitors of prolactin secretion, we wanted to ascertain whether the prolactin-secreting cells were permanently effected by the in vitro presence of the drug. The data presented in Figure 8 demonstrate that ergocryptine blocked prolactin secretion for 28 hours of incubation. After this time, however, the pituitary gland regained its ability to secrete prolactin at rates similar to those observed in control tissue. That the effect of ergocryptine was reversible and that growth hormone secretion was not altered suggests the drug stimulated a specific biological receptor site. Ergocryptine probably inhibits

TABLE 1

Blockade by perphenazine injection of the apomorphine-mediated inhibition of prolactin release

Treatment	Prolactin (c.p.m./mg pituitary)		
	Incubation medium	Glands	Total
Control	1454 ± 149	531 ± 88	1984 ± 188
Apomorphine 6.4 x 10^{-8} M	546 ± 190*	992 ± 212	1544 ± 407
Perphenazine †	3295 ± 451*	1059 ± 110*	4354 ± 383*
Perphenazine † + apomorphine	3565 ± 174*	1026 ± 59*	4591 ± 125*

† Daily s.c. injections of 1.0 mg perphenazine for 2 days; 3 flasks/group, each containing 4 hemipituitary glands.
*P < 0.01.

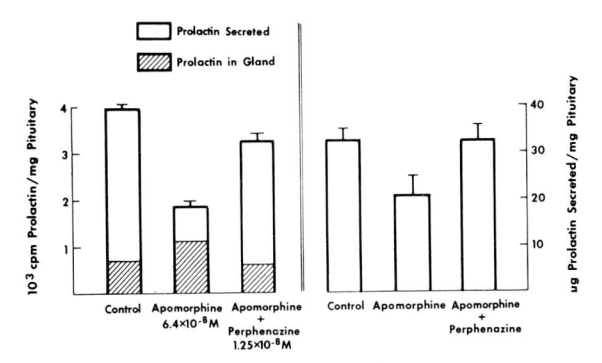

Fig. 6. Effect of perphenazine on the apomorphine-mediated inhibition of prolactin release. The incorporation of ^{3}H-leucine (left side); the amount of radioimmunoassayable prolactin (right side).

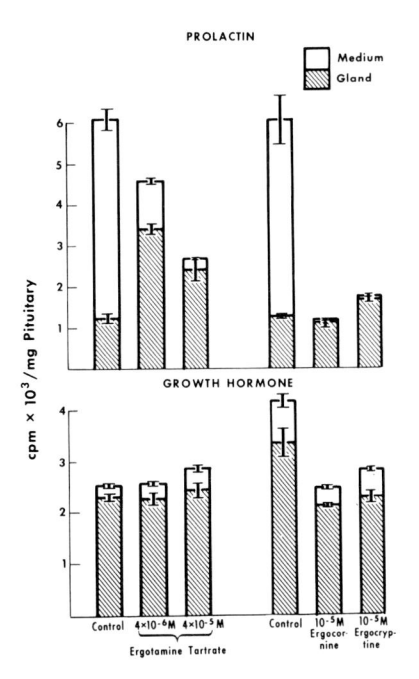

Fig. 7. The in vitro effect of ergot derivatives on the synthesis and release of prolactin and growth hormone.

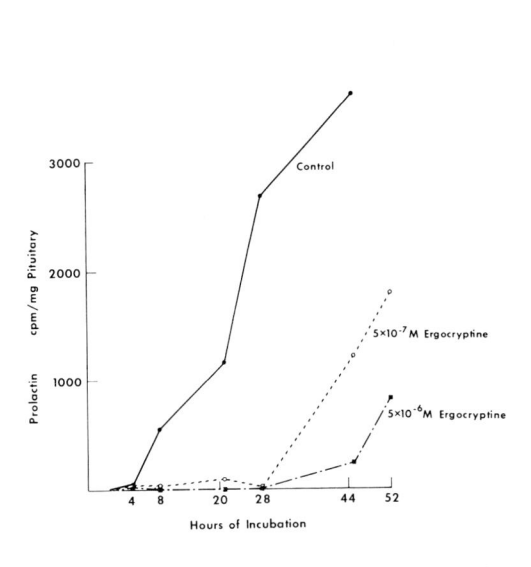

Fig. 8. Inhibition of prolactin release by ergocryptine.

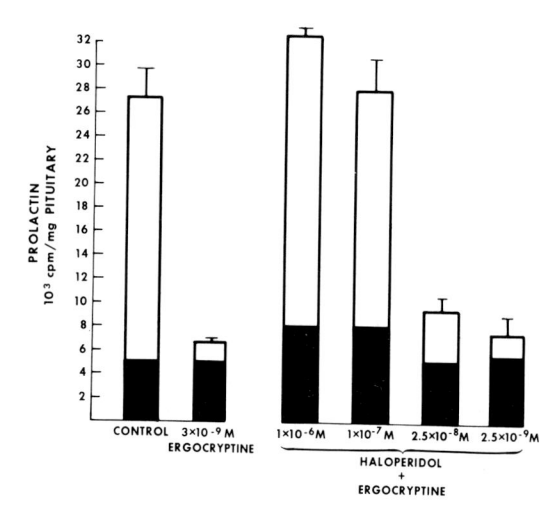

Fig. 9. Blockade of the ergocryptine-mediated inhibition of prolactin release by haloperidol.

prolactin secretion by stimulating the α-adrenergic or dopaminergic receptor sites (Fig. 9). Ergocryptine at 3×10^{-9} M caused a profound inhibition of prolactin secretion. However, the coincubation of ergocryptine with varying concentrations of haloperidol or any of the other known dopaminergic blocking agents produced a graded blockade of the inhibition that ergocryptine produces on prolactin secretion.

ET-495 or Piribedil

In the past few years several antiparkinsonian drugs have appeared which are dopamine agonists. Among the drugs used for this purpose is Piribedil (ET-495), the structure of which is shown in Figure 10. Campbell et al. (1973) demonstrated that ET-495 is metabolized to the catechol form, or S-584. It is uncertain whether the parent compound or S-584 is the active dopaminergic agonist (Consolo et al., 1975). Because of this intriguing chemical structure and because of the dopaminergic activity, we initiated a study to determine whether these compounds were effective in regulating pituitary function, specifically prolactin and growth hormone secretion. Female rats were injected intraperitoneally with ET-495 and the animals sacrificed at varying periods of time (Fig. 11). The data indicate that within 30 minutes after the injection of ET-495 (40 mg/kg) the serum prolactin level fell to approximately 40% of control values and remained suppressed for at least 180 minutes thereafter. Following incubation of the pituitary glands of these animals with radioactive leucine, it was found that within 30 minutes of the in vivo administration of ET-495, the in vitro secretion of ^{3}H-prolactin was almost completely suppressed. It was of great interest to observe that the injection of S-584 (10 mg/kg), the catechol form of the drug, was much less effective in reducing the serum prolactin levels or in altering the in vitro release of ^{3}H-prolactin (Table 2).

A series of experiments was conducted to determine whether ET-495 and S-584 had any direct in vitro effect on prolactin production by the pituitary (Fig. 12). The presence of 5×10^{-7} M ET-495 was extremely effective in decreasing ^{3}H-prolactin or radioimmunoassayable prolactin release, while lower concentrations of the agent were less effective. S-584, on the other hand, had no effect on prolactin release, even when

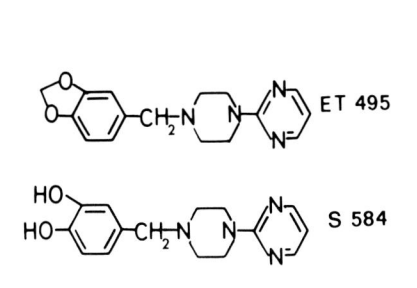

Fig. 10. Chemical structure of Piribedil (ET-495) and its metabolite S-584.

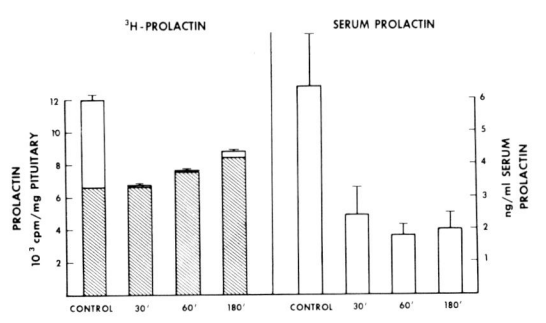

Fig. 11. Effect of ET-495 injection on prolactin secretion. 40 mg/kg body weight of ET-495 injected intraperitoneally. Animals killed at times indicated.

TABLE 2

Effect of S-584 on serum prolactin and the in vitro secretion of ^{3}H-prolactin

Time after injection (min)	Serum prolactin (ng/ml)	^{3}H-prolactin secreted (c.p.m./mg pituitary)
0	11.70 ± 1.98	8281 ± 485
2	10.01 ± 2.11	4868 ± 436*
5	16.68 ± 3.13	5572 ± 278*
10	35.90 ± 14.9	6879 ± 604
20	16.78 ± 3.43	4488 ± 241*
60	5.35 ± 0.74	5269 ± 464*
120	11.02 ± 1.73	6153 ± 910

10 mg/kg S-584 injected intraperitoneally.
* $P < 0.01$.

used in much larger concentrations. These data indicate that the parent compound, and not the metabolite S-584, is the active form of the dopaminergic agonist which affects pituitary hormone secretion. That ET-495 acts via stimulation of the dopaminergic receptor to inhibit prolactin secretion is suggested by the data presented in Figure 13. Once again, 5×10^{-8} M ET-495 caused an approximate 60% decrease in the amount of newly synthesized prolactin released into the incubation medium. However, after coincubation with either 5×10^{-9} M haloperidol or 100 nM pimozide, the secretion of ^{3}H-prolactin was restored to near normal levels. These data provide added support for the concept that ET-495 stimulates a specific dopaminergic receptor in the pituitary which governs prolactin secretion. In contrast to the marked effects that ET-495 has on prolactin secretion, this agent, and its metabolite S-584, are without effect on growth hormone secretion, either by the direct effect on the pituitary gland or after systemic injection of the agents.

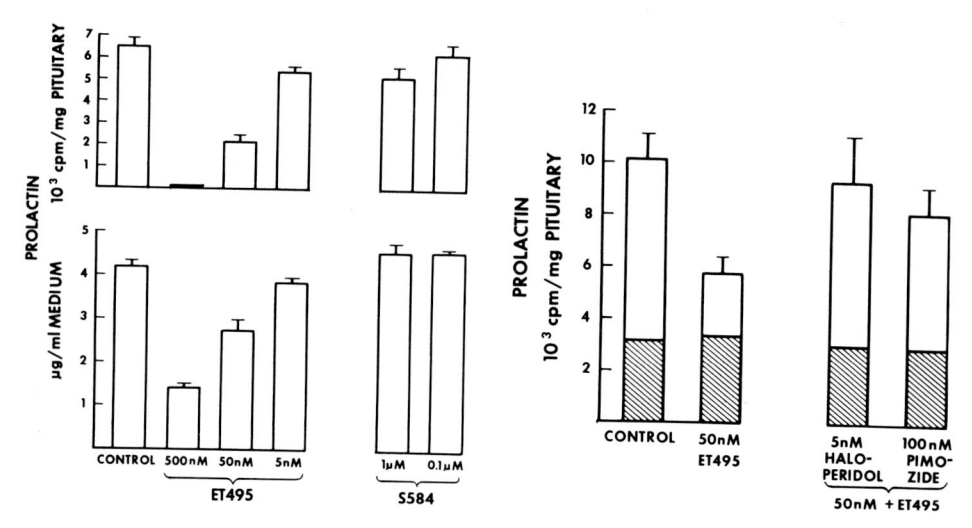

Fig. 12. In vitro effects of ET-495 and S-584 on prolactin secretion.

Fig. 13. Blockade by haloperidol and pimozide of the in vitro inhibitory effects of ET-495 on prolactin secretion.

Pituitary dopamine receptors

Much of the data presented to this point suggest very strongly that dopamine reacts with a dopamine receptor site on the pituitary to inhibit the secretion of prolactin, both in vivo and in vitro. The data also suggest that the dopaminergic agonists such as apomorphine and the ergot derivatives bind to a very similar if not identical receptor on the pituitary gland and thereby affect prolactin secretion. In order to more fully comprehend the mechanism of action of dopamine and its analogs, we have undertaken a systematic investigation of the dopamine receptor in the pituitary gland. This work has principally been carried out by Dr. Hiroko Kimura of our laboratory.

The anterior pituitary was homogenized in 0.25 M sucrose and the subcellular fractions were separated by centrifugation. The binding of ^{3}H-dopamine to these fractions was studied by incubating 50 μg of pituitary protein with 0.15 μM ^{3}H-dopamine for 20 minutes at 25°C. In order to minimize non-specific binding of ^{3}H-dopamine, 0.1 mM ascorbate was added to the incubation medium. The dopamine bound to the pituitary fraction was separated by Millipore filtration. Specific binding was determined by subtracting the net radioactivity obtained with excess unlabeled dopamine from the net radioactivity found in the presence of tracer dopamine alone. Although all subcellular fractions bound ^{3}H-dopamine, the 100,000 × g precipitate, or the microsomal fraction contained the most radioactivity (Table 3). For that reason the microsomal fraction was used in all of the following experiments without further purification.

The data presented in Figure 14 show the time course of ^{3}H-dopamine binding to the microsomal fraction in the absence and presence of 0.1 mM ascorbate. This concentration of ascorbate was found to maximally inhibit ^{3}H-dopamine binding. The data indicate that the binding of ^{3}H-dopamine is rapid and complete within 10 minutes. This bound dopamine, however, could be displaced from the pituitary binding sites by the addition of excess unlabeled dopamine. The binding of ^{3}H-dopamine was found to be complete at 1.2×10^{-6} M and half-maximal at 4.5×10^{-7} M (Fig. 15).

TABLE 3

^{3}H-Dopamine binding to the subcellular fractions of the anterior pituitary gland of female rats

Fraction	Specific binding (pmole/mg protein)
Homogenate	1.54 ± 0.07
800 × g precipitate	0.50 ± 0.01
10,000 × g precipitate	0.73 ± 0.01
105,000 × g precipitate	10.24 ± 0.24
105,000 × g supernatant	0.14 ± 0.004

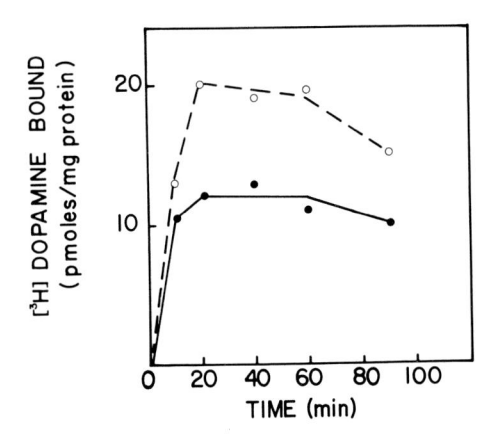

Fig. 14. 50 μg of the microsomal fraction were incubated for various times with 1.5×10^{-7} M ^{3}H-dopamine and 0.1 M Tris, pH 7.4, in the absence and presence of 0.1 mM ascorbate at 25°C. —●—●—, in the presence of 0.1 mM ascorbate; —o—o—, in the absence of 0.1 mM ascorbate.

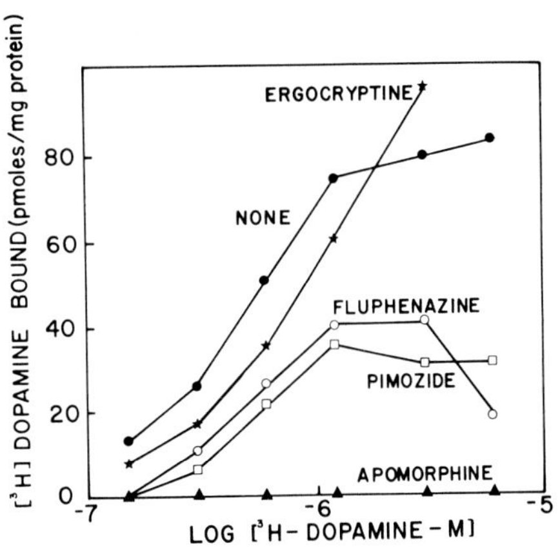

Fig. 15. 50 μg of the microsomal fraction were incubated with various concentrations of ^{3}H-dopamine in the absence and presence of 3×10^{-6} M apomorphine, or 3×10^{-6} M pimozide, 3×10^{-6} M fluphenazine or 3×10^{-6} M ergocryptine in 0.1 M Tris, pH 7.4 for 20 minutes at 25°C.

The addition of 3×10^{-6} M apomorphine, a dopamine agonist, completely inhibited the binding of ^{3}H-dopamine at any concentration to its receptor. Likewise, the addition of the dopaminergic antagonists pimozide and fluphenazine completely blocked the binding of ^{3}H-dopamine to its receptor. At higher concentrations of ^{3}H-dopamine, these antagonists partially blocked its binding. Ergocryptine, at low concentrations, partially inhibited the binding of ^{3}H-dopamine. A Scatchard plot analysis of the data in the absence and presence of 3×10^{-6} M pimozide or haloperidol suggests that the anterior pituitary contains a single type of binding site for dopamine and that dopaminergic antagonists inhibit competitively the binding of ^{3}H-dopamine (Fig. 16).

We extended these studies to investigate the effect of ET-495 and S-584 on the binding of ^{3}H-dopamine to the microsomal fraction. Although ET-495 is a potent inhibitor of prolactin secretion, it is almost entirely devoid of activity to block the

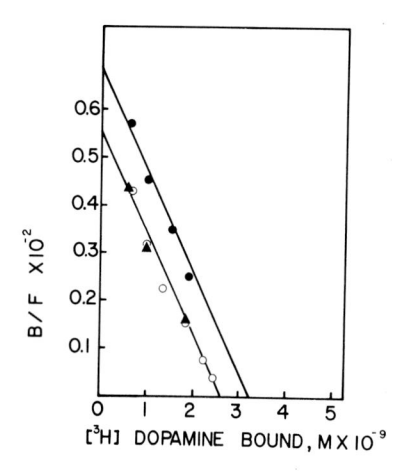

Fig. 16. Scatchard plot of the data presented in Figure 2 on the binding of ³H-dopamine to the microsomal fraction from pituitary. —●—●—●—, no addition; —▲—▲—, 3×10^{-6} M pimozide; —o—o—, 3×10^{-6} M haloperidol.

Fig. 17. 50 µg of the microsomal fraction were incubated with various amounts of ³H-dopamine in the absence and presence of 3×10^{-6} M ET-495 or 3×10^{-6} M S-584 for 20 minutes at 25°C.

binding of ³H-dopamine (Fig. 17). In contrast, the catechol metabolite, S-584, although it is devoid of significant prolactin-inhibitory activity, is a good blocker of ³H-dopamine binding to the microsomal fraction. Although interpretation of these latter data must be undertaken with caution, all of the data strongly suggest that the binding of ³H-dopamine to its pituitary receptor in the presence of ascorbate is a specific binding phenomenon because several known dopamine agonists and antagonists inhibit the binding of the catecholamine. Further work is required to describe in more detail the characteristics of the pituitary dopamine receptor.

CONCLUSIONS

It is increasingly evident that hypothalamic catecholamines, and specifically dopamine, have an important action on the pituitary to inhibit the secretion of prolactin. It is known that the administration of L-dopa causes a decrease in serum prolactin levels after decarboxylation of the catecholamine to dopamine. The administration of dopamine or other catecholamines is ineffective presumably because of their inability to penetrate the blood-brain barrier. However, these agents are excellent inhibitors of the in vitro secretion of prolactin, and their action is blocked directly at the pituitary level by the presence of known dopaminergic blocking agents, specifically perphenazine, haloperidol and pimozide. The known dopaminergic agonists, apomorphine and the ergot alkaloids, also directly inhibit the secretion of prolactin in vitro, and their action is likewise blocked by the dopaminergic blocking agents. Piribedil, or ET-495, is also an effective blocker of prolactin secretion, but its specific mechanism of action is unknown at present. There is suggestive evidence, however, that its action is similar to that of dopamine because dopaminergic blocking agents neutralize its ability to inhibit prolactin secretion. Because of its duration of action and lack of adverse side effects, it may be an effective clinical agent to inhibit the secretion of prolactin.

ACKNOWLEDGMENT

The authors would like to acknowledge the collaborative efforts of Ronald C. Pace.

REFERENCES

Andén, N.E., Butcher, S.G., Corrodi, H., Fuxe, K. and Ungerstedt, U. (1970): *Europ. J. Pharmacol., 11,* 303.
Augustine, E.C. and MacLeod, R.M. (1975): *Proc. Soc. exp. Biol. (N.Y.), 150,* 551.
Ben-David, M., Danon, A. and Sulman, F.G. (1970): *Neuroendocrinology, 6,* 336.
Birge, C.A., Jacobs, L.S., Hammer, C.T. and Daughaday, W.H. (1970): *Endocrinology, 86,* 120.
Campbell, D.B., Jenner, P. and Taylor, A.R. (1973): In: *Advances in Neurology, Vol. III,* p. 199. Editor: D.B. Calne. Raven Press, New York.
Consolo, S., Fanelli, R., Garattini, S., Ghezzi, D., Jori, A., Ladinsky, H., Mare, V. and Samanin, R. (1975): In: *Advances in Neurology, Vol. IX,* p. 257. Editors: D.B. Calne, T.N. Chase and A. Barbeau. Raven Press, New York.
Dickerman, S., Kledzik, G., Gelato, M., Chen, H.J. and Meites, J. (1974): *Neuroendocrinology, 15,* 10.
Hökfelt, T. (1967): *Brain Res., 5,* 121.
Kanematsu, S., Hilliard, J. and Sawyer, C.H. (1963): *Acta endocr. (Kbh.), 44,* 467.
Kimura, H. and MacLeod, R.M. (1975): In: *Program of the Endocrine Society, 57th Annual Meeting,* p. 87.
Kimura, H. and MacLeod, R.M. (1975): In: *Program, VI International Congress of Pharmacology, Helsinki,* p. 88.
Koch, Y., Lu, K.H. and Meites, J. (1970): *Endocrinology, 87,* 673.
Lu, K.H., Amenomori, Y., Chen, C.L. and Meites, J. (1970): *Endocrinology, 87,* 667.
MacLeod, R.M. (1969): *Endocrinology, 85,* 916.
MacLeod, R.M. (1976): In: *Frontiers in Neuroendocrinology — 1975.* Editors: W.F. Ganong and L. Martini. Raven Press, New York.
MacLeod, R.M. and Abad, A. (1968): *Endocrinology, 83,* 799.
MacLeod, R.M. and Lehmeyer, J.E. (1974): *Endocrinology, 94,* 1077.
Meites, J. and Clemens, J.A. (1972): *Vitam. Horm., 30,* 165.
Ojeda, S.R., Harms, P.G. and McCann, S.M. (1974): *Endocrinology, 95,* 1694.
Pasteels, J.L. (1961): *C.R. Acad. Sci. (Paris), 253,* 2140.
Shaar, C.J. and Clemens, J.A. (1974): *Endocrinology, 95,* 1202.
Smalstig, E.B., Sawyer, B.D. and Clemens, J.A. (1974): *Endocrinology, 95,* 123.
Talwalker, P.K., Ratner, A. and Meites, J. (1963): *Amer. J. Physiol., 205,* 213.
Van Maanen, J.H. and Smelik, P.G. (1968): *Neuroendocrinology, 3,* 177.

AUTHOR INDEX

Ahrén, K., 94
Aiello, A., 50
Albertsson-Wikland, K., 94
Arezzini, C., 345
Aynsley-Green, A., 286

Bancroft, F.C., 84
Bazzarre, T.L., 261
Bennett, L.L., 116
Bewley, T.A., 14
Bhattarai, Q., 104
Bialecki, H., 141
Blizzard, R.M., 261
Bolander, F.F., 315
Bornstein, J., 41
Botalla, L., 236
Bryant, G.D., 407, 414

Caufriez, A., 396
Chamley, W.A., 407
Chillemi, F., 50
Chiodini, P.G., 236
Cocchi, D., 236
Compton, P.J., 222
Curchod, A., 369
Curry, D.L., 116

Daughaday, W.H., 169
Davies, R.V., 1
Decedue, C.J., 190
Delvoye, P., 396
D'Ercole, A.J., 190
Drezner, M.K., 202

Efendić, S., 216
Ellis, S., 75
Enberg, G., 178

Falconer, I.R., 433
Felber, J.P., 369
Fellows, R.E., 315
Fioretti, P., 345
Forsyth, I.A., 422
Foushee, D.B., 190

Fruschelli, C., 345
Fryklund, L., 156, 178

Genazzani, A.R., 345
Gorden, P., 127
Gospodarowicz, D., 141
Grindeland, R.E., 75
Gupta, P.D., 104

Hall, K., 156, 178
Handwerger, S., 315
Hanjan, S.N.S., 104
Hart, I.C., 422
Herington, A.C., 169
Hökfelt, T., 216
Hooley, R.D., 407
Hurley, T.W., 315
Huseman, C.A., 261

Isaksson, O., 94

Johanson, A.J., 261
Joostens-Defleur, V., 396

Kahn, C.R., 127
Kidwai, Z., 104
Kimura, H., 443
Knazek, R.A., 386
Kostyo, J.L., 33, 94
Kowalski, K., 327
Kwok, S.C.M., 414

Laron, Z., 297
Lawrence, J.H., 312
Lazarus, L., 222
Lebovitz, H.E., 202
Leopold, N.A., 252
Letchworth, A.T., 380
Lewis, U.J., 64
L'Hermite, M., 396
Li, C.H., 14
Liuzzi, A., 236
Login, I., 443

Luft, R., 216

MacLeod, R.M., 443
Magrini, G., 369
McMurtry, J.P., 414
Méan, F., 369
Megyesi, K., 127
Mehrotra, N.N., 104
Mills, J.B., 33
Mochizuki, M., 334
Moran, J.S., 141
Morikawa, H., 334
Müller, E.E., 236, 345

Neelon, F.A., 202
Neri, P., 345
Neville Jr, D.M., 127

Ohga, Y., 334
Olgiati, V.R., 50

Panerai, A.E., 236
Pecile, A., 50
Pertzelan, A., 297
Peterson, S.M., 64
Phillips, L.S., 169
Podolsky, S., 252
Prader, A., 286

Reagan, C.R., 33
Reilly, T.J., 75
Robyn, C., 396
Roth, J., 127
Rudman, D., 33

Saxena, R., 104
Schneider, A.B., 327
Secchi, C., 236
Sherwood, L.M., 327
Sievertsson, H., 156
Silvestrini, F., 236
Singh, R.N.P., 64
Skottner, A., 156
Skyler, J.S., 386

Author Index

Smythe, G.A., 222
Sussman, P.M., 84

Takano, K., 178
Talwar, G.P., 104
Tojo, S., 334
Tushinski, R.J., 84

Underwood, L.E., 190

Van den Brande, J.L., 271
Van Wyk, J.J., 190
VanderLaan, W.P., 64
Varma, M.M., 261
Vekemans, M., 396

Wallis, M., 1
Wilhelmi, A.E., 33

Yang, S.H., 75

Zachmann, M., 286

SUBJECT INDEX

Prepared by L.M. Boot, D.Sc., Amsterdam

abortion, threatened
 see threatened abortion
acetyl CoA carboxylase
 HGH fragments, 41-49
achondroplasia
 somatomedin, plasma, 281-285
acromegaly
 big somatomedin A and B in serum, 184-189
 HGH secretion, 2-bromo-α ergocryptine, 222-235
 HGH secretion, cyproheptadine, 222-235
 HGH secretion, L-dopa, 222-235
 HGH secretion, dopaminergic drugs, 236-251
 HGH secretion, insulin hypoglycemia, 222-235
 HGH secretion, melatonin, 222-235
 HGH secretion, phentolamine, 222-235
 HGH secretion, L-tryptophan, 5-hydroxy-L-tryptophan, 222-235
 hyperparathyroidism, pheochromocytoma, 224
 hypophysis tumor, heavy particle radiation, 246 cases, 312-314
 hypophysis tumor, hormone production in vitro, 390-395
 NSILA-s, plasma, 127-140
 plasma GH, 75-83
 somatomedin A and B in serum, 180-189
 somatomedin, plasma, 278-285
actinomycin D
 amino acid and sugar transport, diaphragm, rat, 95
adenohypophysis
 dopaminergic compounds, HGH secretion, acromegaly, 245-251
 HGH and RGH, comparison with plasma hormone, 75-83
 HGH secretion, 64-74
 HGH secretion, age, 261-270
 hormone production, artificial capillaries, 386-395
 organ culture, GH synthesis, rat, 6-8
 prolactin secretion, regulation, rat, 443-453
adenosine
 cartilage metabolism, somatomedin, chicken embryo, rat, 202-215
adenosine 3′,5′-cyclic monophosphate
 see AMP, cyclic
adenyl cyclase
 adipose tissue, human placental lactogen, rat, 334-344
 cartilage metabolism, somatomedin, chicken embryo, rat, 202-215
adipose tissue
 GH, acute stimulatory effect, rat, 94-103
 metabolism, human placental lactogen, rat, 334-344
 somatomedin B, metabolic effects, rat, 188
adrenal cortex carcinoma
 hypoglycemia, NSILA-s, 130, 131
adrenalin
 diaphragm, cyclic AMP, rat, 99-103
α-adrenergic blockade
 HGH secretion, 222-235
age
 amino acid and sugar transport, diaphragm, rat, 95-99
 fetus, somatomedin C receptor, guinea pig, 190-201
 HGH secretion, 261-270
 male genital tract, human placental lactogen, prolactin, HGH, mouse, 352-358
 pregnancy, plasma human lactogen, 380-385
 prolactin, human serum, 396-406
 somatomedin A and B in serum, 180-189
 somatomedin, human plasma, 272-285
 thymus, GH, rat, 104-115
aging
 HGH secretion, 261-270
albumin
 sodium transport, mammary gland, rabbit, 433-442
allelic polymorphism
 BGH, 4
amenorrhea
 prolactin, 396-406
amino acid
 sequence, bovine and rat GH, 1-13
 sequence, HGH, plasmin, 14-32
amino acid transport
 cartilage, somatomedin, cyclic AMP, chicken embryo, rat, 202-215
 GH, rat, 94-103
α-aminoisobutyric acid
 cartilage, somatomedin, cyclic AMP, chicken embryo, rat, 202-215
 transport, diaphragm, GH, rat, 94-103
amniotic fluid
 somatomedin A and B, human, 187-189
AMP
 cartilage metabolism, somatomedin, chicken embryo, rat, 202-215
AMP, cyclic
 adipose tissue, human placental lactogen, rat, 334-344
 protein synthesis, membrane

Subject Index

transport, diaphragm, rat, 99-103
somatomedin, cartilage, chicken embryo, rat, 202-215
anabolic steroid
hypopituitarism, growth, 286-296
androgen
HGH, hypopituitarism, growth, 286-296
androgenicity
testosterone, HGH, hypopituitarism, 286-296
anorchia
HGH, testosterone, growth, 286-296
anorexia nervosa
NSILA-s, plasma, 127-140
antibody formation
GH, rat, 104-115
apomorphine
GH secretion, 236-251
HGH secretion, 222-235
HGH secretion, acromegaly, 243-251
prolactin secretion, rat, 443-453
arginine
glucagon release, somatostatin, rat, 218-221
HGH secretion, age, 261-270
insulin release, somatostatin, rat, 218-221
arginine tolerance test
HGH secretion, blood sugar, Huntington disease, 252-260
Argonz-Del Castillo syndrome 396-406
axillary hair
hypopituitarism, HGH, testosterone, 292-296
BGH
carboxymethylated, activity in dwarf mouse, 4-6
chemistry, 1-13
insulin release, glucose, hypophysectomy, rat, 116-126
liver, somatomedin generation, rat, 169-177
methionine residue modification, 4-6
placental lactogen, 315-326
BGH fragments
synthesis, 50-63
big somatomedin A
see also somatomedin A
human serum, 184-189
big somatomedin B
see also somatomedin B
human serum, 184-189
bioassay
GH in rat and human plasma, 75-83
HGH, derivatives and fragments, plasmin, 14-32

HGH fragments, plasmin, 33-40
human placental lactogen and big human placental lactogen, 327-333
somatomedin A and B, 156-168
somatomedin, human, 271-285
somatomedin, rat, 169-177
bleeding
pregnancy, plasma human placental lactogen, 380-385
blood
prolactin, rabbit, 433-442
blood plasma
NSILA-s binding, human, 129-140
somatomedin, rat, human, 169-177
blood pressure
pregnancy, plasma human placental lactogen, 380-385
blood serum
cartilage, somatomedin, cyclic AMP, chicken embryo, rat, 202-215
somatomedin A and B, human, 180-189
somatomedin C, mammals, 190-201
blood sugar
see also glucose
arginine tolerance test, HGH secretion, Huntington disease, 252-260
glucose tolerance test, HGH secretion, Huntington disease, 252-260
HGH fragments, mouse, 41-49
NSILA-s, human, 127-140
body weight
HGH and OGH fragments, hypophysectomized rat, 50-63
HGH secretion, age, 261-270
BOL
HGH secretion, 222-235
prolactin release, 223
bone age
hypopituitarism, HGH, testosterone, 286-296
hypopituitary dwarfism, HGH therapy, 301-311
bovine placental lactogen
isolation, characterization, 315-326
bovril
HGH secretion, age, 261-270
brain
catecholamine, GH secretion, 236-251
catecholamine, HGH secretion, 222-235
mitogenic factor, fibroblast growth factor, cow, 141-155
breast
see mammary gland

breast cancer
prolactin, 396-406
breast feeding
prolactin, relaxin, 409
2-bromo-α-ergocryptine
GH secretion, 236-251
HGH secretion, 222-235
HGH secretion, acromegaly, 243-251
hyperprolactinemia, 404-406
2-bromo-D-lysergic acid diethylamine
see BOL
Buscopan compositum
threatened abortion, premature labor, 10 cases, 359-368
calcium excretion
HGH secretion, age, 266-270
calcium ion
somatostatin, glucose, insulin release, rat, 116-126
calcium uptake
adipose tissue, somatomedin B, rat, 188
cancer
paraneoplastic syndrome, hypoglycemia, NSILA-s, 127-140
cartilage
rib, thymidine incorporation, HGH fragment, rat, 33-40
^{35}S incorporation, NSILA-s, rat, chicken, 127
^{35}S incorporation, somatomedin A and B, rat, 165-168
^{35}S incorporation, somatomedin, rat, 169-177
somatomedin, cyclic AMP, chicken embryo, 202-215
somatomedin bioassay, pig, 271-285
catecholamine
brain, GH secretion, 236-251
prolactin secretion, rat, 443-453
CB-154
see 2-bromo-α-ergocryptine
cell culture
hypophysis, hormone production, artificial capillaries, 386-395
hypophysis tumor, GH synthesis, rat, 84-93
somatomedin assay, 272
cell-free system
GH synthesis, rat, 84-93
cell growth
fibroblast, fibroblast growth factor, 141-155
myoblast, fibroblast growth factor, 141-155
cell membrane
GH synthesis, rat, 84-93
lymphocyte, GH receptor, human, 111-115

Subject Index

thymocyte, GH receptor, rat, 106-115
cell proliferation
fibroblast growth factor, various cell types, 141-155
mitogenic factor in brain and hypophysis, cow, 141-155
myeloblast, fibroblast growth factor, 147-155
NSILA-s, NSILA-s receptor, liver, rat, 127-140
childhood
hypopituitarism, HGH, testosterone, growth, 286-296
hypopituitary dwarfism, HGH therapy, 297-311
somatomedin, plasma, 273-285
chlorpromazine
HGH secretion, 222-235
chondrocyte
fibroblast growth factor, 141-155
somatomedin, cyclic AMP, chicken embryo, 202-215
choriocarcinoma
hormone production in vitro, 386-395
chorionic somatomammotropin
see lactogen
circular dichroism
HGH, plasmin derivatives and fragments, 21-32
CNS
dopaminergic compounds, HGH secretion, acromegaly, 245-251
HGH secretion, Huntington disease, 17 cases, 252-260
congenital malformation
somatomedin, plasma, dysmorphic children, 281-285
corpus luteum
prolactin, women, 396-406
corticosteroid secretion
insulin hypoglycemia, age, human, 261-270
corticotherapy
somatomedin, plasma, 279-285
cortisol
fibroblast growth factor, fibroblasts in vitro, 144-155
somatomedin, children, 274
craniopharyngioma
growth, GH, somatomedin, 174
growth, somatomedin, 276-285
growth, surgery, HGH, androgen, 286-296.
creatinine excretion
HGH secretion, age, 266-270
crop sac assay
HGH, proteinase, 72-74
Cushing syndrome
somatomedin, plasma, 279-285
cyproheptadine

GH secretion, 236-251
HGH secretion, acromegaly, 222-235
cytoplasm
GH synthesis, rat, 84-93
dehydroepiandrosterone
plasma, normal and pathological pregnancy, 359-368
dehydroepiandrosterone sulfate
plasma, normal and pathological pregnancy, 369-379
plasma, threatened abortion, human placental lactogen treatment, 359-368
diabetes mellitus
NSILA-s, human, 133
pregnancy, plasma human placental lactogen, 380-385
somatomedin, human plasma, 279-285
diaphragm
amino acid and sugar transport, GH, rat, 94-103
glucose uptake, HGH fragments, rat, 41-49
diet
HGH secretion, age, 261-270
disulfide bond
HGH, derivatives and fragments, plasmin, 22-32
disulfide dimer
HGH, 64-74
dithiothreitol
HGH, derivatives and fragments, plasmin, 22-32
diurnal rhythm
prolactin, human, 396-406
somatomedin, human plasma, 273-285
DNA synthesis
cell culture, NSILA-s, 127
fibroblast growth factor, 144-155
spleen and thymus cell, GH, rat 104-115
L-dopa
GH secretion, 236-251
HGH secretion, acromegaly, 222-235, 243-251
HGH secretion, age, 261-270
HGH secretion, glucose tolerance test, Huntington disease, 252-260
prolactin release, acromegaly, 222-235
prolactin release, Huntington disease, 252-260
dopamine
brain, HGH secretion, 222-235
brain, HGH secretion, Huntington disease, 252-260
GH secretion, 236-251
prolactin secretion, rat, 443-453
dopamine antagonist

prolactin secretion, rat, 443-453
dopamine-β-hydroxylase
GH secretion, 236-251
dopamine receptor
adenohypophysis, prolactin secretion, rat, 443-453
GH receptor, 236-251
prolactin, human, 396-406
dwarfism
see also Laron dwarfism
GH deficiency, thymus, mouse, 104
hypopituitarism, HGH therapy, 297-311
hypopituitarism, thymus, mouse, 104
male genital tract, human placental lactogen, mouse, 345-358
EGH
amino acid sequences, 8-13
emotional deprivation
short stature, somatomedin, plasma, 276-285
endothelial cell
fibroblast growth factor, 141-155
epidermal growth factor
fetal lung growth, 200
epinephrine
see adrenalin
ergocryptine
prolactin secretion, rat, 443-453
ergot derivative
prolactin secretion, rat, 443-453
erythrocyte cell membrane
somatomedin C receptor, 190-201
estradiol
plasma, human placental lactogen, human pregnancy, 352-358
plasma, normal and pathological pregnancy, 369-379
prolactin, human, 396-406
relaxin release, sheep, 409-413
somatomedin, plasma, 279-285
estriol
plasma, normal and pathological pregnancy, 359-368
estrogen
prolactin, human, 396-406
estrone
plasma, normal and pathological pregnancy, 369-379
estrous cycle
relaxin, sheep, 412
ethinylestradiol
prolactin, human, 396-406
evolution
GH, amino acid sequences, 8-13
fasting
amino acid and sugar transport, diaphragm, rat, 95-99

Subject Index

hypoglycemia, nonpancreatic tumors, NSILA-s, 130, 131
liver, somatomedin generation, rat, 172
fat cell
   insulin binding, HGH fragments, 41-49
fat cell membrane
   somatomedin C receptor, 190-201
fatty acid metabolism
   mother and fetus, human placental lactogen, rat, 334-344
fatty acid synthesis
   HGH fragments, 41-49
fertility
   prolactin, women, 396-406
fetal distress
   plasma human placental lactogen, 380-385
   plasma human placental lactogen and sex steroids, 369-379
fetus
   growth retardation, plasma human placental lactogen, 380-385
   growth retardation, plasma human placental lactogen and sex steroids, 369-379
   metabolism, growth, human placental lactogen, rat, 334-344
fibroblast
   fibroblast growth factor, 141-155
follicle stimulating hormone
   see FSH
Forbes-Albright syndrome 396-406
free fatty acid
   human placental lactogen, pregnant rat, 334-344
FSH
   prolactin, human, 396-406
galactorrhea
   prolactin, 396-406
genetics
   hypopituitary dwarfism, HGH therapy, 297-311
   molecular evolution, GH, prolactin, 8-13
GH
   insulin release, glucose, rat, 116-126
   liver, somatomedin generation, rat, 169-177
   mammary gland, ruminants, 422-432
   mammary gland, sodium transport, rabbit, 433-442
   molecular evolution, 8-13
   somatomedin A and B, 156-168
   somatomedin, cyclic AMP,

cartilage, chicken embryo, rat, 202-215
thymus and lymphoid cells, rat, human, 104-115
GH cell
   hypophysis tumor, acromegaly, 248-251
GH deficiency
   male genital tract, human placental lactogen, dwarf mouse, 345-358
GH inhibiting factor
   see somatostatin
GH receptor
   lymphocyte, human, 111-115
   thymus cell membrane, rat, 104-115
GH releasing factor
   see GRF
GH secretion
   neuroendocrine control, 236-251
   serotonin, rat, 222
gigantism
   cerebral, plasma somatomedin, 281-285
glucagon release
   HGH fragments, mouse, 41-49
   somatostatin, rat, 216-221
glucocorticoid
   fibroblast, DNA synthesis, fibroblast growth factor, 144-155
glucose
   see also blood sugar
   insulin release, GH, rat, 116-126
   insulin release, somatostatin, rat, 218-221
glucose load
   NSILA-s, human, 133
glucose metabolism
   mother and fetus, human placental lactogen, rat, 334-344
glucose oxidation
   adipose tissue, HGH fragment, plasmin, rat, 33-40
   adipose tissue, somatomedin B, rat, 188
glucose tolerance
   HGH fragments, mouse, 41-49
glucose tolerance test
   HGH secretion, age, 261-270
   HGH secretion, blood sugar, Huntington disease, 252-260
glucose uptake
   diaphragm, HGH fragments, rat, 41-49
glycemia
   see blood sugar
glyceraldehyde-3-phosphate dehydrogenase
   HGH fragments, 41-49
glycogen

mother and fetus, human placental lactogen, rat, 334-344
glycogen synthetase
   HGH fragments, 41-49
glycogenolysis
   HGH fragments, 41-49
cyclic GMP
   protein synthesis, membrane transport, diaphragm, rat, 99-103
gonadotropin deficiency
   HGH, testosterone, growth, 286-296
granulosa cell
   relaxin, 407-413
GRF
   GH secretion, 237-251
growth
   craniopharyngioma, somatomedin, 276-285
   fetus, human placental lactogen, rat, 334-344
   GH in rat and human plasma, rat tibia assay, 75-83
   GH, somatomedin, human, 178-189
   HGH fragment, plasmin, rat, 33-40
   HGH and OGH fragments, rat tibia test, 50-63
   HGH, proteinase, 72-74
   hypopituitarism, HGH, testosterone, 286-296
   hypopituitary dwarfism, HGH therapy, 297-311
growth factor
   brain, hypophysis, cow, 141-155
   comparison with NSILA-s, 136-138
   NSILA-s, NSILA-s receptor, rat liver, 127-140
growth hormone
   see GH
growth hormone, bovine
   see BGH
growth hormone, equine
   see EGH
growth hormone, human
   see HGH
growth hormone, ovine
   see OGH
growth hormone, porcine
   see PGH
growth hormone, rat
   see RGH
growth hormone-releasing factor
   see GRF
growth retardation
   fetus, human placental lactogen and sex steroids, plasma, 369-379
   fetus, human placental lactogen, plasma, 380-385
growth velocity

Subject Index

hypopituitarism, HGH, testosterone, 286-296
hypopituitary dwarfism, HGH therapy, 297-311
haloperidol
  prolactin secretion, rat, 443-453
HCG production
  choriocarcinoma in culture, 386-395
heart
  GH, acute stimulatory effects, rat, 94-103
  somatomedin C receptor, fetal pig, 190-201
heavy particle radiation
  hypophysis tumor, acromegaly, 246 cases, 312-314
hepatoma
  hypoglycemia, NSILA-s, 130, 131
hereditary chorea
  HGH secretion, 17 cases, 252-260
HGH
  altered forms, 64-74
  amino acid sequences, 8-13
  human placental lactogen, biological action, rat, 334-344
  hypophysis tumor, acromegaly, heavy particle radiation, 246 cases, 312-314
  insulin release, glucose, rat, 116-126
  male genital tract, mouse, rat, 345-358
  NSILA-s, plasma, 127-140
  plasma, nature of, 75-83
  plasmin derivatives and fragments, rat liver, ornithine decarboxylase, 14-32
  plasmin digestion, 14-32, 33-40
  protein synthesis, hypophysectomized rat, 163-168
  radioimmunoassay, 222-235
  somatomedin A and B, 178-189
  somatomedin, clinical studies, 271-285
  testosterone, hypopituitarism, growth, 286-296
HGH antibody
  hypopituitary dwarfism, HGH therapy, 305-311
HGH, decreased responsiveness
  see Laron dwarfism
HGH deficiency
  big somatomedin A and B in serum, 184-189
  HGH fragment therapy, human, 33-40
  HGH therapy, 297-311
  somatomedin A and B in serum, 180-189
  somatomedin, plasma, 275-285
  testosterone, growth, 286-296

HGH derivative
  plasmin, 14-32
HGH fragments
  plasmin digestion, 14-32, 33-40
  plasmin, insulin release, glucose, rat, 116-126
  synthetic, in vivo and in vitro activity, 41-49
  synthesis, 41-49
HGH production
  hypophysis, capillary culture unit, 386-395
HGH secretion
  age, 261-270
  Huntington disease, 17 cases, 252-260
  metabolic response, age, 266-270
  neuroendocrine control, 236-251
  serotoninergic control, L-dopa, 2-bromo-α-ergocryptine, 222-235
HGH therapy
  hypopituitary dwarfism, 297-311
hormone production
  hypophysis tissue in vitro, rat, human, 386-395
hormone receptor
  lactogen, ruminant, rat, 430
human chorionic gonadotropin
  see HCG
human placental big lactogen
  structure, activity, 327-333
human placental lactogen
  aminoacid sequences, comparison with GH, 8-13
  dimer, 327-333
  insulin release, glucose, rat, 116-126
  male genital tract, mouse, rat, 345-358
  pregnancy, lipolysis, fetal growth, rat, 334-344
  pregnancy, normal and pathological, 359-368, 380-385
  treatment, human pregnancy, threatened abortion, premature labor, 358-368
hydroxyproline excretion
  HGH deficiency and therapy, somatomedin, 275-285
  HGH secretion, age, 266-270
5-hydroxytryptophan
  GH secretion, 236-251
5-hydroxy-L-tryptophan
  HGH and prolactin release, acromegaly, 222-235
hypercorticism
  somatomedin, plasma, 279-285
hyperglycemia
  HGH fragment, 41-49
hyperparathyroidism

acromegaly, pheochromocytoma, 224
hypertension
  pregnancy, plasma human placental lactogen, 380-385
hypoglycemia
  NSILA-s, human plasma, 127-140
  paraneoplastic syndrome, NSILA-s, 127-140
hypoinsulinism
  somatomedin, plasma, 279-285
hypophysectomy
  GH action on thymus and lymphoid cells, rat, 104-115
  GH, acute stimulatory effects, rat, 94-103
  insulin release, GH, rat, 116-126
  liver, somatomedin generation, rat, 169-177
  plasma GH, rat, 75-83
hypophysis
  mitogenic factor, fibroblast growth factor, cow, 141-155
hypophysis extract
  proteinase, HGH alteration, 64-74
hypophysis tumor
  acromegaly, 248-251
  acromegaly, heavy particle radiation, 246 cases, 312-314
  hormone production, capillary culture unit, human, rat, 386-395
  prolactin, 396-406
hypophysis tumor cell
  cell culture, GH synthesis, rat, 75-83
hypopituitarism
  androgen, HGH, growth, 286-296
  big somatomedin A and B in serum, 184-189
  dwarfism, HGH therapy, 297-311
  HGH secretion, age, 261-270
  NSILA-s, human plasma, 127-140
  somatomedin A and B in serum, 180-189
  somatomedin, plasma, 275-285
hypothalamus
  GH secretion, 236-251
  HGH secretion, 222-235
  HGH secretion, Huntington disease, 17 cases, 252-260
  prolactin secretion, rat, 443-453
  somatostatin, 216
hypothalamus extract
  GH excretion, age, rat, 261
hypothyroidism
  prolactin, human, 396-406
  somatomedin A and B in serum, 180-189

Subject Index

somatomedin, plasma, 279-285
hypoxysuprine
  threatened abortion, premature labor, 10 cases, 359-368
insulin
  see also NSILA-s
  adrenalin, cyclic AMP, diaphragm, rat, 99-103
  binding to receptor, HGH fragments, 41-49
  cartilage metabolism, somatomedin, chicken embryo, rat, 202-215
  craniopharyngioma, growth, somatomedin, 276
  liver, somatomedin generation, rat, 169-177
  mammary gland, ruminants, 422-432
  somatomedin binding to organs, 187
insulin hypoglycemia
  HGH secretion, 236-251
  HGH secretion, acromegaly, 222-235
  HGH secretion, age, 261-270
insulin receptor
  somatomedin C, 190-201
insulin release
  arginine, Huntington disease, 252-260
  GH, pancreas, rat, 116-126
  HGH fragments, rat, 41-49
insulin secretion
  somatostatin, rat, 216-221
insulinoma
  NSILA-s, 130-136
islet of Langerhans
  see pancreas islet
kidney
  prolactin binding, rabbit, 433-442
  somatomedin C receptor, fetal pig, 190-201
kidney insufficiency
  somatomedin, plasma, 282-285
Krebs ascites tumor
  cell free system, GH synthesis, 84-93
kwashiorkor
  GH, somatomedin, 174
  somatomedin, 275-285
labor, premature
  see premature labor
lactation
  mammotropic hormones, ruminants, 422-432
  prolactin, human, 396-406
  prolactin, rabbit, 433-442
lactogen, bovine placental
  see bovine placental lactogen
lactogen, human placental
  see human placental lactogen
lactogen, human placental, big

see human placental big lactogen
lactogen, ovine placental
  see ovine placental lactogen
lactogen, placental
  see placental lactogen
lactogenesis
  prolactin, rabbit, 433-442
Laron dwarfism
  see also dwarfism
  somatomedin A and B in serum, 180-189
  somatomedin, plasma, 278-285
LH
  ovary tumor cell line, rat, 141
  ovine, male genital tract, mouse, rat, 345-358
  prolactin, human, 396-406
lipid metabolism
  adipose tissue, human placental lactogen, rat, 334-344
lipolysis
  adipose tissue, somatomedin B, rat, 188
  HGH fragments, 41-49
  human placental lactogen, rat, 334-344
liver
  GH, acute stimulatory effects, rat, 94-103
  ornithine decarboxylase, HGH and plasmin derivatives and fragments, rat, 14-32
  plasma membrane, NSILA-s, NSILA receptor, rat, 127-140
  prolactin binding, rabbit, 433-442
  somatomedin C receptor, fetal pig, 190-201
  somatomedin generation, rat, 169-177
liver cell
  insulin binding, HGH fragments, 41-49
liver cirrhosis
  somatomedin, plasma, 282-285
liver perfusion
  somatomedin generation, rat, 169-177
LSD
  HGH secretion, 222-235
lung
  growth, epidermal growth factor, 200
  somatomedin C receptor, fetal pig, 190-201
luteinizing hormone
  see LH
lymphocyte
  GH, human, 104-115
lymphocyte transformation
  phytohemagglutinin, GH, human, in vitro, 104-115
lymphoid cell

GH, rat, 104-115
D-lysergic acid diethylamine
  see LSD
male genital tract
  human placental lactogen, mouse, rat, 345-358
malnutrition
  somatomedin, human plasma, 275-285
mammary gland
  mammotropic hormones, ruminants, 422-432
  prolactin binding, rabbit, 433-442
marasmus
  somatomedin, plasma, 275-285
median eminence
  somatostatin, 216
melatonin
  GH secretion, 236-251
  HGH secretion, acromegaly, 222-235
  prolactin release, 242
menopause
  prolactin, 396-406
menstrual cycle
  prolactin, 396-406
  relaxin, 412
mesenchymal cell
  fibroblast growth factor, 141-155
mesenchymal tumor
  hypoglycemia, NSILA-s, 130, 131, 134, 135
metabolism
  mother and fetus, human placental lactogen, rat, 334-344
methergoline
  GH secretion, 236-251
3-O-methylglucose
  transport, diaphragm, rat, 94-103
methylthymidine-^{3}H
  incorporation, cartilage, somatomedin assay, pig, 271-285
methysergide
  GH secretion, 236-251
  HGH secretion, 222-235
milking
  prolactin, relaxin, sheep, 409-413
mitogenic factor
  brain, hypophysis, cow, 141-155
  brain, hypophysis, purification, properties, 141-155
myoblast
  growth factor, peptide, 141-155
  growth factor, purification, properties, 141-155
nerve growth factor
  fetal lung, 200
  fibroblast growth factor, 143, 144
neuroendocrine control

462

Subject Index

GH secretion, 236-251
newborn
  somatomedin, human plasma, 272-285
nitrogen excretion
  HGH secretion, age, 266-270
  hypopituitarism, HGH, testosterone, 295, 296
nitrogen metabolism
  mother and fetus, human placental lactogen, rat, 334-344
nonesterified fatty acid
  see free fatty acid
nonsuppressible insulin-like activity
  see NSILA-s
noradrenalin
  brain, HGH secretion, 222-235
  GH secretion, 236-251
norepinephrine
  see noradrenalin
NSILA-s
  see also insulin
  fibroblast growth factor, 143, 144
  radioreceptor assay, normal and pathological values in human plasma, 127-140
NSILA-s receptor
  liver cell plasma membrane, rat, 127-140
5′-nucleotidase
  mammary gland, prolactin, rabbit, 433-442
nucleotide, cyclic
  fibroblast growth factor, 144-155
nutrition
  amino acid and sugar transport, diaphragm, rat, 95-99
  somatomedin, human plasma, 275-285
obesity
  GH, somatomedin, 174
  male genital tract, human placental lactogen, mouse, 345-358
  somatomedin, plasma, 275-285
OGH
  amino acid sequences, 8-13
  placental lactogen, 315-326
OGH fragments
  synthesis, 50-63
oligomenorrhea
  prolactin, 396-406
ontogeny
  somatomedin C receptor, 190-201
orchidectomy
  seminal vesicle, prostate, human placental lactogen, mouse, 352-358
organ culture
  adenohypophysis, GH synthesis, rat, 6-8
  adenohypophysis, prolactin secretion, rat, 443-453
ornithine decarboxylase
  liver, HGH and plasma derivatives and fragments, rat, 14-32
ouabain
  prolactin, sodium transport, rabbit, 433-442
ovary
  prolactin, human, 396-406
  relaxin, 407-421
ovary growth factor
  ovary tumor cell line, rat, 141
ovary tumor
  ovary growth factor, hypophysis, rat, 141
ovine placental lactogen
  isolation, characterization, 315-326
ovine prolactin
  male genital tract, mouse, rat, 345-358
oxytocin
  relaxin release, sheep, 409-413
pancreas islet
  glucagon release HGH fragments, mouse, 41-49
  insulin release, GH, rat, 116-126
  insulin release, HGH fragments, rat, 41-49
  somatostatin, rat, 216-221
pancreas perfusion
  hormone release, somatostatin, 218-221
papaverine
  protein synthesis, membrane transport, diaphragm, rat, 99-103
parachlorophenylalanine
  GH secretion, 236-251
paraneoplastic syndrome
  hypoglycemia, NSILA-s, 127-140
partial hepatectomy
  somatomedin generation, rat, 169-177
parturition
  mammotropic hormones, ruminants, 422-432
  prolactin and relaxin, sheep, 412
peptide
  brain, hypophysis, growth factor, 141-155
perphenazine
  prolactin secretion, rat, 443-453
PGH
  amino acid sequences, 8-13
phentolamine
  HGH secretion, acromegaly, 222-235
pheochromocytoma
  acromegaly, hyperparathyroidism, 224
phosphodiesterase
  cartilage metabolism, somatomedin, chicken embryo, rat, 202-215
physical exercise
  HGH secretion, 236-251
  HGH secretion, age, 261-270
phytohemagglutinin
  lymphocyte, GH, human, in vitro, 104-115
pimozide
  GH secretion, 236-251
  HGH secretion, acromegaly, 245-251
  prolactin secretion, rat, 443-453
Piribedil
  prolactin secretion, rat, 443-453
placenta
  relaxin, 407-413
  somatomedin C receptor, fetal pig, 190-201
placenta cell membrane
  insulin receptor, human, 190-201
  somatomedin assay, human, 272
  somatomedin C receptor, human, 190-201
placenta function
  human placental lactogen and sex steroids in plasma, 369-379
placenta insufficiency
  human placental lactogen and sex steroids in plasma, 369-379
placental lactogen
  introduction, 315
  ruminants, 422-432
placental lactogen, human
  see human placental lactogen
plasma membrane
  liver, NSILA-s receptor, rat, 127-140
  mammary gland, prolactin, rabbit, 433-442
plasma protein
  somatomedin, malnutrition, human, 275
plasmin
  HGH digestion, nature of fragments, 33-40
  HGH and fragments, modification, 14-32
pluriglandular syndrome
  acromegaly, hyperparathyroidism, pheochromocytoma, 224
prednisone
  somatomedin, plasma, 279-285
pregnancy
  fetus, human placental lactogen, rat, 334-344

human placental lactogen and big human placental lactogen, 327-333
mammotropic hormones, ruminants, 422-432
normal and pathological, plasma human placental lactogen, 359-368
NSILA-s, human plasma, 127-140
placental lactogen, bovine and ovine, 315-326
prolactin, human, 396-406
prolactin and relaxin, sheep, 412

pregnancy toxemia
human placental lactogen and sex steroids in plasma, 369-379

pregnenolone
plasma, normal and pathological pregnancy, 369-379

pregnenolone sulfate
plasma, normal and pathological pregnancy, 369-379

pregrowth hormone
GH precursor, 84-93

premature labor
human placental lactogen treatment, 3 cases, 359-368

progesterone
plasma, human placental lactogen, human pregnancy, 352-358
plasma, normal and pathological pregnancy, 369-379
relaxin release, sheep, 409-413
sulpiride, prolactin, women, 402-406

prohormone
HGH, 74

proinsulin
93
somatomedin C receptor, 190

prolactin
amino acid sequences, comparison with GH, 10-13
fertility control, women, 396-406
human, radioimmunoassay, 408
insulin release, glucose, rat, 120-126
liver, somatomedin generation, rat, 172
mechanism of action, rabbit, 433-442
ovine, radioimmunoassay, 408
placental lactogen, bovine and ovine, 315-326
radioimmunoassay, 222-235
relaxin, human, sheep, 407-421
ruminants, 422-432
somatomedin, plasma, 283

prolactin activity

HGH, altered forms, 64-74
HGH and fragments, plasmin, 14-32

prolactin cell
hypophysis tumor, acromegaly, 248-251

prolactin deficiency
male genital tract, human placental lactogen, dwarf mouse, 345-358

prolactin, ovine
see ovine prolactin

prolactin production
hypophysis, capillary culture unit, human, rat, 386-395

prolactin receptor
mammary gland, rabbit, 433-442
mammary gland, ruminants, rat, 430

prolactin release
BOL, 223
2-bromo-α-ergocryptine, 222
L-dopa, acromegaly, 222-235
L-dopa, Huntington chorea, 252-260
dopamine, Piribedil, rat, 443-453
HGH secretion, age, 261-270
5-hydroxy-L-tryptophan, 222-235
melatonin, 242
L-tryptophan, acromegaly, 222-235

proparathyroid hormone
93

prostate
human placental lactogen, prolactin, HGH, mouse, 345-358

protein anabolism
HGH, testosterone, 286-296

protein calorie deficiency
GH, somatomedin, 174

protein kinase
adipose tissue, human placental lactogen, rat, 334-344

protein synthesis
cartilage, somatomedin, cyclic AMP, chicken embryo, rat, 202-215
GH, rat, 94-103
hypophysis tumor cell, cell free system, rat, 84-93
insulin release, glucose, GH, rat, 116-126
somatomedin A and B, hypophysectomized rat, 163-168

proteinase
HGH alterations, 64-74

proteoglycan
cartilage, somatomedin, cyclic AMP, chicken embryo, rat, 202-215

puberty
HGH, testosterone, growth, 286-296

puromycin
amino acid and sugar transport, diaphragm, rat, 95, 101
insulin release, glucose, GH, rat, 116-126

pyruvic dehydrogenase
HGH fragments, 41-49

quinine
protein synthesis, membrane transport, diaphragm, rat, 99-103

radioimmunoassay
GH in rat and human plasma, 75-83
HGH, 222-235
HGH, altered forms, 64-74
HGH and fragments, plasmin, 14-32
human placental lactogen and big human placental lactogen, 327-333
human placental lactogen, pregnancy, 359-368
placental lactogen, bovine, 324
prolactin, human, 222-235, 408
prolactin, ovine, 408, 414-421
relaxin, 408
somatomedin A and B, 156-168
somatomedin B, 178-189, 272

radioreceptor assay
human placental lactogen and big human placental lactogen, 327-333
NSILA-s, human plasma, 127-140
somatomedin A, 178-189, 272
somatomedin A and B, 156-168
somatomedin C, 190-201

radiotherapy
heavy particles, hypophysis tumor, acromegaly, 246 cases, 312-314

relaxin
prolactin, human, sheep, 407-421
radioimmunoassay, sheep, 408, 414-421

reproduction
prolactin, women, 396-406

reserpine
prolactin secretion, rat, 443-453

RGH
chemistry, 1-13
plasma, nature of, 75-83
production, hypophysis, capillary culture unit, 386-395

RGH secretion
regulation, 443-453

RGH synthesis
hypophysis tumor cell, cell free system, rat, 84-93

rhesus isoimmunization

Subject Index

plasma human placental lactogen, 380-385
ribosome
  GH synthesis, rat, 84-93
RNA synthsis
  cartilage, somatomedin, cyclic AMP, chicken embryo, rat, 202-215
  fibroblast growth factor, 144-155
mRNA
  GH synthesis, rat, 84-93
rough endoplasmic reticulum
  GH synthesis, rat, 84-93
secondary sex characteristics
  hypopituitarism, HGH, testosterone, 286-296
seminal vesicle
  human placental lactogen, prolactin, HGH, mouse, 345-358
serotonin
  GH secretion, 236-251
  GH secretion, rat, 222
serotonin receptor
  GH secretion, 236-251
  HGH secretion, 222-235
serotoninergic control
  HGH secretion, 222-235
sex
  HGH secretion, age, 261-270
  hypopituitarism, androgen, HGH, growth, 286-296
  prolactin, human, 396-406
sex hormones
  somatomedin, plasma, 279-285
short stature
  constitutional, plasma somatomedin, 280-285
  emotional deprivation, plasma, somatomedin, 276-285
  somatomedin, plasma, dysmorphic children, 281-285
skeletal growth
  GH, somatomedin, human, 178-189
skeletal muscle
  GH, acute stimulatory effects, rat, 94-103
skin
  sulfate-^{35}S incorporation, somatomedin A and B, rat, 165-168
sleep
  HGH secretion, age, 261-270
sodium transport
  mammary gland, prolactin, rabbit, 433-442
somatomedin
  fibroblast growth factor, 143, 144
  plasma, clinical studies, 271-285
somatomedin A
  see also big somatomedin A
  isolation, chemistry, in vivo effect, 156-168
  NSILA-s, radioreceptor, 132-139
  regulation, pathophysiology, human, 178-189
somatomedin B
  see also big somatomedin B
  isolation, chemistry, in vivo effects, 156-168
  regulation, pathophysiology, human, 178-189
somatomedin C
  NSILA-s, radioreceptor, 132-139
somatomedin C receptor
  specificity, ontogeny, topography, mammalian tissues, 190-201
somatostatin
  fibroblast growth factor, 143, 144
  GH releasing factor, 236-251
  insulin release, glucose, rat, 116-126
  pancreas α_1 cell, glucose, insulin secretion, rat, 216-221
spleen cell
  GH, rat, 104-115
steroid metabolism
  pregnancy, normal and pathological, 369-379
STM
  see GH
suckling
  prolactin, relaxin, sheep, 409-413
sugar transport
  GH, rat, 94-103
sulfate[35]S
  incorporation, cartilage and skin, somatomedin A and B, rat, 165-168
  incorporation, cartilage, somatomedin, pig, 271-285
  incorporation, cartilage, somatomedin, rat, 169-177
sulfation factor
  see somatomedin
sulpiride
  prolactin secretion, human, 396-406
tall stature
  constitutional, plasma somatomedin, 280-285
teat stimulation
  prolactin, relaxin, sheep, 409-413
testis
  human placental lactogen, mouse, rat, 345-358
testosterone
  HGH, hypopituitarism, growth, 286-296
  somatomedin, plasma, 279-285
testosterone propionate
  male genital tract, mouse, rat, 345-358
testosterone secretion
  insulin hypoglycemia, age, human, 261-270
theophylline
  cyclic AMP, somatomedin, cartilage, 202-215
  protein synthesis, membrane transport, diaphragm, rat, 99-103
threatened abortion
  human placental lactogen treatment, 7 cases, 358-368
thymocyte
  GH, rat, 104-115
thymus
  GH, rat, 104-115
thyroid hormone
  somatomedin, plasma, 279-285
thyroxine
  hypopituitarism, HGH, testosterone, growth, 292-296
  somatomedin, plasma, 279-285
tibia cartilage
  human placental lactogen, rat, 345-358
tissue culture
  polypeptide hormone production, artificial capillaries, 386-395
toxemia of pregnancy
  see pregnancy toxemia
TRF
  GH secretion, acromegaly, 247-251
  hypophysis tumor, hormone production in vitro, human, 390-395
  prolactin release, human, 396-406
triglyceride
  mother and fetus, human placental lactogen, rat, 334-344
L-tryptophan
  GH secretion, 236-251
  HGH and prolactin release, acromegaly, 222-235
TSH releasing factor
  see TRF
Turner syndrome
  somatomedin, plasma, 281-285
  somatomedin A and B in serum, 180-189
udder
  mammotropic hormones, ruminants, 422-432
umbilical cord blood
  somatomedin, human, 272-285
uremia
  somatomedin, plasma, 282-285
  somatomedin A and B in serum, 180-189

Subject Index

urine
somatomedin A and B, 187-189
uterus
relaxin, 407-413
vaginal smear
threatened abortion, premature labor, human placental lactogen, treatment, 359-368
wheat germ cell
cell free system, GH synthesis, 84-93

wound healing
fibroblast growth factor, 141-155
xiphoid cartilage
human placental lactogen, rat, 345-358